Harmonic Analysis in Hypercomplex Systems

Mathematics and Its Applications

Volume 434

Harmonic Analysis in Hypercomplex Systems

by

Yu. M. Berezansky

and

A. A. Kalyuzhnyi

Department of Functional Analysis,
Institute of Mathematics,
Ukrainian Academy of Sciences,
Kiev, Ukraine

KLUWER ACADEMIC PUBLISHERS

DORDRECHT / BOSTON / LONDON

A C.I.P. Catalogue record for this book is available from the Library of Congress.

ISBN 978-90-481-5022-9

Published by Kluwer Academic Publishers,
P.O. Box 17, 3300 AA Dordrecht, The Netherlands.

Sold and distributed in North, Central and South America
by Kluwer Academic Publishers,
101 Philip Drive, Norwell, MA 02061, U.S.A.

In all other countries, sold and distributed
by Kluwer Academic Publishers,
P.O. Box 322, 3300 AH Dordrecht, The Netherlands.

This is a completely revised and updated translation of
the original Russian work of the same title,
Kiev, Naukova Dumka, 1992. Copyright by the authors.
Translated by P. Malyshev.

Printed on acid-free paper

CONTENTS

PREFACE TO THE ENGLISH EDITION

First works related to the topics covered in this book belong to J. Delsarte and B. M. Levitan and appeared since 1938. In these works, the families of operators that generalize usual translation operators were investigated and the corresponding harmonic analysis was constructed. Later, starting from 1950, it was noticed that, in such constructions, an important role is played by the fact that the kernels of the corresponding convolutions of functions are nonnegative and by the properties of the normed algebras generated by these convolutions. That was the way the notion of hypercomplex system with continuous basis appeared. A hypercomplex system is a normed algebra of functions on a locally compact space Q—the "basis" of this hypercomplex system. Later, similar objects, hypergroups, were introduced, which have complex-valued measures on Q as elements and convolution defined to be essentially the convolution of functionals and dual to the original convolution (if measures are regarded as functionals on the space of continuous functions on Q).

However, until 1991, the time when this book was written in Russian, there were no monographs containing fundamentals of the theory (with an exception of a short section in the book by Yu. M. Berezansky and Yu. G. Kondratiev [BeKo]). The authors wanted to give an introduction to the theory and cover the most important subsequent results and examples. We now came to realizing that the first chapters of the book should have been written in more detail, but such a change into the English edition of the book would entail its complete revision, and so we only made small additions and clarifications.

But we still think that this book can serve as an introduction to the theory. We should also mention a recently published book by W. R. Bloom and H. Heyer "Harmonic Analysis of Probability Measures on Hypergroups," which contains interesting and important applications of the theory of hypergroups to problems in probability theory. It also contains an introduction to the theory of hypergroups and a large number of examples.

There are many mathematicians who currently work in this area of functional analysis. In 1993, a representative conference devoted to this range of problems was held in Seattle. The proceedings of this conference (Contemporary Mathematics, **183**, 1995) give a good idea about the scope and depth of recently obtained results in this area.

Of course, we were not able to describe in this English edition the works published after 1991; we only confined ourselves to supplementing the bibliography with works in neighboring areas written by the authors or their students and colleagues. We also include a supplement written on the basis of our joined communication at the mentioned conference. In particular, it may be considered as a survey of principal results presented in the first two chapters of the present book. Moreover, it contains a detailed exposition of the cases of compact and discrete bases, a brief description of the basic examples of hypercomplex systems, and the theory of almost periodic functions.

In this communication, we also suggested to introduce a certain uniformity into the terminology and to replace the term "hypercomplex system with locally compact basis" by the term "L_1-hypergroup". From this point of view, it would probably be more appropriate to call this translation of the book "Harmonic Analysis in Hypergroups."

The authors express their deep gratitude to the Kluwer Academic Publishers for publishing this monograph and to P. V. Malyshev, D. V. Malyshev, and Yu. A. Chapovsky who translated it and performed much editorial work that improved the clarity of the exposition.

April 1997 *Yu. M. Berezansky, A. A. Kalyuzhnyi*

INTRODUCTION

In a series of works originated as early as in 1938, J. Delsarte [Del1]–[Del4] and then Levitan [Lev1]–[Lev10] noticed that some facts of classical harmonic analysis can be generalized by replacing exponential functions $e^{i\lambda q}$ $(q, \lambda \in \mathbb{R}^1)$ by some family of complex-valued functions $\chi(q, \lambda)$ which inherit the following property of the indicated exponential functions: The exponential functions are connected with the family of ordinary translation operators R_p $(p \in \mathbb{R}^1)$ acting upon complex-valued functions $f(q)$ $(q \in \mathbb{R}^1)$ according to the rule $(R_p f)(q) = f(q+p)$, i.e.,

$$(R_p e^{i\lambda\bullet})(q) = e^{i\lambda p} e^{i\lambda q} \tag{1}$$

for any λ.

A collection of functions $\chi(q, \lambda)$, where q varies within a set Q and λ takes values from another set $\hat{Q}$, must also be associated with a family of linear operators R_p $(p \in Q)$ of "generalized translation" acting upon functions of a point $q \in Q$ and satisfying an equality similar to (1) ("multiplication formula"), namely,

$$(R_p \chi(\cdot, \lambda))(q) = \chi(p, \lambda)\chi(q, \lambda).$$

It is natural that the collection of operators R_p must also have some additional properties imitating ordinary translations.

As became clear from the first works of J. Delsarte and Levitan, an important role is played not only by the translation itself but also by the convolution of functions generated by this translation. Thus, by analogy with ordinary convolution

$$\begin{aligned}(f * g)(q) &= \int_{\mathbb{R}^1} f(q-p)g(p)\,dp \\ &= \int_{\mathbb{R}^1} f(p)g(q-p)\,dp = \int_{\mathbb{R}^1} f(p)(R_{-p}g)(q)\,dp \quad (q \in \mathbb{R}^1),\end{aligned} \tag{2}$$

it is possible to introduce generalized convolution $*$ associated with the generalized translation operator in exactly the same way as in (2), i.e.,

$$(f * g)(q) = \int_Q f(p)(R_{p*} g)(q)\,dm(p) \quad (q \in Q), \tag{3}$$

where $Q \ni p \to p^* \in Q$ is an involution given in Q which replaces the transition to the inverse element in $\mathbb{R}^1$ and m is a fixed measure in Q inheriting some properties of the Lebesgue measure.

In 1950, Gelfand suggested Berezansky and S. Krein to investigate the properties of *a priori* given convolutions (3) of functions defined on Q which enable one to construct meaningful harmonic analysis. As a result, Berezansky and S. Krein introduced the concept of hypercomplex systems with continual basis (see [BKr1] and [BKr2]).

Recall that an ordinary hypercomplex system is defined as a d-dimensional $(d < \infty)$ associative algebra (i.e., a d-dimensional complex vector space with operation of multiplication $f * g$ of its vectors $f, g, \ldots$ with standard properties) with a fixed basis Q consisting of d points $p, q, \ldots$. By identifying a vector f with its coordinates, we can understand this system as the space of functions $Q \ni q \mapsto f(q) \in \mathbb{C}^1$ with the operations of addition and multiplication of functions by a scalar and with multiplication

$$(f * g)(r) = \sum_{p,\, q \in Q} f(p) g(q) c(p, q, r) \quad (r \in Q),$$

where $c(p, q, r)$ is a function ("cubic matrix of structural constants") which determines multiplication and possesses certain properties guaranteeing the assosiativity (and, if necessary, the commutativity) of multiplication. It is clear that

$$c(p, q, r) = (p * q)(r).$$

A generalization of such hypercomplex systems proposed by Berezansky and S. Krein is connected with the transition from a finite basis Q to a certain locally compact space Q. According to already constructed examples, it seems reasonable to replace c with a "structure measure" $c(A, B, r)$ $(A, B \subset Q;\, r \in Q)$ rather than by a function defined on $Q \times Q \times Q$. The most complete results of harmonic analysis are obtained in the case of commutative hypercomplex systems with nonnegative measure c whose properties are similar to the properties of the group algebra of a locally compact group (the so-called normal hypercomplex system with basis unity).

In 1950–1953, the theory of hypercomplex systems with continual basis was extensively developed by Berezansky and S. Krein [BKr1]–[BKr4] and Berezansky [Ber1]–[Ber9]. They mainly considered the cases of compact and discrete (countable) bases. The detailed presentation of these results can be found in the Candidate's Degree Thesis of Berezansky [Ber9] which, unfortunately, was not published as a book. The development of this theory was significantly influenced by the Gelfand theory of commutative normed algebras (see [GRS]) and the works of M. Krein [Kre1], [Kre2] and Raikov [Rai] on harmonic analysis on groups and homogeneous spaces. It is now reasonable to emphasize that the group algebra of a locally compact commutative group G is one of the most simple examples of commutative hypercomplex systems with locally compact basis $Q = G$. The construction of general hypercomplex systems was mainly performed as a generalization of this classical object of harmonic analysis.

Since 1953, the interest of Berezansky and S. Krein to this and related subjects became much weaker, basically due to the small number of mathematicians working in the field and the lack of new examples. The situation changed only after the appearance, since 1973, of numerous works devoted to the theory of hypergroups which can be regarded as objects close to hypercomplex systems (Dunkl, Jewett, Spector, Ross, Bloom, Heyer, etc.). It became clear that the axiomatics of hypercomplex systems is broader than the axiomatics of hypergroups and, therefore, some results obtained in the theory of hypergroups were actually contained in older works devoted to the theory of hypercomplex systems. Since this time, the theory of hypercomplex system acquired renewed life. Since 1982, many works by Berezansky, Kalyuzhnyi, Vainerman, Podkolzin, and other authors have been devoted to the investigation of hypercomplex systems.

Note that, up to now, despite a vast amount of data accumulated in the theory of hypercomplex systems, hypergroups, and related subjects, there is no book with systematic presentation of all basic results obtained in this branch of functional analysis. In the present monograph, we try to fill this gap. This book contains the description of basic constructions and detailed analysis of applications. The monograph includes practically the entire thesis [Ber9] mentioned above.

At present, a similar book by Bloom and Heyer is prepared for publication. In this book, main applications deal with the problems of probability theory and random processes (we do not consider applications of this sort). One should also mention two well-known monographs by Levitan [Lev9] and [Lev10] devoted to the Lie theory of generalized translation operators. A detailed presentation of the fundamentals of the theory of hypercomplex systems with locally compact basis, their representations by commuting operators, and the theory of generalized functions on the basis of a hypercomplex systems can be found in the book by Berezansky and Kondrat'ev [BeKo]. Principal results in the theory of hypergroups are presented in the monographic paper by Jewett [Jew].

Definitions and constructions of hypercomplex systems (both commutative and noncommutative) with locally compact basis are given in Chapter 1. In Chapter 1, we also present elements of harmonic analysis in the commutative case, i.e., generalizations of the Bochner and Plancherel theorems and duality theory. We do not dwell upon the theory of almost periodic functions to make the book more concise. In Section 5, we consider the theory of quantization of hypercomplex systems related to the concept of Kac algebra and harmonic analysis in the noncommutative case. In Section 2, we establish relationships between hypercomplex systems and the theory of generalized translation operators and hypergroups.

In Chapter 2, we present basic examples of hypercomplex systems with locally compact (in particular, compact and discrete) bases: the center of the group algebra of a compact group, Gelfand pairs, systems associated with J. Delsarte translations, with orthogonal polynomials, and with the Sturm–Liouville operator. It is shown that general facts presented in Chapter 1 imply principal results of harmonic analysis for these objects. In particular, in this chapter, we present the theory of representations of compact groups, including the Tannaka–Krein duality theory.

Chapter 3 contains principal notions and facts of the Lie theory of hypercomplex systems which deals with the investigation of infinitesimal objects associated with such systems. The idea of relevant constructions, clearly, goes back to the construction and investigation of the Lie algebra for a given Lie group G. However, unlike the case of Lie groups, an infinitesimal object given at a single point of the basis Q of a hypercomplex system generally speaking, does not reconstruct the entire system because homogeneity is absent in this case (for groups, the character of the group operation in a neighborhood of a single point, e.g., in a neighborhood of the identity element, determines (after translations) the character of this operation in neighborhoods of all other points). In Chapter 3, we also present the theory of duality of generators of one-dimensional compact and discrete systems.

Let us now dwell upon some notions whose definitions available in the literature are ambiguous and introduce necessary notation. We denote the boundary of a set A by ∂A. The notions of compactness and bicompactness are equivalent. For convenience, we assume that all locally compact spaces used in what follows are σ-compact, i.e., they can be represented as countable unions of compact sets (although all results remain true in the general case). We write

$$\int_Q f(r)\, d_r c\,(E_r, B, s)$$

to denote the integral of a function $f(r)$ with respect to a measure $\mu(A) = c(A, B, s)$ ($B \subset Q$ and $s \in Q$ are fixed). The indicator of a set A is defined as

$$\kappa_A(r) = \begin{cases} 1, & r \in A \\ 0, & r \notin A \end{cases}$$

The superscript $\mathbb{R}\,(+)$ in the designation of a class of functions or measures means that this is the collection of all real-valued (positive) functions or measures from the indicated class. The linear and closed linear spans of a subset M of a linear topological space are denoted by l.s. (M) and c.l.s. (M), respectively. The space X' dual to a space X is defined as the collection of all linear continuous functionals. Unless otherwise stated, $\langle l, x\rangle$ denotes the action of a functional $l \in X'$ upon a vector $x \in X$. The topology of simple convergence on a set $M \subset X$ in X' is defined as the weakest topology in which each function $E' \ni l \mapsto \langle l, x\rangle \in \mathbb{C}$ $(x \in M)$ is continuous. The domain of definition of an operator A in a Hilbert space H is denoted by $\mathfrak{D}(A)$. An operator A with dense domain of definition $\mathfrak{D}(A)$ in the Hilbert space H is called Hermitian if

$$(Ax, y)_H = (x, Ay)_H \qquad (x, y \in \mathfrak{D}(A)),$$

self-adjoint if $A^* = A$, and essentially self-adjoint if the closure of the operator A is self-adjoint. We say that A is a $*$-algebra if it is an algebra over the field $\mathbb{C}$ with involution, i.e., with an antilinear mapping $A \ni x \mapsto x^* \in A$ such that (1) $(x^*)^* = x$ and (2) $(xy)^* = x^* y^*$ $(x, y \in A)$. A Banach algebra A with involution $x \mapsto x^*$ such that $\| x^* \| = \| x \|$ $(x \in A)$ is called a Banach $*$-algebra. A weakly closed $*$-subalgebra of the algebra of linear continuous operators in a Hilbert space is called a W^*-algebra (or a von Neumann algebra). A C^*-algebra is defined as a Banach $*$-algebra A such that $\| x \|^2 = \| x^* x \|$ for any $x \in A$.

References are given both directly in the text and in Bibliographical Notes at the end of the book. The list of references should in no case be regarded as complete. At the same time, it contains principal works even in those branches of the theory of hypergroups, generalized translations, convolution algebras, etc. which are not included in the book. It has already been noted that we do not dwell upon applications to probability theory, the theory of functions almost periodic with respect to generalized translations, problems of spectral analysis and synthesis, the theory of generalized functions on the basis of a hypercomplex system, algebraic combinatorics, and the theory of multi-valued groups and related problems in algebraic topology.

The authors express their deep gratitude to B. M. Levitan, S. G. Krein and L. I. Wainerman for fruitful discussions and to G. L. Litvinov, D. I. Gurevich, B. P. Osilenker, and Yu. S. Samoilenko who read some parts of this manuscript and made useful critical remarks.

November 1991 *Yu. M. Berezansky, A. A. Kalyuzhnyi*

1. GENERAL THEORY OF HYPERCOMPLEX SYSTEMS

In 1950, when extending the Delsarte–Levitan theory of generalized translation operators, Yu. M. Berezanskii and S. G. Krein introduced the concept of (commutative) hypercomplex system with continuous basis and developed harmonic analysis for such systems. Each hypercomplex system is a Banach $*$-algebra of functions on a locally compact space (the basis of a hypercomplex system). It generalizes the concept of hypercomplex system with finite basis and the concept of locally compact group algebra. The role of the group translation operators is played by generalized translation operators, which are naturally associated with a hypercomplex system. Note that it is possible to completely characterize hypercomplex systems in terms of such generalized translation operators (Theorem 2.1) and, hence, a hypercomplex system with continuous basis can be considered as a class of generalized translation operators that admit the construction of rich harmonic analysis and duality theory.

In Section 1, we introduce the required class of hypercomplex systems and establish their properties. In Section 2, we generalize the concept of hypercomplex system to the case where the "structure measure" c is real and establish the relationship between hypercomplex systems and related objects such as generalized translation operators, hypergroups, and convolution algebras. Section 3 is devoted to the construction of harmonic analysis on hypercomplex systems. We study representations of hypercomplex systems and prove analogs of the approximation theorem, the Plancherel theorem, the inversion formula, and the Bochner theorem. We also establish an analog of the Pontryagin duality theory for locally compact commutative groups. In Section 4, we study the properties of hypercomplex subsystems generalizing the concept of subgroups to the case of hypercomplex systems and introduce the concept of homomorphism for hypercomplex systems.

To construct duality theory for locally compact groups, Kac [Kac] introduced the concept of Hilbert bialgebras. Thus, functions defined on a group form a bialgebra with respect to the ordinary pointwise multiplication (commutative operation) and convolution (generally speaking, noncommutative operation). One can understand the transition from this special bialgebra to the general bialgebra with two noncommutative operations as a quantization of the group, moreover, the "conformity principle" is true, i.e., under certain conditions, the commutativity of one operation allows the bialgebra to be realized as a bialgebra of functions on a group. The development of this approach and its application to Lie groups leads to the quantization of these groups in terms of deformations of algebras (Drinfeld [Dri]). In Section 5, we realize this construction for hypercomplex systems To do this, we introduce the concept of quantized hypercomplex system whose axiomatics is simpler and less restrictive than that of Kac bialgebras. If one operation is commutative, a quantized hypercomplex system can be realized as an object that generalizes a hypercomplex system with locally compact basis to the case where the measure c is real. Such hypercomplex systems also admit the construction of harmonic analysis and duality theory. At the end of Section 5, we present a generalization of the concept of hypercomplex systems to the case where the measure c is, generally speaking, not positive.

1. Fundamental Concepts of the Theory of Hypercomplex Systems with Locally Compact Basis

A hypercomplex system with locally compact basis Q is a set $L_1(Q, m)$ with generalized convolution, which can be defined in terms of a structure measure $c(A, B, r)$ (A, $B \subset Q$, $r \in Q$) and generalizes the concept of locally compact group algebra. The structure measure $c(A, B, r)$ is equal to the convolution of the characteristic functions of the sets A and B, whereas the positive measure m is an analog of the invariant measure. In this section, we prove the existence of this measure if the convolution is commutative. We extract a subclass of hypercomplex systems (the so-called normal hypercomplex systems with basis unity) which possesses many good properties which enable one to construct substantive harmonic analysis. In particular, it is established that every hypercomplex system of this sort is semisimple. In the last subsection, we present some technical facts about integration with respect to a measure depending on a parameter frequently used in this section. Examples of hypercomplex systems presented in Section 1 are pure illustrative. More important examples of hypercomplex systems are studied in Chapter 2.

1.1. Definition of Hypercomplex Systems. Characters. Assume that Q is a complete separable locally compact space, $\mathcal{B}(Q)$ is the σ-algebra generated by its Borel sets, $\mathcal{B}_0(Q)$ is the subring of $\mathcal{B}(Q)$ consisting of sets with compact closure. In what follows, we consider regular Borel measures, i.e., measures nonnegative on Q and finite on compact sets; a measure ρ of this sort is *regular* in a sense that $\rho(A) = \sup \rho(\Delta)$ for all $A \in \mathcal{B}(Q)$, where supremum is taken over all compact sets $\Delta \subset A$. A function $Q \ni r \mapsto f(r) \in \mathbb{C}$ is called *finite* if it is equal to zero outside a certain compact set. The support of a function $f(r)$ is denoted by $\operatorname{supp} f$. The spaces of continuous functions, of finite continuous functions, of continuous functions vanishing at infinity, and of bounded continuous functions are denoted by $C(Q)$, $C_0(Q)$, $C_\infty(Q)$, and $C_b(Q)$, respectively. The spaces $C_\infty(Q)$ and $C_b(Q)$ are Banach spaces with norm

$$\|\cdot\|_\infty = \sup_{r\in Q} |(\cdot)(r)|,$$

$C_0(Q)$ is endowed with the topology of inductive limit

$$C_0(Q) = \liminf_n C_0(Q_n),$$

where Q_n is an increasing sequence of compact sets whose union is equal to Q, and $C_0(Q_n)$ is the space with norm $\|\cdot\|_\infty$ of functions from $C(Q)$ with support in Q_n. Thus, convergence in $C_0(Q)$ is actually the uniform convergence of functions whose supports belong to one of Q_n. Recall that any continuous linear functional defined on the space $C_0(Q)$ with inductive topology is called a (complex) Radon measure. We denote the space of Radon measures by $M(Q)$. Let $M_b(Q) = (C_\infty(Q))'$ be the Banach space of bounded Radon measures with norm

$$\|\mu\|_\infty = \sup\{|\mu(f)| \,|\, f\in C_\infty(Q),\ |f|_\infty \le 1\},$$

and let $M_c(Q)$ be the space of Radon measures with compact support. The Riesz–Markov theorem establishes a one-to-one correspondence between positive Radon measures $M^+(Q)$ (bounded positive Radon measures $M_b^+(Q)$) and regular Borel (finite regular Borel) measures on Q. By $M_1(Q) \subset M_b^+(Q)$ we denote the set of Radon probability measures. The topology of simple convergence on functions from $C_0(Q)$ in the space of Radon measures $M(Q)$ is called *vague* topology or v*-topology*. In the space $M_b(Q)$ of bounded Radon measures, one can also introduce the topology of simple convergence on functions from $C_b(Q)$. This topology is called *weak;* we write $\mu_\alpha \Rightarrow \mu$ to denote that the net $\mu_\alpha \in M_b(Q)$ weakly converges to the measure $\mu \in M_b(Q)$.

A hypercomplex system with basis Q is determined by its *structure measure* $c(A, B, r)$ $(A, B \in \mathcal{B}(Q); r \in Q)$. The structure measure $c(A, B, r)$ is a nonnegative regular Borel measure with respect to $A(B)$ with fixed $B \in \mathcal{B}(Q)$, $r \in Q(A, r)$, and satisfies the following requirements:

(H1) for each $A, B \in \mathcal{B}_0(Q)$, we have $c(A, B, r) \in C_0(Q)$;

(H2) for each $A, B \in \mathcal{B}_0(Q)$, $C \in \mathcal{B}(Q)$ and $s \in Q$, the following associativity relation holds:

$$\int c(A, B, r)\, d_r c(E_r, C, s) = \int c(B, C, r)\, d_r c(A, E_r, s) \tag{1.1}$$

(if the domain of integration is not specified explicitly, this always means that integration is carried out over the entire Q).

A structure measure is called *commutative* whenever

(H3) $c(A, B, r) = c(B, A, r) \quad (A, B \in \mathcal{B}_0(Q);\ r \in Q).$

A regular Borel measure m positive on open sets

$$\int c(A, B, r)\, dm(r) = m(A)m(B) \quad (A, B \in \mathcal{B}_0(Q)) \tag{1.2}$$

is called *multiplicative.*

In what follows:

(H4) we assume that at least one multiplicative measure exists.

In the next section, we prove the existence theorem for a multiplicative measure under certain restrictions imposed on the commutative structure measure. The problem of whether the multiplicative measure exists for a noncommutative structure measure remains open.

We fix a multiplicative measure and denote the integration with respect to it by dr (the notions such as almost everywhere, orthogonality, etc. are regarded with respect to this measure). We also denote

$$\int_A x(r)\, dr = x(A), \quad c(A, B, C) = \int_C c(A, B, r)\, dr \ (A, B, C \in \mathcal{B}(Q)).$$

It is sometimes convenient to consider, together with the measure $c(A, B, r)$, its extension to the sets from $Q \times Q$. For this purpose, we fix r and, for any $(A, B \in \mathcal{B}(Q))$, put $m_r(A \times B) = c(A, B, r)$. By additivity, the function of sets m_r can be uniquely extended to the algebra $\mathfrak{R}$ spanned by the rectangles $A \times B$ $(A, B \in \mathcal{B}(Q))$. The regularity of structure measure implies the regularity of m_r. Indeed, it suffices to check the indicated property for a rectangle $E = A_1 \times B_1$. Since $\mathcal{B}(Q) \ni A \mapsto c(A, B_1, r)$ is regular for fixed B_1, there exists, for any $\varepsilon > 0$, a compact set $F_1 \subset A_1$ such that $c(A_1, B_1, r) - c(F_1, B_1, r) < \varepsilon$. Similarly, we find a compact set $G_1 \subset B_1$ such that $c(F_1, B_1, r) - c(F_1, G_1, r) < \varepsilon$. This relation implies that $m_r(A_1 \times B_1) - m_r(F_1 \times G_1) < 2\varepsilon$, whence it follows that m_r is regular which means, as is known, that m_r is σ-additive on $\mathfrak{R}$. If we extend this measure by using the well-known procedure to all Borel sets from $Q \times Q$, then we obtain a nonnegative regular Borel measure m_r on $Q \times Q$. For fixed $\mathfrak{A} \in \mathcal{B}(Q \times Q)$, the function of $r \in Q$ $m_r(\mathfrak{A})$ is, in any case, measurable because it is obtained from continuous functions of the form $c(A, B, r)$ by finite summations and limit transitions.

Consider the space $L_1(Q, m) = L_1$ of functions on Q integrable with respect to the multiplicative measure m.

This space with the *convolution*

$$(x * y)(r) = \int \int x(p) y(q) d m_r(p, q), \tag{1.3}$$

is called a hypercomplex system with basis Q. The following theorem shows that this definition is correct.

Theorem 1.1. *The space $L_1(Q, m)$ with the norm*

$$\|x\|_1 = \int |x(r)| dr$$

and multiplication (1.3) (the integral in (1.3) exists for almost all r) is a Banach algebra and $\|x * y\|_1 \leq \|x\|_1 \|y\|_1$. If, moreover, the structure measure is commutative, then the Banach algebra $L_1(Q, m)$ is also commutative.

Proof. First, we note that, by the Fubini theorem, the integral in (1.3) can be rewritten in the form

$$\int \int x(p) y(q) d m_r(p, q) = \int x(p) d_p \Big(\int y(q) d_q c(E_p, E_q, r) \Big). \tag{1.4}$$

Let us show that, for any $x, y \in L_1(Q, m)$, integral (1.4) exists for almost all r, belongs to $L_1(Q, m)$, and

$$\int \Big| \int x(p) d_p \int y(q) d_q c(E_p, E_q, r) \Big| dr \leq \int |x(p)| dp \int |y(q)| dq. \tag{1.5}$$

To do this, we note that the set Q contains a set Q_0 of full measure for every point r of which, the integral

$$\int y(q) d_q c(A, E_q, r) \quad (A \in \mathcal{B}_0(Q))$$

exists and is a measure with respect to A. Indeed, let $Q_1 \subset Q_2 \subset \ldots$ be an increasing sequence of compact sets such that

$$Q = \bigcup_{n=1}^{\infty} Q_n.$$

By virtue of Lemma 1.5 presented below, for each Q_n, we have

$$\int \Big(\int |y(q)| d_q c(Q_n, E_q, r) \Big) dr = \int |y(q)| d_q \Big(\int c(Q_n, E_q, r) dr \Big)$$

$$= \int |y(q)| dq m(Q_n) < \infty.$$

This implies the existence of a set of full measure $Q_0 \subset Q$ such that

$$\int |y(q)| d_q c(Q_n, E_q, r) < \infty$$

for every $n = 1, 2, \ldots$ and $r \in Q_0$. Applying Lemma 1.7 presented below, we obtain the required result.

We can assume that the outer integration in (1.5) is carried out over Q_0. By using Lemmas 1.4 and 1.5, we get

$$\int_{Q_0} \left| \int x(p) d_p \left(\int y(q) d_q c(E_p, E_q, r) \right) \right| dr$$

$$\leq \int_{Q_0} \left(\int |x(p)| d_p \left(\int |y(q)| d_q c(E_p, E_q, r) \right) \right) dr$$

$$= \int |x(p)| d_p \left(\int_{Q_0} \int |y(q)| d_q c(E_p, E_q, r) dr \right)$$

$$= \int |x(p)| d_p \left(\int |y(q)| d_q \int_{Q_0} c(E_p, E_{qv}, r) dr \right)$$

$$= \int |x(p)| dp \int |y(q)| dq.$$

Hence, convolution (1.3) exists for almost all r, belongs to L_1, and

$$\|x * y\|_1 \leq \|x\|_1 \|y\|_1,$$

i.e., multiplication is continuous and, in addition, clearly distributive. Therefore, it suffices to prove the relation of associativity (and commutativity relation if the structure measure is commutative) for a total set of functions from L_1, e.g., for the indicators of sets from $\mathcal{B}_0(Q)$. The associativity and commutativity relations are true for these functions by virtue of (H2) and (H3) because $(\kappa_A * \kappa_B) = c(A, B, r)$.

■

We say that a hypercomplex system $L_1(Q, m)$ is *commutative* if its structure measure is commutative. Unless otherwise stated, we consider general noncommutative hypercomplex systems. In what follows, the results related to the commutative case will be specially indicated. If a hypercomplex system has no unity, then we formally add the unity element e and consider the algebra $\mathfrak{L}_1(Q, m)$ of expressions of the form $\lambda e + x$, where λ is a scalar and $x \in L_1$ with multiplication

$$(\lambda_1 e + x_1) * (\lambda_2 e + x_2) = \lambda_1 \lambda e + (\lambda_1 x_2 + \lambda_2 x_1 + x_1 * x_2).$$

An essentially bounded measurable complex-valued function $\chi(r)$ $(r \in Q)$ not identically equal to zero on a set of positive measure is called a *character* of a hypercomplex system if the equality

$$\int c(A, B, r)\chi(r)\,dx = \chi(A)\chi(B) \tag{1.6}$$

holds for any $A, B \in \mathcal{B}_0(Q)$. Every hypercomplex system has at least one character, namely, the function identically equal to 1. We note that if $\chi(r)$ $(r \in Q)$ is a character, then $\overline{\chi(r)}$ is also a character.

The following definition proves to be very helpful: a measurable complex-valued function $\chi(r)$ $(r \in Q)$ not identically equal to zero on a set of positive measure, is called a *generalized character* of a hypercomplex system if (1.6) holds for any $A, B \in \mathcal{B}_0(Q)$.

Theorem 1.2. *The set of characters of a commutative hypercomplex system is in one-to-one correspondence with the set of maximal ideals of the Banach algebra $L_1(Q, m)$ (or of the algebra $\mathfrak{L}_1(Q, m)$ if $L_1(Q, m)$ has no unity; in this case, one must consider only the ideals other than L_1). The maximal ideal M and the character χ are connected by the relation*

$$x(M) = \int x(r)\chi(r)\,dr \quad (x \in L_1). \tag{1.7}$$

Proof. Indeed, consider a character $\chi(r)$. We associate each $\lambda e + x \in \mathfrak{L}_1(Q, m)$ (if $L_1(Q, m)$ has unity, the proof is similar) with a number

$$\lambda + \int x(r)\chi(r)\,dr.$$

Since $\chi(r)$ is bounded almost everywhere, this correspondence generates a linear multiplicative functional in L_1:

$$\int (x * y)(r)\chi(r)\,dr = \int \Big(\int x(p)\,d_p\Big(\int y(q)\,d_q c(E_p, E_q, r)\Big)\chi(r)\,dr\Big)$$

$$= \int x(p)\,d_p \int y(q)\,d_q\Big(\int c(E_p, E_q, r)\chi(r)\,dr\Big)$$

$$= \int x(p)\,d_p \int y(q)\,d_q(\chi(E_p)(\chi(E_q))$$

$$= \int x(p)\chi(p)\,dp \int y(q)\chi(q)\,dq.$$

Clearly, this multiplicative functional generates a maximal ideal M associated with the character $\chi(r)$ by relation (1.7). The constructed ideal $M \neq L_1$ since, otherwise, $\chi(r) = 0$ almost everywhere.

Conversely, assume that a maximal ideal $M \neq L_1$ is given. The correspondence $x \mapsto x(M)$ is a linear functional in L_1. By using the theorem on the form of linear functionals in L_1, we obtain

$$x(M) = \int x(r)\chi(r)\,dr \quad (x \in L_1)$$

where $\chi(r)$ is a function essentially bounded by one (because the norm of the functional is equal to one). The function $\chi(r)$ satisfies (1.6) since

$$\int c(A, B, r)\chi(r)\,dr = (\kappa_A * \kappa_B)(M) = \kappa_A(M)\kappa_B(M) = \chi(A)\chi(B).$$

In view of the fact that $M \neq L_1$, the function $\chi(r)$ cannot be zero almost everywhere, and, hence, χ is a character.

■

It follows from the proof of Theorem 1.2 that every character of a commutative hypercomplex system is bounded almost everywhere by 1.

We note that the *measure* $c(A, B, r)$ *is absolutely continuous with respect to the measure* m *for any fixed* $B \in \mathcal{B}_0(Q)$ *and* $r \in Q$. Indeed, if A is a set of measure zero, then it follows from (1.2) that $c(A, B, r) = 0$ almost everywhere. Since $c(A, B, r)$ is continuous in r and the measure m is positive on open sets, we have $c(A, B, r)$ for all $r \in Q$.

■

In contrast to the measure $c(A, B, r)$, the measure $m_r(\mathfrak{A})$ is, in general, not absolutely continuous with respect to the square of the measure m.

Let us present several simple examples.

Example 1. Let $Q = G$ be a locally compact group and let λ be its left Haar measure. We set $c(A, B, r) = \lambda(B^{-1} r \cap A)$ and $m(A) = \lambda(A)$. One can easily see that $c(A, B, r)$ satisfies all axioms of structure measure and m is its multiplicative measure. Convolution (1.4) becomes an ordinary convolution of functions on the group

$$(x * y)(r) = \int x(p)y(p^{-1} r)\,dp$$

and the hypercomplex system $L_1(G, \lambda)$ is a group algebra. If G is abelian, then (1.6) implies that the characters of the hypercomplex system $L_1(G, \lambda)$ satisfy the standard equality for characters of the group G, namely, $\chi(pq) = \chi(p)\chi(q)$ almost everywhere.

In particular, if $G = \mathbb{R}^1$ is an additive group of all real numbers, then we arrive at the ordinary convolution on the real axis

$$(x * y)(r) = \int x(p)y(r-p)\,dp \quad (x, y \in L_1(\mathbb{R}^1, dt))$$

and the functions $\chi(t) = e^{its}$ $(s \in \mathbb{R}^1)$ are characters.

Example 2. Let $Q := [0, \infty)$ and let λ be the Lebesgue measure. We set

$$c(A, B, r) = \frac{1}{2}[\lambda((A+r)\cap B) + \lambda(|A-r| \cap B)] \quad (A, B \in \mathcal{B}_0([0, \infty)); r \geq 0),$$

where $|A-r| = \{p \,|\, p = |s-r|, \ s \in A\}$. It is easy to check that $c(A, B, r)$ is a commutative structure measure and the Lebesgue measure is its multiplicative measure. This structure measure generates the convolution

$$(x * y)(r) = \frac{1}{2}\int_0^\infty [x(p+r) + x(|p-r|)]y(p)\,dp \quad (r \geq 0),$$

and relation (1.6) implies that the characters of the related hypercomplex system satisfy the functional equation

$$\chi(p)\chi(q) = \frac{1}{2}[\chi(p+q) + \chi(|p-q|)] \quad (p, q \in [0, \infty)).$$

This yields $\chi(p) = \cos sp$ $(s \geq 0)$ almost everywhere.

Example 3. Let $Q = [0, \infty)$,

$$c(A, B, r) = \lambda((r-A)\cap B), \quad \text{and} \quad m(A) = \lambda(A) \quad (A, B \in \mathcal{B}_0([0, \infty)); r \geq 0).$$

Then $L_1([0, \infty)), m)$ with the Duhamel convolution

$$(x * y)(r) = \int_0^r x(r-p)y(p)\,dp$$

is a commutative hypercomplex system The functions $\chi(p) = e^{zp}$ $(z \in \mathbb{C}, \ \operatorname{Re} z \leq 0; \ p \leq 0)$ are characters of this hypercomplex system.

1.2. Theorem of Existence of a Multiplicative Measure. In this subsection, we present the proof of the existence of a multiplicative measure mentioned in Subsection 1.1. Most of the conditions of Theorem 1.3 are trivial for the compact case.

Theorem 1.3. *Assume that the structure measure* $c(A, B, r)$ *is commutative and satisfies the following conditions:*

(i) $c(Q, A, r) \in C_b(Q)$ *for all* $A \in \mathcal{B}_0(Q_0)$;

(ii) for all $A \in \mathcal{B}_0(Q_0)$, *there exists the limit*

$$\lim_{r\to\infty} c(Q\, A, r) = c(Q, A, \infty)$$

where the function $\mathcal{B}_0(Q) \ni A \mapsto c(Q, A, \infty)$ *is a regular Borel premeasure;*

(iii) for all open $O \in \mathcal{B}_0(Q)$, *the following inequalities hold:* $c(Q, O, r) > 0$ $(r \in Q)$ *and* $c(Q, O, \infty) > 0$. *Then there exists at least one multiplicative measure.*

Proof. First of all, we note that, for any $A \in \mathcal{B}_0(Q)$, any $B \in \mathcal{B}(Q)$, and any compact set K, there exists a sequence of open sets $O_n \supset A$ such that

$$c(O_n, B, r) \xrightarrow[n\to\infty]{} c(A, B, r)$$

uniformly on K. Indeed, by virtue of the regularity of the structure measure for fixed r, one can find an open set $O_{n,r} \supset A$ such that $c(O_{n,r}, B, r) - c(A, B, r) < 1/n$. Since the structure measure is continuous in r, there exists a neighborhood $U(r)$ such that, $c(O_{n,r}, B, p) - c(A, B, p) < 1/n$ for any $p \in U(r)$. We cover K by the neighborhoods $U(r)$ and take the finite covering $U(r_1)$, $U(r_2), \ldots,$ $U(r_k)$. Then

$$c(O_{n,r}, B, r) - c(A, B, r) < 1/n$$

for the open set $O_n = \bigcap_{i=1}^k O_{n,r_i}$ and all $r \in K$. This yields the required assertion. Note that the regularity of the structure measure enables us to choose this sequence so that all O_n lie in a compact neighborhood of the set A. If $B \in \mathcal{B}_0(Q)$, then we take $\operatorname{supp} c(B, O_1, r)$ as K and conclude that $c(B, O_n, r)$ converges to $c(B, A, r)$ uniformly in r.

Let us show that, for any $A \in \mathcal{B}_0(Q)$ and $\varepsilon > 0$, one can find an open set $O \in \mathcal{B}_0(Q)$, $O \supset A$, such that $\| c(Q, O, r) - c(Q, A, r) \|_\infty < \infty$. In fact, by virtue of the regularity of the premeasure $c(Q, A, \infty)$, one can indicate an open set $O_\infty \supset A$ such that $| c(Q, O_\infty, \infty) - c(Q, A, \infty) | < \varepsilon/5$.

Let $K \in \mathcal{B}_0(Q)$ be a compact set such that, for all $r \in Q \backslash K$,

$$| c(Q, O_\infty, r) - c(Q, O_\infty, \infty) | < \frac{\varepsilon}{5} \quad \text{and} \quad | c(Q, A, r) - c(Q, A, \infty) | < \frac{\varepsilon}{5}.$$

For every open $O \subset O_\infty$ that contains A and every $r \in Q \setminus K$, we have

$$c(Q, A, r) - c(Q, O_\infty, \infty) \leq c(Q, O, r) - c(Q, O, \infty) \leq c(Q, O_\infty, r) - c(Q, A, r).$$

This implies that $|c(Q, O, r) - c(Q, O, \infty)| < 2\varepsilon/5$ for all $r \in Q \setminus K$. Finally, by using the assertion established at the beginning of the proof, we find an open set $O_K \supset A$ such that

$$\sup_{r \in K} |c(Q, O_K, r) - c(Q, A, r)| \leq \varepsilon/5.$$

By combining the arguments presented above, for the open set $O = O_K \cap O_\infty$, we obtain

$$\sup_{r \in Q} |c(Q, O, r) - c(Q, A, r)|$$

$$\leq \sup_{r \in K} |c(Q, O, r) - c(Q, A, r)| + \sup_{r \in Q \setminus K} |c(Q, O, r) - c(Q, A, r)|$$

$$\leq \frac{\varepsilon}{5} + \sup_{r \in Q \setminus K} |c(Q, O, r) - c(Q, O, \infty)|$$

$$+ \sup_{r \in Q \setminus K} |c(Q, A, r) - c(Q, A, \infty)| + |c(Q, A, \infty) - c(Q, O, \infty)| \leq \varepsilon.$$

Hence, for any $A \in \mathcal{B}_0(Q)$, one can find a sequence of open sets $O_n \in \mathcal{B}_0(Q)$, $O_n \supset A$, such that $\| c(Q, O_n, r) - c(Q, A, r) \|_\infty \to 0$ as $n \to \infty$.

Consider the space $C_b^{\mathbb{R}}(Q)$ of continuous real bounded functions with uniform norm. An operator T_B $(B \in \mathcal{B}_0(Q))$ is defined by the formula

$$(T_B f)(r) = \int f(p)\, d_p c(E_p, B, r) \quad (f \in C_b^{\mathbb{R}}(Q)).$$

Let us show that the operator T_B maps the space $C_b^{\mathbb{R}}(Q)$ into itself. For any $r \in Q$, we consider a sequence of points $r_n \in Q$ such that $r_n \to r$ as $n \to \infty$. Since $c(Q, B, r) \in C_b(Q)$, by virtue of (H1), we conclude that

$$\int f(p)\, d_p c(E_p, B, r_n) \to \int f(p)\, d_p c(E_p, B, r) \quad (f \in C_0(Q))$$

and $c(Q, B, r_n) \to c(Q, B, r)$ as $n \to \infty$. This means that the sequence of measures $\mu_n = c(\cdot, B, r_n)$ weakly converges to the measure $\mu = c(\cdot, B, r)$. Hence, for any function $f \in C_b^{\mathbb{R}}(Q)$, we have

$$\int f(p)\,d_p c(E_p, B, r_n) \xrightarrow[n\to\infty]{} \int f(p)\,d_p c(E_p, B, r)$$

and $T_B\colon C_b^{\mathbb{R}}(Q) \to C(Q)$. The function $(T_B f)(r)$ is bounded. Indeed,

$$\sup_{r\in Q} |(T_{Bf})(r)| \le \sup_{r\in Q} \int |f(p)|\,d_p c(E_p, B, r) \le \|f\|_\infty \sup_{r\in Q} c(Q, B, r).$$

The operator T_B is clearly linear and continuous. If the function $f(r)$ belongs to the interior of the cone of nonnegative functions $C_b^{+,\mathbb{R}}(Q)$ in $C_b^{\mathbb{R}}$, i.e., there is $\rho > 0$ such that $\inf_{r\in Q} f(r) \ge \rho$, then $(T_O f)(r) \ge \rho c(Q, O, r)$ for any open set $O \in \mathcal{B}_0(Q)$. The function $c(Q, O, r)$ is separated from zero, i.e., $c(Q, O, r) \ge c > 0$. In fact, one can find a compact set K such that

$$\sup_{r\in Q\setminus K} |c(Q, O, r) - c(Q, O, \infty)| \le \frac{1}{2} c(Q, O, \infty)$$

($c(Q, O, \infty) > 0$ by condition 3)). Let $c_1 = \min_{r\in K} c(Q, O, r)$. Then the required

$$c = \min\left\{c_1, \frac{1}{2} c(Q, O, \infty)\right\} > 0.$$

This implies that, for any open set $O \in \mathcal{B}_0(Q)$, the operators T_O map the interior of the cone of nonnegative functions into itself.

The operators T_B $(B \in \mathcal{B}_0(Q))$ are commuting. Indeed, by virtue of (H2), (H3), and Lemma 1.5, we get

$$\begin{aligned}
(T_A(T_B f))(r) &= \int \left(\int f(p)\,d_p c(E_p, B, q)\right) d_q c(E_q, A, r) \\
&= \int f(p)\,d_p\left(\int c(E_p, B, q)\,d_q c(E_q, A, r)\right) \\
&= \int f(p)\,d_p\left(\int c(B, E_p, q)\,d_q c(E_q, A, r)\right) \\
&= \int f(p)\,d_p\left(\int c(E_p, A, q)\,d_q c(B, E_q, r)\right) \\
&= \int f(p)\,d_p\left(\int c(E_p, A, q)\,d_q c(E_q, B, r)\right) = T_B(T_A f))(r).
\end{aligned}$$

Thus, we have the collection $\{T_O \mid O \in \mathcal{B}_0(Q) \text{ is an open set}\}$ of linear continuous commuting operators mapping the interior of the cone of all nonnegative continuous functions bounded on Q into itself. By the Krein theorem (M. Krein and Rutman

[KrR]), the adjoint operators T_0^* have a common eigenvector $l \in (C_b^{\mathbb{R}}(Q))'$, i.e., $T_0^* l = \lambda_O l$ ($\lambda_O > 0$), where the functional l is positive (note that the space $(C_b^{\mathbb{R}}(Q))'$ is isometrically isomorphic to the space of regular finitely additive measures on the algebra induced by closed sets; therefore we may regard l as a regular positive finitely additive measure). Thus, we have

$$\langle l, T_O f\rangle = \lambda_O \langle l, f\rangle \tag{1.8}$$

for all $f \in C_b^{\mathbb{R}}(Q)$ and $O \in \mathcal{B}_0(Q)$. By setting $f(r) \equiv 1$ in (1.8), we obtain $\lambda_O = \langle l, c(Q, O, \cdot)\rangle / \langle l, 1\rangle$.

For any $A \in \mathcal{B}_0(Q)$, we define a finitely additive measure m' as follows:

$$m'(A) = \frac{\langle l, c(Q, A, \cdot)\rangle}{\langle l, 1\rangle}.$$

Since, for any $A \in \mathcal{B}_0(Q)$, there is a sequence of open sets $\mathcal{B}_0(Q) \ni O_n \supset A$ such that $c(Q, O_n, r) \in C_b^{\mathbb{R}}(Q)$ converges uniformly to $c(Q, A, r) \in C_b^{\mathbb{R}}(Q)$ and the functional l is continuous, the pre-measure m' is regular. By the well-known theorem, there exists a unique regular Borel measure m such that $m(A) = m'(A)$ for all $A \in \mathcal{B}_0(Q)$. Let us show that this is the required measure. We fix $A \in \mathcal{B}_0(Q)$ and an arbitrary open set $O \in \mathcal{B}_0(Q)$. As known, for the continuous finite function $c(A, O, r)$, there exists a sequence of uniformly finite simple functions $f_n(p)$ that converges uniformly to $c(A, O, p)$. Let S be a compact set that contains $\operatorname{supp} f_n$ ($n = 1, 2, \ldots$) and $\operatorname{supp} c(A, O, r)$. We have

$$\begin{aligned} &\left| \int c(A, O, p)\, d_p c(Q, E_p, r) - \int f_n(p)\, d_p c(Q, E_p, r) \right| \\ &\qquad \leq \| c(A, O, p) - f_n(p) \|_\infty c(Q, S, r) \\ &\qquad \leq \| c(A, O, p) - f_n(p) \|_\infty \| c(Q, S, r) \|_\infty. \end{aligned} \tag{1.9}$$

It follows from (1.9) that

$$\int f_n(p)\, d_p c(Q, E_p, r)$$

converges to

$$\int c(A, O, p)\, d_p c(Q, E_p, r)$$

uniformly in r. This fact, the Lebesgue theorem, the continuity of the functional l, axiom (H2), and (1.8) yield

$$\begin{aligned}\int c(A,O,r)\,dm(r) &= \lim_{n\to\infty}\int f_n(r)\,dm(r)\\ &= \lim_{n\to\infty}\frac{1}{\langle l,1\rangle}\Big\langle l,\int f_n(p)\,d_pc(Q,E_p,r)\Big\rangle\\ &= \Big\langle l,\int c(A,O,p)\,d_pc(Q,E_p,r)\Big\rangle\Big/\langle l,1\rangle\\ &= \frac{1}{\langle l,1\rangle}\Big\langle l,\int c(Q,A,p)\,d_pc(O,E_p,r)\Big\rangle\\ &= \frac{\langle l,(T_O c)(Q,A,\cdot)\rangle}{\langle l,1\rangle} = \frac{\langle l,c(Q,O,r)\rangle}{\langle l,1\rangle}\frac{\langle l,c(Q,A,r)\rangle}{\langle l,1\rangle}\\ &= m(O)m(A).\end{aligned} \tag{1.10}$$

Assume that B is an arbitrary open set in $\mathcal{B}_0(Q)$. We choose a sequence of open sets $O_n \supset B$ so that $c(A,O_n,r) \xrightarrow[n\to\infty]{} c(A,B,r)$ uniformly in r from $Q_N \supset \operatorname{supp} c(A,O_n,r) \supset \operatorname{supp} c(A,B,r)$. Since the measure m is regular, there is a sequence of open sets $U_n \supset B$ such that $m(U_n) \xrightarrow[n\to\infty]{} m(B)$. We set $V_n = O_n \cap U_n$. Obviously, $m(V_n) \xrightarrow[n\to\infty]{} m(B)$ and $c(A,V_n,r) \xrightarrow[n\to\infty]{} c(A,B,r)$ uniformly in r from Q_N. By setting $O = V_n$ in (1.10) and passing to the limit, we get

$$\int c(A,B,r)\,dm(r) = m(A)m(B).$$

■

1.3. Normal Hypercomplex Systems. Below, we consider only hypercomplex systems with a certain symmetry property possessed by many important examples. For a hypercomplex system, this condition is an analog of the transition to an inverse element in a group. For simplicity, we consider the "unimodular" case; in the general case, the normality condition must be rewritten so that, in Theorem 1.8 presented below, modular bialgebras are used instead of Hilbert algebras (Vainerman and Kac [VaK]).

(H5) A hypercomplex system is called *normal* if there is an involutive homeomorphism $Q \ni p \mapsto p^* \in Q$ such that $m(E^*) = m(E)$, $c(A,B,C) = c(A^*,C,B) = c(C,B^*,A)$ $(E \in \mathcal{B}(Q),\ A,B,C \in \mathcal{B}_0(Q))$.

If $r^* = r$ for all $r \in Q$, a normal hypercomplex system is called *Hermitian*. *Every*

Hermitian hypercomplex system is commutative. In fact, for any $A, B, C \in \mathcal{B}_0(Q)$, we have $c(A, B, C) = c(A, C, B) = c(C, B, A) = c(B, A, C)$, whence $c(A, B, r) = c(B, A, r)$ almost everywhere by virtue of the arbitrariness of $C \in \mathcal{B}_0(Q)$. According to (H1), both functions in this equality are continuous.

■

In a normal hypercomplex system, the correspondence $L_1 \ni x = x(p) \mapsto \overline{x(p^*)} = x^* \in L_1$ *is an involution.* Indeed, all properties of involution are evident except the equality $(x * y)^* = y^* * x^*$ $(x, y \in L_1)$. It suffices to prove this equality for the indicators of sets from $\mathcal{B}_0(Q)$; here, this is equivalent to the relation $c(A, B, C^*) = c(B^*, A^*, C)$ $(A, B, C \in \mathcal{B}_0(Q))$, which can be established by using (H5) as follows:

$$c(A, B, C^*) = c(A^*, C^*, B) = c(B, C, A^*) = c(B^*, A^*, C). \tag{1.11}$$

■

If a hypercomplex system does not contain the unity, then one can extend the involution into the algebra $\mathfrak{L}_1(Q, m)$ by setting $(\lambda e + x)^* = \bar{\lambda} e + x^*$.

A normal hypercomplex system has a unique multiplicative measure. Indeed, let $r \in Q$ be an arbitrary fixed point and let $V_1 \supset V_2 \supset \ldots$ be a sequence of bounded neighborhoods of r that contract to this point. We fix an arbitrary set $A \in \mathcal{B}_0(Q)$ and take an increasing sequence of compact sets $Q_1 \supset Q_2 \supset \ldots$ such that $Q_1 \supset \operatorname{supp} c(V_1, A, r)$ and $Q = \bigcup_{n=1}^{\infty} Q_n$. Then

$$\begin{aligned} c(Q, A^*, r) &= \lim_{n\to\infty} c(Q_n, A^*, r) = \lim_{n\to\infty}\lim_{k\to\infty} \frac{1}{m(V_k)} \int_{V_k} c(Q_n, A^*, s)\,ds \\ &= \lim_{n\to\infty}\lim_{k\to\infty} \frac{1}{m(V_k)} \int_{Q_n} c(V_k, A, s)\,ds \\ &= \lim_{n\to\infty}\lim_{k\to\infty} \frac{1}{m(V_k)} \int c(V_k, A, s)\,ds \\ &= \lim_{k\to\infty} \frac{1}{m(V_k)} m(V_k) m(A) = m(A). \end{aligned} \tag{1.12}$$

In particular, this implies that $c(Q, A, r) = \text{const} < \infty$ for any $A \in \mathcal{B}_0(Q)$.

■

For a normal hypercomplex system, the following inequality is true:

$$m_r(A \times B) = c(A, B, r) \le m(A) \quad (A, B \in \mathcal{B}(Q); r \in Q). \tag{1.13}$$

Indeed, it suffices to prove (1.13) for $A, B \in \mathcal{B}_0(Q)$. By virtue of (1.12), we have $c(A, B, r) = c(B^*, A^*, r^*) \le c(Q, A^*, r) = m(A)$.

■

Denote $L_\alpha(Q, m) = L_\alpha$, $\|\cdot\|_{L_\alpha} = \|\cdot\|_\alpha$, $(\cdot,\cdot)_{L_2} = (\cdot,\cdot)_2$, and $L'_\alpha = L_{\alpha'}$, where $\alpha^{-1} + \alpha'^{-1} = 1 \quad (\alpha \in [1, \infty])$.

Lemma 1.1. *For $f \in L_\alpha$ and $g \in L_{\alpha'}$ $(\alpha \in [1, \infty])$, the convolution $f * g$ exists, is a bounded function, and satisfies the inequality $\|f * g\|_\infty \le \|f\|_\alpha \|f\|_{\alpha'}$. This function belongs to the space $C_\infty(Q)$ for $\alpha \in (1, \infty)$. If $\alpha = 1$, then, for the convolution to be continuous, it suffices to require, in addition, that $g \in L_\beta$ for some $\beta \in [1, \infty)$.*

Proof. Let $\alpha \in (1, \infty)$ and let a and b be step functions (i.e., linear combinations of the indicators of sets from $\mathcal{B}_0(Q)$). Relation (1.3), the Hölder inequality, and (1.13) yield

$$\begin{aligned} |(a * b)(r)| &= \left| \iint a(p) b(q)\, dm_r(p, q) \right| \\ &\le \left(\iint |a(p)|^\alpha dm_r(p, q) \right)^{1/\alpha} \left(\iint |b(q)|^{\alpha'} dm_r(p, q) \right)^{1/\alpha'} \\ &\le \left(\int |a(p)|^\alpha dp \right)^{1/\alpha} \left(\int |b(q)|^{\alpha'} dq \right)^{1/\alpha'} \\ &= \|a\|_\alpha \|b\|_{\alpha'} \quad (r \in Q), \end{aligned} \tag{1.14}$$

where the function $(a * b)$ is continuous by virtue of (H1). Approximating the functions f and g in the metrics of L_α and $L_{\alpha'}$ by the functions a and b and using estimate (1.4), we conclude that $(a * b)(r)$ uniformly converges to $(f * g)(r)$. Hence, $(f * g)(r)$ is a continuous function satisfying estimate (1.4). Lemma 1.1 is proved in the case $\alpha \in (1, \infty)$.

Now assume that $\alpha = 1$, $f \in L_1$, and $g \in L_\infty$. By analogy with (1.14), we obtain

$$\begin{aligned} |(f * g)(r)| &= \left| \iint f(p) g(q)\, dm_r(p, q) \right| \\ &= \|g\|_\infty \iint |f(p)|\, dm_r(p, q) \le \|g\|_\infty \|f\|_1, \end{aligned}$$

which implies that $f * g(r)$ exists and is bounded.

Assume, in addition, that $g \in L_\beta$ for some $\beta \in [1, \infty)$. It suffices to show that one can find sequences a_n and b_n of finite step functions such that $\| a_n * b_n - f * g \|_\infty \to 0$ as $n \to 0$. We choose sequences a_n and b_n such that

$$\| a_n - f \|_1 < n^{-1} \quad \text{and} \quad \| b_n - g \|_\beta < n^{-1} \| a_n \|_{\beta'}^{-1} \quad (\beta^{-1} + \beta'^{-1} = 1).$$

By using the Hölder inequality and (1.13), we get

$$\begin{aligned} &| (f * g)(r) - (a_n * b_n)(r) | \\ &\quad \leq | (f * g)(r) - (a_n * g)(r) | + | (a_n * g)(r) - (a_n * b_n)(r) | \\ &\quad = \left| \iint (f - a_n)(p) g(q) dm_r(p, q) \right| + \left| \iint a_n(p)(g - b_n)(q) dm_r(p, q) \right| \\ &\quad \leq n^{-1}(\| g \|_\infty + 1) \xrightarrow[n \to \infty]{} 0. \end{aligned}$$

■

It follows from Lemma 1.1 that *if a normal hypercomplex system contains the left unity, then its basis Q consists of at most countably many points with discrete topology*. Indeed, let $e \in L_1$ be a unity. By virtue of Lemma 1.1, for $\alpha = 1$ and any $g \in L_1 \cap L_\infty$, the function $e * g$ is continuous and coincides almost everywhere with g. If an arbitrary $g \in L_1 \cap L_\infty$ coincides almost everywhere with a continuous function and the measure is positive on open sets, then the topology of Q must be discrete. It remains to take into account the separability of Q.

■

A similar assertion is true if the left unity is replaced by the right unity.

Lemma 1.2. *For $x \in L_1$ and $f \in L_2$, the convolutions $(x * f)(r)$ and $(f * x)(r)$ exist for almost all $r \in Q$ and belong to L_2. Furthermore,*

$$\| x * f \|_2 \leq \| x \|_1 \| f \|_2,$$

$$\| f * x \|_2 \leq \| f \|_2 \| x \|_1. \tag{1.15}$$

Proof. It suffices to prove the first inequality in (1.15); the second inequality can be proved by analogy. Let $A, B, C \in \mathcal{B}_0(Q)$. Then, by virtue of (H5), we have

$$(\kappa_A * \kappa_B, \kappa_C)_2 = c(A, B, C) = c(A^*, C, B) = c(C, B^*, A) = (\kappa_A, \kappa_C * \kappa_{B^*})_2.$$

For step functions a, b, and c, this yields

$$(a * b, c) = (a, c * b^*) = (b, a^* * c). \tag{1.16}$$

By using the estimate in Lemma 1.1 with $\alpha = 2$, we obtain

$$|(a * b, c)_2| \le \|a\|_1 \|c * b^*\|_\infty \le \|a\|_1 \|b^*\|_2 \|c\|_2 = \|a\|_1 \|b\|_2 \|c\|_2$$

and, therefore, $\|a * b\|_2 \le \|a\|_1 \|b\|_2$ in view of the arbitrariness of c. Approximating x and f by the functions a and b in the metrics of L_1 and L_2, respectively, we complete the proof of the lemma.

■

It follows from Lemma 1.2 that, for fixed $x \in L_1$, one can define the operators of left and right convolution in the space L_2 as follows: $L_2 \ni f \mapsto x * f = L(x) f \in L_2$ and $L_2 \ni f \mapsto f * x = R(x) f \in L_2$. These operators are bounded and satisfy the inequalities $\|L(x)\| \le \|x\|_1$ and $\|R(x)\| \le \|x\|_1$. It is clear that $L(\lambda x + \mu y) = \lambda L(x) + \mu L(y)$ and $L(x * y) = L(x) L(y)$ $(x, y \in L_1,\ \lambda, \mu \in \mathbb{C})$. The adjoint operator satisfies the equality $L^*(x) = L(x^*)$ $(x \in L_1)$. This is equivalent to the relation

$$(x * f, g)_2 = (f, x^* * g)_2 \qquad (f, g \in L_2)$$

obtained by passing to the limit in (1.16). The right convolution $R(x)$ possesses similar properties. If a hypercomplex system $L_1(Q, m)$ is commutative, then $L(x) = R(x)$ and $L(x)$ is a normal operator:

$$L^*(x) L(x) = L(x^* * x) = L(x * x^*) = L(x) L(x^*) = L(x) L^*(x).$$

In the space L_2, in addition to the convolutions $L(x)$ and $R(x)$ $(x \in L_1)$, we define one more operator family associated with hypercomplex systems, namely, left (right) generalized translation operators L_p (R_p), $p \in Q$. The operator L_p is given in L_2 by the bilinear form $(L_p f, g)_2 = (f * g^*)(p)$ $(f, g \in L_2;\ p \in Q)$. The operator R_p is defined similarly: $(R_p f, g)_2 = (g^* * f)(p)$, $f, g \in L_2$, $p \in Q$ (recall that we consider the case of unimodular hypercomplex systems). It follows from Lemma 1.1 with $\alpha = 2$ that these forms are continuous and, hence, the operators L_p and R_p are well defined and satisfy the inequalities $\|L_p\| \le 1$ and $\|R_p\| \le 1$. Since

$$(f, L_{p^*} g)_2 = \overline{(g * f^*)(p^*)} = (g * f^*)^*(p) = (f * g^*)(p) = (L_p f, g)_2,$$

we have $L_p^* = L_{p^*}$ $(p \in Q)$. Similarly, $R_p^* = R_{p^*}$. In the case where Q is a locally compact group and the hypercomplex system L_1 is its group algebra, we have $L_p f(q) = f(pq)$ and $R_p f(q) = f(qp)$ $(p, q \in Q)$. The operator families L_p and R_p $(p \in Q)$ possess all properties of the operator family of Delsarte–Levitan generalized translations (Levitan [Lev9]); in Section 1.2, the relationship between hypercomplex systems and generalized translation operators is discussed in detail.

Obviously,

$$L(x) = \int x(p) L_{p^*} dp \qquad (x \in L_1) \tag{1.17}$$

i.e., the operator of left convolution is the adjoint operator of left translation "averaged with function x". Let us show that Lemma 1.1 with $\alpha = 2$ implies the weak continuity of the function $Q \ni p \mapsto L_p \in \mathfrak{L}(L_2, L_2)$, where $\mathfrak{L}(E_1, E_2)$ is the space of linear continuous operators from E_1 to E_2. Note that the integral in (1.17) should be understood in the weak sense.

Let us prove (1.17). Assume that $f, g \in L_2$. Then

$$\left(\left(\int x(p) L_{p^*} dp\right) f, g\right)_2 = \int x(p) (L_{p^*} f, g)_2 \, dp$$

$$= \int x(p)(f * g^*)(p^*) \, dp = \langle x, \overline{g * f^*} \rangle. \tag{1.18}$$

Passing to the limit in (1.16) as a tends to x in L_1 and b and c tend to f and g, respectively, in L_2, we get $(x * f, g)_2 = \langle x, \overline{g * f^*} \rangle$. Therefore, (1.18) can be continued as follows:

$$\left(\left(\int x(p) L_{p^*} dp\right) f, g\right)_2 = (x * f, g)_2 = (L_x f, g)_2 \quad (f, g \in L_2).$$

■

Let $L_1(Q, m)$ be a commutative normal hypercomplex system. A character $\chi(p)$ is called *Hermitian* if $\chi(p^*) = \overline{\chi(p)}$. *Hermitian characters of an Hermitian hypercomplex system are real-valued:* $\overline{\chi(p)} = \chi(p^*) = \chi(p)$ $(p \in Q)$.

Generally speaking, the Banach algebra $L_1(Q, m)$ is not symmetric, i.e., the equality $M^* = M$ holds not for all maximal ideals of it (or, in view of (1.7), not all characters of it are Hermitian). Denote the set of all characters (generalized characters) and the set of all Hermitian characters (generalized Hermitian characters) by X (X_g) and X_h $(X_{g,h})$, respectively.

Theorem 1.4. *The characters of a commutative hypercomplex system are continuous functions (more precisely, coincide with such functions almost everywhere). If* Q

is compact, then these characters are Hermitian; the set of such characters is at most countable, they are mutually orthogonal and form, up to constant factors, the system of idempotents of the algebra L_1, *i.e.*,

$$(\chi_j, \chi_k) = \delta_{jk}\|\chi_k\|_2^2, \qquad \chi_j * \chi_k = \delta_{jk}\chi_j\|\chi_j\|_2^2$$

$$(\chi = \chi_h = (\chi_j)_{j=1}^{\leq\infty}, \qquad j, k = 1, 2, \ldots).$$

Proof. Let $A, B \in \mathcal{B}_0(Q_0)$, with $\chi(B) \neq 0$. We construct a compact set F so that $c(A, B, r) = 0$ for $r \notin F$ and set $\chi_F(r) = \chi(r)$ for $r \in F$ and $\chi_F(r) = 0$ at all other points. If $E \subset A$ and $\alpha = 1$, according to (1.6), (H5), and Lemma 1.1, we obtain

$$\begin{aligned}\chi(E)\chi(B) &= \int c(E, B, r)\chi(r)\,dr = \int c(E, B, r)\chi_F(r)\,dr \\ &= \int \chi_F(r)\,d_r c(E, B, E_r) = \int \chi_F(r)\,d_r c(E_r, B^*, E) \\ &= \int_E (\chi_F * \kappa_{B^*})(p)\,dp.\end{aligned}$$

Since E is arbitrary, we have

$$\chi(r) = \frac{1}{\chi(B)}(\chi_F * \kappa_{B^*})(r) \tag{1.19}$$

for almost all $r \in A$. The arbitrariness of A, this equality, and Lemma 1.1 imply the continuity of χ.

Let Q be compact. The Hermitian property for χ follows from the relation

$$\begin{aligned}\chi(A)\|\chi\|_2^2 &= \int \overline{x(p)}\,d_p\Big(\int c(E_p, A, r)\chi(r)\,dr\Big) \\ &= \int\Big(\overline{\chi(p)}\,d_p\Big(\int \chi(r)\,d_r c(E_p, A, E_r)\Big) \\ &= \int \overline{\chi(p)}\,d_p\Big(\int \chi(r)\,d_r c(E_r, A^*, E_p)\Big) \\ &= \int \chi(r)\,d_r\Big(\int \Big(\overline{\chi(p)}\,d_p c(E_r, A^*, E_p)\Big) \\ &= \overline{\chi(A^*)}\,\|\chi\|_2^2 \qquad (A \in \mathcal{B}(Q)).\end{aligned} \tag{1.20}$$

To prove the orthogonality of two characters χ' and χ'' $(\chi' \neq \chi'')$, we take $B \in$

$\mathcal{B}(Q)$ such that $\chi'(B) \neq \chi''(B) = \overline{\chi''(B^*)}$. In view of (1.19), we have

$$\left(\chi'(B) - \overline{\chi''(B^*)}\right) \int \chi'(p)\overline{\chi''(p^*)}\,dp$$

$$= \int (\chi' * \kappa_{B^*})(p)\chi''(p)\,dp - \int \chi'(p)\overline{(\chi'' * \kappa_B)(p)}\,dp$$

$$= (\chi' * \kappa_{B^*}, \chi'')_2 - (\chi', \chi'' * \kappa_B)_2 = 0, \tag{1.21}$$

whence $(\chi', \chi'')_2 = 0$.

The separability of L^2 and the orthogonality of characters imply that there are at most countably many characters and we can set $X = (\chi_j)_{j=1}^{\leq\infty}$. The last assertion of the theorem easily follows from the relation (see (1.19))

$$(\chi_j * \chi_k, \kappa_A)_2 = (\chi_k, \chi_j^* * \kappa_A)_2 = (\chi_k, \chi_j)\chi_j(A) \quad (A \in \mathcal{B}(Q)).$$

■

We introduce a topology in X by using the topology of the space of maximal ideals of the commutative Banach algebra $L_1(Q, m)$. It is easy to check that the topology in X coincides with the topology of uniform convergence on compact sets. The set X_h of Hermitian characters is clearly closed in X.

Consider the general case of noncommutative hypercomplex systems An element a of the algebra K is called an *annihilator* if $a * x = 0$ for all $x \in K$.

Theorem 1.5. *The radical of a normal hypercomplex system coincides with the collection of its annihilators.*

Proof. If x is an annihilator, then it belongs the radical, because $x * x = 0$. We show the inverse inclusion. Assume that $u \in L_1$ is a fixed function such that $u * x = u^* * x$ for any $x \in L_1$. In the Hilbert space L_2, we consider the convolution operator L_u with the function u, i.e., $L_2 \in f \mapsto L_u f = u * f \in L_2$.

Let us show that the operator L_u is selfadjoint. In fact, $L_u^* = L_{u^*}$. But $L_{u^*} f = u^* * f = u * f = L_u f$ for an arbitrary function $f \in L_1 \cap L_\infty$. Therefore, the selfadjointness of L_u follows from the denseness of $L_1 \cap L_\infty$ in L_2 and the fact that the operator L_u is bounded.

Now let $u(r)$ belong, in addition, to the radical $\mathfrak{R}$ of the algebra $L_1(Q, m)$. Then, clearly, the element $0e + u$ belongs to the radical of the algebra $\mathfrak{L}_1(Q, m)$. We show that the operator $L_u = 0$. Indeed, let R_z be the resolvent of the operator L_u. It is easy to see that $R_z f = (u - ze)^{-1} * f$ $(f \in L_2)$. Since $u(r)$ belongs to the radical, the function $(u - ze)^{-1}$ is defined for all $z \neq 0$ and belongs to $\mathfrak{L}_1(Q, m)$. Let $(u - ze)^{-1} =$

$\lambda e + y$ ($y \in L_1(Q, m)$). By virtue of Lemma 1.2., for $f \in L_2$, we have $(u - ze)^{-1} * f = \lambda f + y * f \in L_2$. Hence, the resolvent R_z is defined for all $z \neq 0$. Therefore, the spectrum of the self-adjoint operator L consists of zero and, consequently, $L_u = 0$. In particular, $L_u f = u * f = 0$ for every function $f \in L_1 \bigcap L_\infty$. In view of the continuity of multiplication in the algebra $L_1(Q, m)$ and the denseness of $L_1 \bigcap L_\infty$ in L_∞, we conclude that $u * f = 0$ for all $f \in L_1$. The theorem is proved for the special case under consideration ($u * x = u^* * x$, $x \in L_1$).

Let $v \in \mathfrak{R}$ be an arbitrary element of the radical. Since the radical of a $*$-algebra is a symmetric two-sided ideal, v^* also belongs to the radical $\mathfrak{R}$ of the algebra $L_1(Q, m)$. In this case, $u = v + v^* \in \mathfrak{R}$. Therefore, $u * x = 0$ ($x \in L_1$) according to the already proved result, i.e., $v * x = -v^* * x$ ($f \in L_1$). Consequently, the element iv satisfies the condition of the already proved part of the theorem and, thus, it is an annihilator. together with v.

■

Corollary 1.1. *For a normal hypercomplex system to be semisimple, it is necessary and sufficient that it do not contain annihilators other than zero.*

We illustrate the ideas discussed in this section by several examples presented at the end of Section 1.1. It is easy to see that the hypercomplex system in Example 1 is normal if we set $r^* = r^{-1}$ and the group G is unimodular. The hypercomplex system in Example 2 is Hermitian for $r^* = r$. The hypercomplex system in Example 3 is not normal. The generalized translation operators associated with the hypercomplex system in Example 2 are given by the formula

$$(L_p x)(q) = \frac{1}{2}[x(p+q) + x(|p-q|)] \quad (p, q \in [0, \infty)).$$

1.4. Normal Hypercomplex Systems with Basis Unity. In this subsection, we discuss generalizations of finite-dimensional normal hypercomplex systems with unity element e in its basis to the continual case. We say that a normal hypercomplex system possesses a *basis unity* if

(H6) there exists a point e in Q such that $e^* = e$ and $c(A, B, E) = m(A^* \bigcap B)$ ($A, B \in \mathcal{B}(Q)$).

The basis unity of a normal hypercomplex system is called *strong* if

(H7) for each neighborhood U of an arbitrary set $F \in \mathcal{B}_0(Q)$, there is a neighborhood V of the basis unity e such that $\operatorname{supp} c(F, V, r) \subset U$.

A sequence of functions $e_n \in L_1$ is called an *approximative unit* of a hypercomplex system if $e_n(p) \geq 0$, $e_n(p) = e_n(p^*)$ $(p \in Q)$, $\|e\|_1 = 1$ and, for all $x \in L_1$, $e_n * x \xrightarrow[n\to\infty]{} x$ weakly in L_1.

If a normal hypercomplex system possesses the approximative unit $\{e_n\}_{n=1}^{\infty}$, *then* $e_n * f \to f$ *strongly in* L_2 *for all* $f \in L_2$ *as* $n \to \infty$.

In fact, assume that $g \in L_2$ and a and b are step functions. Then

$$|(e_n * f, g)_2 - (f, g_2)| \leq |(e_n * a, b)_2 - (a, b)_2| + 2\|a\|_2\|g - b\|_2 + 2\|g\|_2\|f - a\|_2$$

by virtue of (1.15). Since $(e_n)_{n=1}^{\infty}$ is an approximative unit, this implies that $e_n * f \to f$ weakly in L_2. Strong convergence now follows from the relation

$$\|e_n * f - f\|_2^2 = \|e_n * f\|_2^2 - 2\operatorname{Re}(e_n * f, f)_2 + \|f\|_2^2 \leq 2\|f\|_2^2 - 2\operatorname{Re}(e_n * f, f)_2 \xrightarrow[n\to\infty]{} f,$$

(we have used (1.15) and weak convergence just proved).

■

Lemma 1.3. *A normal hypercomplex system with basis unity possesses an approximative unit.*

Proof. Assume that $(O_n)_{n=1}^{\infty}$, $O_n \supset O_{n+1}$, is a sequence of bounded neighborhoods of the basis unity e contracting to this point. We show that the sequence of functions $e_n \in L_1$ such that $e_n(p) \geq 0$ $(p \in Q)$, $\operatorname{supp} e_n \subset O_n$, and $\|e_n\|_1 = 1$ is an approximative unit of the hypercomplex system (specifically, one can take $e_n(p) = (m(O_n))^{-1}\kappa_{O_n}(p)$ $(p \in Q;\, n = 1, 2, \dots)$. If necessary, one may take $e_n(p) \in C_0(Q)$ (the existence of a function of this sort for all s is guaranteed by the Uryson lemma). Let us show that $x * e_n \xrightarrow[n\to\infty]{} x$ weakly in L_1. Let $l \in L_1' = L_\infty$ be a linear functional in L_1. We must prove that $l(e_n * x) \to l(x)$ as $n \to \infty$. It suffices to consider the case $x = \kappa_A$ $(A, B \in \mathcal{B}(Q))$. Then the required assertion is also true for step functions which are dense in L_1 and the proof is completed by using the inequality

$$|l(e_n * x) - l(x)| \leq |l(e_n * a) - l(a)| + 2\|l\|_\infty\|x - a\|_1$$

(we have used the fact that $\|e_n * (x - a)\|_1 \leq \|e_n\|_1\|x - a\|_1 = \|x - a\|_1$). Thus, let $x = \kappa_A$. By $F \subset Q$ we denote a sufficiently large compact set such that $F \supset O_n$ $(n = 1, 2, \dots)$. By virtue of (H1), there is a compact set $G \supset A$ such that $c(A, B, E) = 0$ on the complement of G. Let l_G be a function equal to $l(r)$ for $r \in G$ and equal to zero at the other points. By using (1.16), we obtain

$$l(e_n * \kappa_A) = \int (e_n * \kappa_A) l(r)\, dr$$

$$= \int (e_n * \kappa_A)(r) l_G(r)\, dr = \int e_n(r)(l_G * \kappa_{A^*})(r)\, dr.$$

The function $l_G * \kappa_{A^*}$ is continuous for $\alpha = 1$ by virtue of Lemma 1.1. Therefore, by applying the standard argument and (H6) to the last inequality, we obtain

$$l(e_n * \kappa_A) \xrightarrow[n \to \infty]{} (l_G * \kappa_{A^*})(e) = \int l_G(p)\, d_p c(E_p, A^*, e)$$

$$= \int l_G(p)\, dm\, (E_p \cap A) = l_G(A) = \langle l, \kappa_A \rangle.$$

■

Corollary 1.2. *A normal hypercomplex system with approximative unit is semisimple. In particular, a normal hypercomplex system with basis unity is semisimple.*

Indeed, in view of Theorem 1.5, it is necessary to show the absence of nonzero annihilators in the hypercomplex system. Let x be an annihilator. Then

$$x = \lim_{n \to \infty} x * e_n = 0$$

in a sense of weak convergence in L_1, i.e., $x(r)$ almost everywhere.

■

Clearly, the approximative unit $e_n(p)$ constructed in proving Lemma 1.3 has the following property:

$$\int e_n(p) f(p)\, dp \to f(e) \quad \text{as} \quad n \to \infty \qquad (f \in C_b(Q)). \tag{1.22}$$

We now establish several properties of the collection of characters of a commutative normal hypercomplex system with basis unity. First, we recall the following definition: the system $\mathfrak{A}$ of functions on Q is called *complete* in a class $\mathfrak{B}$ of functions on Q if the fact that the equality

$$\int x(q) \varphi(q)\, dq = 0$$

is true for all $\varphi \in \mathfrak{A}$ and some $x \in \mathfrak{B}$ (all $x(q)\varphi(q)$ are supposed to be summable) implies that $x(q) = 0$ almost everywhere.

Theorem 1.6. *In order for a point* $e \in Q$ *to be the basis unity of a commutative normal hypercomplex system, it is necessary and sufficient that*

1) the system of all characters be complete in L_1;

2) for each character, $\chi(e) = 1$.

Proof. *Necessity.* The fact that a normal hypercomplex system with basis unity is semisimple yields the completeness of the system of all characters. We show that $\chi(e) = 1$ for every character. We take a compact set A such that $\chi(A) \neq 0$ and denote by $U_1 \supset U_2 \supset \ldots$ a sequence of balls contracting to e. It follows from Lemma 1.3 that

$$\chi(A) \lim_{n\to\infty} \frac{\chi(U_n)}{m(U_n)} = \lim_{n\to\infty} \frac{1}{m(U_n)} \int c(A, U_n^*, r)\chi(r)\,dr$$

$$= \lim_{n\to\infty} \int (\kappa_A * e_n)(r)\chi(r)\,dr = \chi(A).$$

Hence,

$$\lim_{n\to\infty} \frac{\chi(U_n)}{m(U_n)} = 1.$$

Together with the continuity of the character χ, this yields the equality $\chi(e) = 1$.

Sufficiency. Let B be a compact set, and let U be a precompact neighborhood of the point e. Consider a compact set F such that $c(U, B^*, r) = 0$ for $r \notin F$ and define $\chi_F(r)$ as in the proof of Theorem 1.4. For $E \subset U$, $\chi \in X$, we have

$$\int_E (\chi_F * \kappa_B)(r)\,dr = \int \chi_F(p)\,d_p c(E_p, B, E)$$

$$= \int \chi_F(p)\,d_p c(E, B^*, E_p) = \int c(E, B^*, p)\chi(p)\,dp$$

$$= \chi(E)\chi(B^*).$$

Since $E \subset U$ is arbitrarily and the functions $(\chi_F * \kappa_B)(r)$ and $\chi(r)$ are continuous, we obtain

$$(\chi_F * \kappa_B)(e) = \chi(e)\chi(B^*) = \chi(B^*). \tag{1.23}$$

Consider the measure $c(A, B, e)$ $(A \in \mathcal{B}_0(Q))$. For $A \subset Q \backslash F$, we have $c(A, B, e) = 0$ since

$$c(A, B, U) = c(U, B^*, A) = \int_A c(U, B^*, r)\,dr = 0.$$

Moreover, it follows from (1.13) that the magnitude of the derivative of the measure $c(A, B, e)$ with respect to $m(A)$ is bounded by the unit element. This implies that there exists a bounded function $c(r)$ vanishing outside F such that

$$c(A, B, e) = \int c(r)\kappa_B(r)\,dr \quad (A \in \mathcal{B}_0(Q)).$$

For any character, by using (1.23), we obtain

$$\int [c(r) - \kappa_{B^*}(r)]\chi(r)\,dr = \int_F \chi(r)c(r)\,dr - \chi(B^*)$$

$$= \int \chi_F(r)\,d_r c(E_r, B, e) - \chi(B^*)$$

$$= (\chi_F * \kappa_B)(e) - \chi(B^*) = 0.$$

Since $c(r) - \kappa_{B^*}(r)$ is bounded and finite, we have $c(r) = \kappa_{B^*}(r)$ almost everywhere because the system of characters is complete in L_1. Hence, for all $A \in \mathcal{B}_0(Q)$,

$$c(A, B, e) = \int_A c(r)\,dr = \int_A \kappa_{B^*}(r)\,dr = \int_A \kappa_B(r)\,d = m(A^* \cap B).$$

■

As follows from the proof, for a point $e \in Q$ to be the basis unity of a commutative normal hypercomplex system, it suffices to check the equality $\chi(e) = 1$ and the fact that the system of characters is complete in the class of essentially bounded finite functions.

It is clear that the hypercomplex systems considered in Examples 1 and 2 (Subsection 1.1) are systems with basis unity: In Example 1, the basis unity e coincides with the unit element of the group, whereas in Example 2, $e = 0$.

1.5. Even Subsystem of a Normal Hypercomplex System. In this subsection, we construct an Hermitian hypercomplex system which is a subalgebra of the initial normal hypercomplex system.

Let $x(r)$ be a function on Q. We set $x'(r) = x(r^*)$ and call the function $x(r)$ *even (odd)* if $x' = x$ $(x' = -x)$. Clearly, every function on Q can be uniquely represented as the sum of its even and odd components, namely,

$$x = \frac{1}{2}(x + x') + \frac{1}{2}(x - x').$$

Consider the mapping $x \to x'$ in the space L_1. This mapping is linear, involutive, and $(x * y)' = y' * x'$. Therefore, the set of all even functions from L_1 forms a commutative subalgebra of $L_1(Q, m)$. We denote this subalgebra by $H_1(Q, m)$ and call it the *even subsystem. It can be regarded as an Hermitian hypercomplex system whose basis is a locally compact space* $\boldsymbol{Q}$ *obtained from the initial basis* Q *by uniting the points* r *and* r^* *into the class* $\boldsymbol{r}$.

In fact, we denote the preimages of the sets $\boldsymbol{A}, \boldsymbol{B}, \boldsymbol{E} \ldots \in \mathcal{B}(\boldsymbol{Q})$ under the mapping $r \to \boldsymbol{r}$ $(r \in \boldsymbol{r})$ by $A, B, E, \ldots \in \mathcal{B}(Q)$. These preimages are invariant under $*$. We introduce a measure $\boldsymbol{m}$ on $\boldsymbol{Q}$ by setting $\boldsymbol{m}(\boldsymbol{E}) = m(E)$ $(\boldsymbol{E} \in \mathcal{B}(\boldsymbol{Q}))$. Denote $d\boldsymbol{m}(\boldsymbol{r}) = d\boldsymbol{r}$. Every function $x(\boldsymbol{r}) \in L_1(\boldsymbol{Q}, \boldsymbol{m})$ induces an even function $x(r) \in L_1(Q, m)$, and vice versa. In addition,

$$\int_{\boldsymbol{E}} x(\boldsymbol{r})\, d\boldsymbol{r} = \int_{E} x(r)\, dr.$$

This readily implies that the convolution of two functions from $H_1(Q, m)$ can be represented in the form

$$x * y(\boldsymbol{r}) = \int x(\boldsymbol{p})\, d\boldsymbol{p} \int x(\boldsymbol{q})\, d\boldsymbol{q}\, \boldsymbol{c}(E_{\boldsymbol{p}}, E_{\boldsymbol{q}}, \boldsymbol{r}),$$

where the measure $\boldsymbol{c}$ is defined by the equality $\boldsymbol{c}(\boldsymbol{A}, \boldsymbol{B}, \boldsymbol{r}) = c(A, B, r)$. Thus, $H_1(Q, m)$ is a hypercomplex system for which the Hermitian property follows from the relations $\boldsymbol{c}(\boldsymbol{A}, \boldsymbol{B}, \boldsymbol{C}) = c(A, B, C)$ and $B = B^*$.

■

One can easily show that *if* $\chi(r)$ *is an Hermitian character of a commutative normal hypercomplex system, then the function*

$$v(\boldsymbol{r}) = \frac{1}{2}[\chi(r) + \chi(r^*)] = \operatorname{Re}\chi(r) \quad (r \in \boldsymbol{r}) \tag{1.24}$$

is an Hermitian character of its even subsystem.

Indeed, let M be a maximal ideal of the algebra $\mathfrak{L}_1(Q, m)$ associated with the character χ (for definiteness, we assume that $L_1(Q, m)$ does not contain the unit element). The mapping $x \to x(M)$ $(x \in H_1(Q, m))$ is a homeomorphism of the algebra $H_1(Q, m)$ in the field of complex numbers. This homeomorphism does not map all $H_1(Q, m)$ into zero because if we assume the opposite, then it possible to find an odd function $x \in L_1$ such that $x(M) \neq 0$. But then $(x^* * x)(M) = |x(M)|^2 \neq 0$ in contradiction with our hypothesis because $x^* * x$ is even. Thus, the homeomorphism $x \to x(M)$ $(x \in H_1(Q, m))$ generates a character of the even subsystem. We denote it by $v(\boldsymbol{r})$. For every $x \in H_1(Q, m)$, we have

$$\int x(\boldsymbol{r})v(\boldsymbol{r})d\boldsymbol{r} = \int x(r)\chi(r)dr,$$

which gives the required equality.

■

If the even subsystem of a commutative hypercomplex system is a symmetric algebra, then every its character admits representation (1.24).

Indeed, let $v(\boldsymbol{r})$ be a character of the hypercomplex system $H_1(Q, m)$ and let M' be the corresponding maximal ideal. By virtue of the Shilov theorem on the extensions of maximal ideals [Shi], the maximal ideal M can be extended as a maximal ideal of the symmetric subalgebra to the maximal ideal M of the entire algebra. Since the mapping $x \to x(M)$ $(x \in L_1)$ clearly does not map all $L_1(Q, m)$ into zero, M is generated by a character $\chi(r)$ of the hypercomplex system $L_1(Q, m)$. This is just the required character.

■

One can show that if the hypercomplex system $L_1(Q, m)$ is semisimple (with basis unity), then the hypercomplex system $H_1(Q, m)$ is also semisimple (with basis unity). The converse assertion is weaker: If the even subsystem $H_1(Q, m)$ of a commutative hypercomplex system $L_1(Q, m)$ possesses a basis unity and the hypercomplex system $L_1(Q, m)$ is semisimple, then $L_1(Q, m)$ possesses a basis unity

The hypercomplex system in Example 2 (Section 1.1) is the even subsystem of the hypercomplex system $L_1(\mathbb{R}, d)$ with ordinary convolution. In this case, equality (1.24) takes the form $\cos \lambda x = \operatorname{Re} e^{i\lambda x}$.

1.6. The Case of a Discrete Basis. In this subsection, we study basic notions of the theory of hypercomplex systems in the case where the basis of a hypercomplex system is a countable set Q with discrete topology The structure measure $c(A, B, r)$ transforms into the collection of structure constants $c(p, q, r)$ $(p, q, r \in Q)$ with the following properties: (α) $c(p, q, r) \geq 0$, (β) for any fixed $p, q \in Q$, the sequence $(c(p, q, r))_{r \in Q}$ is finite, and (γ) the associativity relation

$$\sum_r c(p, q, r)c(r, l, s) = \sum_r c(q, l, r)c(p, r, s) \tag{1.25}$$

holds.

The multiplicative measure turns into the multiplicative weight, i.e., the sequence of positive numbers $m(p)$ $(p \in Q)$ such that

$$\sum_r c(p, q, r)m(r) = m(p)m(q) \quad (p, q \in Q) \tag{1.26}$$

The hypercomplex system with basis Q is the space $l_1(m)$ of sequences of complex numbers summable with respect to the multiplicative weight with convolution

$$(x * y)(r) = \sum_{p,q} x(p)y(q)c(p,q,r) \quad (x, y \in l_1(m)).$$

Any bounded sequence $\chi(p) \in \mathbb{C}^1$ $(p \in Q)$ not identically equal to zero and such that

$$\sum_r c(p,q,r)\chi(r)m(r) = \chi(p)m(p)\chi(q)m(q) \tag{1.27}$$

is a character.

This hypercomplex system is normal if there exists an involutive mapping $Q \ni p \mapsto p^* \in Q$ such that $m(p) = m(p^*)$ and

$$c(p,q,r)m(r) = c(r,q^*,p)m(p) = c(p^*,r,q)m(q).$$

The involution in the hypercomplex system $l_1(m)$ is given in the standard way:

$$l_1(m) \ni x = (x(p))_{p \in Q} \mapsto \overline{(x(p^*))}_{p \in Q} = x^* \in l_1(m).$$

The point $e \in Q$ is the basis unity of the hypercomplex system $l_1(m)$ provided that $e^* = e$ and $c(p,q,e) = \delta_{pq^*}$ (δ_{pq} is the Kronecker symbol and $p, q \in Q$).

If the hypercomplex system $l_1(m)$ possesses a basis unity, then there is a two-sided identity element with respect to the convolution in this hypercomplex system. The sequence $e(p)$, equal to $1/m(e)$ for $p = e$ and to zero for $p \neq e$, is the required element.

In fact,

$$(e * x)(r) = \frac{1}{m(e)} \sum_q x(q)c(e,q,r) = \frac{1}{m(r)} \sum_q x(q)c(r,q^*,e)$$

$$= \frac{1}{m(r)} \sum_q x(q)\delta_{r,q}m(q) = x(r) \quad (p,q,r \in Q).$$

The generalized translation operator in a normal hypercomplex system has the form

$$L_p x(q) = \frac{1}{m(p)m(q)} \sum_q x(r)c(p,q,r)m(r). \tag{1.28}$$

We give a simple example of a discrete hypercomplex system. Let $Q = \{0, 1, \dots\}$.

We set $m(0) = 1/\pi$ and $m(p) = 2/\pi$ $(p \neq 0)$ and define the structure measure as follows:

$$c(p, q, r) = \frac{m(p)m(q)}{2m(r)}[\delta_{|p-q|,r} + \delta_{|p+q|,r}].$$

The obtained hypercomplex system is Hermitian and possesses the basis unity $e = 0$. The functions $\chi(p) = \cos pt$ $(t \in [0, \pi])$ are characters of this hypercomplex system. One can easily show that the hypercomplex system $l_1(m)$ is isometrically isomorphic to the algebra of continuous functions on $[0, \pi]$ with ordinary algebraic operations and absolutely convergent Fourier cosine expansions, i.e., it is isomorphic to the algebra of functions

$$\hat{x}(t) = \frac{1}{\pi}x(0) + \frac{2}{\pi}\sum_{p=1}^{\infty} x(p)\cos pt$$

such that

$$\sum_{n=0}^{\infty} |x(p)| < \infty.$$

Theorem 1.7. *The following conditions are equivalent:*

(i) the basis Q of a normal hypercomplex system with basis unity is discrete;

(ii) the point e is isolated;

(iii) $m(e) > 0$.

Proof. The implication (i) $\Rightarrow$ (ii) is evident. The implication (ii) $\Rightarrow$ (iii) holds due to the fact that the multiplicative measure is positive on open sets. We now show that (iii) implies (i). First, we prove that the function

$$g(p) = \frac{1}{m(e)}\kappa_{\{e\}}(p)$$

is the unit element in the algebra $L_1(D, m)$. By virtue of Lemma 1.1, $(g * f)(r) \in C_\infty(Q)$ for all $f \in L_1$. We assume that $f(r) \in C_b(Q) \cap L_1$ and E_n is a sequence of neighborhoods of the point $r \in Q$ contracting to this point. Then

$$f * g(r) = \lim_{n\to\infty} \frac{1}{m(E_n)} \int_{E_n} (f * g)(r)\,dr$$

$$= \lim_{n\to\infty} \frac{1}{m(\{e\})} \frac{1}{m(E_n)} \int f(p)\, d_p c(E_p, \{e\}, E_n)$$

$$= \lim_{n\to\infty} \frac{1}{m(E_n)} \int f(p)\, d_p c(E_p^*, E_n, \{e\})$$

$$= \lim_{n\to\infty} \frac{1}{m(E_n)} \int f(p)\, d_p m(E_n \cap E_p)$$

$$= \lim_{n\to\infty} \frac{1}{m(E_n)} \int_{E_n} f(p)\, dp = f(r).$$

by virtue of Lemma 1.1 (see below). By approximating an arbitrary function $f \in L_1$ by functions from $L_1 \cap C_b(Q)$ in the metric of L_1 and applying Lemma 1.1, we conclude that $(f * g)(r) = (g * f)(r) = f(r)$ almost everywhere.

■

1.7. Hilbert Algebras. Let us show that a normal hypercomplex system with basis unity is closely related to the theory of double Hilbert algebras (Hilbert bialgebras). Recall some definitions (for simplicity, we consider the unimodular case). The lineal $\mathcal{D}$ dense in a Hilbert space H is called a *Hilbert algebra* if $\mathcal{D}$ is a $*$-algebra with the following properties:

(i) $(a, b)_H = (b^*, a^*)_H$ $(a, b \in \mathcal{D})$;

(ii) $(a * b, c)_H = (a, c * b^*)_H$ $(a, b, c \in \mathcal{D})$;

(iii) for each $a \in \mathcal{D}$, the operators $L(a)b = a * b$ $(b \in \mathcal{D})$ can be extended to continuous operators in the entire space H;

(iv) $\mathcal{D} * \mathcal{D}$ is dense in H.

Suppose that the following two pairs of operations are defined on a lineal $\mathcal{D}$ dense in the Hilbert space H: the multiplication $a \cup b$ and involution $a^{\cup}$ (the first pair) and the multiplication $a \cap b$ and involution $a^{\cap}$ (the second pair). In this case, the lineal $\mathcal{D}$ is called a *Hilbert bialgebra* if

(i) for each pair of operations, $\mathcal{D}$ is a Hilbert algebra in H;

(ii) the operator W uniquely defined in $H \otimes H$ by the equality

$$(a \otimes b'^{\cap}, W(a'^{\cup} \otimes b))_{H \otimes H} = (a \cup a', b \cap b')_H \quad (a, b, a', b' \in \mathcal{D}),$$

is continuous (W is called a *binding operator*).

Theorem 1.8. *Let $\mathfrak{L}_1(Q, m)$ be a normal hypercomplex system with basis unity. Then the lineal $\mathcal{D} = L_1 \cap L_\infty$ dense in the Hilbert space $H = L_2$ is a Hilbert bialgebra with respect to the operations $f \cup g = f * g$ and $f \cap g = f \cdot g$ and the involutions $f(r) \mapsto \overline{f(r^*)}$ and $f(r) \mapsto \overline{f(r)}$, respectively.*

Proof. By virtue of Theorem 1.1, Lemma 1.1, and the normality condition, $\mathcal{D}$ is a $*$-algebra with respect to convolution and the involution f^*. It is clear that $\mathcal{D}$ is a $*$-algebra with respect to pointwise multiplication of functions and complex conjugation. The fact that $(\mathcal{D}, \cdot)$ is a Hilbert algebra is also obvious.

Let us show that $(\mathcal{D}, *)$ is a Hilbert algebra. The first condition is satisfied by virtue of the fact that $m(A) = m(A^*)$ $(A \in \mathcal{B}(Q))$, the second one follows from (1.16), the third condition is a consequence of Lemma 1.2, and, finally, the fourth condition follows from Lemma 1.3 and the fact that the existence of an approximative unit $e_n(r)$ in a hypercomplex system allows one to approximate an arbitrary function $f \in \mathcal{D}$ in L_2 by convolutions of the form $e_n * f$ (it should be noted that the functions $e_n(r)$ constructed in Lemma 1.3 lie in $\mathcal{D}$). Therefore, it remains to show that the binding operator is continuous. Indeed, for all $f_1, f_2, g_1, g_2 \in \mathcal{D}$, by virtue of Lemma 1.2 and the Schwartz inequality, we have

$$\begin{aligned}
|(f_1 * f_2, g_1 \cdot g_2)_{L_2}| &= |(f_1, (g_1 \cdot g_2) * f_2^*)_{L_2}| \\
&\le \|f_1\|_2 \|(g_1 \cdot g_2) * f_2^*\|_2 \le \|f_1\|_2 \|g_1 \cdot g_2\|_1 \| f_2^*\|_2 \\
&\le \|f_1\|_2 \|g_1\|_2 \|\overline{g_2}\|_2 \| f_2^*\|_2 \\
&= \|f_1 \otimes \overline{g_2}\|_{L_2 \otimes L_2} \| f_2^* \otimes g_1\|_{L_2 \otimes L_2}
\end{aligned}$$

and, thus, $\|W\| \le 1$.

■

The fact that normal hypercomplex systems with basis unity have the structure of Hilbert bialgebra implies that these systems possess many important properties (see Vainerman and Litvinov [VaL]). However, the presence of such a structure does not give a complete description of a hypercomplex system. In Section 5, we formulate axioms that distinguish hypercomplex systems from the class of Hilbert bialgebras.

1.8. Parameter Depending Measures. In the construction of hypercomplex systems with locally compact basis, we have used several auxiliary lemmas concerning the integration over measures depending on a parameter. Though these assertions are rather simple, in this subsection, we give their proofs for completeness of presentation.

Let Q be a locally compact space. Recall that a nonnegative measure $\rho_m(E)$ $(E \in \mathcal{B}(Q))$ is called a majorant of a measure $\rho(E)$ (generally speaking, complex-valued) if $|\rho(E)| \leq \rho_m(E)$ $(E \in \mathcal{B}(Q))$. The obvious relation $|\rho_m|(E) = (\operatorname{Var} \rho)(E) \leq \rho_m(E)$ $(E \in \mathcal{B}(Q))$ implies the following statement:

Lemma 1.4. *The inequality*

$$\left| \int f(r)\, d\rho(r) \right| \leq \int |f(r)|\, d\rho_m(r) \quad (\rho \in M(Q))$$

holds for any measurable function $f(r)$ $(r \in Q)$.

Lemma 1.5. *Let $\sigma(E, r)$ and $\rho(E)$ $(r \in Q)$ be nonnegative measures with respect to $E \in \mathcal{B}(Q)$ and let $\sigma(E, r)$ and $f(r)$ be nonnegative measurable functions of $r \in Q$ such that*

$$\int \sigma(E, r)\, d\rho(r) < \infty$$

for any $E \in \mathcal{B}_0(Q)$. Then the existence of at least one of the integrals

$$\int \left(\int f(r)\, d_r \sigma(E_r, s) \right) d\rho(s) \tag{1.29}$$

or

$$\int f(r)\, d_r \left(\int \sigma(E_r, s)\, d\rho(s) \right) \tag{1.30}$$

implies that the other integral exists and they are equal. (The integral in (1.30) makes sense because one can easily verify that

$$\int \sigma(E, r)\, d\rho(r) \quad (E \in \mathcal{B}_0(Q))$$

is a nonnegative measure).

Proof. For a characteristic function $f(t)$ (and, hence, for a step function), the statement is obvious.

Let $f(t)$ be an arbitrary nonnegative measurable function for which integral (1.29) exists. Let us show that (1.30) also exists and is equal to (1.29). Indeed, it follows from the existence of (1.29) that $f(t)$ is almost everywhere finite with respect to the measure

$$\psi(E) = \int \sigma(E, r)\, d\rho(r).$$

Then one can construct a sequence of step functions $0 \leq f_1(r) \leq f_2(r) \leq \ldots \leq f_n(r) \leq \ldots \leq f(r)$ that converges to $f(r)$ almost everywhere with respect to $\psi(E)$. This implies that this sequence converges to $f(t)$ almost everywhere with respect to $\sigma(E, r)$ for almost all r with respect to $\rho(E)$. For such r, by virtue of the Lebesgue theorem on majorized convergence, we have

$$\int f_n(t)\, d_t\sigma(E_t, r) \xrightarrow[n\to\infty]{} \int f(t)\, d_t\sigma(E_t, r). \tag{1.31}$$

By using the Lebesgue theorem on the integration of monotone sequences and the finiteness of integral (1.29), we get

$$\begin{aligned}\int f(t)\, d\psi(t) &= \lim_{n\to\infty} \int f_n(t)\, d\psi(t) \\ &= \lim_{n\to\infty} \int f_n(t)\, d_t\Big(\int \sigma(E_t, r)\, d\rho(r)\Big) \\ &= \lim_{n\to\infty} \int \Big(\int f_n(t)\, d_t\sigma(E_t, r)\Big) d\rho(r) \\ &= \int \Big(\int f(t)\, d_t\sigma(E_t, r)\Big) d\rho(r) < \infty.\end{aligned}$$

Let $f(t)$ be a nonnegative measurable function for which integral (1.30) exists. Then one can construct a sequence of step functions $0 \leq f_1(t) \leq f_2(t) \leq \ldots \leq f(r)$ that converges to $f(t)$ almost everywhere with respect to $\psi(E)$. But then this sequence converges to $f(t)$ $\sigma(E, r)$-almost everywhere for $\rho(E)$-almost every r. For such r, relation (1.31) is valid. Therefore, by virtue of the Lebesgue theorem on the integration of monotone sequences, we obtain

$$\begin{aligned}\int \Big(\int f(t)\, d_t\sigma(E_t, r)\Big) d\rho(r) &= \lim_{n\to\infty} \int \Big(\int f_n(t)\, d_t\sigma(E_t, r)\Big) d\rho(r) \\ &= \lim_{n\to\infty} \int f_n(t)\, d_t\Big(\int \sigma(E_t, s)\, d\rho(s)\Big) \\ &= \int f(t)\, d_t\Big(\int \sigma(E_t, s)\, d\rho(s)\Big).\end{aligned}$$

■

Lemmas 1.4 and 1.5 readily yield the following statement:

Lemma 1.6. *Let* $\sigma(E, t)$ *and* $\rho(E)$ $(t \in Q)$ *be measures with respect to* $E \in \mathcal{B}(Q)$ *and let* $\sigma(E, r)$ *be measurable with respect to* $r \in Q$. *Suppose that the measures* $\sigma(E, r)$ *and* $\rho\cdot(E)$ *have majorants* $\bar{\sigma}(E, r)$ *and* $\bar{\rho}(E)$, *the first of which is measurable with respect to* r *and*

$$\int \bar{\sigma}(E, r)\, d\bar{\rho}(r) < \infty \quad (E \in \mathcal{B}(Q)).$$

Then the existence of at least one of the integrals

$$\int \Big(\int |f(t)|\, d_t \bar{\sigma}(E, r) \Big) d\bar{\rho}(r)$$

or

$$\int |f(t)|\, d_t \Big(\int \bar{\sigma}(E, r)\, d\bar{\rho}(r) \Big),$$

where $f(t)$ *is a measurable function, implies that integrals (1.29) and (1.30) exist and coincide.*

Lemma 1.7. *If* $\psi(A, B)$ $(A, B \in \mathcal{B}(Q))$ *is a nonnegative regular measure with respect to* A (B) *with fixed* $B \in \mathcal{B}_0(Q)$ $(A \in \mathcal{B}_0(Q))$ *and* $f(t)$ $(t \in Q)$ *is a measurable function summable with respect to the measure* $d_r \psi(E_r, A)$ *for all* $A \in \mathcal{B}_0(Q)$, *then the function*

$$\rho(A) = \int f(t) d\psi(E_t, A)$$

can be extended to a unique regular σ*-additive measure on* $\mathcal{B}(Q)$.

Proof. By virtue of Theorem 50_3 in [Schw, p. 584], it suffices to consider a nonnegative function $f(t)$. The function $\rho(A)$ is finitely additive, $\rho(\varnothing) = 0$, and, evidently, $\rho(A_1) \le \rho(A_2)$ if $A_1 \subset A_2$. Therefore, $\rho(A)$ is a compact volume. It follows from the regularity of the measure $\psi(A, B)$ in B that the compact volume is regular. This implies that $\rho(A)$ can be uniquely continued to a regular σ-additive measure on $\mathcal{B}(Q)$.

∎

2. Hypercomplex Systems and Related Objects

Generalized translation operators were first introduced by Delsarte [Del1]–[Del3] as an object that generalizes the idea of translation on a group. Later, they were systematically studied by Levitan [Lev1]–[Lev10]; for some classes of generalized translation operators, he obtained generalizations of harmonic analysis, the Lie theory, the theory of almost periodic functions, the theory of group representations, etc. The fact that generalized translation operators arise in various problems of analysis is explained by the following result of Vainerman and Litvinov [VaL]: Transformations of Fourier type for which the Plancherel theorem and the inversion formula hold, as a rule, are closely related to families of generalized translation operators. As shown in Section 1, each hypercomplex system can be associated with a family of generalized translation operators. In Subsection 2.1, we present conditions that distinguish the class of hypercomplex systems from the class of generalized translation operators. Then we clarify the relationship between hypercomplex systems and hypergroups, which were independently introduced (with slight differences in axiomatics) by Dunkl [Dun2], [Dun3], Spector [Spe1]–[Spe3], and Jewett [Jew] and are extensively studied. A hypergroup is a locally compact space Q on which the Banach $*$-algebra $M_b(Q)$ of bounded Radon measures on Q with the identity element δ_e $(e \in Q)$ is defined; the operation of multiplication in $M_b(Q)$ (in what follows, called "convolution") satisfies certain conditions, the most important of which is that the convolution of probability measures is also a probability measure. Actually, hypercomplex systems and hypergroups describe the same algebraic object. The only difference is that operation of convolution in a hypergroup satisfies stronger topological conditions than in a hypercomplex system Note that the concepts of discrete hypercomplex system and discrete hypergroup are equivalent. It seems that the language of hypercomplex systems is more suitable for the construction of duality theory, whereas the language of hypergroups is preferable for studying the structure of an object itself (see, e.g., Section 4). At the end of this section, we present one of possible extensions of hypercomplex systems to the case of complex structure measure and establish the relationship between such hypercomplex systems and the generalized convolution algebras introduced by Ionescu Tulcea and Simon [IoS].

2.1. Generalized Translation Operators and Hypercomplex Systems. Let us establish the relationship between the hypercomplex systems and Delsarte–Levitan generalized translation operators outlined in Subsection 1.3. Below, for convenience, we present some necessary facts from the theory of generalized translation operators (see Levitan [Lev9], [Lev10]).

Let Q be an arbitrary set and let Φ be a space of complex-valued functions on Q. Assume that an operator-valued function $Q \ni p \mapsto R_p\colon \Phi \to \Phi$ is given such that the function $g(p) = (R_p f)(q)$ belongs to Φ for any $f \in \Phi$ and any fixed $q \in Q$. Denote by L_q the operator $\Phi \ni f(p) \mapsto (L_q f)(p) = (R_p f)(q) \in \Phi$. The operators R_p are called *generalized translation operators,* provided that the following axioms are satisfied:

(T1) *Associativity axiom.* The equality $\big(R_p^q(R_q f)\big)(r) = \big(R_q^r(R_p f)\big)(r)$ holds for any elements $p, q \in Q$ (the notation $\big(R_p^q(R_q f)\big)(r)$ means that the operator R_p acts on the function $(R_q f)(r)$ of two variables q and r in the variable q).

(T2) There exists an element $e \in Q$ such that $R_e = I$, where I is the identity operator in Φ.

The associativity axiom can be rewritten in a more transparent form (Delsarte [Del1]–[Del4]):

(T1′) $L_q R_p = R_p L_q$ $(p, q \in Q)$.

Indeed,

$$\big(R_p^r(L_q f)\big)(r) = \big(R_p^r(R_r f)\big)(q) = \big(R_r^q(R_p f)\big)(q) = \big(L_r^q(R_p f)\big)(r),$$

i.e., (T1) implies (T1′). Properly rewriting this chain of equalities, one can obtain the converse statement.

The associativity axiom (T1) also holds for the operators L_p.

Indeed,

$$\big(L_p^q(L_q f)\big)(r) = \big(L_p^q(R_r f)\big)(q) = \big(R_q^p(R_r f)\big)(p) = \big(R_r^q(R_q f)\big)(p)$$

$$= \big(R_r^q(L_p f)\big)(q) = \big(L_q^r(L_p f)\big)(r) \quad (p, q, r \in Q).$$

■

Furthermore, the operators R_p and L_q commute:

$$\big(R_p^r(L_q f)\big)(r) = \big(R_p^r(R_r f)\big)(q) = \big(R_r^q(R_p f)\big)(q) = \big(L_r^q(R_p f)\big)(r).$$

The operator L_e is a projector, i.e., $L_e^2 = L_e$ (the subspace $\tilde{\Phi} = L_e\Phi$ is invariant under the action of L_e). Indeed,

$$\left(L_e^2 f\right)(r) = (R_p(L_e f))(e) = (L_e(R_p f))(e)$$

$$= (R_e(R_p f))(e) = (R_p f)(e) = (L_e f)(p) \quad (p \in Q).$$

The operator L_e is not necessarily identical. Therefore, generally speaking, *the operators L_p are not generalized translation operators.* To avoid this discrepancy between R_p and L_p, one should replace Φ by $\tilde{\Phi} = L_e \Phi$ and consider, instead of Q, the set $\tilde{Q}$ obtained from Q by identifying all points p' and p'' such that $f(p') = f(p'')$ for all $f \in \tilde{\Phi}$ into the class $\tilde{p}$. Since

$$(R_{p'} f)(q) = (L_q f)(p') = (L_q f)(p'') = (R_{p''} f)(q) \quad (p', p'' \in \tilde{p} \in \tilde{Q};\ q \in Q),$$

for all $f \in \tilde{\Phi}$, the operators $(R_{\tilde{p}} f)(\tilde{q}) = (R_p f)(q)$ $(p \in \tilde{p}, q \in \tilde{q})$ are well defined on $\tilde{Q}$. Clearly, they form the family of generalized translation operators. The operators $L_{\tilde{p}} = L_p$ $(p \in \tilde{p} \in \tilde{Q})$ associated with $R_{\tilde{p}}$ are, in fact, generalized translation operators because $L_{\tilde{e}} \tilde{\Phi} = L_e \tilde{\Phi} = \tilde{\Phi}$. The described process is called *reduction* and $\tilde{\Phi} = L_e \Phi$ is called *the principal space.* As a result of reduction, a smooth manifold may be transformed into a manifold with edge. For example, for any value of a parameter q, the operator

$$(L_q f)(p) = (R_p f)(q) = \frac{1}{2}[f(p+q) + f(q-p)] \quad (p, q \in \mathbb{R}^1; f \in C(\mathbb{R}^1))$$

transforms an arbitrary function into an even function and, therefore, as a result of reduction, the manifold $\mathbb{R}^1$ turns into the half line $[0, \infty)$. In the general case, $\tilde{Q}$ may have no smooth structure even if Q is a manifold.

Below, we consider only reduced generalized translation operators, i.e., we assume that reduction is already performed and $\tilde{\Phi} = \Phi$ or, which is the same, $L_e = R_e = I$.

Generalized translation operators are called *commutative* if, for any $p, q \in Q$, we have $\left(R_p^r(R_q f)\right)(r) = \left(R_q^r(R_p f)\right)(r)$. *For commutative generalized translation operators R_p, the equality*

$$(R_p f)(q) = (R_q f)(q) \quad (p, q \in Q), \tag{2.1}$$

holds and, hence, the operators R_p and L_p coincide. Indeed, if the operators R_p commute, the associativity axiom implies that

$$\left(R_q^p(R_p f)\right)(r) = \left(R_p^r(R_q f)\right)(r) = \left(R_q^r(R_p f)\right)(r) = \left(R_p^q(R_q f)\right)(r) \quad (p, q, r \in Q).$$

By setting $r = e$ and using the fact that generalized translation operators are reduced, we obtain $(R_q f)(p) = (R_p f)(q)$.

■

The equality $R_{p'} = R_{p''}$ implies that $p' = p''$. Indeed,

$$(L_q f)(p') = (R_{p'} f)(q) = (R_{p''} f)(q) = (L_q f)(p'') \quad (f \in \tilde{\Phi})$$

for all $q \in Q$. This, in particular, implies that $f(p') = (L_e f)(p') = (L_e f)(p'') = f(p'')$ $(f \in \tilde{\Phi})$ and $p' = p''$ because the generalized translation operators are reduced.

We are interested in the case where Q is a locally compact space with regular Borel measure m positive on open sets and bounded generalized translation operators R_p act in the space of functions $\Phi = L_2(Q, m) = L_2$.

Note that this choice of Φ disagrees with the formal definition of generalized translation operators (because functions from the space $L_2(Q, m)$ have no values at points). We modify this definition by assuming that the relation of associativity for generalized translation operators R_p holds for almost all p, q, and r with respect to the measure $m \times m \times m$. The case under consideration is not exhaustive. Thus, Vainerman [Vai6] studied unbounded generalized translation operators in $L_2([0, \infty), dt])$. As usual, we denote integration with respect to the measure m by dp.

Consider an involutive homeomorphism $Q \ni p \mapsto p^* \in Q$. The generalized translation operators R_p are *involutive* if the equalities

$$(R_p f)(q) = \overline{\left(R_{q^*} f^*\right)(p^*)} \quad (f \in L_2;\ f^*(p) = \overline{f(p^*)})$$

and $e^* = e$ hold for almost all $p, q \in Q$ with respect to the measure $m \times m$. The generalized translation operators R_p are *real* if they commute with complex conjugation, i.e., $(R_p \bar{f})(q) = \overline{(R_p f)(q)}$ for almost all $p, q \in Q$. Generalized translation operators R_p *preserve positivity* if $(R_p f)(q) \geq 0$ almost everywhere in m whenever $f(q) \geq 0$. Clearly, *generalized translation operators are real if they preserve positivity.* The family of operators R_p is called *weakly continuous* if the operator-valued function $Q \ni p \mapsto L_p$ is weakly continuous. We say that the measure m *unimodular* if $m(A) = m(A^*)$ for all $A \in \mathcal{B}(Q)$.

By virtue of the fact that the functions from $L_2(Q, m)$ do not have values at points, the left generalized translation operators L_p cannot be defined by using the equality $(L_p f)(r) = (R_r f)(p)$ $(p, r \in Q)$. However, one can define the family of operators L_p $(p \in Q)$ for a weakly continuous family of generalized translation operators R_p $(p \in Q)$ in the space $L_2(Q, m)$ with unimodular measure m by the bilinear form

$$(L_p f, g) = (g^*, R_{p^*} f^*) \quad (f, g \in L_2; p \in Q).$$

Since R_p is involutive and m is unimodular, we have

$$(L_p f, g) = \int \overline{R_{p^*} f^*(r)\, g(r^*)}dr = \int R_{r^*} f(p)\, \overline{g(r^*)}dr = \int R_r f(q) g(r)\, dr,$$

whence $L_p f(r) = R_r f(p)$ for almost all p and q with respect to the measure $m \times m$, i.e., the *operators* L_p $(p \in Q)$ *form a family of left generalized translation operators* (it follows from the equality $e^* = e$ that L_p are indeed generalized translation operators). It is easy to understand that *left generalized translation operators are involutive (real, weakly continuous, and preserving positivity) if and only if the relevant right generalized translation operators have the same properties.*

Let R_p^* be the operator adjoint to R_p. The operators R_p^* form a family of *adjoint generalized translation operators. The left adjoint generalized translation operators* L_p^* are defined similarly. The measure m is called *strongly right-invariant (strongly left-invariant)* if $R_p^* = R_{p^*}$ $(L_p^* = L_{p^*})$ for all $p \in Q$.

If weakly continuous generalized translation operators L_p *are commutative, then relation (2.1) holds for almost all* p *and* q.

Indeed, for any $f, g, h \in L_2$, we have the following chain of equalities true for almost all r:

$$\int\int \big(L_r^q(R_p f)\big)(q)\overline{g(p)h(q)}\, dp\, dq$$

$$= \int\int \big(R_p^q(R_q f)\big)(r)\overline{g(p)h(q)}\, dp\, dq = \int\int \big(R_q^r(R_p f)\big)(r)\overline{g(p)h(q)}\, dp\, dq$$

$$= \int\int \big(R_p^r(R_q f)\big)(r)\overline{g(p)h(q)}\, dp\, dq = \int\int \big(R_q^p(R_p f)\big)(r)\overline{g(p)h(q)}\, dp\, dq$$

$$= \int\int \big(L_r^p(R_q f)\big)(p)\overline{g(p)h(q)}\, dp\, dq. \tag{2.2}$$

Since L_r are weakly continuous, the first and the last expressions are continuous in r. Hence, they coincide for all $r \in Q$. By setting $r = e$, we complete the proof.

■

Assume that the generalized translation operators L_p satisfy the *finiteness condition:*

(F): For any $A, B \in \mathcal{B}_0(Q)$, there is a compact set F so large that $(L_p f)(q) = 0$ for almost all $p \in A$ and $q \in B$ provided that $\operatorname{supp} f \cap F = \varnothing$.

It is obvious that if L_p satisfy (F), then right generalized translations also satisfy (F). By using condition (F), one can define the action of generalized translation operators L_p

upon all functions f from $C(Q)$. Namely, for any sufficiently small neighborhoods V, $W \in \mathcal{B}_0(Q)$ of arbitrary points $p, q \in Q$, we take a set F according to condition (F) and put $(L_p f)(q) = (L_p f_F)(q)$, where the function $f_F(r)$ is equal to $f(r)$ for $r \in F$ and equal to zero, otherwise (here, $f(r)$ is an arbitrary function from $C(Q)$). We say that generalized translation operators L_p *preserve the function identically equal to one* if their extensions to $C(Q)$ satisfy the equality $(L_p 1)(q) \equiv 1$ for all $p \in Q$ and almost all $q \in Q$.

If the measure m *is strongly left-invariant and the finiteness condition (F) holds, then the subspace* $L_{2,0}$ *of finite functions from* L_2 *is invariant with respect to the action of generalized translation operators* L_p $(p \in Q)$.

In fact, we fix some $p \in Q$ and a function $f \in L_{2,0}$, $\operatorname{supp} f = A \in \mathcal{B}_0(Q)$. Let B be a compact set which contains p and p^*. For given sets A and B, we choose the set F from condition (F). By virtue of (F), we have

$$\int L_p f(q)\overline{g(q)}\,dq = \int f(q)\overline{(L_{p^*}g)(q)}\,dq = \int_A f(q)\overline{(L_{p^*}g)(q)}\,dq = 0$$

for any function $g \in L_2$ whose support does not contain points from F. Hence, $\operatorname{supp} L_p f(q) \subset F$.

■

We now can explain the definition of strongly left-invariant measure.

Let m *be a measure strongly left-invariant with respect to the generalized translation operators* L_p $(p \in Q)$ *which preserve the unit element and satisfy the finiteness condition (F). Then*

$$\int (L_p f)(q)\,dq = \int f(q)\,dq \quad (p \in Q). \tag{2.3}$$

for any $f \in L_{2,0}$.

Indeed, assume that p is a fixed point from Q, $A = \operatorname{supp} f$, B is a compact set containing p and p^*, and F is a compact set constructed from the sets A and B according to condition (F). Since $\operatorname{supp}(L_p f)(q) \subset F$ and $(L_p \kappa_F)(q) = (L_p 1)(q)$ for all $p \in B$ and $q \in A$, we have

$$\int (L_p f)(q)\,dq = \int (L_p f)(q)\kappa_F(q)\,dq$$

$$= \int f(q)(L_{p^*}\kappa_F)(q)\,dq = \int_A f(q)(L_{p^*}\kappa_F)(q)\,dq = \int f(q)\,dq.$$

■

A similar assertion is true for the strongly right-invariant measure.

We say that a measure $m \in M^+(Q)$ satisfying (2.3) is *left-invariant*. The *right-invariant* measure is defined similarly. In the next subsection, we present conditions for an invariant measure to be strongly invariant. If a left-invariant (strongly left-invariant) measure is, at the same time, right-invariant (strongly right-invariant), then it is called invariant (strongly invariant).

Theorem 2.1. *There exists a one-to-one correspondence between normal hypercomplex systems $L_1(Q, m)$ with basis unity e and weakly continuous families of bounded involutive generalized translation operators L_p satisfying the finiteness condition (F), preserving positivity in the space $L_2(Q, m)$ with unimodular strongly invariant measure m, and preserving the unit element. Convolution in the hypercomplex system $L_1(Q, m)$ and the corresponding family of generalized translation operators L_p satisfy the relation*

$$(f * g)(p) = \int (L_p f)(q) g(q^*) dq = (L_p f, g^*)_2 \quad (f, g \in L_2) \tag{2.4}$$

Moreover, the hypercomplex system $L_1(Q, m)$ is commutative if and only if the generalized translation operators L_p $(p \in Q)$ are commutative.

Proof. Let L_p be a family of generalized translation operators satisfying the conditions of the theorem. By using (2.4), we define the function $c(A, B, r) = \kappa_A * \kappa_B(r)$ $(A, B \in \mathcal{B}_0(Q))$ and show that $c(A, B, r)$ is a structure measure. The fact that $c(A, B, r)$ is nonnegative follows from (2.4) and the fact that positivity is preserved by the generalized translation operators. The equalities

$$c(A, B, r) = \int (L_r \kappa_A)(q) \kappa_B(q^*) dq = \int_{B^*} (L_r \kappa_A)(q) dq = \int_A (L_{r^*} \kappa_{B^*})(q) dq$$

imply that the function $c(A, B, r)$ can be uniquely extended to a regular Borel measure with respect to $B(A)$ with fixed $A \in \mathcal{B}_0(Q)$, $r \in Q$ $(B \in \mathcal{B}_0(Q))$. The weak continuity of the family of generalized translation operators L_r implies that $c(A, B, r)$ is continuous in r.

Let us show that $c(A, B, r)$ is finite in r for all fixed $A, B \in \mathcal{B}_0(Q)$. For this purpose, we take a compact set F from condition (F) such that $L_p f(q) = 0$ for almost all $p \in A$ and $q \in B$ provided that $\operatorname{supp} f \cap F = \emptyset$ and show that $c(A, B, r) = 0$ for any $r \notin F$. Indeed, consider a point $r \notin F$ such that $c(A, B, r) > 0$. Since $c(A, B, p)$ is continuous, one can indicate a neighborhood V of the point r such that $V \cap F = \emptyset$ and $c(A, B, p) > 0$ for all $p \in V$. Then

$$
\begin{aligned}
0 < \int_V c(A,B,p)dq &= \int\int (L_p\kappa_A)(q)\kappa_{B^*}(q)\kappa_V(q)dqdp \\
&= \int\int (L_p\kappa_A)(q^*)\kappa_B(q)\kappa_V(q)dqdp \\
&= \int\int (L_q\kappa_{A^*})(p)\kappa_{V^*}(p)dp\kappa_B(q)dq \\
&= \int\int \kappa_{A^*}(p)\kappa_B(q)(L_{q^*}\kappa_{V^*})(p)dpdq \\
&= \int\int \kappa_A(p)\kappa_B(q)(L_p\kappa_V)(q)dpdq.
\end{aligned}
$$

By virtue of (F), the last integral is equal to zero.

Let us now prove the associativity relation (H2). Since the axiom of associativity holds for generalized translation operators and the measure m is invariant, we obtain

$$
\begin{aligned}
&\int c(A,B,p)d_pc(E_p,C,s) \\
&\qquad = \int\int (L_p\kappa_A)(q)\kappa_{B^*}(q)(L_{s^*}\kappa_{C^*})(q)dp\,dq \\
&\qquad = \int \kappa_{B^*}(q)\int \big(L_s^p(L_p\kappa_A)\big)(q)\kappa_{C^*}(p)dp\,dq \\
&\qquad = \int \kappa_{B^*}(q)\int \big(L_p^q(L_s\kappa_A)\big)(q)\kappa_{C^*}(p)dp\,dq \\
&\qquad = \int\int (L_{p^*}\kappa_{B^*})(q)(L_s\kappa_A)(q)\kappa_{C^*}(p)dp\,dq \\
&\qquad = \int\int (L_q\kappa_B)(p)\kappa_{C^*}(p)(L_s\kappa_A)(q^*)dp\,dq \\
&\qquad = \int c(B,C,q)d_qc(A,E_q,s).
\end{aligned}
$$

Hence, the function $c(A,B,r)$ is a structure measure. It follows from the Fubini theorem, the invariance of m, and (2.4) that

$$
\begin{aligned}
\int c(A,B,p)dp &= \int\int (L_p\kappa_A)(q)\kappa_B(q^*)dpdq \\
&= \int \kappa_{B^*}(q)\int (R_q\kappa_A)(p)dpdq = m(A)m(B^*) = m(A)m(B),
\end{aligned}
$$

i.e., the measure m is multiplicative. To show that $c(A,B,r)$ is normal, we use the following equalities, which are true for all $A, B, C \in \mathcal{B}_0(Q)$:

$$c(A, B, C) = \int\int (R_q \kappa_A)(p)\kappa_{B^*}(q)\kappa_C(p)\,dp\,dq$$

$$= \int\int \kappa_A(p)(R_{q^*} \kappa_C)(p)\kappa_{B^*}(q)\,dp\,dq$$

$$= \int\int \kappa_A(p)(L_p \kappa_C)(q^*)\kappa_{B^*}(q)\,dp\,dq$$

$$= \int\int \kappa_A(p)(L_p \kappa_C)(q)\kappa_{B^*}(q^*)\,dp\,dq = c(C, B^*, A).$$

The equality $c(A, B, C) = c(A^*, C, B)$ is proved similarly.

Now let the generalized translation operators L_p $(p \in Q)$ be commutative. To check the commutativity relation (H3), we show that $(f * g)(p) = (g * f)(p)$ $(f, g \in L_2)$. In fact, since the generalized translation operators are involutive, the measure m is unimodular and invariant, and relations (2.1) and (2.4) are true, we have

$$(f * g)(p) = \int (L_p f)(q)\overline{g^*(q)}\,dq$$

$$= \int f(q)\overline{(L_{p^*} g^*)(q)}\,dq = \int f(q)(L_{q^*} g)(p)\,dq$$

$$= \int f(q)(L_p g)(q^*)\,dq = \int f(q^*)(L_p g)(q)\,dq = (g * f)(p).$$

Let us prove the converse assertion. Assume that $L_1(Q, m)$ is a normal hypercomplex system with basis unity e and structure measure $c(A, B, r)$. We define a family of operators L_p $(p \in Q)$ by using the bilinear form (2.4). In Subsection 1.3, it was shown that this definition is correct, the multiplicative measure is invariant, L_p are weakly continuous, and $\|L_p\| \le 1$. We now define the operators R_p $(p \in Q)$ by the equality $(R_p f, g) = (g^* * f)(p)$ $(f, g \in L_2)$ and show that

$$(L_p f)(q) = (R_q f)(p) \quad (f \in L_2) \tag{2.5}$$

for almost all p and q. Indeed, if we approximate arbitrary functions $f, g \in L_2$ in L_2 and $h \in C_0(Q)$ in L_1 by step functions, then, in view of (1.16), we obtain

$$\int\int (R_p f)(q)\overline{g(q)}\overline{h(p)}\,dp\,dq = (g^* * f, h)_2 = (f, g * h)_2$$

$$= (f * h^*, g)_2 = \int\int (L_q f)(p)\overline{h(p)}\overline{g(q)}\,dp\,dq$$

but this yields (2.5). In its turn, relation (2.5) implies that the function $\varphi(p) = (L_p f)(q) \in L_2$ for almost all $q \in Q$. It follows from the strong invariance of the multiplicative measure and the associativity and involutivity of convolution in the hypercomplex system that

$$\int\int L_p^q(L_q f)(r)\overline{g(r)}h(q)\,dr\,dq = \int\int (L_q f)(r)\overline{g(r)}\,dr\,\overline{(L_{p^*}h)(q)}\,dq$$

$$= \int (f*g^*)(q)\overline{(L_{p^*}h)(q)}\,dq = (L_p(f*g^*), h)_2$$

$$= (f*g^*)*h^*(p) = f*(h*g)^*(p)$$

$$= \int\int (L_p f)(r)\overline{(L_{q^*}g)(r)h(q)}\,dr\,dq$$

$$= \int\int \big(L_p^r(L_p f)\big)(r)\overline{g(r)}\,\overline{h(q)}\,dr\,dq \tag{2.6}$$

for all $f \in L_2$, $g, h \in C_0(Q)$, and $p \in Q$. This means that $\big(L_p^q(L_q f)\big)(r) = \big(L_q^r(L_p f)\big)(r)$ for almost all p, q, and r, i.e., the operators L_p are generalized translation operators. The generalized translation operators L_p are involutive because the hypercomplex system $L_1(Q, m)$ is normal, i.e.,

$$\int\int (L_p f)(q)\overline{g(q)}\,\overline{h(p)}\,dp\,dq = (f*g^*, h)_2$$

$$= (f, h*g)_2 = (h^* * f, g)_2$$

$$= \int\int (L_q h^*)(p)\overline{f^*(p)}g(q)\,dp\,dq$$

$$= \int\int h^*(p)\overline{(L_{q^*}f^*)(p)g(q)}\,dp\,dq$$

$$= \int\int \overline{(L_{q^*}f^*)(p^*)}\,\overline{h(p)}\,\overline{g(q)}\,dp\,dq$$

$$(f \in L_2;\ g, h \in C_0(Q)).$$

The fact that the generalized translation operators L_p are finite follows from the relation

$$\int_C (L_{\kappa_A}\kappa_B)(r)\,dr = c(A, B, C) = c(C, B^*, A) = 0$$

which is true for all $A, B, C \in \mathcal{B}_0(Q)$, whenever A is located far enough from B

and C.

Positivity is preserved because $c(A, B, r)$ is nonnegative and the following chain of equalities holds for any $A \in \mathcal{B}_0(Q)$, $r \in Q$, and an arbitrary sequence of neighborhoods V_n of the point r contracting to this point:

$$\int_A (L_r f)(q)\,dq = \lim_{n\to\infty} \frac{1}{m(V_n)} \int\int (L_p f)(q)\kappa_A(q)\kappa_{V_n}(p)\,dp\,dq$$

$$= \lim_{n\to\infty} \frac{1}{m(V_n)} \int\int f(q)(R_q\kappa_A)(p^*)\kappa_{V_n}(p)\,dp\,dq$$

$$= \lim_{n\to\infty} \frac{1}{m(V_n)} \int f(q)c(V_n, A, q)\,dq \geq 0 \quad (f \in L_2, f(q) \geq 0).$$

Let L_p be a family of generalized translation operators associated with a hypercomplex system $L_1(Q, m)$. The inequality

$$|(L_p f)(q)| \leq \|f\|_\infty \quad (f \in C_b(Q)) \tag{2.7}$$

holds for almost all $q \in Q$.

Indeed, we fix $p \in Q$, choose a decreasing sequence of neighborhoods V_n of the point p contracting to this point, and set

$$g_n(r) = \frac{\kappa_{V_n}(r)}{m(V_n)}.$$

Let $g \in L_1(Q, m)$ be an arbitrary finite function and let $A = \operatorname{supp} g$. Then $(L_p f)(q) = (L_p f_F)(q)$ for all $p \in V_1$ and $q \in A$, where F is taken from axiom (F). We have

$$\int (L_p f)(q)g(q)\,dq = \left(f_F * \bar{g}^*\right)(p)$$

$$= \lim_{n\to\infty} \frac{1}{m(V_n)} \int_{V_n} \left(f_F * \bar{g}^*\right)(q)\,dq = \lim_{n\to\infty} \left(f_F * \bar{g}^*, g_n^*\right)_2$$

$$= \lim_{n\to\infty} \left(\bar{g}^*, f_F^* * g_n\right)_2 = \lim_{n\to\infty} \left(g_n^* * f_F, \bar{g}\right)_2$$

$$= \lim_{n\to\infty} \int \left(g_n^* * f_F\right)(q)g(q)\,dq.$$

Since $\|g_n^*\|_1 = \|g_n\|_1 = 1$, we have

$$\left|\int (L_p f)(q) g(q)\, dq\right| = \lim_{n\to\infty} \left|\int \left(g_n^* * f_F\right)(q) g(q)\, dq\right|$$

$$\leq \lim_{n\to\infty} \|g_n^* * f_F\|_\infty \|g\|_1 \leq \|f\|_\infty \|g\|_1,$$

which implies the required inequality.

We now show that the generalized translation operators L_p preserve the function identically equal to one. Let Q_n be an increasing sequence of compact sets which covers all Q. By virtue of the definition of $L_p 1(q)$ and (1.12), we have

$$\int_A (L_p 1)(q)\, dq = \lim_{n\to\infty} \int (L_p \kappa_{Q_n})(q) \kappa_A(q)\, dq$$

$$= \lim_{n\to\infty} c(Q_n, A^*, p) = c(Q, A^*, p) = m(A)$$

for any $A \in \mathcal{B}_0(Q)$ and $p \in Q$. Since positivity is preserved, it follows from (2.7) that $0 \leq L_p 1(q) \leq 1$, whence $(L_p 1)(q) \equiv 1$.

Assume that the hypercomplex system $L_1(Q, m)$ is commutative. Let us prove equality (2.1). Since the generalized translation operators L_p are involutive for all f, $g \in L_2$ and $p \in Q$, we obtain

$$\int (L_p f)(q) \overline{g(q)}\, dq = (f * g^*)(p)$$

$$= (g^* * f)(p) = \int (L_p g^*)(q) \overline{f^*(q)}\, dq$$

$$= \int \overline{\left(L_{p^*} f^*\right)(q^*)}\, \overline{g(q)}\, dq = \int (L_q f)(p)\, \overline{g(q)}\, dq$$

which means that relation (2.1) is true for almost all p and q. The commutativity of L_p now follows from (2.1) and the commutativity of the hypercomplex system $L_1(Q, m)$:

$$\int\int L_q^r L_p f(r) g(r^*) h(q^*)\, dr\, dq = \int\int L_p^q L_q f(r) g(r^*) h(q^*)\, dr\, dq$$

$$= \int f * g(p) \overline{L_{p^*} h^*(q)}\, dq$$

$$= f * g * h(p) = f * h * g(p)$$

$$= \int\int L_p^r L_r f(q) g(r^*) h(q^*)\, dr\, dq$$

$$= \int\int L_p^r L_q f(r) g(r^*) h(q^*)\, dr\, dq \quad (p \in Q).$$

Since L_q is weakly continuous,

$$\int L_p^r L_q f(r) g(r^*)\, dr = (L_q f, L_{p^*} g^*)$$

and

$$\int L_q^r L_p f(r) g(r^*)\, dr = (L_p f, L_{q^*} g^*)$$

are continuous in q for every p. This implies that $L_p^r L_q f(r) = L_q^r L_p f(r)$ for almost all r, p, q

■

According to Theorem 2.1 and the results obtained in Subsection 1.3, every family of generalized translation operators L_p satisfying the conditions of Theorem 2.1 is uniformly bounded in $L_2(Q, m)$: $\|L_p\| \le 1$ $(p \in Q)$.

Inequality (2.7) allows one to extend the generalized translation operators L_p $(p \in Q)$ to operators continuous in $L_\infty(Q, m)$. This result, together with Theorem 2.1, can be regarded as a formalization (for the case of the duality $\langle L_\infty(Q, m), L_1(Q, m)\rangle$) of Levitan's heuristic ideas (see [8, Section 2, Subsection 2.1]). Levitan conjectured that a linear topological space Φ of functions (measures, functionals, germs, etc.) on a locally compact space Q is an associative topological algebra with right identity element if and only if the action of the right generalized translation operators R_s $(s \in Q)$ is defined on the dual space Φ' and, moreover, the operation $*$ in Φ and the corresponding generalized translation operators R_s in Φ' satisfy the relation

$$\langle F * G, f\rangle = \langle F, \langle G^s, R_s f\rangle\rangle \quad (f \in \Phi'; F, G \in \Phi),$$

where the symbol G^s denotes the action of a functional G in a variable s). A formalization of these ideas was proposed by Litvinov [Lit1]–[Lit3] (see Chapter 3, Section 1) for the cases $\langle M_c(Q), C(Q)\rangle$, $\langle M_b(Q), C_\infty(Q)\rangle$ (the hypergroup case), $\langle \mathcal{D}(Q), C^\infty(Q)\rangle$ (in this case, Q is a smooth manifold, $C^\infty(Q)$ is the space of infinitely differentiable functions on Q equipped with the topology of uniform (including all derivatives) convergence on compact sets, and $\mathcal{D}(Q)$ is the space of generalized functions with compact support), $\langle \mathcal{A}(Q), \mathcal{H}(Q)\rangle$ (here, Q is a complex analytic manifold, $\mathcal{H}(Q)$ is the space of functions holomorphic on Q with the topology of uniform convergence on compact sets, and $\mathcal{A}(Q)$ is the dual space which consists of analytic functionals), and for the case of the duality "real analytic functions–hyperfunctions with compact support".

In order that a function $\chi(r) \in C(Q)$ *be a generalized character of a normal commutative hypercomplex system* $L_1(Q, m)$ *with basis unity* e, *it is necessary and sufficient that the equality*

$$(L_p\chi)(q) = \chi(p)\chi(q) \tag{2.8}$$

hold for almost all $p, q \in Q$.

Indeed, assume that a function $\chi(p) \in X_g$ is a generalized character, i.e., equality (1.6) holds. Then

$$\begin{aligned}\chi(A)\chi(B) &= \int c(A, B, r)\chi(r)\,dr \\ &= \int\int_{B^*} (L_r\kappa_A)(s)\,ds\,\chi(r)\,dr = \int_B\int (L_{s^*}\kappa_A)(r)\chi(r)\,dr\,ds \\ &= \int_B\int_A (L_s\chi)(r)\,dr\,ds\end{aligned}$$

for any $A, B \in \mathcal{B}_0(Q)$, which yields (2.8). The converse statement can be proved by analogy.

∎

Note that if a hypercomplex system $L_1(Q, m)$ satisfies the condition of separate continuity (see Subsection 2.2 below), then (2.8) holds for all $p, q \in Q$.

Note that all results obtained in this section can easily be extended to the case of nonunimodular hypercomplex systems (with nonunimodular multiplicative measure).

2.2. Strong Invariance of Invariant Measures. Below, we present a condition that guarantees the strong invariance of an invariant measure.

Let L_s $(s \in Q)$ be generalized translation operators acting in the space of bounded functions on a locally compact space Q equipped with an involutive homeomorphism $Q \ni p \mapsto p^* \in Q$. We say that the generalized translation operators L_s satisfy the *condition of separate continuity* if the function $(L_s f)(t)$ is continuous in each variable for any $f \in C_0(Q)$.

Lemma 2.1. *Let* L_s $(s \in Q)$ *be a family of involutive generalized translation op-*

erators in $C_b(Q)$ that preserve finiteness and satisfy inequality (2.7) and the condition of separate continuity, and let $\mu \in M_b^+(Q)$. Then the equalities

$$L_s \int (L_t f)(r)\, d\mu(r) = \int \left(L_s^t (L_t f)\right)(r)\, d\mu(r) \tag{2.9}$$

and

$$L_s^t \int (L_{t^*} f)(r)\, d\mu(r) = \int \left(L_{s^*}^r (L_{t^*} f)\right)(r)\, d\mu(r) \tag{2.10}$$

hold for all $t, s \in Q$ and $f \in C_0(Q)$.

Proof. Equality (2.9) obviously holds for any Dirac measure. Note that the cone $\mathcal{K}$ generated in $M_b(Q)$ by Dirac measures is dense in $M_b^+(Q)$ in the υ-topology (the topology of simple convergence on functions from $C_0(Q)$). Let the net of $\mu_\alpha \in \mathcal{K}$ be such that $\upsilon - \lim \mu_\alpha = \mu$. Then

$$\lim_\alpha \int \left(L_s^t (L_t f)\right)(r)\, d\mu_\alpha(r) = \lim_\alpha \int \left(L_t^r (L_s f)\right)(r)\, d\mu_\alpha(r)$$

$$= \int \left(L_t^r (L_s f)\right)(t)\, d\mu(r) = \int \left(L_s^t (L_t f)\right)(r)\, d\mu(r)$$

for $f \in C_0(Q)$ and $t, s \in Q$ by virtue of the fact that the generalized translation operators L_s $(s \in Q)$ are separately continuous and preserve finiteness. Let us show that

$$\lim_\alpha \int L_s \int (L_t f)(r)\, d\mu_\alpha(r) = L_s \int (L_t f)(r)\, d\mu(r).$$

Denote

$$h_\alpha(t) = \int (L_t f)(r)\, d\mu_\alpha(r).$$

Obviously, for any t, we have

$$\lim_\alpha h_\alpha(t) = h(t),$$

where

$$h(t) = \int (L_t f)(r)\, d\mu(r).$$

Since generalized translation operators are separately continuous, the functions $h_\alpha(t)$ and $h(t)$ are continuous. Therefore, $h_\alpha(t)$ uniformly converges to $h(t)$ on every compact set. We fix $t, s \in Q$. By virtue of the finiteness condition (F), there exists a compact set $F_{t,s}$ such that $(L_s f)(t) = 0$ if $\operatorname{supp} f \cap F_{t,s} = \varnothing$. It is clear that, for all α, the function $(h_\alpha - h) \mid_{F_{ts}}(t)$ can be extended to the function $(h_\alpha - h)^{\sim}(t)$ so that the net $(h_\alpha - h)^{\sim}(t)$ uniformly converges to zero. Then, by using (2.7) and the finiteness condition, we get

$$\lim_\alpha |(L_s(h_\alpha - h))(t)| = \lim_\alpha |(L_s(h_\alpha - h)^{\sim})(t)| \leq \lim_\alpha \| (h_\alpha - h)^{\sim} \|_\infty = 0.$$

Let us prove (2.10). Assume that $\mu = \delta_r$ $(r \in Q)$. Then

$$\begin{aligned} L_s^t((L_{t^*} f)(r)) &= \overline{L_s^t\big((L_{r^*} f^*)(t)\big)} = \overline{\big(L_s^t(L_{r^*} f^*)(t)\big)} \\ &= \overline{\big(L_{r^*}^s(L_s f^*)(t)\big)} = \overline{\big(L_{r^*}^s(R_t f^*)(s)\big)} \\ &= \big(L_{s^*}^r\big(R_t f^*\big)^*\big)(r) = \big(L_s^r(L_{t^*} f)\big)(r) \end{aligned}$$

because

$$\big(R_t f^*\big)^*(t) = \overline{(R_t f^*)(r^*)} = (R_r f)(t^*) = (L_{t^*} f)(r).$$

The proof of relation (2.10) can now be completed by analogy with the proof of (2.9).

■

Consider a family of generalized translation operators L_t $(t \in Q)$ in $L_2(Q, m)$ with respect to a left-invariant measure m. Let O_n, $O_n \supset O_{n+1}$, be a sequence of bounded neighborhoods of the identity element contracting to this element. Let us construct a sequence of functions $e_n(p) \in C_0(Q)$ such that $e_n(p) \geq 0$, $\operatorname{supp} e_n \subset O_n$, $e_n^* = e_n$, and

$$\int e_n(p)\, dm(p) = 1.$$

Lemma 2.2. *Let L_t be a family of involutive generalized translation operators in the space of functions square integrable with respect to a left-invariant measure; furthermore, we assume that these operators preserve finiteness and positivity and satisfy inequality (2.7), the condition of separate continuity, and the following condition:*

(C) For any neighborhood W of a point $s \in Q$, one can find a neighborhood U

of the identity element e *such that* $\operatorname{supp} L_{s^*}\varphi(\cdot) \subset W$ *for every* $\varphi \in C_0(Q)$ *with* $\operatorname{supp}\varphi \subset U$.

Then every function $f(s) \in C_0(Q)$ *can be uniformly approximated by the sequence*

$$f_n(s) = \int (L_{s^*}e_n)(t)f(t)dt.$$

Proof. Note that, by virtue of inequality (2.7), the continuity of functions $f_n(s)$ follows from the Lebesgue theorem and the fact that generalized translation operators are separately continuous. For any $\varepsilon > 0$, one can find a neighborhood W_t of an arbitrary point $t \in \operatorname{supp} f$ such that $|f(t)-f(s)| < \varepsilon$ for all $s \in W_t$. We cover $\operatorname{supp} f$ by these neighborhoods W_t and take a finite subcovering $W_{t_1}, \ldots, W_{t_n}$. Denote by $U_1, \ldots, U_n$ the neighborhoods of the identity element e that correspond to $W_{t_1}, \ldots, W_{t_n}$ according to condition (C). Denote

$$U = \bigcap_{k=1}^{\infty} U_k.$$

Then, for any $\varphi \in C_0(Q)$ and $s \in W_{t_k}$, we have $\operatorname{supp}(L_{s^*}\varphi)(\cdot) \subset W_{t_k}$ whenever $\operatorname{supp}\varphi \subset U$. There exists n_0 such that $\operatorname{supp} e_n \subset U$ for all $n > n_0$. By using (2.3), we obtain

$$\begin{aligned}|f_n(s)-f(s)| &= \left|\int (L_{s^*}e_n)(t)f(t)dt - f(s)\right| \\ &= \left|\int (L_{s^*}e_n)(t)(f(t) - f(s))dt\right| \le \varepsilon \int (L_{s^*}e_n)(t)dt = \varepsilon\end{aligned}$$

for all $s \in \operatorname{supp} f$.

Since the generalized translation operators are involutive, we have

$$f_n(s) = \int (L_{t^*}e_n)(s)f(t)dt$$

and, obviously, $f_n(s) = 0$ for

$$s \notin \bigcup_{k=1}^{n} W_{t_k}.$$

Hence, $\|f_n - f\|_\infty \to 0$ as $n \to \infty$.

Theorem 2.2. *Let* L_s $(s \in Q)$ *be an involutive family of generalized translation* *operators in the space $L_2(Q, m)$ with left-invariant measure m such that*

(i) *these operators are separately continuous;*

(ii) *inequality (2.7) holds;*

(iii) *the operators L_s preserve positivity;*

(iv) *the finiteness condition (F) is satisfied;*

(v) *condition (C) in Lemma 2.2 is satisfied, i.e., for any neighborhood W of a point $s \in Q$, one can find a neighborhood U of the identity element e such that $\operatorname{supp} L_{s^*}\varphi(\cdot) \subset W$ for every $\varphi \in C_0(Q)$ with $\operatorname{supp}\varphi \subset U$.*

Then $L_s^* = L_{s^*}$ $(s \in Q)$, i.e., the measure m is strongly left-invariant.

Proof. Since the generalized translation operators are continuous and preserve positivity, it suffices to show that $(L_s f, g)_2 = (f, L_{s^*} f)_2$ for any $f, g \in C_0^+(Q)$, $s \in Q$. By virtue of (2.7) and the condition of separate continuity, the generalized translation operators L_s act continuously in $C_b(Q)$. Therefore, by virtue of Lemmas 2.1 and 2.2, we obtain

$$\int (L_s f)(t)\overline{g(t)}\,dt = \lim_\alpha \int L_s^t \int (L_{t^*} e_n)(r) f(r)\,dr\,\overline{g(t)}\,dt$$

$$= \lim_\alpha \int \int \left(L_{s^*}^r (L_{t^*} e_n)\right)(r) f(r)\,dr\,\overline{g(t)}\,d\cdot$$

$$= \lim_\alpha \int f(r) \int \overline{\left(L_{r^*}^s (L_{t^*} e_n)^*\right)(s)}\;\overline{g(t)}\,dt\,dr$$

$$= \lim_\alpha \int f(r) \int \left(L_{r^*}^s (L_s e_n^*)\right)(t)\;\overline{g(t)}\,dt\,dr$$

$$= \lim_\alpha \int f(r) \int \left(L_s^t (L_{r^*} e_n)\right)(t)\;\overline{g(t)}\,dt\,dr$$

$$\doteq \lim_\alpha \int f(r) L_{s^*}^r \int (L_{r^*} e_n)(t)\overline{g(t)}\,dt\,dr = \int f(r)\overline{(L_{s^*} g)(r)}\,dr$$

for any $f, g \in C_0^+(Q)$ and $s \in Q$.

∎

2.3. Hypergroups and Hypercomplex Systems. In this subsection, we establish the relationship between hypercomplex systems and hypergroups.

Let Q be a locally compact Hausdorff space. The space Q is called a *hypergroup* if the following conditions are satisfied:

(G1) There is a separately continuous mapping $Q \times Q \ni (p, q) \mapsto \delta_p * \delta_q \in M^1(Q)$, which is called *convolution,* where $M^1(Q)$ is equipped with weak topology and δ_p is the Dirac measure. This mapping can be extended to $M_b(Q)$ according to the formula

$$\langle \mu * \nu, f \rangle = \int \int \langle \delta_p * \delta_q, f \rangle d\mu(p) d\nu(q). \tag{2.11}$$

The operation of convolution is assumed to be associative, i.e., $\delta_p * (\delta_q * \delta_r) = (\delta_p * \delta_q) * \delta_r$.

(G2) $\operatorname{supp}(\delta_p * \delta_q)$ is compact.

(G3) There is an involutive homeomorphism $Q \ni p \mapsto p^* \in Q$ such that $(\delta_p * \delta_q)^* = \delta_{q^*} * \delta_{p^*}$; here, μ^* denotes the measure

$$\langle \mu^*, f \rangle = \int f(p^*) d\mu(p).$$

(G4) There exists a point $e \in Q$, called the identity element, such that $\delta_e * \delta_p = \delta_p * \delta_e = \delta_p$ $(p \in Q)$.

(G5) $e \in \operatorname{supp} \delta_p * \delta_{q^*}$ if and only if $p = q$.

(G6) The mapping $(p, q) \mapsto (\operatorname{supp} \delta_p * \delta_q)$ continuously maps $Q \times Q$ into the space of compact sets $\mathfrak{K}(Q)$ equipped with the Michael topology [Mic]; this topology is generated by the pre-base $\{K \in \mathfrak{K}(Q) \mid K \cap U \neq \emptyset, K \subseteq V\}$, where U and V are open sets in Q.

A locally compact space Q that satisfies (G1)–(G4) is called a *weak hypergroup.* If $\delta_p * \delta_q = \delta_q * \delta_p$ for any $p, q \in Q$, the hypergroup Q is called *commutative.* We define the action of generalized translation operators R_q $(q \in Q)$ upon an arbitrary Borel function f on Q by the formula $(R_q f)(p) = \langle \delta_p * \delta_q, f \rangle$ if the last integral exists (but is not necessarily finite). In any case, the action of R_q is defined on bounded functions on Q. By virtue of (G1) and (G4), the operators R_q form a family of generalized translation operators and are reduced. Obviously, the generalized translation operators R_p and L_p $(p \in Q)$ are involutive, preserve positivity and identity element, and satisfy the conditions of separate continuity and finiteness and inequality (2.7).

A regular Borel measure m positive on open sets is called the *left Haar measure of a hypergroup* Q if it is weakly invariant with respect to the family of generalized translation operators L_p $(p \in Q)$. The right Haar measure is defined in a similar way. If a Haar measure exists, then it is unique up to multiplication by a constant ([Jew], see also Theorem 4.1) Spector [Spe4] showed that any commutative hypergroup possesses the Haar measure. Jewett [Jew] proved the existence of the Haar measure for noncommutative hypergroups in compact and discrete cases. At present, the problem of existence of the Haar measure for general locally compact hypergroups is still open and the presence of this measure is postulated. We fix a left Haar measure m. As usual, we denote integration with respect to the measure $dm(p)$ by dp.

Let us introduce the *convolution of sets* $A, B \in \mathcal{B}(Q)$:

$$A * B = \bigcup_{\substack{p \in A \\ q \in B}} \operatorname{supp}(\delta_p * \delta_q).$$

Clearly, this operation of convolution is associative, i.e., $A * (B * C) = (A * B) * C$, and involutive, i.e., $(A * B)^* = B^* * A^*$ $(A, B, C \in \mathcal{B}(Q))$.

We now show that $(A * B) \cap C$ *is nonempty if and only if* $B \cap (A^* * C)$ *is nonempty.*

Indeed, let $(A * B) \cap C$ be nonempty. Then, by virtue of (G5), $C^* * (A * B)$ contains the identity element e. At the same time, $C^* * (A * B) = (A^* * C)^* * B$. Since the last set contains e, we have $(A^* * C) \cap B \neq \emptyset$ by virtue of (G5). Note that we have, in fact, also proved the converse statement (it suffices to the rename sets in a proper way).

■

Let us show that *the generalized translation operators* L_p *associated with the hypergroup* Q *satisfy condition (C) in Lemma 2.2.*

By virtue of (G6), for any neighborhood W of an arbitrary point $s \in Q$, one can find a neighborhood of the unit element $U \ni e$ such that $\{s\} * U \subset W$. Condition (C) means that $\operatorname{supp}(\delta_{s^*} * \delta_t) \cap U \neq \emptyset$ for any $t \notin W$. Assume the contrary: $(\{s^*\} * (Q \setminus W)) \cap U \neq \emptyset$. Then we have $(Q \setminus W) \cap (\{s\} * U) \neq \emptyset$ by virtue of the assertion proved above. The last inequality contradicts the choice of U.

■

Thus, the generalized translation operators L_p $(p \in Q)$ satisfy the conditions of Theorem 2.2 and, hence, the left Haar measure of the hypergroup Q is strongly left-invariant with respect to the family of generalized translation operators L_p associated with the hypergroup Q, i.e., $L_p^* = L_{p^*}$ in the space $L_2(Q, m)$ (by virtue of (2.7), the generalized translation operators L_p are bounded in $L_2(Q, m)$). By applying the Lebesgue theorem and the fact that the generalized translation operators L_p are separately continuous, we show that the generalized translation operators L_p are weakly continuous in $L_2(Q, m)$. Thus, all conditions of Theorem 2.1 hold for the generalized translation operators L_p. By this theorem, the space $L_1(Q, m)$ is a normal hypercomplex system with basis unity e with respect to the convolution

$$f * g(p) = \int (L_p f)(q) g(q^*) dq \quad (f, g \in L_2(Q, m)), \tag{2.12}$$

and involution p^*.

We have shown that *it is possible to construct a normal hypercomplex system with basis unity* e *for any unimodular hypergroup.* Let us now analyze the possibility of the reverse procedure. Let $L_1(Q, m)$ be a normal hypercomplex system with basis unity In order to define the convolution of measures, we assume that the generalized translation operators L_p $(p \in Q)$ defined in Subsection 1.3 and acting in the space L_2 can be extended to a family of separately continuous operators in $C_\infty(Q)$. One can now define the convolution of measures as $\langle \delta_p * \delta_q, f\rangle = (L_p f)(q)$ $(f \in C_\infty(Q))$ and extend it to $M_b(Q)$ by using (2.11). The associativity (and, if necessary, commutativity) of this convolution follows from the associativity of the generalized translation operators L_p and relation (2.11). Indeed,

$$\langle \delta_p * (\delta_q * \delta_r), f\rangle = \left(L_q^r(L_p f)\right)(r) = \left(L_p^q(L_q f)\right)(r)$$

$$= \left(L_p^q(R_r f)\right)(q) = \langle (\delta_p * \delta_q) * \delta_r, f\rangle;$$

$$\langle \delta_p * \delta_q, f\rangle = (L_p f)(q) = (L_q f)(p) = \langle \delta_q * \delta_p, f\rangle \quad (p, q, r \in Q; f \in C_\infty(Q)).$$

Since L_p preserve the function identically equal to one, $\delta_p * \delta_q$ is a probability measure: $\langle \delta_p * \delta_q, 1\rangle = L_p 1(q) = 1$. Condition (G2) readily follows from the finiteness condition (F) for generalized translation operators, (G4) follows from (T2), and (G3) results from the fact that the generalized translation operators are involutive. Thus, if the operators L_p can be extended to continuous operators from $C_\infty(Q)$ into $C_\infty(Q)$, a weak hypergroup can easily be associated with any hypercomplex system. It is *a priori* clear that condition (G6) is, generally speaking, not satisfied. It is also impossible to check the validity of (G5). Nevertheless, *in the case of a discrete base* Q, *the concepts of normal hypercomplex system with basis unity and hypergroup are equivalent* because there is no need to check (G6) and (G5) is equivalent to the equality $c(p, q, e) = \delta_{p,q^*} m$, where $p, q \in Q$ and δ_{p,q^*} is the Kronecker symbol.

Let us now present a formula which establishes the relationship between the structure measure $c(p, q, r)$ of a hypercomplex system with discrete base Q and the coefficients $g(p, q, r)$ of the expansion of the measure $\delta_p * \delta_q$ in the basis δ_r, i.e., $\delta_p * \delta_q = \sum_r g(p, q, r)\delta_r$ ($\sum_r g(p, q, r) = 1$ because $\delta_p * \delta_q$ is a probability measure):

$$g(p, q, r) = \frac{c(p, q, r)m(r)}{m(p)m(q)}. \tag{2.13}$$

2.4. Hypercomplex Systems Whose Structure Measure Is Not Necessarily Nonnegative. In this subsection, we present a generalization of the concept of hypercomplex systems to the case of, generally speaking, complex-valued structure measures and reestablish the results of Sections 1 and 2 for these hypercomplex systems.

First, we introduce the following definition:

A complex-valued function $c(A, B, r)$ $(A, B \in \mathcal{B}(Q))$ is called a structure measure if it is a bounded Radon measure in A (B) for fixed B and r $(A$ and $r)$ and satisfies the conditions

$(H1)'$ the function $c(A, B, r)$ is continuous in r for fixed $A, B \in \mathcal{B}_0(Q)$ and for all $E_1 \subset A$ and $E_2 \subset B$, there exists a compact set $F \subset Q$ such that $\operatorname{supp} c(E_1, E_1, r) \subset F$ $(A, B \in \mathcal{B}_0(Q))$;

$(H2)'$ the relation of associativity

$$\int c(A, B, r)\, d_r c(E_r, C, s) = \int c(B, C, r)\, d_r c(A, E_r, s)$$

holds for all $A, B, C \in \mathcal{B}_0(Q)$ and $s \in Q$;

$(H3)'$ the commutativity relation $c(A, B, r) = c(B, A, r)$ holds for all $A, B \in \mathcal{B}_0(Q)$ and $r \in Q$.

As in Subsection 1.1, we define a measure $m_r(\mathfrak{A})$ $(\mathfrak{A} \in \mathcal{B}(Q \times Q))$ by setting $m_r(Q \times Q) = c(A, B, r)$ $(A, B \in \mathcal{B}_0(Q))$ and extending the measure m_r to all Borel subsets of $Q \times Q$ (in view of its regularity, this extension is always possible).

We say that a regular Borel measure $dm(r) = dr$ defined on Q and positive on open sets is a multiplicative measure if

$$\int V(A, B, r)\, dm(r) \leq m(A)m(B) \tag{2.14}$$

where $V(A, B, r)$ is a variation of the measure $m_r(\mathfrak{A})$ on the set $A \times B$ for any $A, B \in \mathcal{B}_0(Q)$.

If the base Q is discrete, then, clearly, inequality (2.14) has the form

$$\sum_r |c(p,q,r)|m(r) \le m(p)m(q) \quad (p,q \in Q). \tag{2.15}$$

$(H4)'$ We assume that at least one multiplicative measure exists.

However, in the case of a complex-valued structure measure, it is impossible to prove the existence theorem similar to Theorem 1.3 for multiplicative measure and the problem of existence of multiplicative measure remains open.

Consider the space $L_1(Q,m) = L_1$ of functions on Q summable with respect to a multiplicative measure m. This space equipped with convolution

$$(x*y)(r) = \int x(p)d_p \int y(q)d_q c(E_p, E_q, r),$$

is called a *hypercomplex system with complex-valued structure measure.* This definition is meaningful because even in the case of a complex-valued structure measure one can show that this hypercomplex system is also a Banach algebra with $\|x*y\|_1 \le \|x\|_1 \|y\|_1$. The definition of a character remains unchanged. Note that, for a complex-valued structure measure, the function identical to one is not, generally speaking, a character of a hypercomplex system. As for the positive structure measure, one can show that, in the commutative case, the set of characters is in one-to-one correspondence with the set of maximal ideals of the Banach algebra $L_1(Q,m)$ (or $\mathfrak{L}_1(Q,m)$ if L_1 has no unit element; in this case, one must consider ideals other than L_1). A maximal ideal and the corresponding character satisfy relation (1.7).

If $(H5)$ is satisfied, a hypercomplex system with complex-valued structure measure is called normal and if $(H6)$ is satisfied, then the hypercomplex system possesses a basis unity

Lemma 2.3. *In a normal hypercomplex system,*

$$V(A,B,r) \le m(A) \tag{2.16}$$

for all $A, B \in \mathcal{B}(Q)$ *and* $r \in Q$.

Proof. In fact, it suffices to prove (2.16) for $A, B \in \mathcal{B}_0(Q)$. We fix $r \in Q, A, B \in \mathcal{B}_0(Q)$. For any $\varepsilon > 0$, one can find finitely many disjoint sets $\mathfrak{A}_j \subset Q \times Q$ constructed from finitely many rectangles such that

$$V(A,B,r) - \varepsilon < \sum_{j=1}^{n} |m_r(\mathfrak{A}_j)|, \quad \bigcup_{j=1}^{n} \mathfrak{A}_j = A \times B.$$

We decompose the sets A and B into small sets A_α and B_β ($\alpha = 1, 2, \dots, k$; $\beta = 1, 2, \dots, l$) such that each $\mathfrak{A}_j$ can be represented as a sum of disjoint rectangles $A_\alpha \times B_\beta$. Then

$$V(A, B, r) - \varepsilon < \sum_{\alpha=1}^{k} \sum_{\beta=1}^{l} |c(A_\alpha, B_\beta, r)|.$$

The lemma will be proved if the relationship

$$\sum_{\beta=1}^{l} |c(A_\alpha, B_\beta, r)| \le m(A_\alpha)$$

is established. For any $C \in \mathcal{B}_0(Q)$, we have by (2.14),

$$\sum_{\beta=1}^{l} \left| \int_C c(A_\alpha, B_\beta, r)\, dr \right| = \sum_{\beta=1}^{l} |c(A_\alpha^*, C, B_\beta)|$$

$$\le \int |c(A_\alpha^*, C, r)|\, dr \le m(C) m(A_\alpha^*) = m(C) m(A_\alpha)$$

This and $(H1)$ imply the required inequality.

■

By using Lemma 2.3, one can show that all assertions in Section 1 and Subsection 3.1 remain true for normal hypercomplex systems with complex-valued structure measure and basis unity. The proofs are practically the same as in the case of nonnegative structure measure. It is easy to see that Theorem 3.4 is also true for such hypercomplex systems, although the Plancherel measure $d\chi$ is not, in general, a multiplicative measure of the dual hypercomplex system Indeed,

$$\int c(\Phi, \Psi, \chi)\, d\chi = \int (x_\Phi \cdot x_\Psi)^\wedge(\chi)\, d\chi = (x_\Phi \cdot x_\Psi)(e) = m(\Phi) m(\Psi).$$

for any $\Phi, \Psi \in \mathcal{B}_0(Q)$. Then it is clear that inequality (2.14) is, generally speaking, not true for the measure $d\chi$ (the corresponding example was given by Gasper [Gas3]). Hence, in this axiomatics, it is impossible to construct an acceptable duality theory for hypercomplex systems with complex-valued structure measure. In Section 5, we present another approach (based on the ideas of Kac [Kac]) to the investigation of hypercomplex systems with complex-valued structure measure.

The results established in Subsection 3.3 also remain valid for hypercomplex systems with complex-valued structure measure if the definition of A-systems introduced in the next subsection is properly changed. An A-system is now defined as an orthogonal system $\Sigma = \{\varphi_p(t)\}_{p \in Q}$ of continuous functions complete in $L_2(E, \sigma)$ and such that (A1)′ the sequence $\{\int \varphi_p(t)\varphi_q(t)\varphi_r(t)d\sigma(t)\}_{r \in Q}$ is finite for all fixed $p, q \in Q$; (A2)′ if $\varphi_p(t) \in \Sigma$, then $\overline{\varphi_p(t)} \in \Sigma$; (A3)′ there is a sequence $\{m(r)\}_{r \in Q}$ of positive numbers such that

$$\sum_r \left|\int \varphi_p(t)\varphi_q(t)\overline{\varphi_r(t)}d\sigma(t)\right| m(r) \leq 1$$

for any fixed $p, q \in Q$, and (A4)′ there is a point $e \in Q$ such that $\varphi_e(t) \equiv 1$ $(t \in E)$.

2.5. Convolution Algebras and Hypercomplex Systems. In this subsection, we establish the relationship between hypercomplex systems and convolution algebras introduced for the first time by Ionescu Tulcea and Simon [IoS] (see also [Mal1] and [Mal2]).

Let Q be a locally compact space and let m_r be a real Radon measure on $Q \times Q$ for all $r \in Q$. For any $f, g \in C_0(Q)$, we define their convolution $f * g$ as

$$(f * g)(r) = \int \int f(p)g(q)dm_r(p, q).$$

Suppose that the following conditions are satisfied:

(C1) for any $A \subset Q$, there exists a compact set $K_A \subset Q$ such that $\operatorname{supp} f * g \subset K_A$ whenever $\operatorname{supp} f \subset A$ and $\operatorname{supp} g \subset A$;

(C2) $C_0(Q)$ is a commutative associative algebra with respect to convolution;

(C3) there exists a Radon measure m on Q strictly positive on nonempty open sets and such that

$$\int |(f * g)(r)| dm(r) \leq \int |f(p)| dm(p) \int |g(q)| dm(q)$$

for any $f, g \in C_0(Q)$;

(C4) there is an involutive homeomorphism $r \to r^*$ of the space Q such that

$$\int f(r)dm(r) = \int f(r)dm(r^*)$$

and

$$\int (g * \bar{h})(r)k(r)dm(r) = \int g(r)(h^* * k)(r)dm(r)$$

for any $f, g, h, k \in C_0(Q)$ (here, as usual, $f^*(r) = \overline{f(r^*)}$);

(C5) there is a point $e \in Q$ (the identity element) such that $e^* = e$ and

$$\int f(p)\overline{g(q)}\,dm_e(p,q) = \int f(p)g^*(p)\,dm(p)$$

for all $f, g \in C_0(Q)$.

Clearly, the maps $(f, g) \to f * g$ and $f \to f^*$, extended by continuity to $L_1(Q, m)$, transform $L_1(Q, m)$ into a commutative Banach algebra with involution. By L we denote the set of all finite functions summable with respect to the measure m. We also impose the following condition:

(C6) for any function $g \in L$ and any compact set $A \subset Q$, the family of maps $s \to (\check{g} * f)(s)$, where $f \in C_0(Q)$, $\operatorname{supp} f \subset A$, and $\|f\|_\infty \le 1$, is equicontinuous. Here, $\check{g}(s) = g(s^*)$.

The space $L_1(Q, m)$ satisfying (C1)–(C6) is called a *convolution algebra.*

Let us show that *any convolution algebra is a normal hypercomplex system with, generally speaking, complex-valued structure measure and basis unity.*

Indeed, we set $c(A, B, r) = m_r(A \times B)$ and show that $c(A, B, r)$ is a structure measure. Since m_r is a real Radon measure on $Q \times Q$, $c(A, B, r)$ is a real measure on Q for fixed B and r. Further, since $m_r(A \times B) = (\kappa_A * \kappa_B)(r)$, by using the Lebesgue theorem, we conclude from (C6) that $c(A, B, r)$ is continuous in r for fixed $A, B \in \mathcal{B}_0(Q)$. Moreover, it follows from (C1) that there exists a compact set $F \subset Q$ such that $\operatorname{supp} c(E_1, E_2, r) \subset F$ if $E_1 \subset A$ and $E_2 \subset B$. The relation of associativity now follows from (C2) and the following chain of relations valid for all $A, B, C \in \mathcal{B}_0(Q)$ and $r \in Q$:

$$\int c(A, B, p)\,d_p c(E_p, C, r)$$

$$= \int\int (\kappa_A * \kappa_B)(p)\kappa_C(s)\,dm_r(p,s) = ((\kappa_A * \kappa_B) * \kappa_C)(r)$$

$$= ((\kappa_B * \kappa_C) * \kappa_A)(r) = \int c(B, C, p)\,d_p c(A, E_p, r)$$

The relation of commutativity for $c(A, B, r)$ is a consequence of (C2). Condition (C3) implies that the measure m is a multiplicative measure. Thus, $L_1(Q, m)$ is a hypercomplex system This hypercomplex system is normal by virtue of (C4), (C6), and the

Lebesgue theorem. Condition (C5) implies that e is the basis unity of this hypercomplex system.

■

On the other hand, (C1)–(C5) are satisfied for all commutative normal hypercomplex systems with basis unity as follows from Section 1 (as indicated in Subsection 2.4, all results established in Section 1 remain valid for hypercomplex systems with complex-valued structure measure) but condition (C6) is, in general, not true.

3. Elements of Harmonic Analysis for Normal Hypercomplex Systems with Basis Unity

In this section, we establish some principal theorems of the harmonic analysis on locally compact groups for hypercomplex systems. In Subsection 3.1, we introduce an analog of the Fourier transformation, prove analogs of the Bochner and Plancherel theorems, and establish the inversion formula for commutative normal hypercomplex systems with basis unity (the case of noncommutative hypercomplex systems will be considered in Section 5). Then we generalize the Pontryagin duality theory to hypercomplex systems and describe an analog of the Fourier algebra. Further, we consider representations of normal, generally speaking, noncommutative hypercomplex systems with basis unity and prove the approximation theorem.

3.1. Fourier Transformation and the Plancherel Theorem. Let us show that analogs of the Bochner and Plancherel theorems and the inversion formula hold for commutative normal hypercomplex systems. The relevant results for noncommutative hypercomplex systems are presented in Section 5.

Let $\mathfrak{L}_1(Q, m)$ be a commutative normal hypercomplex system with basis unity, i.e., axioms (H1)–(H6) are satisfied. Let X and X_h be the sets of characters and Hermitian characters, respectively, endowed with the topology of the space of maximal ideals. We denote the spaces of generalized and Hermitian generalized characters by X_g and $X_{g,h}$, respectively. The space X is compact (locally compact) if L_1 contains the identity element (does not contain). We now consider the second more complicated case. At the end of the subsection, we describe the situation occurring in the first case.

An essentially bounded function $\varphi(r)$ $(r \in Q)$ is called *positive definite* if

$$\int \varphi(r)(x^* * x)(r)\,dr \geq 0 \tag{3.1}$$

for all $x \in L_1$.

We also present another definition of positive definiteness.

A continuous bounded function $\varphi(r)$ $(r \in Q)$ is called positive definite if the inequality

$$\sum_{i,j=1}^{n} \lambda_i \bar{\lambda}_j \big(R_{r_i^*}\varphi\big)(r_j) \geq 0 \tag{3.1$'$}$$

holds for all $r_1, \ldots, r_n \in Q$ and $\lambda_1, \ldots, \lambda_n \in \mathbb{C}$ $(n \in \mathbb{N})$.

If the generalized translation operators R_t *extended to* L_∞ *map* $C_b(Q)$ *into* $C_b(Q \times Q)$, *then the definitions of positive definiteness (3.1) and (3.1)′ are equivalent for the functions* $\varphi(r) \in C_b(Q)$.

Indeed,

$$\begin{aligned}\int \varphi(r)(x^* * x)(t)dt &= \int \varphi(t) \int (L_s x)(t)\overline{x(s)}ds\,dt \\ &= \iint (L_{s^*}\varphi)(t)\overline{x(s)}ds\,x(t)dt \\ &= \iint (R_t\varphi)(s^*)x(t)\overline{x(s)}dt\,ds \geq 0.\end{aligned}$$

By the condition, we have $(R_t\varphi)(s^*) \in C_b(Q \times Q)$, then the last inequality clearly implies (3.1)′. Let us prove the converse assertion. Let Q_n be an increasing sequence of compact sets covering the entire Q. We consider a function $y(r) \in C_0(Q)$ and set $\lambda_i = y(r_i)$ in (3.1)′. This yields

$$\sum_{i,j=1}^{n} \left(R_{r_i^*}\varphi\right)(r_j)y(r_i)\overline{y(r_j)} \geq 0.$$

By integrating this inequality with respect to each $r_1, \ldots, r_n$ over the set Q_k $(k \in \mathbb{N})$ and collecting similar terms, we conclude that

$$nm(Q_k) \int_{Q_k} \left(R_{r^*}\varphi\right)(r)|y(r)|^2 dr + n(n-1) \int_{Q_k}\int_{Q_k} \left(R_{r^*}\varphi\right)(s)y(r)\overline{y(s)}dr\,ds \geq 0$$

Further, we divide this inequality by n^2 and pass to the limit as $n \to \infty$. We get

$$\int_{Q_k}\int_{Q_k} \left(R_{r^*}\varphi\right)(s)y(r)\overline{y(s)}dr\,ds \geq 0$$

for each $k \in \mathbb{N}$. By passing to the limit as $k \to \infty$ and applying Lebesgue theorem, we see that (3.1) holds for all functions from $C_0(Q)$. Approximating an arbitrary function from L_1 by finite continuous functions, we arrive at (3.1).

■

In what follows, we use definition (3.1) which is formally more general. Clearly, the set of functions positive definite in the sense of (3.1) is weakly closed in $L_\infty(Q, m)$. By $P(Q)$ we denote a set of continuous positive definite functions and by $\mathfrak{R}$ we denote the linear span of the functions from $L_1 \cap P(Q)$.

Let $L_{2,0}$ be the set of functions with compact support.

Lemma 3.1. *The linear span of the set of functions from* $\mathfrak{R} \cap C_0(Q)$ *of the form* $a * b$ $(a, b \in L_{2,0})$ *is dense in* L_2 *and* L_1.

Proof. Let e_n be the approximative unit of a hypercomplex system constructed in Lemma 1.3. The sequence $a * e_n \in C_0(Q)$ converges to a in L_2 for any $a \in L_{2,0}$. Thus, the functions of the form $a * b^* \in C_0(Q)$ are dense in $L_{2,0}$ (in the metric of the space L_2). Hence, they are dense in L_2. On the other hand, $a * b^*$ is a linear combination of the positive definite functions $(a \pm b) * (a \pm b)^*$ and $(a \pm ib) * (a \pm ib)^*$ and, hence, it belongs to $\mathfrak{R}$.

Since step functions are dense in L_1, it suffices to prove the second part of the lemma for the indicators κ_A of sets $A \in \mathcal{B}_0(Q)$. Let O_n be the sequence of neighborhoods constructed in Lemma 1.3. Denote $F = \operatorname{supp} c(A, O_1, r)$. Taking into account that $\operatorname{supp} e_n \subset O_n \subset O_1$ and the structure measure is nonnegative, we establish that $\operatorname{supp} \kappa_A * e_n \subset F$. Thus,

$$\begin{aligned}
\| \kappa_A * e_n - \kappa_A \|_1 &= \int_Q |(\kappa_A * e_n)(r) - \kappa_A(r)| dr \\
&= \int \kappa_F(r) |(\kappa_A * e_n)(r) - \kappa_A(r)| dr \\
&\leq \sqrt{m(F)} \| \kappa_A * e_n - \kappa_A \|_2 \to 0, \qquad n \to \infty.
\end{aligned}$$

■

Lemma 3.2. *Assume that the function* $L_s f(t)$ *is separately continuous for each* $f \in C_0(Q)$ *(i.e., the corresponding hypercomplex system satisfies the condition of separate continuity). Then the set of functions from* $\mathfrak{R} \cap C_0(Q)$ *of the form*

$$\varphi(r) = \sum_{k=1}^{N(\varphi)} \lambda_k (a_k * b_k^*)(r) \qquad (\lambda_k \in \mathbb{C};\ a_k, b_k \in C_0(Q))$$

is dense in $C_\infty(Q)$.

Proof. Since $C_0(Q)$ is dense in $C_\infty(Q)$, it suffices to prove this statement for an arbitrary function $x \in C_0(Q)$. First, we show that $x * e_n(r) \to x(r)$ as $n \to \infty$ for all $r \in Q$. Indeed, since $e_n^* = e_n$, we have

$$x * e_n(r) - x(r) = \int_{Q_n} ((L_s x)(r) - x(r)) e_n(s) ds \to 0, \qquad n \to \infty,$$

because $(L_s x)(r)$ is continuous for fixed $r \in Q$. Since $\| x * e_n \|_\infty \leq \| x \|_\infty \| e_n \|_1 = \| x \|_\infty$, by virtue of the Lebesgue theorem, we have $\langle \mu, x * e_n \rangle \to \langle \mu, x \rangle$ as $n \to \infty$ for any Radon measure $\mu \in M_b^+(Q)$. This implies that $x * e_n(r)$ weakly converges to $x(r)$ in $C_\infty(Q)$. Then there exists a sequence of functions from the linear span of $\{ x * e_n \mid n \in \mathbb{N} \}$ that uniformly converges to x.

■

Lemma 3.3. *If* $f \in L_2(Q, m)$, *then* $f^* * f \in P(Q)$.

Proof. By virtue of Lemma 1.1, we have $f^* * f \in C_\infty(Q)$. We fix $x \in L_1(Q, m)$. Passing to the limit in the equality $(a^* * a, b^* * b)_2 = \| a * b \|_2^2$ (here, a and b are step functions) as $a \to f$ in the space L_2 and $b \to x$ in L_1, and using Theorem 1.1 and Lemmas 1.2 and 1.1, we get

$$\langle f^* * f, x^* * x \rangle = \| f * x \|_2^2 \geq 0.$$

■

For any function $x \in L_1$ and any character $\chi \in X$, we set

$$\hat{x}(\chi) = \int x(r) \overline{\chi(r)} dr.$$

This integral exists and $\hat{x}$ is a continuous function on X. It is called a *Fourier transform* of the function $x \in L_1$. By Theorem 1.2, $\hat{x}(\chi)$ is simply $x(M)$, where M is a maximal ideal of the algebra $\mathfrak{L}_1(Q, m)$ associated with the character $\overline{\chi}$.

This implies that

$$(x_1 + x_2)^\wedge(\chi) = \hat{x}_1(\chi) + \hat{x}_2(\chi),$$

$$(\lambda x)^\wedge(\chi) = \lambda \hat{x}(\chi), \qquad (x_1 * x_2)^\wedge(\chi) = \hat{x}_1(\chi) \hat{x}_2(\chi),$$

for all $\chi \in X$. For all $\chi \in X_h$, we have $(x^*)^\wedge(\chi) = \overline{\hat{x}(\chi)}$. In particular, $(x^* * x)^\wedge(\chi) = |\hat{x}(\chi)|^2$. We also note that the set of Fourier transforms of functions from $L_1(Q, m)$ is dense in the space $C_\infty(X_h)$ with uniform norm.

Theorem 3.1 is an analog of the Bochner theorem for hypercomplex systems.

Theorem 3.1. *Every function* $\varphi \in P(Q)$ *admits a unique representation in the form of an integral*

$$\varphi(r) = \int_{X_h} \chi(r)\, d\mu(\chi) \quad (r \in Q), \tag{3.2}$$

where μ *is a nonnegative finite regular measure on the space* X_h. *Conversely, each function of the form (3.2) belongs to* $P(Q)$.

Proof. Let $\varphi(r) \in P(Q)$. Consider a continuous functional Φ in $L_1(Q, m)$ defined as follows:

$$\Phi(x) = \int \varphi(r) x(r)\, dr \quad (x \in L_1).$$

It is clear that this functional is positive. The functional Φ can be extended to a positive functional $\tilde{\Phi}$ in $\mathfrak{L}_1(Q, m)$. To do this, it suffices to show that the functional Φ is real (i.e., $\Phi(x^*) = \overline{\Phi(x)}$ for all $x \in L_1(Q, m)$) and satisfies the inequality $|\Phi(x)|^2 \le C\Phi(x^* * x)$ for all $x \in L_1(Q, m)$, where C is a constant (see, e.g., [Nai1, p. 223]). Let $e_n(r)$ be the approximative unit constructed in the proof of Lemma 1.3. Since Φ is positive, we have

$$\Phi(x^*) = \lim_{n\to\infty} \Phi\left(e_n^* * x^*\right) = \lim_{n\to\infty} \overline{\Phi\left(x * e_n\right)} = \overline{\Phi(x)}$$

for all $x \in L_1(Q, m)$. Further, by using Lemma 1.3, we obtain

$$|\Phi(x)|^2 = \lim_{n\to\infty} |\Phi(e_n * x)|^2 \le \lim_{n\to\infty} \Phi\left(e_n^* * e_n\right) \Phi(x^* * x) \le \|\Phi\| \Phi(x^* * x).$$

Consequently, it is possible to extend Φ to a positive functional $\tilde{\Phi}$ on $\mathfrak{L}_1(Q, m)$.

By virtue of the theorem on representations of positive functionals on commutative Banach $*$-algebras with identity element, the functional $\tilde{\Phi}$ (and, hence, Φ) can be uniquely represented in the form

$$\Phi(x) = \int_{X_h} \int_Q x(r) \chi(r)\, dr\, d\mu(\chi),$$

where μ is a finite regular Borel measure on $\mathcal{B}_0(X_h)$. Consequently,

$$\varphi(r) = \int_{X_h} \chi(r)\, d\mu(\chi)$$

almost everywhere on Q. Since the characters of normal hypercomplex systems are continuous, both functions in this equality are also continuous. This yields (3.2).

The second part of the theorem follows from the relation

$$\int_Q \int_{X_h} \chi(r)\, d\mu(\chi)(x^* * x)(r)\, dr = \int_{X_h} \int_Q (x^* * x)(r)\chi(r)\, dr\, d\mu(\chi)$$

$$= \int_{X_h} |\hat{x}(\chi)|^2 d\mu(\chi) \geq 0$$

and the Lebesgue theorem on the limit transition.

Corollary 3.1. *If the product of any two Hermitian characters is positive definite, then the product of any two continuous positive definite functions is also positive definite.*

Indeed, by virtue of Theorem 3.1,

$$\int f(r)g(r)(x^* * x)(r)\, dr = \int_Q \int_{X_h} \chi(r)\, d\mu(\chi) \int_{X_h} \psi(r)\, d\nu(\psi)(x^* * x)(r)\, dr$$

$$= \int_{X_h} \int_{X_h} \int_Q \chi(r)\psi(r)(x^* * x)(r)\, dr\, d\mu(\chi)\, d\nu(\psi) \geq 0$$

for all $f, g \in P(Q)$ and $x \in L_1$.

■

Corollary 3.2. *Assume that $L_1(Q, m)$ is a commutative hypercomplex system with basis unity. Then a continuous bounded function $\varphi(r)$ is positive definite in the sense of (3.1) if and only if it is positive definite in the sense of (3.1)′. Moreover, it has the following properties:*

(i) $\varphi(e) \geq 0$;

(ii) $\varphi(r^*) = \overline{\varphi(r)}$;

(iii) $|\varphi(r)| \leq \varphi(e)$;

(iv) $|(R_s\varphi)(t)|^2 \leq (R_{s^*}\varphi)(s)(R_{t^*}\varphi)(t)$;

(v) $|\varphi(s) - \varphi(t)|^2 \leq 2\varphi(e)[\varphi(e) - \operatorname{Re}(R_s\varphi)(t^*)]$.

Proof. At the beginning of this subsection, we have shown that positive definiteness in the sense of (3.1)′ implies positive definiteness in the sense of (3.1). Let us show that the converse assertion is also true (i.e., the requirement of continuity of $(R_s\varphi)(r)$ $(f \in C_b(Q))$ is unnecessary). Assume that $\varphi(r) \in C_b(Q)$ is positive definite in the sense of (3.1) and let $\lambda_1, \ldots, \lambda_n \in \mathbb{C}$ and $r_1, \ldots, r_n \in Q$. Relation (3.2) and the fact that the generalized translation operators are continuous in $L_\infty(Q, m)$ imply that

$$\sum_{i,j=1}^{n} \lambda_i \bar{\lambda}_j \left(R_{r_i^*}\varphi\right)(r_j) = \sum_{i,j=1}^{n} \lambda_i \bar{\lambda}_j \int_{X_h} \left(R_{r_i^*}\chi\right)(r_j)\, d\mu(\chi)$$

$$= \int_{X_h} \sum_{i,j=1}^{n} \lambda_i \bar{\lambda}_j\, \overline{\chi(r_i)}\chi(r_j)\, d\mu(\chi) = \int_{X_h} \left| \sum_{k=1}^{n} \lambda_k \overline{\chi(r_k)} \right|^2 d\mu(\chi) \geq 0.$$

It also follows from relation (3.2) that

$$\varphi(e) = \mu(X_h) \geq 0; \ \varphi(r^*) = \int_{X_h} \chi(r^*)\, d\mu(\chi) = \int_{X_h} \overline{\chi(r)}\, d\mu(\chi) = \overline{\varphi(r)};$$

$$|\varphi(r)| \leq \int_{X_h} |\chi(r)|\, d\mu(\chi) \leq \mu(X) = \varphi(e);$$

$$\begin{aligned} |(R_s\varphi)(t)|^2 &= \left| \int_{X_h} \chi(s)\chi(t)\, d\mu(\chi) \right|^2 \\ &\leq \int_{X_h} |\chi(s)|^2 d\mu(\chi) \int_{X_h} |\chi(t)|^2 d\mu(\chi) \\ &= \int_{X_h} \chi(s)\chi(s^*)\, d\mu(\chi) \int_{X_h} \chi(t)\chi(t^*)\, d\mu(\chi) \\ &= (R_{s^*}\varphi)(s)(R_{t^*}\varphi)(t). \end{aligned}$$

Finally,

$$|\varphi(s) - \varphi(t)|^2 \leq \left| \int_{X_h} (\chi(s) - \chi(t))\, d\mu(\chi) \right|^2 \leq \mu(X_h) \int_{X_h} |(\chi(s) - \chi(r))|^2 d\mu(\chi)$$

$$= \varphi(e) \int_{X_h} \left(|\chi(s)|^2 + |\chi(t)|^2 - 2\,\mathrm{Re}\, \chi(s)\overline{\chi(r)} \right) d\mu(\chi)$$

$$\leq \varphi(e) \int_{X_h} 2(1 - \operatorname{Re}(R_{r^*}\chi)(s))\, d\mu(\chi)$$

$$= 2\varphi(e)[\varphi(e) - \operatorname{Re}(R_{r^*}\varphi)(s)].$$

■

Note that *properties (i)–(v) hold for continuous functions positive definite in the sense of (3.1)′ on a noncommutative hypercomplex system.*

Inequality (3.1)′ means that the kernel $K(t, s) = (R_{t^*}\varphi)(s)$ is positive definite. Therefore, this kernel possesses the following properties: $K(t, t) \geq 0$, $|K(t, s)|^2 \leq K(t, t)K(s, s)$, $K(t, s) = \overline{K(s, t)}$, and

$$|K(t, r) - K(s, r)|^2 \leq K(r, r)(K(t, t) + 2\operatorname{Re} K(t, s) + K(s, s))$$

(see, e.g., [BeKo, p. 469]). This immediately yields (ii). Indeed,

$$\varphi(r^*) = (R_e\varphi)(r^*) = K(e, r^*) = \overline{K(r^*, e)} = \overline{(R_r\varphi)(e)} = \overline{(L_e\varphi)(r)} = \overline{\varphi(r)}.$$

Similarly, $(R_r\varphi)(s) = \overline{(R_{s^*}\varphi)(r^*)}(r^*)$. This implies that

$$|(R_t\varphi)(s)|^2 \leq (R_{t^*}\varphi)(t)(R_{s^*}\varphi)(s),$$

i.e., (iv). By setting $s = e$ in (iv), we obtain $|\varphi(r)|^2 \leq \varphi(e)(R_{r^*}\varphi)(r)$. In view of (2.7), we have $(R_{r^*}\varphi)(r) \leq \|\varphi\|_\infty$. Consequently, $|\varphi(r)|^2 \leq \varphi(e)\|\varphi\|_\infty$ which implies (iii) and, hence, (i). Finally, (v) follows from the last inequality for $K(t, s)$, where $s = e$,

$$|\varphi(s) - \varphi(t)|^2 \leq \varphi(e)[(R_{t^*}\varphi)(t) + R_{s^*}\varphi(s) - 2\operatorname{Re}(R_{t^*}\varphi)(s)]$$

$$\leq 2\varphi(e)[\varphi(e) - 2\operatorname{Re}(R_{t^*}\varphi)(s)].$$

■

Remark. For simplicity, we consider only bounded positive definite functions on hypercomplex systems. But omitting the boundedness condition, one may study functions continuous and positive definite in the sense of (3.1)′ on hypercomplex systems with the property of separate continuity. For example, the positive definite functions in Example 2 in Section 1 are even-positive definite functions. The reader can find an analog of the Bochner theorem for such functions, e.g., in [Ber10]. An analog of the Bochner theorem for general unbounded positive definite functions associated with commutative normal hypercomplex systems with basis unity and the property of separate continuity, can easily be obtained by using the theorem on spectral decomposition for representations of commutative hypercomplex systems (see Subsection 4 below). General unbounded positive definite functions and measures on hypergroups were studied by Bloom and Heyer [BlH3] and Voit [Voi7].

Theorem 3.2 (an analog of the inversion formula). *There exists a unique regular Borel measure* $\hat{dm}(\chi) = d\chi$ *on the space* X_h *such that, for any function* $x \in \mathfrak{R}$, *its Fourier transform* $\hat{x}$ *belongs to* L_1 *and satisfies the equality*

$$x(r) = \int_{X_h} \hat{x}(\chi)\chi(r)\,d\chi.$$

Proof. Denote by I the set of all functions $x \in L_1(Q, m) \cap C_b(Q)$. Lemma 1.1 implies that I is a $*$-ideal in L_1. In I, we define a linear functional $F(x) = x(e)$ $(x \in I)$, where e is the basis unity of the hypercomplex system $\mathfrak{L}_1(Q, m)$. The functional $F(x)$ is real (since $e = e^*$) and positive on I because

$$F(x^* * x) = (x^* * x)(e) = \int \overline{x(p^*)}d_p \int x(q)\,d_p c(E_p, E_q, e)$$

$$= \iint \overline{x(p^*)}x(q)\,dm(E_p \cap E_q^*) = \int |x(p)|^2 dm \geq 0 \quad (x \in I).$$

For what follows, we need the Weil–M. Krein–Godeman theorem (see, e.g., [Nai1, p. 336]): *Let* R *be a commutative Banach* $*$*-algebra without identity element. Then, for any real positive functional* f *defined on an ideal* I *which is dense in* R, *there exists a unique measure* μ *on the space* M_h *of symmetric maximal ideals of the algebra* R *such that*

(*i*) *for any* $\varphi \in P$, *we have* $\varphi(t) \in L_1(M_h, \mu)$ *and*

$$f(\varphi x) = \int_{M_h} x(M)\varphi(M)\,d\mu(M);$$

(*ii*) *the mapping* $\varphi \mapsto \varphi(M)$ *can be extended to an isometric mapping of the Hilbert space* H_f' *into the space* $L_2(M_h, \mu)$.

If $I \cdot I$ *is dense in* I *in the sense of* $\|\cdot\|_{H_f}$, *then* $\varphi \mapsto \varphi(M)$ *can be extended to an isometric mapping from the entire* H_f *into* $L_2(M_h, \mu)$.

Here, the symbol P denotes the collection of all elements positive with respect to f (i.e., elements $\varphi \in I$ such that the functional $f_\varphi(x) = f(\varphi x)$ $(x \in R)$ is real and positive and satisfies the condition $|f_\varphi(x)|^2 \le C f_\varphi(x^*x)$ for all $x \in R$), H_f is the Hilbert space generated by elements $x \in I$ with scalar product $(x, y) = f(xy^*)$, and H_f' is the subspace in H_f generated by elements $x \in P$.

By virtue of the denseness of the ideal I in L_1, we can apply this theorem to the functional F. Consequently, there exists a unique regular Borel measure $d\hat{m}(\chi) = d\chi$, given on the space X_h of Hermitian characters of a hypercomplex system such that the Fourier image $\hat{\varphi}$ of any function $\varphi \in I$ positive with respect to F belongs to $L_1(\mathrm{X}_h, \hat{m})$ and

$$F(\varphi * x) = \int_{\mathrm{X}_h} \hat{\varphi}(\chi)\hat{x}(\chi)\,d\chi. \tag{3.3}$$

Let $\varphi \in L_1 \cap P(Q)$. Since

$$F_\varphi(x) = F(\varphi * x) = (\varphi * x)(e) = \int \varphi(r)x(r)\,dr,$$

the functional F_φ coincides with the functional Φ constructed in the proof of Theorem 2.1. The latter can be extended to a positive functional in $\mathfrak{L}_1(Q, m)$ because L_1 contains an approximative unit. Hence, functions $\varphi \in L_1 \cap P(Q)$ are positive with respect to the functional F. Relation (3.3) holds for any function $\varphi \in L_1 \cap P(Q)$ and, hence, for any $\varphi \in \mathfrak{R}$. In (3.3), we replace x by x^*, substitute the corresponding expressions for $F(\varphi * x)$ and $(x^*)^\wedge(\chi) = \overline{\hat{x}(\chi)}$, and change the order of integration on the right-hand side. As a result, we get

$$\int \varphi(r)\overline{x(r)}\,dr = \int \overline{x(r)}\Big[\int \hat{\varphi}(\chi)\chi(r)\,d\chi\Big]dr$$

(here and below, we omit the indices Q and X_h of the domain of integration if this does not lead to misunderstanding). In view of the arbitrariness of $x \in L_1$, this completes the proof of the theorem.

■

The measure $\hat{m}$ in Theorem 3.2 is called the *Plancherel measure*. Theorem 3.2 implies, in particular, that, for any two functions $f, g \in L_1 \cap L_2$, the convolution $f * g$ belongs to $\mathfrak{R}$ and satisfies the equality

$$(f * g)(r) = \int \hat{f}(\chi)\,\hat{g}(\chi)\chi(r)\,d\chi \qquad (r \in Q). \tag{3.4}$$

Denote by $\hat{Q} \subset \mathrm{X}_h$ the support of the Plancherel measure.

Corollary 3.3. *If the Fourier transforms of functions* $f_1, f_2 \in L_1(Q, m)$ *coincide on* $\hat{Q}$, *then* $f_1 = f_2$.

Indeed, by virtue of Lemma 3.1, the convolution $e_n * f$ converges to f in L_1 and, according to (3.4),

$$e_n * f_1(r) = \int_{\hat{Q}} (e_n * f_1)^{\wedge}(\chi)\chi(r)\,d\chi = \int_{\hat{Q}} \hat{e}_n(\chi)\hat{f}_1(\chi)\chi(r)\,d\chi$$

$$= \int_{\hat{Q}} \hat{e}_n(\chi)\hat{f}_2(\chi)\chi(r)\,d\chi = e_n * f_2(r).$$

Passing here to the limit as $n \to \infty$, we obtain $f_1 = f_2$.

■

Corollary 3.4. *A function* $\varphi \in L_1 \cap C_b(Q)$ *is positive definite if and only if* $\hat{\varphi}(\chi) \geq 0$ *for all* $\chi \in \hat{Q}$.

Indeed, let $\varphi \in L_1 \cap P(Q)$. By virtue of Theorem 3.2, $\hat{\varphi} \in L_1(\hat{Q}, \hat{m})$ and

$$\varphi(r) = \int \hat{\varphi}(r)\chi(r)\,d\chi.$$

Therefore,

$$\int |\hat{x}(\overline{\chi})|^2\, \hat{\varphi}(\chi)\,d\chi = \int (x^* * x)^{\wedge}(\chi)\hat{\varphi}(\chi)\,d\chi$$

$$= \iint (x^* * x)(r)\chi(r)\hat{\varphi}(\chi)\,d\chi\,dr = \int \varphi(r)(x^* * x)(r)\,dr \geq 0$$

for any function $x \in L_1(Q, m)$. Therefore, since all functions $\hat{x}$ $(x \in L_1)$ form a dense set in $C_\infty(X_h)$, we get

$$\int \hat{\varphi}(\chi)v(\chi)\,d\chi \geq 0$$

for any nonnegative function $v \in C_\infty(X_h)$. This is possible only if $\hat{\varphi}(\chi) \geq 0$ $(\chi \in \hat{Q})$.

Conversely, let $\varphi \in L_1 \cap C_b(Q)$ and let $\hat{\varphi}(\chi) \geq 0$ for all $\chi \in \hat{Q}$. By using Theorem 3.2 and the fact that $g^* * g \in P(Q)$ for any function $g \in L_1 \cap L_\infty$, we obtain

$$\int \varphi(r)(g^* * g)(r)\,dr = \int \varphi(r) \int (g^* * g)^{\wedge}(\chi)\chi(r)\,d\chi\,dr$$

$$= \int |\hat{g}(\chi)|^2 \int \varphi(r)\chi(r)\,dr\,d\chi = \int |\hat{g}(\chi)|^2 \hat{\varphi}(\overline{\chi})\,d\chi \geq 0.$$

Approximating an arbitrary function $y \in L_1$ by functions from $g \in L_1 \cap L_\infty$ in the metric of L_1 and using the boundedness of $\varphi(r)$ and the continuity of convolution, we get

$$\int \varphi(r) y^* * y(r)\,dr \geq 0,$$

i.e., $\varphi(r) \in P(Q)$.

■

Remark. It follows from Lemma 3.3 and Corollary 3.4 that if $f^* * f \in L_1$ for some function $f \in L_2(Q, m)$, then $(f^* * f)^\wedge(\chi) \geq 0$ $(\chi \in \hat{Q})$.

Applying (3.4) to the function $c(A, B, r) = \kappa_A * \kappa_B(r)$, we obtain

$$c(A, B, r) = \int_{\hat{Q}} \int_{Q} c(A, B, p)\overline{\chi(p)}\,dp\,\chi(r)\,d\chi = \int_{\hat{Q}} \overline{\chi}(A)\overline{\chi}(B)\chi(r)\,d\chi.$$

Let us prove an analog of the Plancherel theorem for hypercomplex systems. The proof is based on a continual analog of the Schur lemma (see, [Nai1, Section 26, Subsection 5]). For convenience, we present the assertions necessary for what follows.

Let T be a locally compact space, let μ be a regular Borel measure on T, and let R be a Banach algebra with involution.

Theorem (a continual analog of the Schur lemma). *Suppose that a representation $x \mapsto A_x$ of the Banach algebra R with involution is given by the direct integral of representations of this algebra in a separable Hilbert space H, namely,*

$$x \mapsto \int_T A_x(t)\,d\mu(t).$$

Furthermore, assume that the following conditions are satisfied:

(i) the function $T \ni t \mapsto \|A_x(t)\| \in C_\infty(T)$ for any $x \in R$;

(ii) the operator $A_x(t)$ is compact for all $x \in R$ and $t \in T$;

(iii) each representation $x \mapsto A_x(t)$ is irreducible;

(iv) the representations $x \mapsto A_x(t_1)$ and $x \mapsto A_x(t_2)$ are not equivalent for any

$t_1 \neq t_2,\ t_1, t_2 \in T.$

Then any bounded operator B *acting in*

$$\mathcal{H} = \int_T^{\oplus} H \, d\mu$$

and commuting with all operators A_x $(x \in R)$ *has the form* $B = \{\beta(t)\mathbb{1}\}$, *where* $\beta(t) \in L_\infty(T, \mu)$. *Furthermore, any subspace* $\mathfrak{M}$ *of* $\mathcal{H}$ *invariant under the action of all operators* A_x *is a collection of vectors* $\xi = \{\xi(t)\} \in \mathcal{H}$ *satisfying the condition* $\xi(t) = 0$ *for almost all* $t \in \Delta$, *where* Δ *is a fixed* μ*-measurable set in* T.

Theorem 3.3 (an analog of the Plancherel theorem). *The Fourier transformation* $x \mapsto \hat{x}$ *is an isometric mapping from a set dense in* $L_2(Q, m)$ *onto a set dense in* $L_2(\hat{Q}, \hat{m})$ *and, hence, it can be uniquely extended to an isometric operator* $\mathcal{F}$ *that maps* $L_2(Q, m)$ *onto* $L_2(\hat{Q}, \hat{m})$.

Proof. Let I, F, and $P(Q)$ be the same as in Theorem 3.2. By setting $x = \psi^*$ in (3.3), where $\psi \in \mathfrak{R}$, we get

$$\int \varphi(r)\overline{\psi(r)}dr = (\varphi * \psi^*)(e) = F(\varphi * \psi^*) = \int \hat{\varphi}(\chi)\overline{\hat{\psi}(\chi)}d\chi^*.$$

Therefore, the Fourier transformation $\varphi \mapsto \hat{\varphi}$ is an isometric mapping from the subspace $\mathfrak{R}$ of $L_2(Q, m)$ into $L_2(\hat{Q}, \hat{m})$. Since $\mathfrak{R}$ is dense in $L_2(Q, m)$, this mapping can be uniquely extended to an isometric operator $\mathcal{F}$ that maps $L_2(Q, m)$ into $L_2(\hat{Q}, \hat{m})$.

It remains to show that $\mathcal{F}$ maps $L_2(Q, m)$ onto $L_2(\hat{Q}, \hat{m})$. It suffices to show that the image of the space $L_2(Q, m)$ under the mapping $\mathcal{F}$ is dense in $L_2(\hat{Q}, \hat{m})$. For this purpose, we use the continual analog of the Schur lemma presented above. If we take $L_1(Q, m)$ as R and set $T = \hat{Q}$, $d\mu = d\chi$, and $H = \mathbb{R}^1$, then $\mathcal{H} = L_2(\hat{Q}, \hat{m})$. Let A_x be the operator of multiplication by the function $\hat{x}(\chi)$ $(\chi \in \hat{Q})$ in $L_2(\hat{Q}, \hat{m})$; then $A_x(t)$ is the operator of multiplication by the number $\hat{x}(\chi)$ in $\mathbb{R}^1$. Clearly, the operators A_x thus chosen satisfy all conditions of the theorem. Let us show that the image $\mathfrak{M}$ of the space $L_2(Q, m)$ under the mapping $\mathcal{F}$ is invariant under the action of the operators A_x $(x \in L_1(Q, m))$. Consider a function $v(\chi) \in \mathcal{F}(L_1(Q, m) \cap L_2)$. Then $v = \mathcal{F}y$, where $y \in L_1(Q, m) \cap L_2(Q, m)$ and $A_x v = \hat{x}(\chi)v(\chi) = (x * y)^{\wedge}(\chi)$. By virtue of Theorem 1.1, $x * y(r) \in L_1(Q, m)$. According to Lemma 1.2, we have $(x * y)(r) \in L_2(Q, m)$. Therefore, the manifold $\mathcal{F}(L_1(Q, m) \cap L_2(Q, m))$ (and, hence, $\mathfrak{M}$) is invariant under the action of the operators A_x $(x \in L_1(Q, m))$. By using the continual analog of the Schur lemma, we find that $\mathfrak{M}$ consists of functions equal to zero almost everywhere on some measurable set Δ in $\hat{Q}$. Let us show that $\hat{m}(\Delta) = 0$. Denote by $U_1 \supset U_2 \supset \ldots$ a sequence of balls that contracts to e and consider the sequence of functions $\hat{e}_n(r) = \kappa_{U_n}(r) / m(U_n)$. Clearly, $e_n(r) \in L_1 \cap L_2$ and $e_n(\chi) = \chi(U_n) / m(U_n)$. As shown in Theorem 1.6,

$$\lim_{n\to\infty} \frac{\chi(U_n)}{m(U_n)} = 1,$$

which implies that $\hat{m}(\Delta) = 0$ by virtue of the continuity of the functions $\hat{e}_n(\chi)$.

■

Denote by S the Fourier image of the set $\mathfrak{R} \cap C_0(Q)$.

Lemma 3.4. *The set* S *is dense in* $L_1(\hat{Q}, \hat{m})$ *and* $L_2(\hat{Q}, \hat{m})$.

Proof. Consider the set $S_1 \subset L_2(\hat{Q}, \hat{m})$ of Fourier transforms of functions from $C_0(Q)$. Since $C_0(Q)$ is dense in $L_2(Q, m)$, the set S_1 is dense in $L_2(\hat{Q}, \hat{m})$. Let $x \in L_1(\hat{Q}, \hat{m})$. Then $x = y \cdot z$, where $y, z \in L_2(\hat{Q}, \hat{m})$. Let us find $y_1, z_1 \in C_0(Q)$ such that

$$\| y - \hat{y}_1 \|_{L_2(\hat{Q}, \hat{m})} < \varepsilon, \qquad \| z - \hat{z}_1 \|_{L_2(\hat{Q}, \hat{m})} < \varepsilon.$$

Theorem 3.2 implies that the function $y_1 * z_1(r)$ belongs to $\mathfrak{R} \cap C_0(Q)$. Hence, the function $(y_1 * z_1)^{\wedge}(\chi) = \hat{y}_1 \hat{z}_1(\chi)$ belongs to S and, therefore,

$$\| yz - \hat{y}_1\hat{z}_1 \|_{L_1} \le \| y \|_{L_2} \| z - \hat{z} \|_{L_2} + \| \hat{z}_1 \|_{L_2} \| y - \hat{y}_1 \|_{L_2} \le \varepsilon(\| y \|_{L_2} + \| z \|_{L_2} + \varepsilon).$$

Finally, by virtue of Theorem 3.3 and Lemma 3.1, we conclude that S is dense in $L_2(\hat{Q}, \hat{m})$.

■

By virtue of Theorem 3.2, the relation

$$f(r) = \int_{\hat{Q}} f(\chi)\chi(r)\, d\chi$$

determines the inverse Fourier transformation of the set S dense in $L_2(\hat{Q}, \hat{m})$ onto the set $\mathfrak{R} \cap C_0(Q)$ dense in $L_2(Q, m)$. Consequently, the operator $\mathcal{F}^* = \mathcal{F}^{-1}$ is given by the equality

$$\check{f}(r) = (\mathcal{F}^{-1} f)(r) = \int_{\hat{Q}} f(\chi)\chi(r)\,d\chi \qquad (f(\chi) \in S). \tag{3.5}$$

Relation (3.5) can be extended to the entire space $L_1(\hat{Q}, \hat{m})$ by passing to the limit as $f(\chi)$ tends to $x(\chi) \in L_1(\hat{Q}, \hat{m})$ in the same way as in the proof of Lemma 3.4. By Theorem 3.3 and Lemma 1.1, the functions $\check{f}(r) = (y_1 * z_1)(r) \in C_0(Q) \cap \mathfrak{R}$ (we use the notation of Lemma 3.4) uniformly converge to some function $\varphi(r) \in C_\infty(Q)$. Hence, $\mathcal{F}^{-1}$ maps $L_1(\hat{Q}, \hat{m})$ into $C_\infty(Q)$.

3.2. Duality of Commutative Hypercomplex Systems. Recall the Pontryagin duality principle for commutative locally compact groups: The group of characters constructed for the group of characters of some group is isomorphic to the original group. If the groups in this chain are regarded as bases of the corresponding group algebras regarded, in turn, as hypercomplex systems, then the Pontryagin duality principle can be reformulated as follows: On a given group, we construct a hypercomplex system that coincides with its group algebra. Then, by using the set of characters of this hypercomplex system as a basis, one can construct a hypercomplex system whose characters coincide with the elements of the original group. In this subsection, we show that this construction can be realized in the case of general commutative hypercomplex systems. This duality can be explained by the fact that, along with associative convolution (or, which is the same, generalized translation), there is another operation—ordinary multiplication of functions on Q. Thus, we have a double topological algebra (see {Lit2] and [Lit3]); in the case of hypercomplex systems, this algebra is a Hilbert bialgebra (see Subsection 1.7). As we pass to the dual object $\hat{Q}$ (the space of multiplicative functionals), the operations in the double topological algebra interchange, namely, convolution (or generalized translation) transforms into multiplication of functions defined on $\hat{Q}$ while multiplication of functions defined on Q turns into the dual operation (convolution of functions on $\hat{Q}$).

Let $\mathfrak{L}_1(Q, m)$ be a commutative hypercomplex system satisfying conditions (H1)–(H6). Denote sets from $\mathcal{B}_0(\hat{Q})$ by the letters $\Phi, \Psi, \Theta, \ldots$ and the characters of the hypercomplex system $\mathfrak{L}_1(Q, m)$ by $\chi, \varphi, \psi, \theta, \ldots$. For what follows, we need the functions

$$x_\Phi = \int_{\hat{Q}} \kappa_\Phi(\chi)\chi(r)\,d\chi.$$

By virtue of (3.5), $x_\Phi(r) \in L_2(Q, m)$ and $\hat{x}_\Phi(\chi) = \kappa_\Phi(\chi)$. Theorem 3.1 implies that

$x_\Phi(r) \in P(Q)$. Obviously, $|x_\Phi(r)| \le x_\Phi(e) = \hat{m}(\Phi)$ $(\Phi \in \mathcal{B}_0(\hat{Q}))$. For all $\Phi, \Psi \in \mathcal{B}_0(\hat{Q})$ and $\chi \in \hat{Q}$, we set

$$\hat{c}(\Phi, \Psi, \chi) = \int_Q \overline{\chi(r)} \int_\Phi \varphi(r)\, d\varphi \int_\Psi \psi(r)\, d\psi\, dr. \tag{3.6}$$

Clearly, the integral on the right-hand side converges. We now show that, under certain conditions, $\hat{c}(\Phi, \Psi, \chi)$ is a structure measure which defines convolution in the dual hypercomplex system.

We say that an Hermitian character χ is *essential* if $\chi \in \hat{Q}$.

Theorem 3.4. *In order that $\hat{c}(\Phi, \Psi, \chi)$ be a structure measure, it is necessary and sufficient that its support is compact for fixed $\Phi, \Psi \in \mathcal{B}_0(\hat{Q})$ and that the product of any two essential characters $\chi, \psi \in \hat{Q}$ be positive definite.*

Proof. *Necessity.* By virtue of Corollary 3.4 and the fact that $\hat{c}(\Phi, \Psi, \chi) = (x_\Phi x_\Psi)^\wedge(\chi) \ge 0$ for all $\chi \in \hat{Q}$, the function $x_\Phi(r) x_\Psi(r)$ is positive definite. Let us show that the product of any two essential characters $\varphi(r)$ and $\psi(r)$ of the hypercomplex system $\mathfrak{L}_1(Q, m)$ is positive definite. Let $(\Phi_n)_{n=1}^\infty$ and $(\Psi_n)_{n=1}^\infty$ be sequences of neighborhoods in $\hat{Q}$ that contract to φ and ψ, respectively. Since

$$\left| \frac{x_{\Phi_n}(r)}{\hat{m}(\Phi_n)} \right| \le 1,$$

by passing to the limit as $n \to \infty$ and using the fact that the function $r(\chi) = \chi(r)$ is continuous, we conclude that

$$\int \frac{x_{\Phi_n}(r)}{\hat{m}(\Phi_n)} \cdot \frac{x_{\Psi_n}(r)}{\hat{m}(\Psi_n)} x^* * x(r)\, dr \xrightarrow[n\to\infty]{} \int \varphi(r)\psi(r)(x^* * x)(r)\, dr$$

for any function $x \in L_1(Q, m)$. The nonnegativity of the limit follows from the positive definiteness of the function $x_{\Phi_n}(r)\, x_{\Psi_n}(r)$ for all n.

Sufficiency. Let $\Phi, \Psi \in \mathcal{B}_0(\hat{Q})$. Since $x_\Phi(r) x_\psi(r) \in L_1(Q, m)$, we have $\hat{c}(\Phi, \Psi, \chi) \in C_b(X_h)$. Moreover, $x_\Phi(r) x_\Psi(r) \in P(Q)$ by Corollary 3.1. Hence, $\hat{c}(\Phi, \Psi, \chi) = (x_\Phi x_\Psi)^\wedge(\chi) \ge 0$ $(\chi \in \hat{Q},\ \Phi, \Psi \in \mathcal{B}_0(\hat{Q}))$ by Corollary 3.4. Clearly, $\hat{c}(\Phi, \Psi, \chi)$ is a regular Borel measure with respect to Ψ for any fixed $\Phi \in \mathcal{B}_0(\hat{Q})$ and $\chi \in \hat{Q}$. Clearly, (H3) is also satisfied. It remains to check the associativity relation (H2). Indeed, Theorems 3.2 and 3.3 yield

$$\int \hat{c}(\Phi, \Psi, \chi)\, d_\chi \hat{c}(E_\chi, \Omega, \varphi) = \int_Q \int_{\hat{Q}} (x_\Phi x_\Psi)^\wedge(\chi)\chi(r)\, d\chi\, x_\Omega(r)\overline{\varphi(r)}\, dr$$

$$= \int_Q x_\Phi(r) x_\Psi(r) x_\Omega(r)\overline{\varphi(r)}\, dr$$

$$= \int \hat{c}(\Psi, \Omega, \chi)\, d_\chi \hat{c}(\varphi, E_\chi, \varphi)$$

$$(\Phi, \Psi, \Omega \in \mathcal{B}_0(\hat{Q}),\ \varphi \in \hat{Q}).$$

■

Assume that the conditions of Theorem 3.4 are satisfied.

The Plancherel measure can be regarded as the multiplicative measure for the structure measure $\hat{c}(\Phi, \Psi, \chi)$.

In fact,

$$\int \hat{c}(\Phi, \Psi, \chi)\, d\chi = \int (x_\Phi \cdot x_\Psi)^\wedge(\chi)\, d\chi = x_\Phi(e) x_\Psi(e) = \hat{m}(\Phi)\hat{m}(\Psi)$$

by virtue of Theorems 1.6 and 3.2.

■

It is easy to see that the dual hypercomplex system is normal if the complex conjugation $\chi^*(r) = \overline{\chi(r)}$ *is taken as the involutive homeomorphism. Every point* $r \in Q$ *generates an Hermitian character* $r(\chi) = \chi(r)$ $(\chi \in \hat{Q})$ *of the dual hypercomplex system.*

Indeed,

$$\hat{c}(\Phi, \Psi, \Omega) = \int_\Omega \int_Q x_\Phi(r) x_\Psi(r)\overline{\chi(r)}\, dr\, d\chi = \int x_\Phi(r) x_\Psi(r) x_{\Omega^*}(r)\, dr$$

$$= \int x_\Phi(r^*) x_\Psi(r^*)\, x_{\Omega^*}(r^*)\, dr$$

$$= \int_Q \int_\Phi \chi(r^*)\, d\chi \int_\Psi \psi(r^*)\, d\Psi \int_\Omega \overline{\omega(r^*)}\, d\omega\, dr$$

$$= \int x_{\Phi^*}(r)(x)_{\Psi^*}(r) x_\Omega(r)\, dr = \hat{c}(\Phi, \Psi^*, \chi)$$

for all $\Phi, \Psi, \Omega \in \mathcal{B}_0(\hat{Q})$, whence we conclude that the dual hypercomplex system is normal.

By using Theorem 3.2, we can show that the function $r(\chi)=\chi(r)$ $(\chi\in\hat{Q})$ is a character of the dual hypercomplex system for each $r\in Q$:

$$\int \hat{c}(\Phi,\Psi,\chi)\chi(r)\,d\chi = \int (x_\Phi x_\Psi)^\wedge(\chi)\chi(r)\,d\chi = (x_\Phi x_\Psi)(r)$$

$$= x_\Phi(r)x_\Psi(r) = \int_\Phi \chi(r)\,d\chi \int_\Psi \psi(r)\,d\psi = r(\Phi)r(\Psi).$$

■

Each character of a compact hypercomplex system is essential.

Indeed, by Theorem 1.4, every character χ_n is an idempotent. Hence, $\chi_n\in\mathfrak{R}$. The characters are orthogonal and, therefore,

$$\hat{\chi}_n(p) = \int \chi_n(r)\,\overline{\chi_p(r)}\,dr = \delta_{np}\|\chi_n\|_2^2.$$

This and Theorem 2.2 imply that

$$\chi_n(r) = \sum_{p=0}^{\infty} \hat{\chi}_n(p)\chi_p(r)\hat{m}(p) = \chi_n(r)\|\chi_n\|_2^2\,\hat{m}(n).$$

By setting $r=e$, we obtain

$$\hat{m}(n) = \frac{1}{\|\chi_n\|_2^2} \quad (\chi_n\in \mathrm{X} = (\chi_j)_{j=1}^{\le\infty}), \tag{3.7}$$

whence it follows that $\hat{m}(n)>0$.

■

If a character identically equal to one is essential, then it is a basis unity of the dual system.

Indeed, by Theorem 3.3, for all $\Phi,\Psi\in\mathcal{B}_0(\hat{Q})$, we have

$$\hat{c}(\Phi,\Psi,1) = \int_Q \int_\Phi \varphi(r)\,d\varphi \int_\Psi \psi(r)\,d\psi\,dr = \int_Q \int_\Phi \varphi(r)\,d\varphi\,\overline{\int_{\Psi^*} \psi(r)d\psi}\,dr$$

$$= \int_Q x_\Phi(r)\overline{x_{\Psi^*}(r)}dr = \int_{\hat{Q}} \kappa_\Phi(r)\overline{\kappa_{\Psi^*}(\chi)}d\chi = \hat{m}(\Phi\cap\Psi^*).$$

■

To prove duality, we suppose (throughout the remaining part of this subsection) that the *identity character is essential and the conditions of Theorem 3.4 are satisfied.* In addition, we require the following condition to be satisfied:

(P) If φ and ψ are continuous positive definite functions on Q which are the characteristic functions of measures concentrated on $\hat{Q}$, then the measure associated with the positive definite function $\varphi \cdot \psi$ is also concentrated on $\hat{Q}$.

By Q_h we denote the space of Hermitian characters of the dual hypercomplex system and let $\hat{\hat{Q}} \subset Q_h$ be the support of the Plancherel measure $\hat{\hat{m}}$ constructed according to the dual hypercomplex system. We have shown that $Q \subset Q_h$. It is therefore reasonable to ask under what conditions we have $Q \subset \hat{\hat{Q}}$ or $Q = \hat{\hat{Q}}$. In the second case, we say that *duality* takes place.

Let us describe the range of the Fourier transformation.

Lemma 3.5. *The equality*

$$(fg)^{\wedge}(\chi) = \hat{f} * \hat{g}(\chi) \quad (\chi \in \hat{Q})$$

holds for all functions $f, g \in L_2(Q, m)$.

Proof. Let $f, g \in L_2(Q, m)$ be arbitrary fixed functions. Consider sequences $f_n \in \mathfrak{R}$ and $g_n \in \mathfrak{R}$ which converge in $L_2(Q, m)$ to the indicated functions (the fact that such sequences exist follows from Lemma 3.1). By Theorem 3.2,

$$\begin{aligned}(f_n g_n)^{\wedge}(\chi) &= \int f_n(r) g_n(r) \overline{\chi(r)} dr \\ &= \int_Q \int_{\hat{Q}} \hat{f}_n(\varphi)\varphi(r) d\varphi \int_{\hat{Q}} \hat{g}_n(\psi)\psi(r) d\psi \overline{\chi(r)} dr \\ &= \int_{\hat{Q}} \hat{f}_n(r) d_{\varphi} \int_{\hat{Q}} \hat{g}_n(\psi) d_{\psi} \hat{c}(E_{\varphi}, E_{\psi}, \chi) = (\hat{f}_n * \hat{g}_n)(\chi) \quad (\chi \in \hat{Q}).\end{aligned}$$

On the other hand, by virtue of Theorem 3.3,

$$\begin{aligned}\| (f_n g_n)^{\wedge}(\chi) - (fg)^{\wedge}(\chi) \|_{\infty} &= \sup_{\chi \in \hat{Q}} \left| \int [f_n g_n(p) - fg(p)] \overline{\chi(p)} dp \right| \\ &\le \|f_n\|_2 \|g_n - g\|_2 + \|g\|_2 \|f_n - f\|_2 \xrightarrow[n \to \infty]{} 0,\end{aligned}$$

$$\|(\hat{f}_n * \hat{g}_n)(\chi) - (\hat{f} * \hat{g})(\chi)\|_\infty \leq \|\hat{f}_n\|_2 \|\hat{g}_n - \hat{g}\|_2 + \|\hat{g}\|_2 \|\hat{f}_n - \hat{f}\|_2$$

$$= \|f_n\|_2 \|g_n - g\|_2 + \|g\|_2 \|f_n - f\|_2 \xrightarrow[n\to\infty]{} 0.$$

■

Corollary 3.5. *The rank of the Fourier transformation $\mathcal{F}(L_1(Q, m))$ is equal to $L_2(\hat{Q}, m) * L_2(\hat{Q}, \hat{m})$. In particular, $L_2(\hat{Q}, \hat{m}) * L_2(\hat{Q}, \hat{m})$ is a linear space.*

Indeed, by representing a function $f \in L_1(Q, m)$ as $f = yz$ $(y, z \in L_2(Q, m)$ (e.g., one may set $y = |f|^{1/2}$, $z = |f|^{-1/2} f$ if $f \neq 0$ and 0 if $f = 0$) and applying Lemma 3.5, we conclude that $\mathcal{F}(L_1(Q, m)) \subset L_2(\hat{Q}, \hat{m}) * L_2(\hat{Q}, \hat{m})$. To prove the inverse inclusion, for any two functions $f, g \in L_2(\hat{Q}, \hat{m})$, we find (according to Theorem 3.3) their Fourier preimages $\check{f}, \check{g} \in L_2(Q, m)$. Then $\check{f} \cdot \check{g} \in L_1(Q, m)$ and $(\check{f}\check{g})(\hat{\chi}) = (f * g)(\chi)$ by virtue of Lemma 3.5.

■

Let $P(Q, \hat{Q})$ be the set of continuous positive definite functions on Q which are characteristic functions of measures concentrated on $\hat{Q}$ and let $\mathfrak{R}(Q, \hat{Q})$ be the linear span of $P(Q, \hat{Q}) \cap L_1$. Consider the manifold $R = \mathfrak{R}(Q, \hat{Q}) \cap C_0(Q)$. If $x(r) \in R$, then $|x(r)|^2 \in R$ because, by virtue of condition (P), the product of any two functions from $P(Q, \hat{Q})$ is also an element of this set.

Throughout the remaining part of this subsection, we suppose that the *hypercomplex system $L_1(Q, m)$ satisfies the condition of separate continuity.*

By virtue of Theorem 3.2, the linear span of the set $\{ a * b \mid a, b \in L_{2,0} \}$ lies in R. Consequently, R *is dense* in $C_\infty(Q)$ according to Lemma 3.2.

Let us show that R *majorize* $C_0(Q)$, i.e., for any function $x(r) \in C_0(Q)$, there exists a function $x'(r) \in R$ such that $|x(r)| \leq x'(r)$. In fact, consider the function $y(r) \in C_\infty(Q)$ such that $y(r) > \max\{|x(r)| + 1, 2\}$ for all $r \in \operatorname{supp} x$. We find a function $z(r) \in R$ such that $\|y - z\|_\infty < 1/2$. Then $|z(r)| > \max\{|x(r)| + 1/2, 1\}$ $(r \in \operatorname{supp} x)$ and one can take $|z(r)|^2 = z(r)\overline{z(r)} \in R$ as $x'(r)$.

■

By using the property proved above, we may conclude that every positive linear functional given on R can be extended to a positive linear functional given on the entire space $C_0(Q)$ (see, e.g., [KVP]).

We define the *convolution of a measure* $\mu \in M_b^+(Q)$ *with a function* $f \in C_0(Q)$ as follows:

$$(f * \mu)(r) = \int (L_r f)(s^*)\, d\mu(s) = \int (L_r f)(s)\, d\mu^*(s),$$

where $\mu^*(A) = \mu(A^*)$. By virtue of (2.7), we have

$$|(f * \mu)(r)| \leq \|f\|_\infty \|\mu\|.$$

The function $(f * \mu)(r)$ *is continuous.* Indeed, consider a sequence $r_n \in Q$ which converges to $r \in Q$. Since $(L_r f)(s)$ is separately continuous, we have $(L_{r_n} f)(s) \to (L_r f)(s)$ for each $s \in Q$. The continuity of $(f * \mu)(r)$ follows from the Lebesgue theorem.

■

Let us show that $(f * \mu)(r)$ *is integrable.* In fact, since generalized translation operators are linear, it suffices to check this assertion for nonnegative $f \in C_0(Q)$. By using the fact that L_s preserves positivity, the invariance of the multiplicative measure, and relation (2.1), we obtain

$$\int |(f * \mu)(r)|\,dr \leq \iint |(L_r f)(s^*)|\,d\mu(s)\,dr = \iint (L_{s^*} f)(r)\,dr\,d\mu(s)$$

$$= \iint f(r)\,dr\,d\mu(s) = \|\mu\|\,\|f\|_1.$$

■

Hence, $(f * \mu)(r) \in C_b(Q) \cap L_1$ $(f \in C_0(Q),\ \mu \in M_b^+(Q))$.

We define the *Fourier transform of a measure* $\mu \in M_b^+(Q)$ *by the relation*

$$\hat{\mu}(\chi) = \int \overline{\chi(t)}\,d\mu(t) \quad (\chi \in X).$$

Clearly, $\hat{\mu}(\chi) \in C_\infty(X)$.

If the Fourier transforms of two measures from $M_b^+(Q)$ *coincide on* $\hat{Q}$, *then these measures coincide.*

Indeed, it suffices to show that $\hat{\mu}(\chi) \not\equiv 0$ $(\chi \in \hat{Q})$ for any nonnegative measure $\mu \in M_b^+(Q)$. Since

$$(f * \mu)(e) = \int f(s^*)\,d\mu(s),$$

for each $\mu \in M_b^+(Q)$, one can find $f \in C_0(Q)$ such that $f * \mu \not\equiv 0$ $(r \in Q)$. Since $f * \mu \in L_1$, in view of Corollary 3.3, there exists a character $\chi \in \hat{Q}$ such that $(f * \mu)\hat{}(\chi) \neq 0$. At the same time,

$$(f*\mu)^\wedge(\chi) = \iint (L_r f)(s^*)\,d\mu(s)\overline{\chi(r)}\,dr = \iint (L_{s^*}f)(r)\overline{\chi(r)}\,dr\,d\mu(s)$$

$$= \iint f(r)\overline{(L_s\chi)(r)}\,dr\,d\mu(s) = \int \hat{f}(\chi)\overline{\chi(s)}\,d\mu(s) = \hat{f}(\chi)\hat{\mu}(\chi) \neq 0$$

but this implies that $\hat{\mu}(\chi) \neq 0$.

■

The characters from $\hat{Q}$ separate points of the set Q, i.e., for any points $p', p'' \in Q$, one can find $\chi \in Q$ such that $\chi(p') \neq \chi(p'')$.

Suppose the contrary. By virtue of Lemma 3.2, there exists a function $f \in R$ such that $f(p') \neq f(p'')$. Then Theorem 3.2 implies that

$$f(p') = \int_{\hat{Q}} \hat{f}(\chi)\chi(p')\,d\chi = \int_{\hat{Q}} \hat{f}(\chi)\chi(p'')\,d\chi = f(p'')$$

and we arrive at a contradiction.

■

Theorem 3.5. *Let $L_1(Q, m)$ be a commutative normal hypercomplex system with basis unity which possesses the following properties:*

(i) the identity character is essential;

(ii) the product of essential characters is a positive definite function and the measure that corresponds to this function according to Theorem 2.1 is concentrated on the set of essential characters;

(iii) the structure measure $\hat{c}(\Phi, \Psi, \chi)$ of the dual hypercomplex system has compact support for fixed $\Phi, \Psi \in \mathcal{B}_0(\hat{Q})$;

(iv) the hypercomplex system $L_1(Q, m)$ satisfies the condition of separate continuity.

Then the dual hypercomplex system also satisfies conditions (i)–(iv) and duality takes place.

Proof. We have already shown that each point $r \in Q$ generates an Hermitian character $r(\chi) = \overline{\chi(r)}$ $(\chi \in \hat{Q})$ of the dual hypercomplex system. In view of the fact that characters $\hat{Q}$ separate the points of Q, we have $Q \subset Q_h$, where Q_h is the set of Hermitian characters of the dual hypercomplex system.

Consider a family of functions R. Functions from $C_\infty(Q)$ separate points in Q. For any point $r \in Q$, there exists $f \in C_\infty(Q)$ such that $f(r) \neq 0$. By Lemma 3.2, R is dense in $C_\infty(Q)$. Therefore, R also possesses the indicated properties. It is easy to see that the weak topology in Q generated by the family of functions R coincides with the original topology in Q (see, e.g., [Nai1]). By Theorem 3.2, this implies that the sets

$$U(r;\varepsilon;\hat{x}_1,\ldots,\hat{x}_n) = \left\{ s : \left| \int_{\hat{Q}} \hat{x}_i(\chi)[\chi(r)-\chi(s)] \right| d\chi < \varepsilon,\ i=1,\ldots,n \right\},$$

$$(x_1,\ldots,x_n \in R,\ r\in Q,\ \varepsilon>0)$$

form a base of neighborhoods in Q. Since $r(\chi) = \overline{\chi(r)}$, by the definition of the topology in the space of characters, $U(r;\varepsilon;\hat{x}_1,\ldots,\hat{x}_n)$ are neighborhoods in the space Q regarded as a subspace of Q_h. In Lemma 3.4, it was actually proved that the Fourier image of the linear span of $\{y * z \mid y, z \in C_0(Q)\} \subset R$ is dense in $L_1(\hat{Q},\hat{m})$. Therefore, $U(r;\varepsilon;\hat{x}_1,\ldots,\hat{x}_n)$ form a base of neighborhoods in $Q\subset Q_h$.

We show that $\hat{\hat{Q}} \subset Q$. Let $\tau \in \hat{\hat{Q}}$ be an essential character of the hypercomplex system $\mathfrak{L}_1(\hat{Q},\hat{m})$. Consider a linear functional

$$I(x(r)) = \int \hat{x}(\chi)\tau(\chi)\,d\chi \quad (x\in R).$$

By Lemma 3.5 and Theorem 3.2,

$$I(x(r)y(r)) = \int (xy)^\wedge(\chi)\tau(\chi)\,d\chi = \int (\hat{x} * \hat{y})(\chi)\tau(\chi)\,d\chi$$

$$= \int \hat{x}(\chi)\tau(\chi)\,d\chi \int \hat{y}(\theta)\tau(\theta)\,d\theta = I(x(r))I(y(r)) \quad (x,y\in R),$$

i.e., the functional is multiplicative on R. Let us now show that I is positive on R. Let $x(r)\geq 0$ $(x\in R)$ and $y(r) = \sqrt{x(r)}$. Then $y(r)\in C_0(Q)$ and, hence, $y\in L_2(Q,m)$. Consequently, by virtue of Lemma 3.5,

$$I(x(r)) = \int (y^2)^\wedge(\chi)\tau(\chi)\,d\chi = \int (\hat{y}^* * \hat{y})(\chi)\tau(\chi)\,d\chi.$$

Since $x(r)\in R$, in view of Theorem 3.2, we have $\hat{x}(\chi) = \hat{y}^* * \hat{y}(\chi)\in L_1(\hat{Q},\hat{m})$. By applying the remark to Theorem 3.2 to the function $\hat{y}(\chi)$, we obtain

$$\int (\hat{y}^* * \hat{y})(\chi)\tau(\chi)\,d\chi \geq 0$$

i.e., $I(x(r))\geq 0$. Since R majorizes $C_0(Q)$, the functional $I(x(r))$ can be extended to a positive linear functional $\tilde{I}$ on the entire $C_0(Q)$. By the Riesz–Markov theorem, there exists a regular Borel measure μ on Q such that

$$\tilde{I}(x(r)) = \int x(r)\,d\mu(r).$$

for all $x \in C_0(Q)$. This and the multiplicativity of the functional I imply that

$$\int x(r)\,d\mu(r)\int y(r)\,d\mu(r) = \int x(r)y(r)\,d\mu(r) \quad (x, y \in R). \tag{3.8}$$

We show that the measure μ is finite. Indeed, let Q_n be a sequence of compact sets such that $\bigcup_{n=1}^{\infty} Q_n = Q$ and $Q_n \subset Q_{n+1}$. For each open $A \in \mathcal{B}_0(Q)$, we consider a sequence of functions

$$x_n(r) = \frac{1}{m(A)}c(Q_n, A, r).$$

Clearly, $x_n \in R$. By virtue of (1.13), we have $0 \le x_n(r) \le 1$. In view of (1.12), $x_n(r) \to 1$ as $n \to \infty$ for all $r \in Q$. Since $x_n^2(r) \le x_n(r)$, we find

$$\left(\int x_n(r)\,d\mu(r)\right)^2 = \int x_n^2(r)\,d\mu(r) \le \int x_n(r)\,d\mu(r).$$

Hence,

$$\int x_n(r)\,d\mu(r) \le 1.$$

By using the Fatou lemma, we obtain

$$\mu(Q) = \int \lim_{n\to\infty} x_n(r)\,d\mu(r) \le \lim_{n\to\infty}\int x_n(r)\,d\mu(r) \le 1.$$

If we approximate arbitrary functions $f, g \in C_\infty(Q)$ by functions $x, y \in R$ in the uniform norm, then it follows from (3.8) that

$$\int f(r)\,d\mu(r)\int g(r)\,d\mu(r) = \int f(r)g(r)\,d\mu(r).$$

This relation immediately implies that there exists a unique point $s_0 \in Q$ such that

$$I(x(s)) = \int x(s)\,d\mu(s) = x(s_0).$$

Let us show that $Q = \hat{Q}$. By Theorem 3.3, $L_2(Q, m)$ is isomorphic to $L_2(\hat{Q}, \hat{m})$

which is, in turn, isomorphic to $L_2(\hat{\hat{Q}}, \hat{\hat{m}})$. Consequently, $L_2(Q, m)$ is isomorphic to $L_2(\hat{\hat{Q}}, \hat{\hat{m}})$. Denote this isomorphism by $\hat{\mathcal{F}} \circ \mathcal{F}$. By Theorem 3.2, we have

$$f(r) = \int_{\hat{Q}} \hat{f}(\chi)\chi(r)d\chi = \int_{\hat{Q}} \hat{f}(\chi)\overline{r(\chi)}d\chi \quad (r \in \hat{\hat{Q}})$$

for all $f \in R$, i.e., under the isomorphism $\hat{\mathcal{F}} \circ \mathcal{F}$, the function $f \in R$, is mapped into $f \mid \hat{\hat{Q}}$. Since the topology in Q is induced by the topology from $\hat{X}$, the space $\hat{\hat{Q}}$ is closed, and R is uniformly dense in $C_\infty(Q)$, we conclude that, for each $x \in C_0(Q)$, its image $\hat{\mathcal{F}} \circ \mathcal{F}x = x \mid \hat{\hat{Q}}$. Since $\hat{\hat{Q}}$ is closed in Q, $U = Q \backslash \hat{\hat{Q}}$ is an open subset of Q. If $U \neq \varnothing$, then. by virtue of the regularity of m, there exists a compact set $A \in \mathcal{B}_0(Q)$ such that $A \subset U$ and $m(A) > 0$. Let the function $f \in C_0(Q)$ be such that $f(r) = 1$ on A, $0 \leq f(r) \leq 1$, and $f(r) = 0$ on $\hat{\hat{Q}}$. Since $m(A) > 0$, we have $f \neq 0$ in $L_2(Q, m)$. Since $\hat{\mathcal{F}} \circ \mathcal{F}f = f \mid \hat{\hat{Q}} = 0$, we arrive at a contradiction with the assertion that $L_2(Q, m)$ is isomorphic to $L_2(\hat{\hat{Q}}, \hat{\hat{m}})$. Thus, $U = \varnothing$ and $Q = \hat{\hat{Q}}$.

Consider functions of the form $(f * f^*)(\chi)$ $(f(\chi) \in C_0(\hat{Q}))$. Clearly, $f * f^* \in \mathfrak{R}(\hat{Q}) \cap C_0(\hat{Q})$. By applying Lemma 3.5 and Theorem 3.2 to the dual hypercomplex system, we get

$$\int_Q \left|\check{f}(r)\right|^2 \overline{\chi(r)}dm(r) = f * f^*(\chi) = \int_Q \left|\hat{f}(r)\right|^2 r(\chi)d\hat{\hat{m}}(r).$$

It follows from the relations $\hat{\mathcal{F}} \circ \mathcal{F}x = x \mid \hat{\hat{Q}}$ and $Q = \hat{\hat{Q}}$ that $\hat{f}(r) = \check{f}(r)$. Therefore, in view of the uniqueness of the Fourier transforms of measures from $M_b^+(Q)$, we obtain

$$d\mu(r) = \left|\hat{f}(r)\right|^2 dm(r) = \left|\hat{f}(r)\right|^2 d\hat{\hat{m}}(r).$$

Since the image of $\mathfrak{R}(\hat{Q}) \cap C_0(\hat{Q})$ is dense in $L_2(Q, \hat{\hat{m}})$, we have $m = \hat{\hat{m}}$.

Relations (3.4) and (3.6) imply that the structure measure $\hat{\hat{c}}(A, B, r)$ $(A, B \in \mathcal{B}_0(Q))$ of the hypercomplex system dual to $L_1(\hat{Q}, \hat{m})$ coincides with the initial structure measure $c(A, B, r)$.

By virtue of the fact that $e(\chi) = \overline{\chi(e)} = 1$ $(\chi \in \hat{Q})$, the basis unity $e \in Q$ generates the identity character of the dual hypercomplex system. Since $Q = \hat{\hat{Q}}$, this character is essential.

Let us show that the hypercomplex system dual to $L_1(\hat{Q}, \hat{m})$ satisfies condition (ii) of the theorem. Since $Q = \hat{\hat{Q}}$, it suffices to show that the function $p^*(\chi)q^*(\chi) = \chi(p)\chi(q)$ $(\chi \in \hat{Q})$ is positive definite for any $p, q \in Q$ and the support of the measure associated by Theorem 3.1 with $\chi(p)\chi(q)$ lies in Q. By virtue of (1.19), for any $B \in \mathcal{B}_0(Q)$, one can find a compact set $F \in \mathcal{B}_0(Q)$ such that

$$\chi(p)\chi(B) = \int \chi_F(r)\, d_r c(E_r, B^*, p).$$

In this equality, we set $B = U_n$, where U_n is a sequence of neighborhoods of the point q contracting to q. Clearly, we can take the same compact F for all n. Then

$$\chi(p)\frac{\chi(U_n)}{m(U_n)} = \int \chi_F(r)\, d\mu_n(r), \tag{3.9}$$

where

$$\mu_n(E) = \frac{c\left(E, U_n^*, p\right)}{m(U_n)}.$$

Clearly, $\mu_n(E) \geq 0$ and $\mu_n(Q) = 1$. Let us show that the sequence of probability measures μ_n is uniformly dense, i.e., for any $\varepsilon > 0$, one can indicate a compact set K such that $\mu_n(K) > 1 - \varepsilon$. In fact, let F_n be a sequence of neighborhoods of the point p contracting to p. We set $K = \operatorname{supp} c(F_1, U_1, \cdot)$. Then, in view of the continuity of the structure measure, we have

$$\mu_n(K) = \frac{c\left(K, U_n^*, p\right)}{m(U_n)} = \lim_{i\to\infty} \frac{1}{m(F_i)m(U_n)} \int_{F_i} c\left(K, U_n^*, p\right) dp$$

$$= \lim_{i\to\infty} \frac{c\left(K, U_n^*, F_i\right)}{m(F_i)m(U_n)} = \lim_{i\to\infty} \frac{c(F_i, U_n, K)}{m(F_i)m(U_n)}$$

$$= \lim_{i\to\infty} \frac{\int_Q c(F_i, U_n, p)\, dp}{m(F_i)m(U_n)} = 1.$$

Hence, by virtue of the Prokhorov theorem, one can choose a subsequence μ_{n_k} of the sequence μ_n weakly convergent to a nonnegative regular measure $d\mu(r) = d\mu_{p,q}(r)$. By passing in (3.9) to the limit in the subsequence μ_{n_k}, we obtain

$$\chi(p)\chi(q) = \int \chi_F(r)\, d\mu_{p,q}(r) = \int_F \chi(r)\, d\mu_{p,q}(r). \tag{3.10}$$

It is easy to show that the function on the right-hand side of (3.10) is positive definite. Indeed,

$$\int_{\hat{Q}}\int_{F}\chi(r)\,d\mu_{p,q}(r)(x^{*}*x)(\chi)\,d\chi = \int_{F}\int_{\hat{Q}}\chi(r)(x^{*}*x)(\chi)\,d\chi\,d\mu_{p,q}(r)$$

$$= \int_{F}|\check{x}(r)|^{2}d\mu_{p,q}(r)$$

for all $x \in L_1(\hat{Q}, \hat{m})$.

Finally, let us show that condition (iv) is satisfied for the hypercomplex system $L_1(\hat{Q}, \hat{m})$. Let $\chi, \psi \in \hat{Q}$. By virtue of (ii), the function $\chi(r)\psi(r)$ is positive definite. By Theorem 3.1, this yields

$$\chi(r)\psi(r) = \int_{\hat{Q}}\theta(r)\,d\mu_{\chi,\psi}(\theta),$$

where $\mu_{\chi,\psi}$, by virtue of (ii), is concentrated on $\hat{Q}$. Let us show that

$$(L_{\chi}f)(\psi) = \int_{\hat{Q}}f(\theta)\,d\mu_{\chi,\psi}(\theta) \quad (\chi,\psi \in \hat{Q}) \tag{3.11}$$

for all $f \in C_{\infty}(\hat{Q})$.

Let $f \in R$, then, clearly, $\hat{f} \in L_1(\hat{Q}, \hat{m}) \cap C_{\infty}(\hat{Q})$ and for arbitrary $g \in L_2(\hat{Q}, \hat{m})$, we obtain

$$\left(L_{\chi}\hat{f}, g\right)_{L_2(\hat{Q},\hat{m})} = \int_{\hat{Q}}\int_{\hat{Q}}\left(\int_{Q}f(r)\overline{\theta(r)}\,dr\right)d\mu_{\chi,\psi}(\theta)\overline{g(\psi)}\,d\psi$$

$$= \int_{\hat{Q}}\left(\int_{Q}f(r)\int_{\hat{Q}}\overline{\theta(r)}\,d\mu_{\chi,\psi}(\theta)\,dr\right)\overline{g(\psi)}\,d\psi$$

$$= \int_{\hat{Q}}\int_{Q}f(r)\overline{\chi(r)}\,\overline{\psi(r)}\,dr\,\overline{g(\psi)}\,d\psi.$$

On the other hand, by virtue of Theorem 3.2 and (3.6), we have

$$(\hat{f} * g^{*})(\chi) = \int_{\hat{Q}}\int_{\hat{Q}}\hat{f}(\theta)\overline{g(\psi)}\,d\hat{c}(E_{\theta}, E_{\psi}, \chi)$$

$$= \int_Q \int_{\hat{Q}} \hat{f}(\theta)\theta(r)d\theta \int_{\hat{Q}} \overline{g(\psi)}\,\overline{\psi(r)}d\psi\,\overline{\chi(r)}dr$$

$$= \int_{\hat{Q}} \int_Q f(r)\overline{\chi(r)}\,\overline{\psi(r)}dr\,\overline{g(\psi)}d\psi.$$

Thus, relation (3.11) holds for all $\hat{f}(\chi)$ $(f \in R)$. By Lemma 3.1, R is dense in L_1. Hence, for any $x \in L_1$, one can choose a sequence $f_n \in R$ that converges to x in L_1. Then the relevant Fourier transforms $\hat{f}_n$ uniformly converge to $\hat{x}$ in $C_\infty(\hat{Q})$. In view of (2.7), the generalized translation operators L_χ are continuous in $L_\infty(\hat{Q}, \hat{m})$. By passing to the limit on both sides of relation (3.11), we conclude that (3.11) holds for all $\hat{x}(\chi)$ $(x \in L_1)$. Since the Fourier image of $L_1(Q, m)$ is dense in $C_\infty(\hat{Q})$, we arrive at (3.11) by passing to the limit.

We fix $\psi \in \hat{Q}$, assume that $\chi_n \to \chi$ in $\hat{Q}$, and choose $\Delta \in \mathcal{B}_0(Q)$ such that $\chi_n \in \Delta$ for all $n \in \mathbb{N}$. The structure measure $\hat{c}(\Phi, \Psi, \chi)$ has a compact support with respect to χ by the condition of the theorem. Thus, by virtue of Theorem 2.1, the generalized translation operators L_χ $(\chi \in \hat{Q})$ associated with the dual hypercomplex system satisfy the finiteness condition (F). Recall that, according to condition (F), there exists a compact set $K \subset \hat{Q}$ such that $(L_\chi f)(\psi) = 0$ for all $\chi \in \Delta$, provided that $\operatorname{supp} f \cap K = \varnothing$. By virtue of (3.11), this means that $\operatorname{supp} \mu_{\chi,\psi} \subset K$ for all $\chi \in \Delta$. In addition, it is clear that $\sup_n \mu_{\chi_n,\psi}(\hat{Q}) = 1$. Thus, the sequence of measures $\mu_{\chi_n,\psi}$ is uniformly dense and, by the Prokhorov theorem, weakly compact, i.e., any its subsequence $\mu_{\chi_{n'},\psi}$ contains a weakly convergent subsequence $\mu_{\chi_{n''},\psi}$. Assume that $\mu_{\chi_{n''},\psi}$ weakly converges to a measure ν ($\mu_{\chi_{n''},\psi} \Rightarrow \nu$). This enables us to conclude that, for all $r \in Q$,

$$\chi_{n''}(r)\psi(r) = \int \theta(r)\,d\mu_{\chi_{n''},\psi}(\theta) \xrightarrow[n\to\infty]{} \int \theta(r)\,d\nu(\theta).$$

On the other hand, since $\chi_n \to \chi$, we have

$$\chi_{n''}(r)\psi(r) \xrightarrow[n\to\infty]{} \chi(r)\psi(r) = \int \theta(r)\,d\mu_{\chi,\psi}(\theta)$$

for all $r \in \hat{Q}$. This yields

$$\int \theta(r)\,d\mu_{\chi,\psi}(\theta) = \int \theta(r)\,d\nu(\theta).$$

Each integral in this equality is a continuous positive definite function on Q. By Theorem 3.1, the integral representation (3.2) is unique. By using this fact, we conclude that $\nu = \mu_{\chi,\psi}$, whence $\mu_{\chi_n,\psi} \Rightarrow \mu_{\chi,\psi}$. In view of (3.11), this immediately yields the continuity of $(L_\chi f)(\psi)$ in χ for fixed ψ. The continuity of $(L_\chi f)(\psi)$ in ψ can be proved analogously.

■

Let us now describe an analog of the Fourier algebra. Let $L_1(Q, m)$ be a normal hypercomplex system with basis unity satisfying the conditions of Theorem 3.5. By virtue of this theorem, all results established in 3.1 and 3.2 hold for the dual hypercomplex system $L_1(\hat{Q}, \hat{m})$. Denote

$$A(Q) = L_2(Q, m) * L_2(Q, m) = \{f_1 * f_2 \mid f_1, f_2 \in L_2\}.$$

By Lemma 1.1, $A(Q) \subset C_\infty(Q)$. Since $L_1(\hat{\hat{Q}}, \hat{\hat{m}}) = L_1(Q, m)$, by applying Corollary 3.5 to the hypercomplex system $L_1(\hat{Q}, \hat{m})$, we can show that $A(Q)$ is a linear space. A norm in $A(Q)$ is introduced by setting

$$\|f_1 * f_2\|_{A(Q)} = \|(f_1 * f_2)^\wedge\|_{L_1(\hat{Q}, \hat{m})}.$$

By the same Corollary 3.5, $A(Q)$ with this norm is complete.

Theorem 3.6. *Assume that* $L_1(Q, m)$ *is a commutative normal hypercomplex system with basis unity satisfying the conditions of Theorem 3.5, i.e.,*

(i) *the identity character is essential;*

(ii) *the product of essential characters is a positive definite function and the measure that corresponds to this function according to Theorem 3.1 is concentrated on the set of essential characters;*

(iii) *the structure measure* $\hat{c}(\Phi, \Psi, \chi)$ *of the dual hypercomplex system has compact support for fixed* $\Phi, \Psi \in \mathcal{B}_0(\hat{Q})$*;*

(iv) *the hypercomplex system* $L_1(Q, m)$ *satisfies the condition of separate continuity.*

Then $A(Q) = L_2(Q, m) * L_2(Q, m)$ *is a Banach algebra with respect to the ordinary multiplication of functions and the norm* $\|\cdot\|_{A(Q)}$ *The algebra* $A(Q)$ *is called the Fourier algebra of the hypercomplex system* $L_1(Q, m)$.

Proof. First, we show that the set $C_0(Q) * C_0(Q) = \{u_1 * u_2 \mid u_1, u_2 \in C_0(Q)\}$ is dense in $A(Q)$. We fix arbitrary $f_1, f_2 \in L_2$ and take $u_1, u_2 \in C_0(Q)$ such that

$$\|f_1 - u_1\|_2 \le \frac{\varepsilon}{2\|f_2\|_2}, \qquad \|f_2 - u_2\|_2 \le \frac{\varepsilon}{2\|f_1\|_2}.$$

By applying Theorem 3.3, we obtain

$$\|f_1 * f_2 - u_1 * u_2\|_{A(Q)} = \left\|\hat{f}_1\hat{f}_2 - \hat{u}_1\hat{u}_2\right\|_{L_1(\hat{Q},\hat{m})}$$

$$\le \|\hat{f}_1\|_2\|\hat{f}_2 - \hat{u}_2\|_2 + \|\hat{f}_1 - \hat{u}_1\|_2\|\hat{u}_2\|_2$$

$$= \|f_1\|_2\|f_2 - u_2\|_2 + \|f_1 - u_1\|_2\|u_2\|_2 \le \varepsilon\left(1 + \frac{\varepsilon}{\|f_2\|_2}\right).$$

Now let $u_1, u_2, v_1, v_2 \in C_0(Q)$. Denote $\varphi_1 = u_1 * v_1$ and $\varphi_2 = u_2 * v_2$. Then $\varphi_1, \varphi_2, \varphi_1\varphi_2 \in L_1(Q, m) \cap C_\infty(Q)$. By virtue of Lemma 3.5, we have $(\varphi_1\varphi_2)^\wedge(\chi) = \hat{\varphi}_1 * \hat{\varphi}_2(\chi)$. Since $\varphi_i = u_i * v_i \in \mathfrak{R}$ $(i = 1, 2)$, by virtue of Theorem 3.2, we conclude that $\hat{\varphi}_i(\chi) \in L_1(\hat{Q}, \hat{m})$ $(i = 1, 2)$. At the same time, by Theorem 3.4, $L_1(\hat{Q}, \hat{m})$ is a Banach algebra. Therefore, $(\varphi_1\varphi_2)^\wedge(\chi) = (\hat{\varphi}_1 * \hat{\varphi}_2)(\chi) \in L_1(\hat{Q}, \hat{m})$. Consequently, one can find elements $h_1, h_2 \in L_2(\hat{Q}, \hat{m})$ such that $(\varphi_1\varphi_2)^\wedge(\chi) = (\hat{\varphi}_1 * \hat{\varphi}_2)(\chi) = h_1(\chi)h_2(\chi)$. Denote by $\check{h}_1, \check{h}_2 \in L_2(Q, m)$ the preimages of the functions h_1 and h_2 under the Fourier transformation. By applying Corollary 3.5 to the hypercomplex system $L_1(\hat{Q}, \hat{m})$, we obtain

$$(u_1 * v_1)(u_2 * v_2)(r) = (\varphi_1\varphi_2)(r) = (\mathcal{F}^{-1}(h_1h_2))(r) = \check{h}_1 * \check{h}_2(r) \in A(Q).$$

Moreover, in view of the fact that $L_1(\hat{Q}, \hat{m})$ is a Banach algebra, we get

$$\|\varphi_1\varphi_2\|_{A(Q)} = \left\|(\varphi_1\varphi_2)^\wedge\right\|_1 = \|\hat{\varphi}_1 * \hat{\varphi}_2\|_1 \le \|\hat{\varphi}_1\|_1\|\hat{\varphi}_2\|_1$$

$$= \|\varphi_1\|_{A(Q)}\|\varphi_2\|_{A(Q)}.$$

To complete the proof, it remains to use the fact that $C_0(Q) * C_0(Q)$ is dense in $A(Q)$.

■

In passing, we have proved the fact that the Fourier algebra $A(Q)$ is isomorphic to the algebra $L_1(\hat{Q}, \hat{m})$. By passing to the dual object, we conclude that the Fourier algebra $A(\hat{Q})$ of the dual hypercomplex system is isomorphic to $L_1(Q, m)$.

3.3. The Case of Discrete Hypercomplex Systems. Let us study hypercomplex systems with discrete bases in more details. A bounded numerical sequence $\eta(p)$ $(p \in Q)$ is called positive definite if

$$\sum_{k,l=1}^{n} \left(R_{r_k^*}\eta\right)(r_l)\lambda_k\bar{\lambda}_l \geq 0 \tag{3.12}$$

for any $\lambda_1, \dots, \lambda_n \in \mathbb{C}$ and $r_1, \dots, r_n \in Q$. This definition coincides with the definition presented in Subsection 3.1.

The spaces of characters X and Hermitian characters X_h are compact sets because the hypercomplex system $l_1(m)$ possesses an identity element. In view of relation (3.4), we have

$$c(p, q, r) = m(p)m(q) \int\limits_{X_h} p(\chi)q(\chi)\overline{r(\chi)}d\chi, \tag{3.13}$$

where $d\chi$ is the Plancherel measure and $p(\chi) = \overline{\chi(p)}$ $(p \in Q)$.

Let us show that hypercomplex systems with compact and discrete bases are dual to each other, namely, the following theorem is true:

Theorem 3.7. *Let $l_1(m)$ be a commutative normal hypercomplex system with discrete basis and basis unity. In order that the function $\hat{c}(\Phi, \Psi, \chi)$ defined by relation (3.6) be a structure measure of the dual hypercomplex system with compact basis, it is necessary and sufficient that the product of any two essential characters be positive definite. If, in addition, the measure corresponding to this product according to Theorem 3.1 is concentrated on the set of essential characters and the identity character is essential, then duality takes place, the product of any two continuous positive definite functions $f(\chi)$ and $g(\chi)$ from the dual hypercomplex system is positive definite, and the dual hypercomplex system satisfies the condition of separate continuity.*

Proof. The fact that the basis of the dual hypercomplex system is compact follows from the fact that the hypercomplex system $l_1(m)$ contains the identity element. The space of characters of this compact hypercomplex system is discrete by virtue of Theorem 1.4 and each character in this space is essential. The proof is completed by applying Theorem 3.5.

■

It is easy to see that a similar result is true for hypercomplex systems with compact bases.

We introduce an important concept of A-system of functions. Let E be a compact space with nonnegative Borel measure $d\sigma(t)$ such that supp $\sigma = E$, let Q be a countable set, and let $\Sigma = \{\varphi_p(t)\}_{p \in Q}$ be a system of continuous functions orthogonal and complete in $L_2(E, d\sigma)$. This system is called an *A-system* if

(A1) for all fixed $p, q \in Q$, the sequence

$$\left(\int \varphi_p(t)\varphi_q(t)\overline{\varphi_r(t)}\,d\sigma(t)\right)_{r\in Q}$$

is finite and consists of nonnegative members;

(A2) if $\varphi_p(t) \in \Sigma$, then $\overline{\varphi_p(t)} \in \Sigma$;

(A3) there is a point $e \in Q$ such that $\varphi_e(t) \equiv 1$ $(t \in Q)$;

(A4) there is a point $a \in E$ such that $\varphi_p(a) = 1$ $(p \in Q)$.

For every normal hypercomplex system $l_1(m)$ with basis unity and discrete basis Q whose identity character belongs to $\hat{Q}$, it is possible to construct an A-system of functions.

Namely, we set

$$\sigma = \hat{m}, \quad E = \operatorname{supp} \sigma = \hat{Q}, \quad \Sigma = \{p(\chi)\}_{p\in Q}$$

and show that Σ is an A-system. First, we note that the function $p(\chi) = \overline{\chi(p)}$ is the Fourier transform of the sequence $\kappa_p(q) = \delta_{p,q}/m(p)$ ($q \in Q$ and $\delta_{p,q}$ is the Kronecker symbol). Therefore, the functions $p(\chi)$ are continuous and, in view of Theorem 3.3, form an orthogonal system in $L_2(\hat{Q}, \hat{m})$, i.e.,

$$\int_{\hat{Q}} p(\chi)\overline{q(\chi)}\,d\chi = \sum_{r\in Q} \kappa_p(r)\kappa_{q^*}m(r) = \frac{1}{m(p)}\delta_{p,q} \quad (p, q \in Q). \qquad (3.14)$$

The completeness of the system Σ follows from the fact that the set of sequences $(\kappa_p(r))_{r\in Q}$ $(p \in Q)$ is total in l_2. Condition (A1) follows from (3.13) and the fact that $c(p, q, r)$ is nonnegative and finite. Since $\overline{p(\chi)} = \chi(p) = \overline{\chi(p^*)} = p^*(\chi)$, (A2) is satisfied. Condition (A3) follows from the equality $\chi(e) = 1$ $(\chi \in \hat{Q})$, and condition (A4) is a consequence of the fact that the identity element is an essential character of the hypercomplex system $l_1(m)$.

■

Note that the normal hypercomplex system $l_1(m)$ with basis unity is isometrically isomorphic to the algebra of continuous functions on $E = \hat{Q}$ with ordinary algebraic operations whose expansions in the complete orthogonal system of functions $p(\chi)$ $(\chi \in \hat{Q},\ p \in Q)$ are absolutely convergent, i.e., to the algebra of functions

$$x(\chi) = \sum_p x(p)p(\chi)m(p)$$

such that

$$\sum_p |x(p)|m(p) = \|x\| < \infty.$$

The validity of the converse statement is established by the following theorem:

Theorem 3.8. *For any A-system of functions $\Sigma = \{\varphi_p(t)\}_{p \in Q}$, it is always possible to construct a commutative normal hypercomplex system with discrete basis Q and basis unity. Each point $t \in E$ generates, generally speaking, a generalized Hermitian character of the derived hypercomplex system $\chi_t(p) = \varphi_p(t)$ $(p \in Q)$. If the sequence $\varphi_p(t)$ $(p \in Q)$ is bounded, then the relevant Hermitian character is ordinary.*

Proof. We set

$$m(p) = \left(\int |\varphi_p(t)|^2 d\sigma(t)\right)^{-1} = \|\varphi_p\|_2^{-2},$$

$$c(p, q, r) = m(p)m(q)\int \varphi_p(t)\varphi_q(t)\overline{\varphi_r(t)}d\sigma(t). \tag{3.15}$$

By virtue of (A1), $c(p, q, r)$ is nonnegative and finite along r for fixed $p, q \in Q$. The commutativity relation is obvious, and the associativity relation (1.25) follows from the Parseval equality:

$$\begin{aligned}\sum_r c(p, q, r)c(r, l, s) &= m(p)m(q)m(l)\sum_r (\varphi_p\varphi_q, \varphi_r)_2(\varphi_r, \overline{\varphi}_l\varphi_s)_2\|\varphi_r\|_2^{-2} \\ &= m(p)m(q)m(l)(\varphi_p\varphi_q, \overline{\varphi}_l\varphi_s)_2 \\ &= m(p)m(q)m(l)(\varphi_q\varphi_l, \overline{\varphi}_p\varphi_s)_2 \\ &= m(p)m(q)m(l)\sum_r (\varphi_q\varphi_l, \varphi_r)_2(\varphi_r, \overline{\varphi}_p\varphi_s)_2\|\varphi_r\|_2^{-2} \\ &= \sum_r c(q, l, r)c(p, r, s).\end{aligned}$$

Thus, $c(p, q, r)$ is a structure measure. We expand the continuous function $\varphi_p(t)\varphi_q(t)$ into the Fourier series in the orthogonal system $\varphi_r(t)$, i.e.,

$$\varphi_p(t)\varphi_q(t) = \sum_r \int \varphi_p(t)\varphi_q(t)\overline{\varphi_r(t)}d\sigma(t)\varphi_r(t)/\|\varphi_r\|_2^2. \tag{3.16}$$

By virtue of (A1), the sum in (3.16) is finite. Hence, both functions in this equality are continuous and (3.16) holds at every point $t \in E$. By setting $t = a$ in (3.16), we obtain

$$1 = \sum_r \int \varphi_p(t)\varphi_q(t)\overline{\varphi_r(t)}d\sigma(t)/\|\varphi_r\|_2^2.$$

Multiplying the last equality by $\|\varphi_p\|_2^{-2}\|\varphi_q\|_2^{-2}$, we get (1.26). Hence, $\|\varphi_p\|_2^{-2}$ is a multiplicative measure. We introduce involution in Q as follows: Let p^* be the index of the function $\overline{\varphi_p(t)} \in \Sigma$. Then

$$m(p^*) = \|\varphi_{p^*}\|_2^{-2} = \|\overline{\varphi}_p\|_2^{-2} = m(p)$$

and

$$\begin{aligned} c(p, q, r)m(r) &= m(p)m(q)m(r)\int \varphi_p(t)\varphi_q(t)\overline{\varphi_r(t)}d\sigma(t) \\ &= m(p)m(q)m(r)\overline{\int \overline{\varphi_q(t)}\varphi_r(t)\overline{\varphi_p(t)}d\sigma(t)} \\ &= m(p)m(q)m(r)\int \varphi_r(t)\varphi_{q^*}(t)\,\overline{\varphi_p(t)}d\sigma(t) \\ &= c(r, q^*, p)m(p), \end{aligned}$$

i.e., the hypercomplex system is normal; recall that, in view of (A1), we have

$$\int \varphi_p(t)\varphi_q(t)\overline{\varphi_r(t)}d\sigma(t) \geq 0.$$

The point $e \in Q$ is the basis unity of the derived hypercomplex system. Indeed, $e = e^*$ and

$$c(p, q, e) = m(p)m(q)\int \varphi_p(t)\overline{\varphi_{q^*}(t)}d\tau(t) = m(p)\delta_{p,q^*}.$$

Multiplying equality (3.16) by $m(p)m(q)$, we obtain (1.27), which proves the last assertion of the theorem.

■

Let $l_1(m)$ be a normal hypercomplex system with discrete basis Q and basis unity constructed from the A-system of functions $\varphi_p(t)$ $(p \in Q,\ t \in E)$. Assume that the linear span of $\{\varphi_p(t) \mid p \in Q\}$ is dense in $C(E)$. We endow $X_{g,h}$ with the topology of uniform convergence on compact sets. Consider a continuous map $\alpha: E \to X_{g,h}$ defined as follows: $\alpha(t)(p) = \varphi_p(t)$ $(p \in Q,\ t \in E)$. Since the functions $\varphi_p(t)$ $(p \in Q)$ separate the points of E, the map α is injective. Moreover, $\alpha(E)$ is homeomorphic to E and is a compact subset of $X_{g,h}$ (as the image of the compact set E under a continuous map). Thus, without loss of generality, we may consider the functions $\varphi_p(t)$ as given on $\alpha(E)$. As already mentioned, under the condition that the identity character of the hypercomplex system $l_1(m)$ is essential, the set of functions $\varphi_p(\chi) = \chi(p)$ $(\chi \in \hat{Q},$ $p \in Q)$ is also an A-system such that the sequence $(\varphi_p(\chi))_{p \in Q}$ is bounded for each $\chi \in \hat{Q}$ and $\varphi_p(\chi) = \varphi_p(t)$ whenever $\chi = \alpha(t) \in \hat{Q} \cap \alpha(E)$ (note that $\hat{Q} \cap \alpha(E) \neq \varnothing$ because $\alpha(a) \in \hat{Q}$).

Thus, each function $\varphi_p(t)$ from an arbitrary A-system of functions on the compact set E such that the linear span of these functions is dense in $C(E)$ and the identity character of the hypercomplex system $l_1(m)$ constructed from this A-system is essential can be extended to a function φ_p' on, generally speaking, another set E' $(E \cap E' \neq \varnothing)$ so that the functions $\varphi_p' \mid E'$ also form an A-system generating the same hypercomplex system $l_1(m)$, and the sequence $(\varphi_p'(t'))_{p \in Q}$ is bounded for all $t' \in E'$. Hence, without loss of generality, we may consider A-systems of functions $(\varphi_p(t))_{p \in Q}$ $(t \in E)$ such that the sequence $(\varphi_p(t))_{p \in Q}$ is bounded for all $t \in E$. In this case, $\alpha(E) \subset X_h$ and *the image* $\alpha(\sigma)(d\chi)$ *of the measure* $d\sigma(t)$ *under the mapping* α *is a Plancherel measure on* X_h. Hence, $\hat{Q} = \alpha(E)$. Indeed, consider the functions

$$\hat{\kappa}_p = \frac{1}{m(p)}\hat{\delta}_{p,q}(\chi) = \chi(p) \quad (p \in Q,\ \chi \in X_h).$$

In view of orthogonality, we have

$$\int_{X_h} \hat{\kappa}_p(\chi)\overline{\hat{\kappa}_q(\chi)}\,d\alpha(\sigma)(\chi) = \int_E \hat{\kappa}_p(\alpha(t))\overline{\hat{\kappa}_q(\alpha(t))}\,d\sigma(t)$$

$$= \int_E \chi_t(p)\overline{\chi_t(q)}\,d\sigma(t) = \int_E \varphi_p(t)\overline{\varphi_q(t)}\,d\sigma(t)$$

$$= \frac{1}{m(p)}\delta_{p,q} = \sum_r \kappa_p(r)\kappa_q(r)m(r). \tag{3.17}$$

Since the functions $\kappa_p(r)$ $(p \in Q)$ form a total set in $l_2(m)$, the Parseval equality

(3.17) holds for all $f \in l_2(m)$. In view of the uniqueness of the Plancherel measure, we get $d\alpha(\sigma)(\chi) = d\hat{m}(\chi)$.

∎

The most difficult problem encountered in proving that a given orthogonal system of functions $\{\varphi_p(t)\}_{p \in Q}$ is an A-system is to show that the integrals in expansion (3.16) are nonnegative. Below, we present a sufficient condition for the nonnegativity of these integrals which belongs to Koorniwinder [Koo2]).

Let E and F be compact Hausdorff spaces with nonnegative regular Borel measures σ and μ, respectively, where $\operatorname{supp} \sigma = E$ and $\operatorname{supp} \mu = F$. Let $\{\varphi_p(t)\}_{p=1}^{\infty}$ and $\{r_p(\tau)\}_{p=1}^{\infty}$ be families of continuous functions orthogonal with respect to the measures σ and μ, respectively, and let $r_0(\tau) \equiv 1$. Assume that the system $\{\varphi_p(t)\}_{p=1}^{\infty}$ is complete in $L_2(E, \sigma)$ and only finitely many terms in expansion (3.16) differ from zero.

Theorem 3.9. *Assume that there exists a continuous mapping* $\Lambda\colon E \times E \times F \to E$ *such that the functions* $\varphi_p(t)$ *satisfy the additional formula*

$$\varphi_p(\Lambda(t, s, \tau)) = \sum_k c_{p,k} \varphi_p^k(t) \varphi_p^k(s) \frac{r_k(\tau)}{\| r_k \|_2^2}, \tag{3.18}$$

where $\varphi_p^k(t)$ *are functions continuous on* E, $\varphi_p^0(t) = \varphi_p(t)$, $c_{p,k} \geq 0$, $c_{p,0} \geq 0$, *and, for each* n, *only finitely many coefficients* $c_{n,k}$ *differ from zero. Then the integrals in expansion (3.16) are nonnegative.*

Proof. Denote

$$a(p, q, l) = \int \varphi_p(t) \varphi_q(t) \overline{\varphi_l(t)} d\sigma(t).$$

We have

$$\begin{aligned}\varphi_q(\Lambda(t, s, \tau))\, \varphi_q(\Lambda(t, s, \tau)) &= \sum_l a(p, q, l) \frac{\varphi_l(\Lambda(t, s, \tau))}{\| \varphi_l \|_2^2} \\ &= \sum_{l,k} a(p, q, l) c_{l,k} \varphi_l^k(t)\, \overline{\varphi_l^k(s)} \frac{r_k(\tau)}{\| \varphi_l \|_2^2 \| r_k \|_2^2}.\end{aligned}$$

Consequently,

$$\begin{aligned} &a(p, q, l) \\ &= \left(c_{l,0} \| \varphi_l \|_2^2\right)^{-1} \int_E \int_E \int_F \varphi_p(\Lambda(t, s, \tau))\, \varphi_q(\Lambda(t, s, \tau))\, \overline{\varphi_l(t)} \varphi_l(s)\, d\sigma(t)\, d\sigma(s)\, d\mu(\tau)\end{aligned}$$

$$= \left(c_{l,0}\|\varphi_l\|_2^2\right)^{-1} \sum_{i,j} c_{p,i} c_{q,j} \frac{1}{\|r_i\|_2^2 \|r_j\|_2^2} \int_E \int_E \int_F \varphi_p^i(t)\, \overline{\varphi_p^i(s)} r_i(\tau) \varphi_q^j(t)\, \overline{\varphi_q^j(s)} r_j(\tau)$$

$$\times\, \overline{\varphi_l(t)} \varphi_l(s)\, d\sigma(t)\, d\sigma(s)\, d\mu(\tau)$$

$$= \left(c_{l,0}\|\varphi_l\|_2^2\right)^{-1} \sum_{i} c_{p,i} c_{q,i} \frac{1}{\|r_i\|_2^2} \left| \int_E \varphi_p^i(t) \varphi_q^i(t) \overline{\varphi_l(t)} d\sigma(t) \right|^2 \geq 0.$$

■

We note that additional formulas of the form (3.18) are true for some classes of orthogonal polynomials (of single or several variables), in particular, for the Jacobi polynomials $P_n^{(\alpha,\beta)}(t)$ $(\alpha > \beta > -1/2)$ (for details, see [Koo2]). Relations of the form (3.18) appear in studying spherical functions on the homogeneous space G/H, where G is a compact group and H is a subgroup of G (see Section 2 in Chapter 2).

3.4. Representations of Hypercomplex Systems and Approximation Theorem. This subsection contains a brief investigation of representations of hypercomplex systems. The commutativity of hypercomplex systems is not supposed.

Let $L_1(Q, m)$ be a normal hypercomplex system with basis unity e. The family of bounded operators $U = (U_p)_{p\in Q}$ in a separable Hilbert space $\mathcal{H}$ is called *a representation of a hypercomplex system* if

(1) $U_e = 1$;

(2) $U_p^* = U_{p^*}$ $(p \in Q)$;

(3) for each $\xi \in \mathcal{H}$, the vector function $Q \ni p \mapsto U_p\xi \in \mathcal{H}$ is weakly continuous;

(4) for all $A, B \in \mathcal{B}_0(Q)$,

$$\int c(A, B, r)\, U_r dr = \int_A U_p dp \int_B U_q dq. \tag{3.19}$$

The integrals in (3.19) are understood in Bochner's sense with the σ-algebra $\mathcal{B}(Q)$, and the integrands are vector functions $Q \ni p \mapsto A_p \in \mathcal{L}(\mathcal{H} \to \mathcal{H})$ or their products with continuous functions. Their existence follows from (3) and general properties of the Bochner integrals; it should be noted that, in view of the Banach–Steinhaus theorem, condition (3) implies that the function $Q \ni p \mapsto \|U_p\|$ is locally bounded (see, e.g., [Yos]).

Let us show that the *family of generalized translation operators* L_{p^*} $(p \in Q)$ *defines a representation (so-called left regular representation) of the hypercomplex sys-*

tem $L_1(Q, m)$ *in the Hilbert space* $L_2(Q, m)$. Indeed, the generalized translation operators L_{p^*} obviously satisfy conditions (1)–(3). Let us prove (3.19). Denote

$$L_x = \int x(p) L_{p^*} dp \quad (x \in L_1).$$

By virtue of (1.17), L_x is the operator of left convolution with $x \in L_1(Q, m)$. Hence,

$$L_x L_y f(r) = (x * y * f)(r) = L_{x*y} f(r) \quad (x, y \in L_1,\ f \in L_2).$$

In particular, we have

$$L_{\kappa_A} L_{\kappa_B} = L_{\kappa_A * \kappa_B},$$

which is obviously equivalent to (3.19).

∎

Similarly, one can show that the right generalized translation operators R_p $(p \in Q)$ define *a right regular representation.*

In [BeK3] and [BeK4], it was shown that *any representation* U_p *of a commutative normal hypercomplex system with strong basis unity (see Subsection 1.4) admits the following spectral decomposition:*

$$U_p = \int_{X_{g,h}} \chi(p) dE(\chi) \quad (p \in Q). \tag{3.20}$$

Here, E *is a resolution of the identity on the* σ*-algebra*

$$\mathfrak{C}_\sigma(X_{g,h}) = \{\alpha \cap X_h \mid \alpha \in \mathfrak{C}_\sigma(\mathbb{C}^Q)\}$$

and $\mathfrak{C}_\sigma(\mathbb{C}^Q)$ *is the* σ*-algebra generated by cylindrical sets with finite-dimensional Borel supports (see, e.g., [BeKo]). The resolution of the identity* E *is concentrated on the set* $\{\chi \in X_{g,h} \mid |\chi(p)| \le \|U_p\|,\ p \in Q\}$. *If the function* $Q \ni p \mapsto \|U_p\|$ *is bounded, then conditions (1)–(4) are necessary for the existence of representation (3.20); in this case,* E *is concentrated on the set* X_h *and given on the* σ*-algebra* $\mathcal{B}(X_h)$.

This statement remains true for any representation of the hypercomplex system $L_1(Q, m)$ *by unbounded operators. Such a representation is defined as a function* $Q \ni p \mapsto U_p$ *whose values are normal operators (generally speaking, unbounded) in a separable Hilbert space* $\mathcal{H}$ *that satisfy the following conditions:*

(i) *the operators* U_p $(p \in Q)$ *commute in the sense of commutation of their spectral projections;*

(ii) *there exists a lineal* $\mathcal{D} \subset \bigcap_{p \in Q} \mathcal{D}(U_p)$ *dense in* $\mathcal{H}$ *and invariant under the action of the operators* U_p;

(iii) *the space* $\mathcal{H}$ *admits a decomposition*

$$\mathcal{H} = \bigoplus_{k=0}^{\infty} \mathcal{H}_k$$

such that $\mathcal{D} \cap \mathcal{H}_k$ *is invariant under the action of* U_p $(p \in Q)$ *for all* k, *and each* $\mathcal{H}_k$ *contains a vector* $\xi_k \in \mathcal{D} \cap \mathcal{H}_k$ *such that the closure of the linear span* $\{U_p \xi_k \mid p \in Q\}$ *coincides with* $\mathcal{H}_k$ *(this condition means that the representation* $Q \ni p \mapsto U_p$ *can be expanded into a sum of cyclic representations);*

(iv) $U_p^* = U_{p^*}$;

(v) $U_e = I$;

(vi) *the function* $Q \ni p \mapsto (U_p \xi, \eta)_{\mathcal{H}}$ *is continuous for all* $\xi \in \mathcal{D}$ *and* $\eta \in \mathcal{H}$;

(vii) *for all* $A, B \in \mathcal{B}_0(Q)$,

$$\left(\int c(A, B^*, r) U_r dr \xi, \eta \right)_{\mathcal{H}} = \left(\int_A U_r dr \xi, \int_B U_s ds \eta \right)_{\mathcal{H}};$$

here, the operator

$$U_\varphi = \int \varphi(p) U_p dp \quad (\varphi \in L_{2,0})$$

is defined on $\mathcal{D}$ *as follows:*

$$(U_\varphi \xi, \eta)_{\mathcal{H}} = \int \varphi(p) (U_p \xi, \eta)_{\mathcal{H}} dp \quad (\xi \in \mathcal{D},\ \eta \in \mathcal{H}).$$

Furthermore, if there exists at most a countable subset $Q_0 \subset Q$ such that

$$\bigcap_{p \in Q_0} \mathcal{D}(U_p) = \bigcap_{p \in Q} \mathcal{D}(U_p),$$

then condition (iii) is automatically satisfied (this simple sufficient condition was established by Ionescu Tulcea). For a detailed exposition of results related to decomposition (3.20), see the monograph [BeKo].

We now return to the general case of noncommutative hypercomplex systems. Let $Q \ni p \mapsto U_p$ be a representation of the hypercomplex system $L_1(Q, m)$. Below, we consider representations that satisfy conditions (1)–(4) and the following additional condition:

(5) The function $Q \ni p \mapsto \| U_p \|$ is bounded.

Such representations are called *bounded.*

Let $L_1(Q, m) \ni x \mapsto U_x$ be a representation of the Banach algebra $L_1(Q, m)$ in a separable Hilbert space $\mathcal{H}$. Denote $\mathfrak{K} = \{\xi \mid U_x \xi = 0 \text{ for all } x \in L_1(Q, m)\}$. Obviously, $\mathfrak{K}$ is an invariant subspace. Below, we consider the case $\mathfrak{K} = 0$ (i.e., the representation U_x is nondegenerate).

Theorem 3.10. *Let $L_1(Q, m)$ be a normal hypercomplex system with basis unity that satisfies the condition of separate continuity. Then there is a bijection between the set of nondegenerate $*$-representations of the Banach $*$-algebra $L_1(Q, m)$ and the set of bounded representations $Q \ni p \mapsto U_p$ of the hypercomplex system. This bijection is given by the relation*

$$U_x = \int_Q x(p) U_p \, dp \quad (x \in L_1(Q, m)). \tag{3.21}$$

Proof. Since any $*$-representation of a Banach $*$-algebra can be decomposed into a direct sum of cyclic representations (see, e.g., [Nai1]), it suffices to prove the theorem for cyclic representations. Let $L_1(Q, m) \ni x \mapsto U_x$ be a cyclic representation of the Banach $*$-algebra $L_1(Q, m)$ in a separable Hilbert space $\mathcal{H}$ with cyclic vector $\xi_0 \in \mathcal{H}$. Let $\mathcal{H}' = \{\xi \in \mathcal{H} \mid \xi = U_x \xi_0 \text{ for all } x \in C_0(Q)\}$ be a dense lineal in $\mathcal{H}$.

The generalized translation operators L_p $(p \in Q)$ act continuously in $L_\infty(Q, m)$. Furthermore, by virtue of (2.7), we have $\| L_p \| \le 1$. Hence, for any $p \in Q$, we can define an operator $L_p^* : L_1 \to (L_\infty)'$ by restriction of the adjoint operator to the space L_1. It is clear that $L_p^* x = L_{p^*} x \in L_1 \cap C_0(Q)$ and $\| L_{p^*} x \|_1 \le \| x \|_1$ for all $x \in C_0(Q)$.

In $\mathcal{H}'$, we introduce an operator U_p by setting $U_p \xi = U_{L_{p^*} x} \xi_0$ for $\xi = U_x \xi_0$, $x \in C_0(Q)$. The operator U_p thus defined does not depend on the choice of x. Indeed, if $\xi = U_x \xi_0 = U_y \xi_0$ $(x, y \in C_0(Q))$, then $U_{x-y} \xi_0 = 0$ and, hence, $U_z \xi_0 = 0$ for all z from the closed left ideal I in $L_1(Q, m)$ generated by the element $x - y$. Let us show that $L_{p^*}(x - y) \in I$. Indeed, let e_n be the approximative unit of the hypercomplex system $L_1(Q, m)$ constructed in Lemma 1.3, and let $z \in I \cap C_0(Q)$. Then $L_{p^*} e_n * z \in I \cap C_0(Q)$. Since generalized translation operators are associative, we have

$$(L_{p^*} e_n) * x(r) = \int \big(L_r^q (L_{p^*} e_n)\big)(q) x(q^*) dq$$

$$= \int L_{p^*}^r L_r e_n(q) x(q^*) dq = L_{p^*}(e_n * x)(r).$$

Since $e_n * z \to z$ in L_1 for all $z \in I \cap C_0(Q)$ and $e_n * z \in C_0(Q)$ for all n, we have $L_{p^*} e_n * z = L_{p^*}(e_n * z) \to L_p z$ in L_1. Hence, $L_p z \in I \cap C_0(Q)$ because I is closed. By setting $z = x - y$, we obtain $L_{p^*} x - L_{p^*} y \in I \cap C_0(Q)$. Consequently, $U_{L_{p^*} x} \xi_0 = U_{L_{p^*} y} \xi_0 = U_p \xi$.

The operator U_p maps $\mathcal{H}'$ into itself and is bounded in $\mathcal{H}$. Indeed, assume that $\xi = U_x \xi_0 \in \mathcal{H}'$ $(x \in C_0(Q))$. Since $e_n * x \to x$ in L_1, we conclude that $U_{e_n} \xi$ tends to ξ in $\mathcal{H}$. Note that

$$x * L_p y = \int (L_r x)(s)(L_p y)(s^*) ds = \int (L_r x)(s) \overline{(L_s y^*)(p^*)} ds$$

$$= \int (L_r x)(s) \overline{(R_{p^*} y^*)(s)} ds = \int \big(R_p^s (L_r x)\big)(s) \overline{y^*(s)} ds$$

$$= \int \big(L_r^s (R_p x)\big)(s) y(s^*) ds = (R_p x) * y.$$

Therefore,

$$U_p \xi = U_{L_{p^*} x} \xi_0 = \lim_{n \to \infty} U_{e_n} U_{L_{p^*} x} \xi_0 = \lim_{n \to \infty} U_{(R_{p^*} e_n) * x} \xi_0 = \lim_{n \to \infty} U_{R_{p^*} e_n} \xi.$$

But

$$\big\| U_{R_{p^*} e_n} \xi \big\|_{\mathcal{H}} \le \| R_{p^*} e_n \|_1 \, \| \xi \|_{\mathcal{H}} \le \| e_n \|_1 \, \| \xi \|_{\mathcal{H}}.$$

Passing to the limit as $n \to \infty$, we obtain $\| U_p \xi \|_{\mathcal{H}} \le \| \xi \|_{\mathcal{H}}$, i.e., U_p is continuous in $\mathcal{H}$ and $\| U_p \| \le 1$. Since $\mathcal{H}'$ is dense in $\mathcal{H}$, we can extend U_p by continuity to the entire space $\mathcal{H}$.

Let us show that the operator family U_p satisfies conditions (1)–(5). The validity of (1) and (5) is obvious. Condition (2) can easily be verified by direct calculation.

Let us show that condition (3) is satisfied. Since the generalized translation operators L_p in $L_1(Q, m)$ are uniformly bounded with respect to p, by using the Lebesgue theorem and the condition of separate continuity, we conclude that $\| L_{p_n} x - L_p x \|_1 \to 0$ as $n \to \infty$ for each $x \in C_0(Q)$. We set $\xi = U_x \xi_0$ and $\eta = U_y \xi_0$ $(x, y \in C_0(Q))$. Then

$$|(U_{p_n}\xi,\eta)_{\mathcal{H}}-(U_p\xi,\eta)_{\mathcal{H}}| = |(U^*_{y*(L_{p_n^*}-L_{p^*})x}\xi_0,\xi_0)_{\mathcal{H}}|$$

$$\leq \|y\|_1\|(L_{p_n^*}-L_px)\|_1 \to 0, \quad n\to\infty.$$

This relation immediately yields (3) because $\|U_p\|\leq 1$.

Finally, let us verify that condition (4) is satisfied. For all $z\in L_1(Q,m)$, we introduce the operators

$$V_z = \int z(r)\,U_r dr,$$

where the integral on the right-hand side is understood in Bochner's sense. For all $\xi = U_x\xi_0$ and $\eta = U_y\xi_0 \in \mathcal{H}'$ we have

$$(V_z\xi,\eta)_{\mathcal{H}} = \int z(r)(U_rU_x\xi_0,U_y\xi_0)_{\mathcal{H}}dr = \int z(r)(U_{y^**L_{r^*}x}\xi_0,\xi_0)_{\mathcal{H}}dr$$

$$= (U_{y^**\int z(r)(L_{r^*}x)(\cdot)dr}\xi_0,\xi_0)_{\mathcal{H}} = (U_{y^**(z*x)}\xi_0,\xi_0)$$

$$= (U_z\xi,\eta)_{\mathcal{H}} \quad (x,y\in C_0(Q)).$$

Thus, $V_z = U_{z^\cdot}$ $(z\in L_1(Q,m))$. Hence, for any $A,B\in\mathcal{B}_0(Q)$, we obtain

$$\int_A U_p dp\int_B U_q dq = U_{\kappa_A}U_{\kappa_B} = U_{\kappa_A*\kappa_B} = \int c(A,B,r)\,U_r dr,$$

i.e., condition (4) is satisfied. Note that it follows from the equality $V_z = U_{z^\cdot}$ $(z\in L_1)$ that the representation U_p of the hypercomplex system associated with the cyclic representation U_x of the algebra $L_1(Q,m)$ is also cyclic.

Conversely, let $Q\ni p\mapsto U_p$ be a bounded representation of the hypercomplex system $L_1(Q,m)$ in a separable Hilbert space $\mathcal{H}$. For each $x\in L_1(Q,m)$, we define a bounded operator U_x according to (3.21). Clearly, $U_x^* = U_{x^*}$. By virtue of (3.19), the equality $U_xU_y = U_{x*y}$ holds for step functions. Taking into account that $L_1(Q,m)$ is a Banach algebra and passing to the limit, we establish that $L_1\ni x\mapsto U_x$ is a $*$-representation of the Banach $*$-algebra $L_1(Q,m)$. Let us show that this representation is nondegenerate. Let $\xi\in\mathcal{H}$ be an arbitrary nonzero vector. We choose a neighborhood $O\in\mathcal{B}_0(Q)$ of the basis unity so that $m(O)\leq 1$ and

$$|(U_p\xi,\xi)_{\mathcal{H}}-(\xi,\xi)_{\mathcal{H}}| \leq \frac{1}{2}\|\xi\|^2_{\mathcal{H}}$$

for all $p\in Q$. Then

$$|(U_{\kappa_O}\xi,\xi)_{\mathcal{H}} - (\xi,\xi)_{\mathcal{H}}| \le \int_O |(U_p\xi - \xi,\xi)_{\mathcal{H}}|\,dp + (1-m(O))\|\xi\|^2_{\mathcal{H}}$$

$$\le \frac{1}{2}m(O)\|\xi\|^2_{\mathcal{H}} + (1-m(O))\|\xi\|^2_{\mathcal{H}} < \|\xi\|^2_{\mathcal{H}}$$

and, hence, $U_{\kappa_O}\xi \ne 0$.

■

Corollary 3.6. *For any bounded representation U_r of a normal hypercomplex system with basis unity that satisfies the condition of separate continuity the following relations hold:*

$$R_s(U_r\xi,\eta)_{\mathcal{H}} = (U_rU_s\xi,\eta)_{\mathcal{H}}, \tag{3.22}$$

$$L_s(U_r\xi,\eta)_{\mathcal{H}} = (U_sU_r\xi,\eta)_{\mathcal{H}} \quad (r,s\in Q;\ \xi,\eta\in\mathcal{H}).$$

Proof. Let U_x be the representation of the algebra $L_1(Q,m)$ associated with a representation U_p of the hypercomplex system according to Theorem 3.10. Without loss of generality, we can assume that U_x (or U_p) is cyclic. Let ξ_0 be a cyclic vector. According to Theorem 3.10, the representations U_x and U_p satisfy the relation $U_p\xi = U_{L_{p*x}}\xi_0$, where $\xi = U_x\xi_0 \in \mathcal{H}'$. Then $U_{L_{p*y}}\xi = U_{(L_{p*}y)*x}\xi_0 = U_{L_{p*}(y*x)}\xi_0 = U_pU_y\xi$ for all $\xi\in\mathcal{H}'$ by virtue of (2.10).

Clearly, it suffices to prove relations (3.22) for $\xi,\eta\in\mathcal{H}'$. Let $s\in Q$ be fixed. By virtue of the equalities $(R_{s^*}x)^*(p) = \overline{(R_{s^*}x)(p^*)} = (R_px^*)(s) = (L_sx^*)(p)$, for any $x\in L_1$, we get

$$\int x(r)R_s(U_r\xi,\eta)_{\mathcal{H}}dr = \left(\int (R_{s^*}x)(r)U_rdr\xi,\eta\right)_{\mathcal{H}}$$

$$= (U_{R_{s^*}x}\xi,\eta)_{\mathcal{H}} = (\xi, U_{(R_{s^*}x)^*}\eta)_{\mathcal{H}} = (\xi, U_{L_sx^*}\eta)_{\mathcal{H}}$$

$$= (\xi, U_{s^*}U_{x^*}\eta)_{\mathcal{H}} = (U_xU_s\xi,\eta)_{\mathcal{H}}$$

$$= \int x(r)(U_rU_s\xi,\eta)_{\mathcal{H}}dr.$$

Hence, $R_s(U_r\xi,\eta)_{\mathcal{H}} = (U_rU_s\xi,\eta)_{\mathcal{H}}$ for almost all r. By virtue of the condition of separate continuity, the left-hand side of this equality is continuous in r; the right-hand side is continuous by definition. Thus, we have proved the first equality in (3.22). The second equality can be proved similarly.

■

Note that, in the definition of representations of hypercomplex systems, one can replace condition (4) by (3.22). Relations (3.22) can be used for the construction of the representation of the measure algebra $M_b(Q)$ associated with a representation $Q \ni p \mapsto U_p$ of a hypercomplex system $L_1(Q, m)$ (for the definition of the convolution of measures on the basis of hypercomplex systems, see Subsection 2.3). For this purpose, one should set

$$(U_\mu \xi, \eta)_{\mathcal{H}} = \int (U_r \xi, \eta)_{\mathcal{H}} d\mu(r) \qquad (\mu \in M_b(Q)),$$

define $(U_{\delta_r * \delta_s} \xi, \eta)_{\mathcal{H}} = L_r(U_s \xi, \eta)_{\mathcal{H}} = (U_r U_s \xi, \eta)_{\mathcal{H}}$ $(r, s \in Q;\ \xi, \eta \in \mathcal{H})$, and extend this operation to the entire algebra $M_b(Q)$.

In what follows, we assume that hypercomplex systems satisfy the condition of separate continuity. A representation of a hypercomplex system $L_1(Q, m)$ is called *irreducible* if any closed subspace invariant under the action of U_p $(p \in Q)$ coincides with either $\{0\}$ or $\mathcal{H}$. It is easy to see that a bounded representation of a hypercomplex system is irreducible if and only if the corresponding representation of the algebra $L_1(Q, m)$ is irreducible. It is also clear that two representations of a hypercomplex system $L_1(Q, m)$ are unitarily equivalent if and only if the corresponding representations of the algebra $L_1(Q, m)$ are equivalent. We say that a system of representations of a hypercomplex system $L_1(Q, m)$ is *complete* if, for any $p \neq q$ $(p, q \in Q)$, it contains a representation $Q \ni r \mapsto U_r$ such that $U_p \neq U_q$.

Theorem 3.11. *The system of all bounded irreducible representations of any normal hypercomplex system with basis unity that satisfies the condition of separate continuity is complete.*

Proof. First, recall some necessary results concerning representations of Banach algebras. A Banach $*$-algebra R is called *reduced* if, for any $x \in R$, $x \neq 0$, there exists a positive functional Φ such that $\Phi(x^* * x) \neq 0$. The collection of all irreducible $*$-representations of a reduced Banach $*$-algebra forms a complete system. Let us show that the hypercomplex system $L_1(Q, m)$ is reduced. Assume that $x \neq 0$ and L_x is the operator of convolution with x in $L_2(Q, m)$. Also assume that $\Phi(x^* * x) = 0$ for any positive functional Φ. Then, in particular, $\| L_x f \|_2 = 0$ for all $f \in L_2(Q, m)$, whence $x * f = 0$ $(f \in L_2)$. Let $f = e_n$, where e_n is the approximative unit constructed in Lemma 1.3. Then $x * e_n \to x$ in L_1, whence $x = 0$, which contradicts the assumption. Hence, the algebra $L_1(Q, m)$ is reduced.

Since the generalized translation operators L_r satisfy the conditions of finiteness (by virtue of Theorem 2.1) and separate continuity, they map $C_0(Q)$ into itself and, hence, are reduced on $C_0(Q)$. Therefore, for $p \neq q$, one can find $x \in C_0(Q) \subset L_1(Q, m)$ such that $L_{p^*} x \neq L_{q^*} x$. Since the Banach algebra $L_1(Q, m)$ is reduced, there exists an irreducible representation $L_1(Q, m) \ni x \mapsto U_x$ such that $U_{L_{p^*}x - L_{q^*}x} \neq 0$, i.e., $U_{L_{p^*}x}\xi_0 \neq U_{L_{q^*}x}\xi_0$ for some vector $\xi_0 \in \mathcal{H}$. Let $U_x\xi_0 = \xi$. Then $U_{L_{p^*}x}\xi_0 = U_p\xi$, $U_{L_{q^*}x}\xi_0 = U_q\xi$, and the last inequality takes the form $U_p\xi \neq U_q\xi$.

■

The statement below describes the relationship between bounded representations of the hypercomplex system $L_1(Q, m)$ and positive definite functions on Q.

Theorem 3.12. *Let $L_1(Q, m)$ be a normal hypercomplex system with basis unity satisfying the condition of separate continuity. Then there is a bijection between the collection of continuous bounded functions on Q positive definite in the sense of (3.1) and the set of classes of unitarily equivalent bounded cyclic representations of the hypercomplex system $L_1(Q, m)$. This bijection is given by the relation*

$$\varphi(r) = (U_r\xi_0, \xi_0)_{\mathcal{H}} \quad (r \in Q), \tag{3.23}$$

where $\varphi \in P(Q)$ and U_r is the corresponding representation of the hypercomplex system $L_1(Q, m)$ in a Hilbert space $\mathcal{H}$ with cyclic vector ξ_0.

Proof. If U_r is a bounded representation of the hypercomplex system $L_1(Q, m)$ with cyclic vector ξ_0, then the function $\varphi(r) = (U_r\xi_0, \xi_0)_{\mathcal{H}}$ is positive definite in the sense of (3.1). Indeed, let $x \in L_1(Q, m)$. Then

$$\int \varphi(r)(x^* * x)(r)\,dr = \left(\int (x^* * x)(r)U_r\,dr\,\xi_0, \xi_0\right)_{\mathcal{H}}$$

$$= (U_{x^* * x}\xi_0, \xi_0)_{\mathcal{H}} = \|U_x\xi_0\|_2^2 \geq 0$$

by virtue of Theorem 3.10. Assume that there exists another representation U'_r of the hypercomplex system $L_1(Q, m)$ in the space $\mathcal{H}'$ with cyclic vector ξ'_0 such that $\varphi(r) = (U'_r\,\xi'_0, \xi'_0)_{\mathcal{H}'}$. Let U'_x be the corresponding (obviously, cyclic) representation of the algebra $L_1(Q, m)$. It is clear that $(U_x\xi_0, \xi_0)_{\mathcal{H}} = (U'_x\,\xi'_0, \xi'_0)_{\mathcal{H}'}$. For any $x, y \in L_1$, we have

$$(U_x\xi_0, U_y\xi_0)_{\mathcal{H}} = (U_{y^* * x}\xi_0, \xi_0)_{\mathcal{H}} = (U'_{y^* * x}\,\xi'_0, \xi'_0)_{\mathcal{H}'} = (U'_x\,\xi'_0, U'_y\,\xi'_0)_{\mathcal{H}'}.$$

Since $U_x\xi_0$ $(U'_x\,\xi'_0)$ are dense in $\mathcal{H}$ (in $\mathcal{H}'$, respectively), this equality implies that there exists an isomorphism $V\colon \mathcal{H} \to \mathcal{H}'$ such that $V(U_x\xi_0) = U'_x\,\xi'_0$ $(x \in L_1)$. Let us show that V maps U_x into U'_x, i.e., $VU_x = U'_xV$ $(x \in L_1)$. Indeed, for all $y \in L_1$, we have

$$(VU_x)U_y\xi_0 = VU_{x*y}\xi_0 = U'_{x*y}\xi'_0 = U'_x(U'_y\xi'_0) = U'_xV(U_y\xi_0).$$

Since $U_y\xi_0$ are dense in $\mathcal{H}$, we get $VU_x = U'_xV$. On the other hand, for any $x \in L_1$,

$$(\xi'_0, U'_x\xi'_0)_{\mathcal{H}'} = (V\xi_0, VU_x\xi_0)_{\mathcal{H}'} = (V\xi_0, U'_x\xi'_0)_{\mathcal{H}'},$$

whence $\xi'_0 = V\xi_0$. It is obvious that the representations U_p and U'_p of the hypercomplex system $L_1(Q, m)$ that correspond to U_x and U'_x, respectively, are also unitarily equivalent.

For an arbitrary continuous positive definite function $\varphi(r)$ $(r \in Q)$, we construct a cyclic bounded representation U_x of the Banach algebra $L_1(Q, m)$ in a Hilbert space $\mathcal{H}$ with cyclic vector ξ_0 by applying the GNS construction to a positive functional

$$\Phi(x) = \int \varphi(r)x(r)dr \quad (x \in L_1(Q, m)) \tag{3.24}$$

(this is possible because $L_1(Q, m)$ contains an approximative unit). Let U_p be a cyclic bounded representation of the hypercomplex system $L_1(Q, m)$ that corresponds to U_x $(x \in L_1(Q, m))$ by virtue of Theorem 3.10. Then, for all $x \in L_1(Q, m)$, we have

$$\int x(r)(U_r\xi_0, \xi_0)_{\mathcal{H}}dr = (U_x\xi_0, \xi_0)_{\mathcal{H}} = \Phi(x) = \int x(r)\varphi(r)dr.$$

Hence, relation (3.23) holds almost everywhere. Since the functions $(U_r\xi_0, \xi_0)_{\mathcal{H}}$ and $\varphi(r)$ are continuous, relation (3.23) holds for all $r \in Q$.

■

Corollary 3.7. *For a normal hypercomplex system with basis unity that satisfies the condition of separate continuity, the concepts of positive definiteness in the sense of (3.1) and (3.1)′ are equivalent.*

It suffices to show that if $\varphi(r)$ is positive definite in the sense of (3.1), then relation (3.1)′ holds for any $\lambda_1, \dots, \lambda_n \in \mathbb{C}$ and $r_1, \dots, r_n \in Q$. Indeed, by virtue of (3.23) and (3.22), we have

$$\sum_{i,j=1}^{n} \lambda_i\bar{\lambda}_j(R_{r_i^*}\varphi)(r_j) = \sum_{i,j=1}^{n} \lambda_i\bar{\lambda}_j R_{r_i^*}(U_{r_j}\xi_0, \xi_0)_{\mathcal{H}}$$

$$= \sum_{i,j=1}^{n} \lambda_i\bar{\lambda}_j(U_{r_j}U_{r_i^*}\xi_0, \xi_0)_{\mathcal{H}} = \left\| \sum_{i=1}^{n} \lambda_i U_{r_i^*}\xi_0 \right\|_{\mathcal{H}}^2 \geq 0.$$

■

Recall that properties (i)–(v) formulated in Corollary 3.2 hold for continuous functions positive definite in the sense of (3.1)′ and, hence, any function $\varphi(r) \in P(Q)$ also possesses these properties.

Remark. It is easy to see that the GNS construction can also be performed for a positive definite function $\varphi(r)$ that is not necessarily continuous. In this case, equality (3.23) holds almost everywhere. Consequently, *an arbitrary bounded positive definite function* $\varphi(r)$ *coincides almost everywhere with some continuous positive definite function.* Since the linear span of the set of bounded positive definite functions is weakly dense in $L_\infty(Q, m)$, the *linear span of the set* $P(Q)$ *of continuous bounded positive definite functions is weakly dense in* $L_\infty(Q, m)$.

It is clear that relation (3.24) establishes a one-to-one correspondence between the set of positive functionals on $L_1(Q, m)$ and the set of bounded positive definite (and, hence, continuous positive definite) functions on Q.

Functions obtained by generalized translation of any continuous positive definite function can be represented as a linear combination of four continuous positive definite functions. Indeed, assume that $\varphi(r) \in P(Q)$, $p, q \in Q$, and U_r is the representation of a hypercomplex system in the Hilbert space $\mathcal{H}$ with cyclic vector ξ_0 that corresponds to the function φ according to Theorem 3.12. We set $U_p\xi_0 = \xi$ and $U_{q^*}\xi_0 = \eta$. By virtue of (3.22),

$$\begin{aligned} 4\big(L_q^r(R_p\varphi)\big)(r) &= 4L_q^r R_p(U_r\xi_0, \xi_0) = 4(U_r\xi, \eta)_{\mathcal{H}} \\ &= (U_r(\xi+\eta), \xi+\eta)_{\mathcal{H}} - (U_r(\xi-\eta), \xi-\eta)_{\mathcal{H}} \\ &\quad + i(U_r(\xi+i\eta), \xi+i\eta)_{\mathcal{H}} - i(U_r(\xi-i\eta), \xi-i\eta)_{\mathcal{H}}. \end{aligned}$$

■

Let $\varphi_1, \varphi_2 \in P(Q)$. We say that φ_1 *majorizes* φ_2 $(\varphi_1 \geq \varphi_2)$ if $\varphi_1 - \varphi_2 \in P(Q)$. Clearly, this is true for the corresponding positive functionals, i.e., $\Phi_1 \geq \Phi_2$. A positive definite function $\varphi(r)$ is called *elementary* if $\varphi \neq 0$ and all positive definite functions $\varphi_1 \in P(Q)$ majorized by φ have the form $\lambda\varphi$ $(0 \leq \lambda \leq 1)$. In other words, a functional Φ that corresponds to an elementary positive definite function is indecomposable. It is known that the representation of a Banach $*$-algebra associated with a continuous positive functional is irreducible if and only if this functional is indecomposable. This implies that the representation of the hypercomplex system $L_1(Q, m)$ that corresponds to an elementary function $\varphi(r) \in P(Q)$ is irreducible.

Let $\varphi(p) = (U_p\xi_0, \xi_0)_{\mathcal{H}}$, *where* $p \mapsto U_p$ *is a cyclic representation of a normal hypercomplex system with basis unity that satisfies the condition of separate continuity. The function* $\varphi_1 \in P(Q)$ *is majorized by* φ *if and only if* $\varphi_1(p) = (U_pB\xi_0, \xi_0)_{\mathcal{H}}$, *where* B *is a positive operator in* $\mathcal{H}$ *which commutes with all* U_p *and satisfies the condition* $\|B\| \le 1$. Indeed, if Φ_1 is a positive functional on $L_1(Q, m)$ corresponding to φ_1, then $\Phi_1(x) = (U_x B\xi_0, \xi_0)_{\mathcal{H}}$ $(x \in L_1)$, where U_x is the representation of $L_1(Q, m)$ corresponding to U_p, and the operator $B \ge 0$, $\|B\| \le 1$, commutes with all U_x (and, hence, with U_p). This implies that $\varphi_1(r) = (U_r B\xi_0, \xi_0)_{\mathcal{H}}$ almost everywhere. Since both sides of this equality are continuous, we arrive at the required result.

■

If $\varphi_1, \varphi_2 \in P(Q)$ *and* φ_1 *majorizes* φ_2, *then* φ_2 *is the uniform limit of functions of the form*

$$\varphi(r) = \sum_{i,j=1}^{n} \lambda_i \bar{\lambda}_j \left(R^r_{s_i}\left(L^r_{s_j}\varphi_1\right)\right)(r)$$

$$(s_1, \ldots, s_n, r \in Q, \quad \lambda_1, \ldots, \lambda_n \in \mathbb{C}).$$

Indeed, let $\varphi_1 = (U_r \xi_0, \xi_0)_{\mathcal{H}}$. Then

$$\varphi_2(r) = (U_r B\xi_0, \xi_0)_{\mathcal{H}} = (U_r B^{1/2}\xi_0, B^{1/2}\xi_0)_{\mathcal{H}}.$$

Denote $\eta = B^{1/2}\xi_0$. For all $\varepsilon > 0$, there exist $\lambda_1, \ldots, \lambda_n \in \mathbb{C}$ and $s_1, \ldots, s_n \in Q$ such that

$$\left\| \eta - \sum_{i=1}^{n} \lambda_i U_{s_i} \xi_0 \right\|_{\mathcal{H}} < \varepsilon.$$

This inequality and (3.22) imply that

$$\left| \varphi_2(r) - \sum_{i,j=1}^{n} \lambda_i \bar{\lambda}_j \left(R^r_{s_i}\left(L^r_{s_j^*}\varphi_1\right)\right)(r) \right| = \left| (U_r \eta, \eta)_{\mathcal{H}} - \left(U_r\left(\sum_{i=1}^{n} \lambda_i U_{s_i}\xi_0\right), \sum_{i=1}^{n} \lambda_i U_{s_i}\xi_0 \right)_{\mathcal{H}} \right|$$

$$\le \varepsilon \|\eta\|_{\mathcal{H}} + \varepsilon(\|\eta\|_{\mathcal{H}} + \varepsilon)$$

for all $r \in Q$.

■

Lemma 3.6. *Suppose that* A *is a bounded set in* $L_\infty(Q, m)$ *and* $f \in C_0(Q)$. *If* $\varphi_n \in A$ *weakly converges to* $\varphi \in A$, *then* $f * \varphi_n$ *converges to* $f * \varphi$ *in the topology* *of uniform convergence on compact sets.*

Proof. Clearly, $f * \varphi_n = \langle \overline{\varphi}_n^*, L_s f \rangle$. If $\varphi_n \in A$ weakly converges to $\varphi \in A$, then $\langle \overline{\varphi}_n^*, y \rangle$ uniformly converges to $\langle \overline{\varphi}^*, y \rangle$ on each compact subset in $L_1(Q, m)$. Let $s_n \to s \in Q$. For any function $f \in C_0(Q)$, according to the condition of separate continuity, we have $L_{s_n} f(r) \to L_s f(r)$ for all $r \in Q$. Since the generalized translation operators L_p $(p \in Q)$ are uniformly bounded in $L_1(Q, m)$, by the Lebesgue theorem, we have

$$\left\| L_{s_n} f - L_s f \right\|_1 \xrightarrow[n \to \infty]{} 0.$$

Hence, for a given function $f \in C_0(Q)$, the mapping $Q \ni s \mapsto L_s f \in L_1(Q, m)$ is strongly continuous and, therefore, for any compact set $K \subset Q$, the set $\{L_s f \mid s \in K\}$ is strongly compact in $L_1(Q, m)$.

■

Lemma 3.7. *Let $P_1(Q)$ be the set of continuous positive definite functions $\varphi(r)$ on Q such that $\|\varphi\|_\infty = \varphi(e) = 1$. The weak topology $\sigma(L_\infty(Q, m), L_1(Q, m))$ coincides on the set $P_1(Q)$ with the topology of uniform convergence on compact sets.*

Proof. Since $\|\varphi\|_\infty = 1$ for any $\varphi \in P_1(Q)$, it is clear that the topology of compact convergence on $P_1(Q)$ is stronger than the weak topology.

Now assume that $\varphi_0 \in P_1(Q)$, K is a compact set in Q, and $\varepsilon > 0$. Let us show that if $\varphi \in P_1(Q)$ lies in a certain appropriate weak neighborhood of φ_0, then

$$|\varphi(r) - \varphi_0(r)| \le \varepsilon + 4\sqrt{\varepsilon}$$

for all $r \in K$. Denote by $O \in \mathcal{B}_0(Q)$ a neighborhood of the point $e \in Q$ such that $|\varphi(r) - 1| < \varepsilon$ for all $r \in Q$. One can find a function $f \in C_0(Q)$ such that $f(r) \ge 0$, $\operatorname{supp} f \subset \overline{O}$, and $\|f\|_1 = 1$. Consider a weak neighborhood U of the point φ_0 in $P_1(Q)$ such that, for $\varphi \in U$,

$$\left| \int f(r)(\varphi(r) - \varphi_0(r)) dr \right| \le \varepsilon.$$

If $\varphi \in U$, then

$$\left| \int f(r)(1 - \varphi(r)) dr \right| \le \left| \int f(r)(1 - \varphi_0(r)) dr \right|$$

$$+ \left| \int f(r)(\varphi(r) - \varphi_0(r)) dr \right| \le \|f\|_1 \varepsilon + \varepsilon = 2\varepsilon.$$

On the other hand, for $\varphi \in U$ and $r \in Q$, we have

$$|f * \varphi(r) - \varphi| = \left|\int (L_r f)(s)\varphi(s^*)ds - \varphi(r)\right|$$

$$= \left|\int f(s)\overline{(L_{r^*}\varphi^*)(s)}ds - \int f(s)\varphi(r)ds\right|$$

$$\leq \int f(s)|L_{s^*}\varphi(r) - \varphi(r)|ds. \tag{3.25}$$

By applying the inequality

$$|K(s,r) - K(t,r)|^2 \leq K(r,r)[K(t,t) + K(s,s) - 2\operatorname{Re} K(s,t)]$$

to the positive definite kernel $K(s,r) = L_{s^*}\varphi(r)$ and setting $t = e$, we get

$$|(L_{s^*}\varphi)(r) - \varphi(r)|^2 \leq (L_{r^*}\varphi)(r)[\varphi(e) + (L_{s^*}\varphi)(s) - 2\operatorname{Re}(L_{s^*}\varphi)(e)]$$

$$\leq 2\varphi(e)(\varphi(e) - 2\,(\operatorname{Re}\varphi)(s^*)) = 2(1 - 2\operatorname{Re}\varphi(s)).$$

Here, we have used the fact that $|(L_{r^*}\varphi)(r)| \leq \|\varphi\|_\infty = \varphi(e)$ and $\varphi(s^*) = \overline{\varphi(s)}$. Consequently, we can extend (3.25) as follows:

$$|f * \varphi(r) - \varphi| \leq \int \sqrt{2} f(s)(1 - \operatorname{Re}\varphi(s))^{1/2} ds$$

$$\leq \sqrt{2}\left(\int f(s)(1 - \operatorname{Re}\varphi(s))ds\right)^{1/2} \|f\|_1^{1/2} \leq 2\sqrt{\varepsilon}.$$

By Lemma 3.6, one can find a weak neighborhood U' of the point $\varphi_0 \in P_1(Q)$ such that $|(f * \varphi_0)(r) - (f * \varphi)(r)| \leq \varepsilon$ for all $\varphi \in U'$ and $r \in K$. Hence,

$$|\varphi(r) - \varphi_0(r)| \leq |\varphi(r) - f * \varphi(r)| + |(f * \varphi)(r) - (f * \varphi_0)(r)|$$

$$+ |(f * \varphi_0)(r) - \varphi_0(r)| \leq \varepsilon + 4\sqrt{\varepsilon}$$

for $\varphi \in U \cap U'$ and $r \in K$.

■

Theorem 3.13. *Let $L_1(Q, m)$ be a normal hypercomplex system with basis unity that satisfies the condition of separate continuity. Any continuous function on Q can be uniformly approximated on every compact set by linear combinations of elementary positive definite functions.*

Proof. Assume that the theorem is proved for arbitrary functions from $P(Q)$. Since any function of the form $a * b^*$ $(a, b \in C_0(Q))$ is a linear combination of four positive definite functions $(a \pm b) * (a \pm b)^*$, $(a \pm ib) * (a \pm ib)^*$, the statement of the theorem is true for all functions of the form

$$\varphi(r) = \sum_k \lambda_k (a_k * b_k)(r) \qquad (a_k, b_k \in C_0(Q)).$$

According to Lemma 3.2, we can uniformly approximate any function $f \in C_0(Q)$ by functions of this form. This proves the theorem because any continuous function can be uniformly approximated by functions from $C_0(Q)$ on every compact set.

It remains to show that any function $\varphi(r) \in P(Q)$ can be uniformly approximated on every compact set by linear combinations of elementary positive definite functions. Without loss of generality, we can assume that $\varphi(e) = \|\varphi\|_\infty = 1$. As is known, every continuous positive functional defined on a Banach $*$-algebra with approximative unit and having the norm less than 1 is a weak limit of convex combinations of indecomposable functionals and the zero functional. Therefore, the positive functional Φ associated with φ is a weak limit of convex combinations $\Psi = \sum \lambda_i \Phi_i$ of indecomposable positive functionals Φ_i, $\|\Phi_i\| = 1$, and the zero functional. Since $\lim \|\Psi\| = \|\Phi\| = 1$ and $\|\Psi\| \le 1$, we can assume that $\|\Psi\| = 1$ (this can always be realized by multiplying Ψ by proper scalars). If φ_i are elementary positive definite functions corresponding to Φ_i, then $\varphi_i(e) = \|\Phi_i\| = 1$. In this case, Ψ corresponds to the positive definite function

$$\psi(r) = \sum_i \lambda_i' \varphi_i(r) \qquad \left(\lambda_i' \ge 0, \sum \lambda_i' = 1\right).$$

Consequently, the functions $\psi(r)$ weakly converge in $P_1(Q)$ to $\varphi(r)$. To complete the proof it remains to use Lemma 3.7.

■

Corollary 3.8. *Assume that $L_1(Q, m)$ is a commutative normal hypercomplex system with basis unity that satisfies the condition of separate continuity. Then any continuous function can be uniformly approximated on compact sets from Q by linear combinations of Hermitian characters.*

Proof. It suffices to show that every elementary normalized positive definite function φ is an Hermitian character of a commutative hypercomplex system. The indecomposable positive functional

$$\Phi(x) = \int \varphi(r) x(r)\, dr \qquad (x \in L_1(Q, m))$$

corresponds to the function φ. It is well known that a positive functional Φ is indecomposable in the commutative Banach $*$-algebra R if and only if $\Phi(x) = x(M)$ $(x \in R)$, where M is some symmetric maximum ideal. This and Theorem 1.2 imply that

$$\Phi(x) = \int x(r)\chi(r)\,dr.$$

Hence, $\varphi(r) = \chi(r)$ almost everywhere. Taking into account that both functions in this equality are continuous, we arrive at the required statement.

■

Let us introduce the concept of Fourier transformation for noncommutative normal hypercomplex systems with basis unity satisfying the condition of separate continuity. Denote by $\mathfrak{R}$ the set of equivalence classes of bounded irreducible representations of a hypercomplex system. The operator-valued function $\mathfrak{R} \ni U \mapsto \hat{\mu}(U) = \overline{U}_\mu$ is called the *Fourier transform* of a measure $\mu \in M_b(Q)$. If a function f belongs to $L_1(Q, m)$, then its *Fourier transform* is defined as the operator function $\hat{f}(U) = \overline{U}_f$ $(U \in \mathfrak{R})$. By $C_b(\mathfrak{R})$ we denote the $*$-algebra of operator functions on $\mathfrak{R}$ with ordinary multiplication and the norm

$$\|f\|_\infty = \sup_{U \in \mathfrak{R}} \|f(U)\|_{\varphi_\infty}$$

(here, $\|A\|_{\varphi_\infty}$ is the greatest eigenvalue of the operator $|A|$). *Let Q be a compact set. The Fourier transformation $M_b(Q) \ni \mu \mapsto \hat{\mu} \in C_b(\mathfrak{R})$ is a continuous injective $*$-homomorphism of $M_b(Q)$ into $C_b(\mathfrak{R})$.* Indeed, it is obvious that $\|\hat{\mu}\|_\infty \le \|\mu\|$ and the Fourier transformation is a $*$-homomorphism. Assume that $\hat{\mu}(U) = \overline{U}_\mu = 0$ for all $U \in \mathfrak{R}$. Then

$$\int (U_p\xi, \eta)_{\mathcal{H}}\,d\mu(p) = 0 \qquad (U \in \mathfrak{R},\ \ \xi, \eta \in \mathcal{H}).$$

In view of Theorems 3.12 and 3.13, this relation implies that

$$\int f(r)\,d\mu(p) = 0$$

for any $f \in C(Q)$. Thus, $\mu = 0$.

■

4. Hypercomplex Subsystems and Homomorphisms

The notion of normal hypercomplex system with basis unity is a generalization of the notion of group algebra of a locally compact group. As shown above, many facts of harmonic analysis for locally compact groups are true for a hypercomplex system with locally compact basis. It is natural to expect that some other facts from the group theory can be generalized to the case of hypercomplex systems. In this section, we introduce and study an analog of subgroups of a group—the so-called hypercomplex subsystems. Since a subgroup of a locally compact group, generally speaking, cannot be described in terms of subalgebras of a group algebra ($\mathbb{R}^1 = \{0\} \times \mathbb{R}^1$ in $\mathbb{R}^2$ is an example of a subgroup of this sort), the definition of hypercomplex subsystem is given in terms of generalized translation operators associated with the hypercomplex system but not in terms of the structure measure. We also present a brief description of the homomorphisms of hypercomplex systems. At the end of the section, we introduce the notions of the direct and semidirect products and the join of compact and discrete hypercomplex systems (hypergroups). The latter notion is of pure hypergroup nature, since, generally speaking, the join of two groups is a hypercomplex system but not a group.

4.1. Definition of Hypercomplex Subsystems. Let us now introduce an analog of the notion of subgroups for the case of hypercomplex systems Let $L_1(Q, m)$ be a normal (not necessarily commutative) hypercomplex system with basis unity that satisfies the condition of separate continuity. A closed nonempty set $H \subset Q$ is called a *basis of a hypercomplex subsystem* if (a) $e \in H$, (b) $H^* = H$, (c) the equality $(L_s f)(t) = 0$ holds for all $t, s \in H$ and any function $f \in C(Q)$ such that $f(r) = 0$ $(r \in H)$.

We equip H with the induced topology of the space Q: The sets of the form $O' = O \bigcap H$, where O is a neighborhood in Q, are regarded as neighborhoods in H. According to condition (c), the generalized translation operators L_s $(s \in H)$ can be correctly restricted to $C(H)$. Namely, we consider an extension $\tilde{f}(r) \in C(Q)$ of an arbitrary function $f \in C(H)$ and set $(L_s f)(t) = (L_s \tilde{f})(t)$. By virtue of (c), this definition does not depend on the choice of the function $\tilde{f}$. Clearly, if H is a basis of the hypercomplex subsystem, then there exists a family of involution-consistent generalized translation operators acting in $C_b(H)$ preserving positivity and the identity, satisfying the finiteness condition (F), and the condition of separate continuity. The generalized translation operators L_s $(s \in H)$ also satisfy inequality (2.7), i.e., the operators L_s $(s \in H)$ are uniformly bounded in $C_b(H)$.

In what follows, for simplicity, a basis H of a hypercomplex subsystem is called simply a hypercomplex subsystem and we keep in mind that the family of generalized translation operators L_s $(s \in H)$ is given on H. If the generalized translation operators L_s $(s \in H)$ are commutative, then we say that the hypercomplex subsystem is *commutative.*

If the hypercomplex subsystem H *is commutative and compact in the induced topology, then there exists a finite regular positive Borel measure* μ *on* H *invariant under the action of the generalized translation operators* L_p $(p \in H)$:

$$\int_H (L_p f)(q)\,d\mu(q) = \int_H f(q)\,d\mu(q) \quad (p \in H;\ f \in C_b(H)).$$

In fact, we apply the M. Krein–Rutman theorem [KrR]: If the collection $\Gamma = \{A\}$ of commuting linear operators which transform a solid cone K into itself has a common fixed vector $v > 0$, i.e., $Av = v$ $(A \in \Gamma)$, then their adjoint operators A^* have a common fixed vector $\psi \neq 0$ in K^*, i.e., $A^*\psi = \psi$ $(\psi \in K^*)$. Since the generalized translation operators L_p $(p \in H)$ preserve positivity, the commutative operator family $\{L_p\}_{p \in H}$ maps the solid cone $C_b^+(H)$ of nonnegative functions from $C_b(H)$ into itself. The operators L_p have a common fixed vector $1(r) \equiv 1$ $(r \in H)$ inside $C_b^+(H)$. By the Krein–Rutman theorem, their adjoint operators have a common fixed vector μ in $(C_b^+(H))'$, i.e., $L_p^*\mu = \mu$ $(p \in H)$. By the Riesz–Markov theorem, $(C_b^+(H))'$ is isomorphic to the set of finite regular positive Borel measures on H. Thus, we have $\mu \in M_b^+(Q)$ and

$$\int_H (L_p f)(q)\,d\mu(q) = \int_H f(q)\,d\mu(q) \quad (f \ni C_b(H), p \in H).$$

∎

Here, we do not prove the existence of weakly invariant measures on noncommutative locally compact hypercomplex subsystems H (this problem contains the unsolved problem of existence of the invariant measure on a hypergroup) and postulate the existence of a measure μ of this sort on H.

The following argument can be regarded as a justification of this assumption: The convolution of probability measures on H can be defined as follows: Let $\langle \delta_p * \delta_q, f \rangle = (L_p f)(q)$ $(f \in C_b(H),\ p, q \in H)$. Since the generalized translation operators L_p $(p \in H)$ preserve the identity, $\delta_p * \delta_q$ is a probability measure. By virtue of the condition of separate continuity, the operation of convolution can be extended to arbitrary probability measures by the relation

$$\langle \mu * \nu, f \rangle = \int \int \langle \delta_p * \delta_q, f \rangle\,d\mu(p)\,d\mu(q).$$

It is easy to show that this convolution in $M_1(H)$ satisfies axioms (G1)–(G4) in the definition of hypergroups, i.e., H becomes a weak hypergroup. If H is a hypergroup with respect to this operation of convolution (in this case, H is called *subhypergroup),* then the existence of weak invariant measure in compact and commutative locally compact cases follows from the results of Jewett [Jew] and Spector [Spe4]. If the hypercomplex subsystem is discrete, then as shown in Section 2, it is a hypergroup and, by virtue of [Jew], it possesses the invariant measure.

Throughout Section 4.1, we assume that the support of the invariant measure μ coincides with the entire H. *If* H *is a subhypergroup, then* $\operatorname{supp}\mu = H$. Indeed, it is clear that, for any measure $\eta \in M_b^+(H)$ with compact support, there exists convolution $\eta * \mu$ which is a positive regular Borel measure and $\operatorname{supp}(\eta * \mu) = \operatorname{supp}\eta * \operatorname{supp}\mu$ (recall that the convolution of sets $A * B$ $(A, B \in \mathfrak{B}(H))$ is defined as

$$\bigcup_{p \in A : q \in B} \operatorname{supp}(\delta_p * \delta_q),$$

see Subsection 2.3). The property of invariance can be reformulated in terms of the convolution of measures. Indeed, $\delta_p * \mu = \mu$ $(p \in H)$. Thus, it suffices to show that $e \in \operatorname{supp}\mu$. Let $p \in \operatorname{supp}\mu$. Since

$$\operatorname{supp}(\delta_{p^*} * \mu) = \bigcup_{q \in \operatorname{supp}\mu} \operatorname{supp}\delta_{p^*} * \delta_q,$$

we see that $\operatorname{supp}(\delta_{p^*} * \delta_p) \subset \operatorname{supp}\mu$. By using axiom (G5) from the definition of hypergroups, we obtain $e \in \operatorname{supp}(\delta_{p^*} * \delta_p)$.

■

Let us now describe hypercomplex subsystems in terms of the convolution of sets.

A closed nonempty subset $H \subset Q$ *is a hypercomplex subsystem if and only if* (a′) $e \in H$, (b′) $H^* = H$, *and* (c′) $H * H \subset H$.

Indeed, assume that H have properties (a′)–(c′) and let $f \in C(Q)$ be such that $f(r) = 0$ $(r \in H)$. Then $\operatorname{supp}(\delta_p * \delta_q) \subset H$ for any $p, q \in H$ by virtue of (c′), whence $(L_p f)(q) = \langle \delta_p * \delta_q, f \rangle = 0$ and H is a hypercomplex subsystem Conversely, assume that H is a hypercomplex subsystem and $f \in C(Q)$ is a function such that $f(r) = 0$ $(r \in H)$. Then $\langle \delta_p * \delta_q, f \rangle = (L_p f)(q) = 0$ for all $p, q \in H$. Since the function f is arbitrary, this equality means that $\operatorname{supp}(\delta_p * \delta_q) \subset H$.

■

Let us show that the invariant measure is unique to within a constant.

Theorem 4.1. *Assume that* H *is a hypercomplex subsystem and the generalized* *translation operators* L_s $(s \in H)$ *satisfy condition (C) in Lemma 2.2 (i.e., for any neighborhood* W *of the point* $s \in Q$, *there exist a neighborhood* U *of the basis unity such that* $\operatorname{supp}(L_{s^*}\varphi)(\cdot) \subset W$ *for any* $\varphi \in C_0(Q)$ *with* $\operatorname{supp}\varphi \subset U$). *If* μ_1 *and* μ_2 *are two weakly invariant measures on* H, *then one can find a positive real number C such that* $\mu_2 = C\mu_1$.

Proof. Denote by $\|\cdot\|_{2,\mu_i}$ and $(\cdot,\cdot)_{2,\mu_i}$ the norm and the scalar product in $L_2(H, \mu_i)$ $(i = 1, 2)$, respectively. Note that $|(L_pf, g)_{2,\mu_i}| \le \|f\|_{2,\mu_i}\|g\|_{2,\mu_i}$ by (2.7). This implies that the operators L_p $(p \in H)$ form a family of bounded generalized translation operators in $L_2(H, \mu_i)$ $(i = 1, 2)$. Clearly, the generalized translation operators L_p $(p \in H)$ satisfy the conditions of Theorem 2.2. Therefore, $L_p^* = L_{p^*}$ in each of the spaces $L_2(H, \mu_i)$ $(i = 1, 2)$, i.e., μ_i are strongly invariant in the sense of Subsection 2.1.

We take $\varepsilon > 0$ and functions $f, g \in C_0^+(H)$ such that $f \ne 0$ and $g \ne 0$. By virtue of the condition of separate continuity,

$$\lim_{p\to e} ((L_pf)(q) - (R_pf)(q)) = 0$$

for each $q \in H$. In view of (2.7) and the Lebesgue theorem, this yields

$$\lim_{p\to e} \int_H |(L_pf)(q) - (R_pf)(q)|\, d\mu_2(q) = 0.$$

Thus, there is an open neighborhood $U \in \mathfrak{B}_0(H)$ of the basis unity e such that

$$\int_H |(L_pf)(q) - (R_pf)(q)|\, d\mu_2(q) < \frac{\varepsilon}{2}\int_H f(q)\, d\mu_1(q)$$

$$\int_H |(L_pg)(q) - (R_pg)(q)|\, d\mu_2(q) < \frac{\varepsilon}{2}\int_H g(q)\, d\mu_1(q)$$

for all $p \in U$. We now take $h \in C_0^+(H)$ such that

$$\int_H h(q)\, d\mu_1(q) > 0,$$

$h^* = h$, and $h = 0$ outside U. Then

$$\int_H h(q)\, d\mu_2(q) \int_H f(q)\, d\mu_1(q) = \int_H f(p) \int_H (L_{p^*}h)(q)\, d\mu_2(q)\, d\mu_1(p)$$

$$= \int_H \int_H f(p)\overline{(L_{q^*}h^*)(p)}\, d\mu_1(p)\, d\mu_2(q)$$

$$= \int_H \int_H (L_q f)(p)\, \overline{h^*(p)}\, d\mu_1(p)\, d\mu_2(q)$$

$$= \int_H h(p)\Big(\int_H (R_p f)(q)\, d\mu_2(q)\Big) d\mu_1(p).$$

Since $h = 0$ outside U, we have

$$\Big|\int h(p)d\mu_1(p)\int f(q)d\mu_2(q) - \int h(p)d\mu_2(p)\int f(q)d\mu_1(q)\Big|$$

$$= \Big|\int h(p)\int (R_p f)(q) - L_p f(q))d\mu_2(p)d\mu_1(q)\Big|$$

$$\le \frac{\varepsilon}{2}\int h(p)\, d\mu_1(p)\int f(q)\, d\mu_1(q).$$

This yields

$$\left|\frac{\int f(q)d\mu_2(q)}{\int f(q)d\mu_1(q)} - \frac{\int h(q)d\mu_2(q)}{\int h(q)d\mu_1(q)}\right| \le \frac{\varepsilon}{2}.$$

The same argument is applicable to the function g. Hence,

$$\left|\frac{\int f(p)d\mu_2(q)}{\int f(q)d\mu_1(q)} - \frac{\int g(p)d\mu_2(p)}{\int g(q)d\mu_1(q)}\right| \le \varepsilon,$$

whence $\mu_2 = C\mu_1$ for some $C \ge 0$.

■

We fix a weakly invariant measure μ and, in the remaining part of this subsection, assume that the generalized translation operators L_p $(p \in H)$ satisfy the conditions of Theorem 4.1. By virtue of Theorems 2.1 and 2.2, $L_1(H, \mu)$ *becomes a normal hypercomplex systems with basis unity with respect to the convolution*

$$f \overset{H}{*} g(p) = (L_p f, g)_{2,\mu}, \quad (f, g \in L_2(H, \mu)).$$

We define the left-invariant measure $\mu * \delta_{p^*} \in M^+(H)$ as follows:

$$\langle \mu * \delta_{p^*}, f \rangle = \int (R_{p^*} f)(q) d\mu(q) \ (f \in C_0(H)).$$

By virtue of Theorem 4.1, there exists a positive number $\Delta(p)$ such that $\mu * \delta_{p^*} = \Delta(p)\mu$. The function $Q \ni p \mapsto \Delta(p)$ is called *modular.* One can show that the modular function is a continuous generalized character of the hypercomplex system $L_1(H, \mu)$. In addition, $\Delta\Delta^* = 1$ and $\mu = \Delta\mu^*$. Note that all assertions in this section hold for $H = Q$. Wherever possible, for the sake of simplicity, we assume that the measure μ is *unimodular,* i.e., $\Delta(p) \equiv 1$ $(p \in H)$. The corresponding hypercomplex subsystem is also called *unimodular.*

4.2. Fundamental Properties of Hypercomplex Subsystems. Let us establish some properties of hypercomplex subsystems.

Assume that the separately continuous generalized translation operators L_p $(p \in Q)$ associated with the hypercomplex system $L_1(Q, m)$ are such that the corresponding convolution of measures satisfies condition (G6) from the definition of hypergroups. Let H be the connected component of the basis unity e of the hypercomplex system $L_1(Q, m)$. Then H is a hypercomplex subsystem.

In fact, since H^* is connected, we have $H^* = H$. It suffices to show that $H * H$ is connected. Let $H * H \subset V \cup W$, where U and W are open subsets of Q such that $(H * H) \cap V$ and $(H * H) \cap W$ have no common points. Assume that $\{e\} * \{e\} = \{e\} \subset (H * H) \cap V$. The sets $P = \{(t, s) \mid \{t\} * \{s\} \in \mathfrak{B}_0(V)\}$ and $S = \{(t, s) \mid \{t\} * \{s\} \in \mathfrak{B}_0(W)\}$ are open in $Q \times Q$ by virtue of (G6). Moreover, $P \cup S$ contains $H \times H$ and the sets $(H \times H) \cap P$ and $(H \times H) \cap S$ have no common points. Consequently, $H \times H \subset P$, since $(e, e) \in P$ and $H \times H$ is connected.

■

In what follows, we denote the set $\{p\} * \{q\}$ $(p, q \in Q)$ simply by $p * q$. Note that the operation of convolution of sets is associative: $A * (B * C) = (A * B) * C$ $(A, B, C \in \mathfrak{B}_0(Q))$. Denote $G = \{r \in Q \mid r * r^* = r^* * r = e\}$. Clearly, $e \in G$ and $G^* = G$.

The set G is a group with respect to the correctly defined operation of multiplication $rs = r * s$ $(r, s \in G)$. If, in addition, the continuity axiom (G6) from the definition of hypergroups holds, then the group G is locally compact.

Indeed, let $r \in G$ and $p \in Q$. We take $q \in r * p$. Then $r^* * q \subset r^* * (r * p) = e * p = p$, i.e., $r^* * q = p$. Multiplying this equality from the left by r, we obtain $q = r * p$. Hence, multiplication is correctly defined in G. Assume that $r, s \in G$. Then $rs(rs)^* = r * s * s^* * r^* = r * r^* = e$, i.e., $G * G \subset G$. The remaining assertions are obvious.

■

The group G is called a *maximal subgroup* (clearly, the case $G = \{e\}$ is not excluded). We have also proved that, for each $r \in G$, the maps $Q \ni p \mapsto r * p \in Q$ and $Q \ni q \mapsto r^* * q$ are inverse to each other.

Theorem 4.2. *Assume that* $L_1(Q, m)$ *is a normal hypercomplex system with compact base and basis unity satisfying the condition of separate continuity. Let* $Q \ni p \mapsto U_p$ *be a bounded representation of the hypercomplex system Denote by* T *a weakly closed subgroup of unitary operators in the Hilbert space* H *of representation. Then* $S = \{r \in Q \mid U_p \in T\}$ *is a hypercomplex subsystem.*

Proof. The closeness of S follows from the weak continuity of the representation $Q \ni p \mapsto U_p$ and from the coincidence of the weak and strong operator topologies. Clearly, $e \in S$ and $S^* = S$. Let us show that $S * S \subset S$. By applying (3.22) and the definition of the convolution of measures, for any $r, s \in S$ and $\xi \in H$, we obtain

$$\begin{aligned}(\xi, \xi)_H &= (U_r U_s \xi, U_r U_s \xi)_H = L_r(U_s \xi, \eta)_H \\ &= \int (U_p \xi, \eta)_H d(\delta_r * \delta_s)(p) = \int (U_p \xi, U_r U_s \xi)_H d(\delta_r * \delta_s)(p),\end{aligned}$$

where $\eta = U_r U_s \xi$. Since $|(U_p \xi, U_r U_s \xi)_H| \leq (\xi, \xi)_H$ and the mapping $\operatorname{supp} \delta_r * \delta_s \ni p \mapsto (U_p \xi, U_r U_s \xi)_H$ is continuous, we have $(\xi, \xi)_H = (U_p \xi, U_q \xi)_H$ $(p, q \in r * s; \xi \in H)$. In particular, for $p = q$, the last equality implies that that the operator U_p is unitary for all $p \in r * s$, whence it follows that $S * S \subset S$.

∎

Corollary 4.1. *Assume that* $G' = \bigcap \{r \in Q \mid U_p$ *is unitary*$\}$ *(the intersection is taken over the set* $\mathfrak{R}$ *of equivalence classes of irreducible representations). Then* G' *is a maximal subgroup of the hypercomplex system* $L_1(Q, m)$.

Proof. Let G be a maximal subgroup. For $r \in G$, we have

$$U_r U_{r^*} = \int U_p d(\delta_r * \delta_{r^*})(p) = U_b = I$$

(the integral is understood in Bochner's sense). Hence, $G \subset G'$. On the other hand, by virtue of Theorem 4.2, G' is a hypercomplex subsystem of $L_1(Q, m)$ and, for any $r \in G'$ and $\xi, \eta \in H$, we have

$$(U_{\delta_r * \delta_{r^*}} \xi, \eta)_H = L_r(U_{r^*} \xi, \eta) = (U_r U_{r^*} \xi, \eta)_H = (\xi, \eta)_H,$$

whence $U_{\delta_r * \delta_{r^*}} = U_e = I$ for all $U \in \mathfrak{R}$. By using the injectivity of the Fourier transformation, we conclude that $\delta_r * \delta_{r^*} = \delta_e$, i.e., $r \in G$.

■

If Q is a hypergroup and H is its subhypergroup, we can introduce the quotient space Q/H. Indeed, by definition, $rH = \{r\} * H$ is a left coset in H. *For any* $r, s \in Q$, *the sets* rH *and* sH *either do not intersect or coincide.* Indeed, let $p \in rH \cap sH$. Obviously, $pH \subset rH$. Since $p \cap rH \neq 0$, by virtue of (G5), we conclude that $e \in p * (r * H)^* = p * (H^* * r^*) = (p * H) * r^*$, whence, by using (G5) once again, $r \in pH$. Therefore, $rH = pH$. The fact that $pH = sH$ can be proved by using the same argument.

■

As shown in [Jew], the quotient space Q/H is a locally compact Hausdorff space and the map $Q \ni rH \in r \mapsto Q/H$ is continuous and open. Moreover, the same work contains the description of the spaces $Q//H$ of dual cosets $H\,rH = H * \{r\} * H$ of the hypergroup Q with respect to the compact subhypergroup H; it is also shown therein that $Q//H$ is a hypergroup with respect to a naturally defined operation. We do not present here these results but, in Section 2 in Chapter 2, we consider the set $G//H$, where G is a locally compact group and H is its compact subgroup.

A hypercomplex subsystem H is called *normal* if $pH = Hp$ for any $p \in Q$, or *supernormal* if $pHp^* \subset H$ $(p \in Q)$. If Q is a hypergroup and H is its normal subhypergroup, then $Q/H = Q//H$ is called a *quotient hypergroup.* In contrast to groups, the notions of normality and supernormality are not equivalent for hypergroups. *If* H *is a supernormal subhypergroup, then* H *is normal.* Indeed, let $p * H * p^* \subset H$ $(p \in Q)$. Then $p * H * p^* * p \subset H * p$. On the other hand, by virtue of (G5), we have $p * H \subset p * H * p^* * p \subset H * p$, i.e., $pH \subset Hp$. By analogy, we get $Hp \subset pH$.

Generally speaking, the converse statement is not true. For example, the normal hypercomplex subsystem $H = \{e\}$ is not necessarily supernormal because, otherwise, $p * p^* = p * e * p^* = e$ for all $p \in Q$, which implies that Q coincides with the maximal subgroup. In [Vre5], it was shown that if H is a supernormal subhypergroup of the hypergroup Q, then Q/H is a group with respect to the induced convolution; furthermore, the set $p * p^*$ is a group for all $p \in Q$.

4.3. Homomorphisms. Let us define a homomorphism of a hypercomplex system Assume that $L_1(Q_1, m_1)$ and $L_2(Q_2, m_2)$ are two normal hypercomplex systems with basis unity satisfying the condition of separate continuity. A continuous mapping $\varphi: Q_1 \to Q_2$ is called a *homomorphism* of $L_1(Q_2, m_2)$ into $L_1(Q_1, m_1)$ if

1) $L_p(f \circ \varphi)(q) = L_{\varphi(p)} f(\varphi(q))$ for any $p, q \in Q_1$ and $f \in C(Q_2)$;

2) $\varphi(p^*) = (\varphi(p))^*$ $(p \in Q_1)$.

If, in addition, the map φ is a homeomorphism of Q_1 onto Q_2, then the hypercomplex systems $L_1(Q_1, m_1)$ and $L_1(Q_2, m_2)$ are called *isomorphic.* In what follows, we prove that the measure m_1 coincides with m_2 to within a constant.

It is more convenient to consider homomorphisms φ with certain topological properties. We say that the complements of compact sets in the locally compact space Q are *neighborhoods of the infinite point* in Q. We say that a continuous map $\varphi: Q_1 \to Q_2$ is *continuous at infinity* if, for any neighborhood U_∞ of the infinite point in Q_2, there is a neighborhood V_∞ of the infinite point in Q_1 such that $\varphi(V_\infty) \subset U_\infty$. Clearly, if the continuous map $\varphi: Q_1 \to Q_2$ is continuous at infinity, then the preimage of any compact set in Q_2 is compact in Q_1. In what follows, the map φ which realizes the homomorphism of $L_1(Q_2, m_2)$ into $L_1(Q_1, m_1)$ is assumed to be continuous at infinity.

Consider a linear operator $C_\infty(Q_2) \ni f \mapsto Ff = f \circ \varphi \in C_\infty(Q_1)$. It is obvious that $\|F\| \le 1$ and $F(C_0(Q_2)) \subset C_0(Q_1)$. If φ maps Q_1 onto Q_2, then $\operatorname{Ker} F = \{0\}$. Let F^* be the operator adjoint to F. Clearly, F^* transforms $M_b(Q_1)$ into $M_b(Q_2)$, with $\|F^*\| \le 1$ and $F^*(\delta_n) = \delta_{\varphi(p)}$. One can show that F^* *can be extended to the linear operator acting from* $M(Q_1)$ *into* $M(Q_2)$. In fact, for any $\mu \in M(Q_1)$, we can set $\langle F^*\mu, f\rangle = \langle \mu, Ff\rangle$ $(f \in C_0(Q_2))$. In order to check that $F^*\mu$ is a Radon measure, we must show that the inequality $|\langle F^*\mu, f\rangle| \le C_K \|f\|_{\infty, Q_2}$ $(f \in C_0(K))$ holds for any compact set $K \subset Q_2$. Indeed, we have

$$|\langle F^*\mu, f\rangle| = \langle \mu, Ff\rangle \le C_{\varphi^{-1}(K)} \|Ff\|_{\infty, Q_1} \le C_{\varphi^{-1}(K)} \|f\|_{\infty, Q_2}$$

for any $f \in C_0(K)$ because $\varphi^{-1}(K)$ is a compact set (since φ is continuous at infinity), μ is a Radon measure on Q_1, and $\|F\| \le 1$.

■

The equality

$$\operatorname{supp}(F^*\mu) = \varphi(\operatorname{supp}\mu) \tag{4.1}$$

holds for all $\mu \in M_b(Q_1)$.

Indeed, assume that there is $p \in \operatorname{supp}\mu$ such that $\varphi(p) \notin \operatorname{supp}(F^*\mu)$. There exists a nonnegative function $f \in C_\infty(Q_2)$ equal to zero on $\operatorname{supp} F^*\mu$ and such that $f(\varphi(p)) = 1$. We have

$$0 = \langle F^*\mu, f\rangle = \int f(\varphi(p))\, d\mu(p).$$

It is clear that the last integral is strictly positive. Hence, $\varphi(\operatorname{supp}\mu) \subset \operatorname{supp} F^*\mu$. The inverse inclusion $\operatorname{supp}(F^*\mu) \subset \varphi(\operatorname{supp}\mu)$ is also obvious.

■

Lemma 4.1. *Assume that the homomorphism* φ *maps* Q_1 *onto* Q_2 *and the hypercomplex system* $L_1(Q_2, m_2)$ *satisfies the conditions of Theorem 4.1, i.e., for any neighborhood* W *of the point* $s \in Q_2$, *one can indicate a neighborhood* U *of the basis unity such that* $\operatorname{supp} L_{s^*} g(\cdot) \subset W$ *for each* $g \in C_0(Q_2)$ *with* $\operatorname{supp} g \subset U$. *Then there exists a constant* $C > 0$ *such that* $F^* m_1 = C m_2$.

Proof. For any $f, g \in C_0(Q_2)$, we have

$$\begin{aligned}
\langle F^* m_1, (L_{\varphi(p)} f) \cdot \bar{g} \rangle &= \int (L_{\varphi(p)} f)(\varphi(q)) \overline{g(\varphi(q))}\, dm_1(q) \\
&= \int (L_p(f \circ \varphi)(q)) \overline{(g \circ \varphi)(q)}\, dm_1(q) \\
&= \int (f \circ \varphi)(q) \overline{(L_{p^*}(g \circ \varphi))(q)}\, dm_1(q) \\
&= \int f(\varphi(q)) \overline{(L_{\varphi(p^*)} g)(\varphi(q))}\, dm_1(q) \\
&= \langle F^* m_1, f \cdot L_{(\varphi(p))^*}\, \bar{g} \rangle.
\end{aligned}$$

Hence, the measure $F^* m_1$ is strongly invariant. By Theorem 2.1, the generalized translation operators L_q $(q \in Q_2)$ satisfy the finiteness condition (F). Therefore, the measure $F^* m_1$ is invariant, i.e., equality (2.3) holds. It remains to apply Theorem 4.1. ∎

Without loss of generality, we may assume that $C = 1$ (otherwise, we must consider $L_1(Q_2, Cm_2)$ instead of $L_1(Q_2, m_2)$).

Theorem 4.3. *Assume that the homomorphism* φ *maps* Q_1 *onto* Q_2. *The operator* F *can be extended to an isometric* $*$*-homomorphism from the Banach algebra* $L_1(Q_2, m_2)$ *to* $L_1(Q_1, m_1)$, *where* $m_2 = F^* m_1$.

Proof. Since, by Lemma 4.1, the measure m_2 is invariant, we have

$$\begin{aligned}
(F(f * g))(p) &= \int_{Q_2} (L_{\varphi(p)} f)(s) g(s^*)\, dm_2(s) \\
&= \langle F^* m_1, (L_{\varphi(p)} f) \cdot \bar{g}^* \rangle = \int_{Q_1} (L_{\varphi(p)} f)(\varphi(q)) g(\varphi(q^*))\, dm_1(q) \\
&= \int (L_p(f \circ \varphi)(q))(g \circ \varphi)(q^*)\, dm_1(q) = (Ff * Fg)(p)
\end{aligned}$$

for any $f, g \in C_0(Q_2)$ and $p \in Q_1$. It is obvious that $\|Ff\|_{L_1(Q_1, m_1)} = \|f\|_{L_1(Q_2, m_2)}$. Consequently, the operator F can be extended by continuity to a homomorphism of $L_1(Q_2, m_2)$ into $L_1(Q_1, m_1)$. Finally, by virtue of the equality

$$F(f^*)(p) = f^*(\varphi(p)) = \overline{f(\varphi(p^*))} = (Ff)^*(p),$$

the operator F is involutory.

■

If the homomorphism φ *maps* Q_1 *onto* Q_2, *then* $\varphi(e_1) = e_2$, where e_i is the basis unity of the hypercomplex system $L_1(Q_i, m_i)$, $i = 1, 2$. Indeed,

$$(L_{\varphi(e_1)}f)(\varphi(q)) = (L_{e_1}(f \circ \varphi))(q) = f(\varphi(q)) \quad (q \in Q_1, f \in C(Q_2)).$$

As q runs over Q_1, $\varphi(q)$ runs over the entire Q_2. Hence, $L_{\varphi(e_1)} = I$ and $\varphi(e_1) = e_2$ by virtue of the fact that the generalized translation operators L_p $(p \in Q_2)$ are reduced.

■

Suppose that $\varphi(Q_1) = Q_2$.

Let $H_1 \subset Q_1$ *be a compact hypercomplex subsystem of* $L_1(Q_1, m_1)$. *Then* $\varphi(H_1) = H_2$ *is a compact hypercomplex subsystem of* $L_1(Q_2, m_2)$.

Indeed, since $e_1 \in H_1$, we have $\varphi(e_1) = e_2 \in H_2$. Moreover, it is clear that $H_2^* = H_2$. Let $f \in C(Q_2)$ and $f(r) = 0$ for all $r \in H_2$. Then

$$(L_{\varphi(p)}f)(\varphi(q)) = (L_p(f \circ \varphi))(q) = 0 \quad (p, q \in H_1),$$

since H_1 is a hypercomplex subsystem and $f \circ \varphi(p) = 0$ for all $p \in H_1$. Consequently, H_2 is a hypercomplex subsystem.

■

Note that this assertion is not proved for general hypercomplex subsystems because $\varphi(H_1)$ is, generally speaking, not closed. Therefore, in what follows, we suppose that the mapping φ is closed.

As usual, the kernel $\operatorname{Ker}\varphi$ of the homomorphism φ is defined as the set of points $p \in Q_1$ such that $\varphi(p) = e_2$.

Theorem 4.4. *Assume that separately continuous generalized translation operators* L_p $(p \in Q_1)$ *and* L_q $(q \in Q_2)$ *associated with the hypercomplex systems* $L_1(Q_1, m_1)$ *and* $L_1(Q_2, m_2)$, *respectively, are such that the corresponding convolution of* *measures satisfies axiom (G5) from the definition of hypergroups. Then the kernel of the homomorphism* φ *is a normal hypercomplex subsystem of* $L_1(Q_1, m_1)$.

Proof. We denote $H = \operatorname{Ker}\varphi$ and show that H is a hypercomplex subsystem. Clearly, H is closed and $e_1 \in H$. Since $\varphi(p^*) = (\varphi(p))^* = e_2^* = e_2$ for each $p \in H$, we have $H^* = H$. Let $p, q \in H$. Since

$$\langle F^*(\delta_p * \delta_q), f\rangle = \langle \delta_p * \delta_q, f \circ \varphi\rangle$$

$$= (L_p(f\circ\varphi))(q) = (L_{\varphi(p)}f)(\varphi(q)) = \langle \delta_{\varphi(p)} * \delta_{\varphi(q)}, f\rangle$$

for each $f \in C_\infty(Q_2)$, we have $F^*(\delta_p * \delta_q) = \delta_{\varphi(p)^*} * \delta_{\varphi(q)} = \delta_{e_2} * \delta_{e_2} = \delta_{e_2}$. In view of (4.1), this implies that $\varphi(p * q) = e_2$ and $H * H \subset H$.

Let us show that the hypercomplex subsystem H is normal. To do this it suffices to show that

$$rH = \varphi^{-1}(\varphi(r)) = Hr \tag{4.2}$$

for each $r \in Q$. Assume that there is a point $s \in \varphi^{-1}(\varphi(r))$ such that $s \notin rH$. By virtue of what has been proved in the previous subsection, $sH \cap rH = \emptyset$. In view of the fact that $\varphi(s^*) = (\varphi(r))^*$ and axiom (G5), we have $e_2 \in \varphi(s^*) * \varphi(r) = \varphi(s^* * r)$. This implies that $s^* * r \cap H \neq \emptyset$. By virtue of (G5) and the associativity of the convolution of sets established in the previous section, we obtain $e_1 \in (s^* * r) * H^* = (s^* * r) * H = s^* * (r * H)$. In view of (G5), this yields $s \cap rH = \emptyset$. The first equality in (4.2) is proved. The second equality is proved similarly.

■

Let separately continuous generalized translation operators L_p $(p \in Q)$ *associated with the hypercomplex system* $L_1(Q, m)$ *be such that the corresponding convolution of sets satisfies axiom (G6) from the definition of hypergroups. Then, for each* r *from the maximal subgroup* G, *the mapping* $Q \ni p \mapsto \varphi_r(p) = r * p * r^* \in Q$ *is an automorphism of the hypercomplex system* $L_1(Q, m)$.

In fact, in the previous subsection, it was shown that, for each $r \in G$, the mappings $Q \ni p \mapsto rp \in Q$ and $Q \ni p \mapsto r^*p \in Q$ continuous by virtue of (G6) are correctly defined and, moreover, each of these mappings is inverse to the other. Hence, the map φ_r is a homeomorphism of Q onto Q. Further, $(\varphi_r(p))^* = (r * p * r^*)^* = \varphi_r(p^*)$ for each $p \in Q$. One can easily check that $F_r^*\mu = \delta_r * \mu * \delta_{r^*}$, whence

$$(L_{\varphi_r(p)}f)(\varphi_r(q)) = \langle \delta_{\varphi_r(p)} * \delta_{\varphi(q)}, f\rangle = \langle \delta_r * \delta_p * \delta_q * \delta_{r^*}, f\rangle$$

$$= \langle F_r^*(\delta_{\varphi_r(p)} * \delta_{\varphi(q)}), f\rangle = (L_p(f \circ \varphi_r))(q).$$

■

4.4. Direct and Semidirect Products of Hypercomplex Systems. Join of Hypercomplex Systems

1. *Direct product.* Let L_{p_i} $(i = 1, 2)$ be generalized translation operators associated with normal hypercomplex systems with basis unity $L_1(Q_1, m_1)$ and $L_2(Q_2, m_2)$, respectively. We denote $H_1 = L_2(Q_1, m_1)$ and $H_2 = L_2(Q_2, m_2)$. The *direct product* of generalized translation operators L_{p_1} and L_{p_2} $(p_1 \in Q, p_2 \in Q)$ is defined as the operator-valued function

$$Q = Q_1 \times Q_2 \ni (p_1, p_2) \mapsto L_p = L_{p_1} \otimes L_{p_2} : H_1 \otimes H_2 \to H_1 \otimes H_2.$$

It is easy to see that the operators L_p $(p \in Q)$ form in $L_2(Q, m)$ $(m = m_1 \otimes m_2)$ a family of generalized translation operators satisfying the conditions of Theorem 2.1. The hypercomplex system $L_1(Q, m)$ constructed from the generalized translation operators L_p is called the *direct product* of the hypercomplex systems $L_1(Q_1, m_1)$ and $L_1(Q_2, m_2)$. To denote the operation of taking the direct product, we write $L_1(Q, m) = L_1(Q_1, m_1) \otimes L_1(Q_2, m_2)$. For example, if L_{p_i} $(p_i \in [0, \infty))$ are generalized translation operators associated with the hypercomplex system in Example 2 of Subsection 1.1, then, clearly,

$$(L_pf)(q) = (L_{(p_1, p_2)}f)(q_1, q_2)$$

$$= \frac{1}{4}[f(p_1 + q_1, p_2 + q_2) + f(p_1 + q_1, |p_2 - q_2|)$$

$$+ f(|p_1 - q_1|, p_2 + q_2)] + f(|p_1 - q_1|, |p_2 - q_2|)]$$

$$(p_i, q_i \in [0, \infty), \; i = 1, 2, \; f \in C([0, \infty) \times [0, \infty)).$$

2. *Semidirect product.* One can also introduce the concept of *semidirect product of a hypercomplex system and a group*. First, we recall the concept of semidirect product of groups. Let G be a locally compact group and let $\operatorname{Aut} G$ be the group of all automorphisms of the group G equipped with standard topology; the base of neighborhoods of the identity in this topology is formed by the sets

$$\mathfrak{B}(U, F) = \{\beta \in \operatorname{Aut} G \mid \beta(g) \in Ug, \beta^{-1}(g) \in Ug \text{ for all } g \in F\},$$

where U is a neighborhood of the identity in G, and F is a compact subset of G (see, e.g., [HeR]). Let G_A be a subgroup of $\operatorname{Aut} G$. The *semidirect product* $G \rtimes G_A$ of groups G and G_A is the set of ordered pairs (g, β) $(g \in G, \beta \in G_A)$ with group multiplication $(g, \beta)(g', \beta') = (g\beta(g'), \beta\beta')$. The element (e, id) is the identity in $G \otimes G_A$, and the pair $(g, \beta)^{-1} = (\beta^{-1}(g^{-1}), \beta^{-1})$ is the inverse of the element (g, β). The topology in $G \rtimes G_A$ is defined as the product of topologies of the spaces G and G_A. The most common situation is as follows: Let G be a normal subgroup of a group L and let a subgroup H of L be such that $GH = L$ and $G \bigcap H = \{e\}$. Then $L = G \rtimes H$ because

$$x_1 x_2 = g_1 h_1 g_2 h_2 = g_1 h_1 g_2 h_1^{-1} h_1 h_2 = g_1 p_{h_1}(g_2) h_1 h_2$$

$$(x_1, x_2 \in L, \quad g_1, g_2 \in G, \quad h_1, h_2 \in H),$$

where p_{h_1} is the automorphism of the group G generated by the internal automorphism $x \mapsto p_{h_1}(x_1)$ of L.

Let $L_1(Q, m)$ be a normal hypercomplex system with basis unity that satisfies the condition of separate continuity and let G be some topological subgroup of the group of automorphisms of $L_1(Q, m)$. Let us define generalized translation operators corresponding to the semidirect product of $L_1(Q, m)$ and G. Assume that the mapping $Q \times G \ni (p, \varphi) \mapsto \varphi(p) \in Q$ is continuous. For each $f \in C(Q \times G)$, we set

$$(R_{(p', \varphi')} f)(p, \varphi) = \left(R^p_{\varphi(p')} f\right)(p, \varphi\varphi') \quad (p, p' \in Q, \quad \varphi, \varphi' \in G).$$

Let us show that the operators $R_{(p, \varphi)}$ $(p \in Q; \varphi \in G)$ *form a family of generalized translation operators.* Indeed, let $p, p', p'' \in Q$ and $\varphi, \varphi', \varphi'' \in G$. Then

$$\left(R^{(p', \varphi')}_{(p'', \varphi'')}(R_{(p', \varphi')} f)\right)(p, \varphi) = R^{(p', \varphi')}_{(p'', \varphi'')}\left[R^p_{\varphi(p')} f(p, \varphi\varphi')\right]$$

$$= \left(R^{p'}_{\varphi'(p'')}(R^p_{\varphi(p')} f)\right)(p, \varphi\varphi'\varphi'').$$

Denote $h(p') = \left(R^p_{p'} f\right)(p, \varphi\varphi'\varphi'')$. Then

$$\left(R^{p'}_{\varphi'(p'')}(R^{p}_{\varphi(p')}f)\right)(p,\ \varphi\varphi'\varphi'')$$

$$= \left(R_{\varphi'(p'')}(h\circ\varphi)\right)(p') = \left(R_{\varphi\varphi'(p'')}h\right)(\varphi(p'))$$

$$= \left(R^{\varphi(p')}_{\varphi\varphi'(p'')}(R^{p}_{\varphi(p')}f)\right)(p,\ \varphi\varphi'\varphi'') = \left(R^{p}_{\varphi(p')}(R^{p}_{\varphi\varphi'(p'')}f)\right)(p,\ \varphi\varphi'\varphi'').$$

On the other hand,

$$\left(R^{(p,\varphi)}_{(p',\varphi')}(R^{(p,\varphi)}_{(p'',\varphi'')}f)\right)(p,\ \varphi) = R^{(p,\varphi)}_{(\varphi',p')}\left[R^{p}_{\varphi(p'')}f(p,\varphi\varphi')\right]$$

$$= \left(R^{p}_{\varphi(p')}(R^{p}_{\varphi\varphi'(p'')}f)\right)(p,\ \varphi\varphi'\varphi'').$$

The validity of other axioms is obvious.

■

One can easily construct a hypercomplex system corresponding to generalized translation operators $R_{(p,\,\varphi)}$ $(p \in Q, \varphi \in G)$; in this case, one should take the mapping

$$Q\times G \ni (p,\varphi) \mapsto (p,\varphi)^{*} = (\varphi^{-1}(p^{*}), \varphi^{-1}) \in Q\times G$$

as an involution in the new hypercomplex system, and the tensor product of the right invariant measures on Q and G as the new right-invariant measure.

The semidirect product of a hypercomplex system and a group possesses standard properties of semidirect products of groups. Let us illustrate this taking the semidirect product of a three-dimensional hypercomplex system and the group $\mathbb{Z}_2$ as an example. In the discrete case, the concepts of hypercomplex system and hypergroup are equivalent and, therefore, it is convenient to define generalized translation operators in terms of the convolution of Dirac measures. Let $Q_3 = \{e, a, b\}$ be a commutative hypergroup with the involution $a^* = b$ and the multiplication

$$\delta_a * \delta_a = \frac{1}{3}\delta_a + \frac{2}{3}\delta_b, \quad \delta_b * \delta_b = \frac{2}{3}\delta_a + \frac{1}{3}\delta_b, \quad \delta_a * \delta_b = \frac{1}{3}\delta_e + \frac{1}{3}\delta_a + \frac{1}{3}\delta_b.$$

Let the action of the group $\mathbb{Z}_2 = \{1, \gamma\}$ on Q_3 be defined by $\gamma(r) = r^*$ $(r \in Q_3)$. Clearly, γ is an automorphism of Q_3. The semidirect product of Q_3 and $\mathbb{Z}_2$ is the hypergroup $Q = Q_3 \rtimes \mathbb{Z}_2 = \{\bar{e}, \bar{a}, \bar{b}, \bar{c}, \bar{d}, \bar{f}\}$, where $\bar{e} = (e, 1)$, $\bar{a} = (a, 1)$, $\bar{b} = (b, 1)$, $\bar{c} = (e, \gamma)$, $\bar{d} = (a, \gamma)$, and $\bar{f} = (f, \gamma)$, with the involution $\bar{a}^* = \bar{b}$, $\bar{c}^* = \bar{c}$, $\bar{d}^* = \bar{d}$, $\bar{f}^* = \bar{f}$ and the multiplication law given by the following Cayley table:

	$\delta_{\bar e}$	$\delta_{\bar a}$	$\delta_{\bar b}$	$\delta_{\bar c}$	$\delta_{\bar d}$	$\delta_{\bar f}$
$\delta_{\bar e}$	$\delta_{\bar e}$	$\delta_{\bar a}$	$\delta_{\bar b}$	$\delta_{\bar c}$	$\delta_{\bar d}$	$\delta_{\bar f}$
$\delta_{\bar a}$	$\delta_{\bar a}$	$\frac{1}{3}\delta_{\bar a}+\frac{2}{3}\delta_{\bar b}$	$\frac{1}{3}\delta_{\bar e}+\frac{1}{3}\delta_{\bar a}+\frac{1}{3}\delta_{\bar b}$	$\delta_{\bar d}$	$\frac{1}{3}\delta_{\bar d}+\frac{2}{3}\delta_{\bar f}$	$\frac{1}{3}\delta_{\bar c}+\frac{1}{3}\delta_{\bar d}+\frac{1}{3}\delta_{\bar f}$
$\delta_{\bar b}$	$\delta_{\bar b}$	$\frac{1}{3}\delta_{\bar e}+\frac{1}{3}\delta_{\bar a}+\frac{1}{3}\delta_{\bar b}$	$\frac{2}{3}\delta_{\bar a}+\frac{1}{3}\delta_{\bar b}$	$\delta_{\bar f}$	$\frac{1}{3}\delta_{\bar c}+\frac{1}{3}\delta_{\bar d}+\frac{1}{3}\delta_{\bar f}$	$\frac{2}{3}\delta_{\bar d}+\frac{1}{3}\delta_{\bar f}$
$\delta_{\bar c}$	$\delta_{\bar c}$	$\delta_{\bar f}$	$\delta_{\bar d}$	$\delta_{\bar e}$	$\delta_{\bar b}$	$\delta_{\bar a}$
$\delta_{\bar d}$	$\delta_{\bar d}$	$\frac{1}{3}\delta_{\bar c}+\frac{1}{3}\delta_{\bar d}+\frac{1}{3}\delta_{\bar f}$	$\frac{1}{3}\delta_{\bar d}+\frac{2}{3}\delta_{\bar f}$	$\delta_{\bar a}$	$\frac{1}{3}\delta_{\bar e}+\frac{1}{3}\delta_{\bar a}+\frac{1}{3}\delta_{\bar b}$	$\frac{1}{3}\delta_{\bar a}+\frac{2}{3}\delta_{\bar b}$
$\delta_{\bar f}$	$\delta_{\bar f}$	$\frac{2}{3}\delta_{\bar d}+\frac{1}{3}\delta_{\bar f}$	$\frac{1}{3}\delta_{\bar c}+\frac{1}{3}\delta_{\bar d}+\frac{1}{3}\delta_{\bar f}$	$\delta_{\bar b}$	$\frac{2}{3}\delta_{\bar a}+\frac{1}{3}\delta_{\bar b}$	$\frac{1}{3}\delta_{\bar e}+\frac{1}{3}\delta_{\bar a}+\frac{1}{3}\delta_{\bar b}$

As expected, the hypergroup Q is noncommutative. Moreover, the hypergroup Q_3 and group $\mathbb{Z}_2$ are isomorphic to the subhypergroup $H = \{\bar e, \bar a, \bar b\}$ and group $\mathbb{Z}_2 = \{\bar e, \bar c\}$, respectively. Note that $\tilde{\mathbb{Z}}_2$ is the maximal subgroup of the hypergroup Q. Since $\bar c * H * \bar c = \{\bar c, \bar d, \bar f\} * \bar c = H$, $\bar d * H * \bar d = \{\bar c, \bar d, \bar f\} * \bar d = H$, and $\bar f * H * \bar f = \{\bar c, \bar d, \bar f\} * \bar f = H$, the subhypergroup H is supernormal and the quotient group Q/H is isomorphic to $\mathbb{Z}_2$.

3. *Join.* Finally, we define the join of two hypergroups (this construction was proposed by Jewett [Jew]). Let J be a compact hypergroup and let K be a discrete hypergroup. We identify the identities of the hypergroups J and K so that $J \cap K = \{e\}$, where e is the identity for both hypergroups. The *join* of the hypergroups J and K is defined as the hypergroup $J \vee K = J \cup K$ with the topology in which J and K are closed subspaces (note that such a topology is unique) and with the operation $\bullet$ defined by the following equalities:

(i) $\delta_p \bullet \delta_q = \delta_p * \delta_q$ $(p, q \in J)$;

(ii) $\delta_p \bullet \delta_q = \delta_p * \delta_q$ $(p, q \in K;\ p \neq q^*)$;

(ii) $\delta_p \bullet \delta_q = \delta_q \bullet \delta_p$ $(p \in J, q \in K;\ q \neq e)$;

(iv) if $p \in K$, $p \neq e$,

$$\delta_p * \delta_{p^*} = \sum_{r \in K} c_r \delta_r,$$

and σ is a normalized $(\sigma(J) = 1)$ invariant measure on J, then

$$\delta_p \bullet \delta_{p^*} = c_e \sigma + \sum_{r \in K} c_r \delta_r.$$

The validity of the axioms of hypergroups is obvious. Note also that J is a compact subhypergroup of $J \vee K$, but K is a subhypergroup if and only if either J or K is equal to $\{e\}$.

Let us give an example of the join of countably many discrete hypergroups. Let b_n^* $(n \geq 1)$ be a sequence of numbers satisfying the inequalities $0 < b_n \leq 1$. We set $c_0 = 1$ and define c_n $(n \geq 1)$ by the recurrence relations $c_n = (c_0 + c_1 + \ldots + c_{n-1}) / b_n$. For each $n \geq 1$, we consider the hypergroup $K_n = \{0, n\}$ with the identity $e = 0$ and multiplication $\delta_n * \delta_n = b_n \delta_0 + (1 - b_n) \delta_0$. Consider the infinite join $K = K_1 \vee K_2 \vee \ldots =$ $\{0, 1, 2, \ldots\}$. Multiplication in K is defined as follows:

$$\delta_m \bullet \delta_n = \delta_n \bullet \delta_m = \delta_n \quad (0 \leq m < n),$$

and

$$\delta_n \bullet \delta_n = \frac{1}{c_n}(c_0 \delta_0 + c_1 \delta_1 + \ldots + c_{n-1} \delta_{n-1}) + (1 - b_n) \delta_n.$$

The join K is an Hermitian hypergroup with the identity 0 and the invariant measure

$$m = \sum_{k=0}^{\infty} c_k \delta_k.$$

Clearly, each set $H_n = \{0, 1, \ldots, n\}$ is a subhypergroup of K. One can easily establish that the characters of the hypergroup K have the following form: $\chi_0(n) = 1$ $(n \in K)$, $\chi_k(n) = 1$ for $0 \leq n < k - 1$, and $\chi_k(k) = -b_k$ and $\chi_k(n) = 0$ for $n > k$. The set of characters $\mathrm{X} = \mathrm{X}_h$ is countable, but the topology of the space of maximal ideals transforms X into a compact set, e.g., $\chi_k \to \chi_0$, $k \to \infty$ in the topology of X.

Note that, generally speaking, the union of two groups is not a group, but a hypergroup (to verify this, it suffices to consider the case where $J = SO(2)$ and $K = \mathbb{Z}_2$).

5. Further Generalizations of Hypercomplex Systems

In the previous sections, we studied hypercomplex systems with nonnegative structure measure. For hypercomplex systems of this sort, we proved numerous results from harmonic analysis including an analog of Pontryagin's duality for commutative hypercomplex systems. If we omit the condition of commutativity of a hypercomplex system, then we encounter certain difficulties connected with definition and investigation of the dual object. For noncommutative locally compact groups, problems of this sort are solved by using specially introduced objects such as Kac algebras (originally, ring groups) or, equivalently, by using Hilbert bialgebras (see Subsection 1.7) with certain relations between the operations in them (see [Kac]). If one operation in a Hilbert bialgebra is commutative, then this bialgebra is generated by a locally compact group. Here, the main role is played by the fact that comultiplication is a homomorphism, i.e., the Kac algebra is a Hopf algebra or, in the language of Hilbert bialgebras, the binding operator W is unitary and satisfies the equality

$$(W\otimes I)(I\otimes W)(W^{*}\otimes I) = (I\otimes W)(\sim\otimes I)(I\otimes W)(\sim\otimes I),$$

where I is the identity operator in H and $\sim(a\otimes b)=b\otimes a$ $(a, b\in H)$.

In Subsections 5.1 and 5.2, we study a class of bialgebras related to unimodular hypercomplex systems in a similar way. These bialgebras are called quantized hypercomplex systems. In this case, the most important thing is that, generally speaking, the operation of comultiplication (generalized translation) is no longer a homomorphism for hypercomplex systems. The bialgebras associated with unimodular hypercomplex systems were studied in [Vai8]. This case is more complicated and requires special technique for its investigation; therefore, we do not consider it here.

As can be shown by examples, the generalized translation operators associated with differential operators and orthogonal polynomials do not always preserve positivity and finiteness. Hence, it is reasonable to extend the concept of hypercomplex systems to the case of hypercomplex systems with alternating structure measure. A generalization of this sort have already been presented in Subsection 2.3. However, for hypercomplex systems considered in Subsection 2.3, one fails to construct the duality theory because the dual object of a hypercomplex system of this kind no longer satisfies the axioms of Subsection 2.3. Therefore, in Subsection 5.3, we consider hypercomplex systems with real structure measure satisfying the axioms of quantized hypercomplex systems. Fundamental facts from harmonic analysis and duality theory remain true for such hypercomplex systems. At the same time, one can construct natural examples for which the axioms of quantized hypercomplex systems are not satisfied (see Subsection 3.5 in Chapter 2). Therefore, in 5.4, we describe another class of compact or discrete hypercomplex systems with real structure measure (quite good from the viewpoint of harmonic analysis and duality theory) which was introduced by Vainerman [Vai2], [Vai3].

5.1. Properties of Hilbert Bialgebras. Assume that $\mathfrak{D}$ is a Hilbert bialgebra in H with respect to the operations $a \bigcup b$ and $a \bigcap b$ $(a, b \in \mathfrak{D})$ and the corresponding involutions $a^{\cup}$, $a^{\cap}$ $(a \in \mathfrak{D})$. Let $\mathfrak{L}_{\cup}$ and $\mathfrak{L}_{\cap}$ be the left W^*-algebras of the bialgebra $\mathfrak{D}$, i.e., they are the weak closures of the $*$-algebras of the operators $\mathfrak{D} \ni b \mapsto L_{\cup}(a)b = a \bigcup b$ and $\mathfrak{D} \ni b \mapsto L_{\cap}(a)b = a \bigcap b$ $(a \in \mathfrak{D})$, respectively. Denote by $\mathfrak{L}_{\cup *}$ and $\mathfrak{L}_{\cap *}$, the corresponding preduals, i.e., the Banach spaces $\mathfrak{L}_{\cup *}$ and $\mathfrak{L}_{\cap *}$ such that $\mathfrak{L}_{\cup}$ and $\mathfrak{L}_{\cap}$ are dual to $\mathfrak{L}_{\cup *}$ and $\mathfrak{L}_{\cap *}$, respectively. Each element ω of $\mathfrak{L}_{\cup *}$ $(\mathfrak{L}_{\cap *})$ admits the representation

$$\langle \cdot, \omega \rangle = \sum_{k=1}^{n} \langle \cdot\, u_k, v_k \rangle_H,$$

where $u_k, v_k \in H$. Hence, it can be continued to a weakly continuous form on the W^*-algebra of all bounded operators in H.

Let $\mathfrak{D} \bigcup \mathfrak{D}$ $(\mathfrak{D} \bigcap \mathfrak{D})$ be the linear span of all vectors of the form $a \bigcup b$ $(a \bigcap b)$, where $a, b \in \mathfrak{D}$. By $E_{\cup}$ and $E_{\cap}$ we denote the Banach spaces which are the completions of $\mathfrak{D} \bigcap \mathfrak{D}$ and $\mathfrak{D} \bigcup \mathfrak{D}$ with respect to the norms $\|\cdot\|_{\cup} = \|L_{\cup}(\cdot)\|$ and $\|\cdot\|_{\cap} = \|L_{\cap}(\cdot)\|$, respectively.

The inequalities

$$\|a \bigcup b\|_{\cap} = \|L_{\cap}(a \bigcup b)\| \le \|W\| \, \|a\|_H \|b\|_H \tag{5.1}$$

and

$$\|a \bigcap b\|_{\cup} = \|L_{\cup}(a \bigcap b)\| \le \|W\| \, \|a\|_H \|b\|_H \tag{5.2}$$

hold for any $a, b \in \mathfrak{D}$.

In fact, we have

$$\begin{aligned} \|L_{\cap}(a \bigcup b)c\|_H^2 &= (a \bigcup b, (a \bigcup b) \bigcap c \bigcap c^{\cap}) \\ &= (a \otimes c, W(b^{\cup} \otimes (a \bigcup b) \bigcap c)) \\ &\le \|W\| \, \|a \otimes c\|_{H \otimes H} \|b^{\cup} \otimes (a \bigcup b) \bigcap c\|_{H \otimes H} \end{aligned}$$

$$= \|W\| \|a\|_H \|b^{\cup}\|_H \|c\|_H \|L_{\cap}(a \cup b) c\|_H$$

for any $a, b, c \in \mathfrak{D}$. This implies that

$$\|L_{\cap}(a \cup b) c\|_H \leq \|W\| \|a\|_H \|b\|_H \|c\|_H$$

which is equivalent to (5.1). Inequality (5.2) is proved similarly.

■

In view of (5.1) and (5.2), the products $u \cup v$ and $u \cap v$ can be defined by continuity for all $u, v \in H$. They lie in $E_{\cap}$ and $E_{\cup}$, respectively. Clearly, $E_{\cap}$ ($E_{\cup}$) is isometrically isomorphic to the closure of the linear span $\{L_{\cap}(a \cup b) \mid a, b \in \mathfrak{D}\}$ ($\{L_{\cup}(a \cap b) \mid a, b \in \mathfrak{D}\}$) in the uniform norm. We denote this closure by $L_{\cap}(\mathfrak{D} \cup \mathfrak{D})$ ($L_{\cup}(\mathfrak{D} \cap \mathfrak{D})$), and the corresponding isomorphism as $\pi_{\cap}: E_{\cap} \to L_{\cap}(\mathfrak{D} \cup \mathfrak{D})$ ($\pi_{\cup}: E_{\cup} \to L_{\cup}(\mathfrak{D} \cap \mathfrak{D})$). We also denote $L_{\cup}(u \cap v) = \pi_{\cup}^{-1}(u \cap v)$ and $L_{\cap}(u \cup v) = \pi_{\cap}^{-1}(u \cup v)$ $(u, v \in H)$.

Any functional $\omega \in \mathfrak{L}_{\cap *}$ is associated with an element of the Banach space $E_{\cup}$ by the formula

$$\mathfrak{L}_{\cap *} \ni \omega(\cdot) = \sum_{k=1}^{n} (\cdot\, u_k, v_k) \mapsto \rho_{\cup}(\omega) = \sum_{k=1}^{n} u_k \cap v_k^{\cap} \in E_{\cup}.$$

The map $\rho_{\cap}$ (clearly, linear) is defined in a similar way. Let us show that the *maps* $\rho_{\cup}$ *and* $\rho_{\cap}$ *are defined correctly.* In other words, if

$$\sum_{k=1}^{n} (A u_k, v_k)_H = \omega(A) = \sum_{k=1}^{n} (A u_k', v_k')_H,$$

then

$$\sum_{k=1}^{n} u_k \cap v_k^{\cap} = \sum_{k=1}^{n} u_k' \cap v_k'^{\cap}.$$

First, we prove the identity

$$(L_{\cup}(u \cap v) u', v')_H = (u, L_{\cap}(v' \cup u'^{\cup}) v^{\cap})_H \quad (u, v, u', v' \in H). \tag{5.3}$$

Let v_n and u_n' from $\mathfrak{D}$ approximate v and u', respectively, in the norm of H. By virtue of (5.2), the sequence of operators $L_{\cup}(u \cap v_n)$ uniformly converges to

$L_{\cup}(u \cap v)$. Since $u \cap v_n = R_{\cap}(v_n)u \in H \cap E_{\cup}$, where $R_{\cap}(a)b = b \cap a$ $(a, b \in \mathfrak{D})$ is an operator of the right $\cap$-multiplication in H, this yields relation (5.3). Indeed,

$$\begin{aligned}
(L_{\cup}(u \cap v)u', v')_H &= \lim_{n\to\infty}(L_{\cap}(u \cup v_n)\, u_n', v')_H \\
&= \lim_{n\to\infty}(R_{\cup}(u_n')(u \cap v_n), v')_H \\
&= \lim_{n\to\infty}(R_{\cup}(u_n')R_{\cap}(v_n)\,u, v')_H \\
&= \lim_{n\to\infty}(u, R_{\cap}(v_n^{\cap})R_{\cup}(u_n'^{\cup})(v_n)v')_H \\
&= \lim_{n\to\infty}(u, R_{\cap}(v_n^{\cap})(v' \cap u_n'^{\cup}))_H \\
&= \lim_{n\to\infty}(u, L_{\cap}(v' \cap u_n'^{\cup})v_n^{\cap})_H \\
&= (u, L_{\cap}(v' \cup u'^{\cup})v^{\cap})_H .
\end{aligned}$$

Let us return to proving the correctness of definition of the map $\rho_{\cup}$. By virtue of (5.3), we have

$$\begin{aligned}
\left(L_{\cup}\left(\sum_{k=1}^{n} u_k \cap v_k^{\cap}\right)v, u\right)_H &= \sum_{k=1}^{n}(L_{\cup}(u_k \cap v_k^{\cap})v, u)_H \\
&= \sum_{k=1}^{n}(u_k, L_{\cap}(u \cup v^{\cup})v_k)_H = \sum_{k=1}^{n}(L_{\cap}^{*}(u \cup v^{\cup})u_k, v_k)_H \\
&= \sum_{k=1}^{n}(L_{\cap}(u \cup v^{\cup\cap})u_k, v_k)_H = \omega(L_{\cap}(u^{\cap} \cup v^{\cup\cap})) \\
&= \sum_{k=1}^{n'}(L_{\cap}(u^{\cap} \cup v^{\cup\cap})u_k', v_k')_H = \sum_{k=1}^{n'}(u_k', L_{\cap}(u \cup v^{\cup})v_k')_H \\
&= \sum_{k=1}^{n'}(L_{\cup}(u_k' \cup v_k'^{\cap})v, u)_H = \left(L_{\cup}\left(\sum_{k=1}^{n'} u_k' \cap v_k'\right)v, u\right)_H
\end{aligned}$$

for any $u, v \in H$, whence

$$\sum_{k=1}^{n} u_k \cap v_k^{\cap} = \pi_{\cup}^{-1}\left(L_{\cup}\left(\sum_{k=1}^{n} u_k \cap v_k^{\cap}\right)\right) = \pi_{\cup}^{-1}\left(L_{\cup}\left(\sum_{k=1}^{n'} u'_k \cap v'^{\cap}_k\right)\right) = \sum_{k=1}^{n'} u'_k \cap v'^{\cap}_k.$$

The correctness of the definition of the map $\rho_{\cap}$ can be proved similarly.

■

The range $\rho_{\cup}(\mathfrak{B}_{\cap *})$ of the map $\rho_{\cup}$ coincides with the linear span $H \cap H$ of all possible $\cap$-products of elements from H and belongs to $E_{\cup}$. The set $\rho_{\cap}(\mathfrak{B}_{\cup *})$ has similar properties. We summarize the assertions established above in the following diagram:

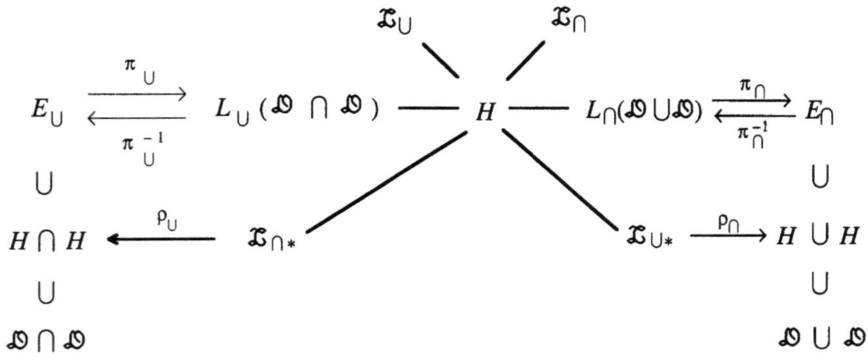

5.2. Quantized Hypercomplex Systems. In this subsection, we define quantized hypercomplex systems as bialgebras $\mathfrak{A}$ with certain relations between the operations $\cup$ and $\cap$ in $\mathfrak{A}$. Quantized hypercomplex systems can be regarded as a generalization of ordinary hypercomplex systems in the sense that they satisfy the theorem on realization (Theorem 5.1): If one operation in $\mathfrak{A}$ is commutative (e.g., $\cap$), then $\mathfrak{B}_{\cap *}$ can be realized as an object close to a hypercomplex system with locally compact base. Then we present axioms on $\mathfrak{A}$ that guarantee the multiplicativity of the measure m, the existence of the basis unity, and the preservation of positiveness for the indicated hypercomplex system, provided that one of operations in $\mathfrak{A}$ is commutative.

Assume that the operations $\cup$ and $\cap$ satisfy the following axioms:

(A1) $a^{\cup\cap} = a^{\cap\cup}$, $(a \cup b)^{\cap} = a^{\cap} \cup b^{\cap}$, $(a \cap b)^{\cup} = a^{\cup} \cup b^{\cup}$ $(a, b \in \mathfrak{A})$;

(A2) $\operatorname{Ker}\rho_{\cup} = \operatorname{Ker}\rho_{\cap} = \{0\}$.

Since

$$\| a^{\cup} \cap b^{\cup} \|_{\cap} = \| L_{\cup}(a^{\cup} \cap b^{\cup}) \| = \| L_{\cup}(a \cap b)^{\cup} \|$$

$$= \|L_{\cup}(a \cap b)\| = \|a \cap b\|_{\cup} \quad (a, b \in \mathfrak{D}),$$

one can introduce involution $\cup$ in the Banach space $E_{\cup}$ by setting $(a \cap b)^{\cup} = a^{\cup} \cap b^{\cup}$ $(a, b \in \mathfrak{D})$, and then extending it by antilinearity onto the entire $\mathfrak{D} \cap \mathfrak{D}$ and by the closure in the norm $\|\cdot\|_{\cup}$ onto the entire $E_{\cup}$. The mapping $\pi_{\cup}: E_{\cup} \to L_{\cup}(\mathfrak{D} \cap \mathfrak{D})$ is, clearly, a $*$-isomorphism; furthermore, $H \cap H$ is invariant under the involution $\cup$ and $(u \cap v)^{\cup} = u^{\cup} \cap v^{\cup}$. Similar assertions are true for $H \cup H$ and $E_{\cap}$.

If the bialgebra $\mathfrak{D}$ satisfies (A1), then in order that (A2) be satisfied, it is necessary and sufficient that the linear span of the operators $L_{\cap}(a \cup b)$ $(a, b \in \mathfrak{D})$ be weakly dense in $\mathfrak{L}_{\cap}$ and that the linear span of the operators $L_{\cup}(a \cap b)$ $(a, b \in \mathfrak{D})$ be weakly dense in $\mathfrak{L}_{\cup}$.

Indeed, it suffices to prove this claim for $\rho_{\cup}$.

Sufficiency. Let

$$\rho_{\cup}(\omega) = \sum_{k=1}^{n} u_k \cap v_k^{\cap} = 0.$$

Then, by virtue of (5.3), for any $a, b \in \mathfrak{D}$, we have

$$\omega(L_{\cap}(a \cup b)) = \sum_{k=1}^{n} (L_{\cap}(a \cup b)u_k, v_k)_H$$

$$= \sum_{k=1}^{n} (a, L_{\cup}(v_k \cap u_k^{\cap})b)_H = \left(a, L_{\cup}\left(\sum_{k=1}^{n} u_k \cap v_k^{\cap}\right)^{\cap} b\right)_H = 0.$$

Hence, $\omega(A) = 0$ for any $A \in \mathfrak{L}_{\cup}$, i.e., $\omega = 0$.

Necessity. We rewrite (5.3) as follows:

$$\langle \rho_{\cup}(\omega_{u \cap v^{\cap}}), \omega_{u' \cap v'^{\cup}} \rangle$$

$$= (L_{\cup}(u \cap v)u', v')_H = (u, L_{\cap}(v' \cup u'^{\cup})v)_H$$

$$= (L_{\cap}((u' \cup v'^{\cup})^{\cup \cap} u, v)_H = \left\langle \rho_{\cap}\left(\omega^{\cup \cap}_{u' \cup v'^{\cup}}\right), \omega_{u \cap v^{\cap}} \right\rangle, \quad (5.4)$$

where $\langle\cdot,\cdot\rangle$ denotes the duality between an W^*-algebra and its predual space. Clearly, (5.4) remains true for arbitrary forms $\omega_\cup \in \mathfrak{L}_{\cup *}$ and $\omega_\cap \in \mathfrak{L}_{\cap *}$. Assume the contrary, i.e., let the set of operators $\rho_\cup(\omega_\cap)$ be nondense in $\mathfrak{L}_\cup$ in the weak operator topology. Then one could find $\omega_\cup^0 \in \mathfrak{L}_{\cup *}$ orthogonal to this set in the sense of the duality $\langle\cdot,\cdot\rangle$. Then it would follow from (5.4) that $\langle\rho_\cap(\omega_\cup^0), \omega_\cap\rangle = 0$ for any $\omega_\cap \in \mathfrak{L}_{\cap *}$, i.e., $\rho_\cap(\omega_\cup^0) = 0$. However, by virtue of $\operatorname{Ker}\rho_\cap = 0$, we have $\omega_\cup^0 = 0$.

■

Since the mappings $\rho_\cup$ and $\rho_\cap$ are bijective, one can identify $\mathfrak{L}_{\cup *}$ and $\mathfrak{L}_{\cap *}$ with $H \cup H$ and $H \cap H$, respectively, i.e., elements $x \in \mathfrak{L}_{\cup *}$ and $y \in \mathfrak{L}_{\cap *}$ can be regarded as linear combinations of elements of the form $u \cup v$ and $u \cap v$ $(u, v \in H)$, respectively. We also identify $E_\cap$ with $L_\cap$ $(\mathfrak{A} \cup \mathfrak{A})$ and $E_\cup$ with $L_\cup$ $(\mathfrak{A} \cap \mathfrak{A})$. Then the diagram presented above takes the form

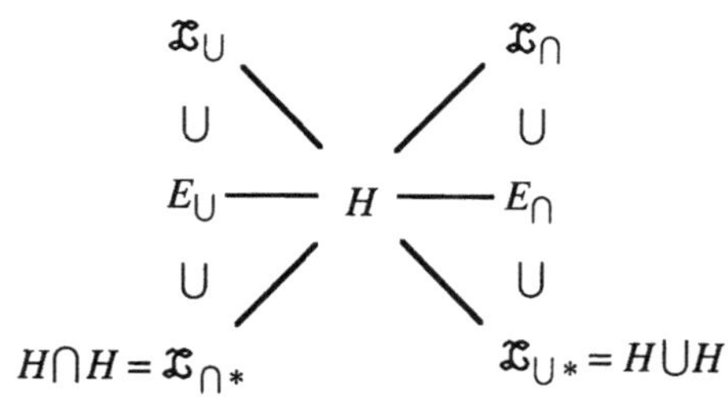

Taking into account that $\mathfrak{L}_{\cap *}$ $(\mathfrak{L}_{\cup *})$ is imbedded into $\mathfrak{L}_\cup$ $(\mathfrak{L}_\cap)$, we introduce the operations of multiplication $\cap$ in $\mathfrak{L}_{\cup *}$ and $\cup$ in $\mathfrak{L}_{\cap *}$.

We say that a Hilbert bialgebra $\mathfrak{A}$ is a *quantized hypercomplex system* if it satisfies (A1), (A2), and

(A3) $\mathfrak{L}_{\cup *}$ and $\mathfrak{L}_{\cap *}$ are subalgebras in $\mathfrak{L}_\cap$ and $\mathfrak{L}_\cup$, respectively.

Also note that, since $H \cap H$ is invariant under the involution $\cup$, it follows from (A3) that $\mathfrak{L}_{\cap *}$ is a $*$-algebra; similarly, $\mathfrak{L}_{\cup *}$ is a $*$-algebra with respect to the operation $\cap$ and involution $\cap$.

A normal hypercomplex system with basis unity $L_1(Q, m)$ that satisfies the conditions of Theorem 3.5 generates a quantized hypercomplex system Indeed, in this case, we have $H = L_2(Q, m)$ and, according to Theorem 1.8, $\mathfrak{A} = L_1 \cap L_\infty$ is a Hilbert bialgebra with respect to convolution (the operation $\cup$), multiplication (the operation $\cap$), and relevant involutions. The W^*-algebra $\mathfrak{L}_\cap$ is isomorphic to $L_\infty(Q, m)$, and $\mathfrak{L}_\cup$ is the weak (in L_2) closure of convolution operators with functions from $L_1 \cap L_\infty$. It is obvious that $\mathfrak{L}_{\cap *} = L_1(Q, m)$ and the range of $\rho_\cap$ coincides with the Fourier algebra $A(Q)$. By virtue of Lemma 3.2, we have $E_\cap = C_\infty(Q)$. Axiom (A1) is evidently satisfied. By virtue of Lemma 3.2, $A(Q)$ is weakly dense in $L_\infty(Q, m)$ and, hence, $\operatorname{Ker} \rho_\cup = 0$. In order to check that $\operatorname{Ker} \rho_\cap = 0$, it suffices to show that the convolution with a function from L_2 can be weakly (or strongly, which is the same) approximated by convolutions in $C_0(Q)$ with functions from $L_1 \cap L_\infty$. This fact was established in Lemma 1.2. It remains to verify (A3). Since $\operatorname{Ker} \rho_\cup = 0$, we have $\mathfrak{L}_{\cup *} = A(Q)$. It follows from Theorem 3.6 that $A(Q)$ is a subalgebra of $\mathfrak{L}_\cap = L_\infty(Q, m)$ and, by virtue of Theorem 1.1, $L_1(Q, m)$ is a subalgebra of $\mathfrak{L}_\cup$.

■

Note that, since $\mathfrak{L}_{\cup *} = A(Q)$ is isomorphic to $L_1(\hat{Q}, \hat{m})$ and $A(\hat{Q})$ is isomorphic to $L_1(Q, m)$, we have $\mathfrak{L}_\cup = L_\infty(\hat{Q}, \hat{m})$ and $E_\cup = C_\infty(\hat{Q})$.

The Banach spaces $E_\cup$ and $E_\cap$ are completions of the $*$-algebras $\mathfrak{L}_{\cup *}$ and $\mathfrak{L}_{\cap *}$ in the operator norm. Therefore, $E_\cap$ and $E_\cup$ are C^*-algebras. Furthermore, *the following inequalities hold:*

$$\|x\|_\cup \leq \|W\| \, \|x\|_{\mathfrak{L}_{\cap *}} \qquad (x \in \mathfrak{L}_{\cap *}), \tag{5.5}$$

$$\|x\|_\cap \leq \|W\| \, \|x\|_{\mathfrak{L}_{\cup *}} \qquad (x \in \mathfrak{L}_{\cup *}). \tag{5.6}$$

Indeed, let $a, b \in H$ and let $x = u \cap v \in \mathfrak{L}_{\cap *}$. Then, by virtue of (5.3) and (5.1), we get

$$\begin{aligned} |(L_\cup(u \cap v) b^{\cap\cup}, a^\cap)_H| &= |(u, L_\cap(a^\cap \cap b^\cap) v^\cap)_H| = |(L_\cap(a \cup b) u, v^\cap)_H| \\ &= |\omega_x(L_\cap(a \cup b))| \leq \|\omega_x\| \, \|L_\cap(a \cup b)\| \\ &= \|x\|_{\mathfrak{L}_{\cap *}} \|a \cup b\|_\cap \leq \|W\| \, \|x\|_{\mathfrak{L}_{\cap *}} \|a\|_H \|b\|_H . \end{aligned}$$

This implies (5.5). Inequality (5.6) can be proved similarly.

■

Lemma 5.1. *The C^*-algebras $E_\cap$ and $E_\cup$ contain left ideals $\mathfrak{D}_\cap$ and $\mathfrak{D}_\cup$, respectively, which belong to the intersection $H \cap E_\cap \cap E_\cup$ and are dense in each of these spaces.*

Proof. Let us prove Lemma 5.1 for $E_\cap$. By $\mathcal{D}_\cap$ we denote the linear span of the set $\{u \cap v \mid u \in E_\cap \cap H$ (here, $\cap$ denotes intersection), $v \in \mathcal{D} \cup \mathcal{D}\}$. Consequently, $\mathcal{D}_\cap$ is a left ideal in $E_\cap$, and $\mathcal{D}_\cap \subset E_\cap \cap H \cap E_\cup$.

Let us show that $\mathcal{D}_\cap$ is dense in $E_\cap$. It suffices to show that $(\mathcal{D} \cup \mathcal{D}) \cap (\mathcal{D} \cup \mathcal{D})$ is dense in $E_\cap$. For this purpose, in turn, it suffices to approximate an arbitrary element $x = u \cup v \in \mathcal{D} \cup \mathcal{D}$ by elements of $(\mathcal{D} \cup \mathcal{D}) \cap (\mathcal{D} \cup \mathcal{D})$ in the norm $\|\cdot\|_\cap$. Let us represent x as $x = x_1 + ix_2$, where

$$x_1 = x_1^\cap = \frac{x + x^\cap}{2} \in \mathcal{D} \cup \mathcal{D} \quad \text{and} \quad x_2 = x_2^\cap = \frac{x - x^\cap}{2i} \in \mathcal{D} \cup \mathcal{D}.$$

Thus, it suffices to consider the case where x is self-adjoint (i.e., $x = x^\cap$). Since $x \in E_\cap$, one can represent this element as the difference of nonnegative elements x_+ and x_-, i.e., $x = x_+ - x_-$, $x_+, x_- \in E_\cap$, $x_+ \geq 0$, $x_- \geq 0$, $x_+ \cap x_- = 0$. Let $y_+ \in E_\cap$ $(y_+ \geq 0)$ be such that $x_+ = y_+ \cap y_+$ and let $y_n \in \mathcal{D} \cup \mathcal{D}$ be a sequence which approximates y_+ in the norm $\|\cdot\|_\cap$. Then $y_n \cap y_n$ approximates x_+ in the norm $\|\cdot\|_\cap$ because

$$\begin{aligned} &\| y_n \cap y_n - y_+ \cap y_+ \|_\cap \\ &\qquad = \| (y_n + y_+) \cap (y_n - y_+) - (y_n \cap y_+ + y_+ \cap y_n) \|_\cap \\ &\qquad \leq \| y_n + y_+ \|_\cap \| y_n - y_+ \|_\cap + \| (y_n - y_+) \cap y_+ \|_\cap + \| y_+ \cap (y_+ - y_n) \|_\cap \\ &\qquad \leq (2 + \varepsilon) \| y_+ \|_\cap \| y_n - y_+ \|_\cap + 2 \| y_n - y_+ \|_\cap \| y_+ \|_\cap \to 0 \end{aligned}$$

as $n \to 0$.

By analogy, we find a sequence $z_n \in \mathcal{D} \cup \mathcal{D}$ which approximates y_- in the norm $\|\cdot\|_\cap$. Then, clearly, the sequence $x_n = y_n \cap y_n - z_n \cap z_n$ approximates x.

Let us show that $\mathcal{D}_\cap$ is dense in $E_\cup$. It suffices to approximate any element $u \cap v \in \mathcal{D} \cup \mathcal{D}$ in the norm $\|\cdot\|_\cup$ by elements from $\mathcal{D}_\cap$. Since $(\mathcal{D}, \cup)$ is a Hilbert algebra, one can find sequences $u_n \in \mathcal{D} \cup \mathcal{D}$ and $v_n \in \mathcal{D} \cup \mathcal{D}$ that approximate u and v, respectively, in the norm $\|\cdot\|_H$. Then, according to inequality (5.2), $u_n \cap v_n$ approximates $u \cap v$ in the norm $\|\cdot\|_\cup$.

Finally, let us show that $\mathcal{D}_\cap$ is dense in H. Note first that the operators $L_\cap(a \cup b)$ $(a, b \in \mathcal{D})$ are strongly dense in $\mathfrak{L}_\cap$. Indeed, since $L_\cap(\mathcal{D} \cup \mathcal{D})$ is dense in $L_\cap$, we conclude that the C^*-algebra $E_\cap$ is weakly (and, hence, strongly) dense in $\mathfrak{L}_\cap$. It remains to note that the operators $L_{\cap}(a \cup b)$ are uniformly dense in $E_{\cap}$. Any element of H can be approximated in the norm $\|\cdot\|_H$ by elements of the form $a \cap b$ $(a, b \in \mathfrak{D})$. As proved above, the element $L_{\cap}(a)$ can be approximated by the operators $L_{\cap}(u \cup v)$ $(u, v \in \mathfrak{D})$. Let us approximate $b \in \mathfrak{D}$ by elements of the form $u' \cup v'$ $(u', v' \in \mathfrak{D})$. We obtain

$$\|(u \cup v) \cap (u' \cup v') - (a \cap b)\|_H$$

$$\leq \|(u \cup v) \cap (u' \cup v') - a \cap (u' \cup v')\| + \|L_{\cap}(a)\|_H \|u' \cup v' - b\|_H \to 0.$$

This implies that $(\mathfrak{D} \cup \mathfrak{D}) \cap (\mathfrak{D} \cup \mathfrak{D})$ is dense in H. It remains to note that $(\mathfrak{D} \cup \mathfrak{D}) \cap (\mathfrak{D} \cup \mathfrak{D})$ is a part of $\mathfrak{D}_{\cap}$.

■

Theorem 5.1. *Let* $\mathfrak{D}$ *be a quantized hypercomplex system with commutative operation* $\cap$. *Then there exist a locally compact space* Q, *a Borel measure* m *regular and positive on open sets, and an involutory homeomorphism* $Q \ni p \mapsto p^* \in Q$ *such that*

(i) the C^**-algebra* $E_{\cap}$ *is isomorphic to the* C^**-algebra* $C_\infty(Q)$ *of continuous functions on* Q *vanishing at infinity;*

(ii) this isomorphism can be continued to an isometry between H *and* $L_2(Q, m)$ *and between* $\mathfrak{L}_{\cap *}$ *and* $L_1(Q, m)$;

(iii) $m(A) = m(A^*)$ *for any Borel set* $A \subset Q$.

Proof. By the Gelfand theorem, the commutative C^*-algebra $E_{\cap}$ is isometrically isomorphic to the algebra $E_{\cap}$ of continuous functions on the Hausdorff locally compact space Q of maximal ideals of the algebra $E_{\cap}$. By $I_{\cap}$ we denote the image of $\mathfrak{D}_{\cap}$ under this isomorphism. Let us introduce involution $*$ in Q by setting $t^*(x) = t(x^{\cap\cup})$ $(t \in Q, x \in E_{\cap})$.

Below, we use the Weil–Krein–Godeman theorem (see Subsection 3.1).

We define a functional f on the ideal $\mathfrak{D}_{\cap}$ as follows:

$$f(x) = \lim_{n \to \infty} (x, f_n) \quad (x \in \mathfrak{D}_{\cap}),$$

where $f_n = f_n^{\cap} \in \mathfrak{D}_{\cap}$ is an approximative unit in the Hilbert algebra $(\mathfrak{D}, \cap)$. Note that

$$x = \sum_k u_k \cap v_k \in \mathfrak{D}_\cap \quad (u_k \in H \cap E_\cap,\ v_k \in \mathfrak{D} \cap \mathfrak{D}).$$

Then

$$\lim_{n\to\infty} (x, f_n)_H = \sum_k \lim_{n\to\infty} (u_k \cap v_k, f_n)_H$$

$$= \sum_k \lim_{n\to\infty} (u_k, f_n \cap v_k^\cap)_H = \sum_k (u_k, v_k^\cap)_H.$$

Therefore, the functional f is well defined. Let us show that f satisfies the conditions of the Weil–Krein–Godeman theorem. Clearly, f is real. Indeed,

$$f(x^\cap) = \lim_{n\to\infty} (x^\cap, f_n)_H = \lim_{n\to\infty} \overline{(f_n, x)} = \overline{f(x)}.$$

Since

$$f(x \cap x^\cap).= \lim_{n\to\infty} (x \cap x^\cap, f_n)_H = \lim_{n\to\infty} (x, f_n \cap x)_H = \|x\|_H^2,$$

the functional f is positive. Hence, one can find a regular Borel measure $dm(p) = dp$ on Q such that

$$\lim_{n\to\infty} (x \cap y, f_n)_H = \int_Q x(p) y(p)\, dp$$

for any $x \in P$, $y \in E_\cap$.

Let us show that $\mathfrak{D}_\cap \subset L_2(Q, m)$. It is easy to see that, for any $x \in \mathfrak{D}_\cap$, the element $x \cap x^\cap$ is positive with respect to f. Hence,

$$(x \cap y, x)_H = \lim_{n\to\infty} (x \cap x^\cap \cap y, f_n)_H = \int_Q |x(p)|^2 y(p)\, dp \quad \forall y \in E_\cap,\ x \in \mathfrak{D}_\cap.$$

Consequently, we get $x(p) \in L_2(Q, m)$ because $y(p) \in C_\infty(Q)$ is arbitrary.

Since

$$(x, y)_{H_f} = f(y^\cap \cap x) = (x, y)_2 \quad (x, y \in \mathfrak{D}_\cap),$$

we have $H_f = H$. By analogy with the proof of Lemma 5.1, one can prove that $\mathfrak{D}_{\cap} \cap \mathfrak{D}_{\cap}$ is dense in H. Therefore, H is isometrically isomorphic to some closed subspace $F(H)$ of $L_2(Q, m)$.

Let us verify that $\dot{m}$ is positive on open sets from Q. Let $O \subset Q$ be an open set such that $m(O) = 0$. Assume that $p_0 \in O$. Obviously, one can find a function $\tilde{x}(p) \in C_\infty(Q)$ such that $\tilde{x}(p_0) = 1$, $0 \le \tilde{x}(p) \le 1$, and $\tilde{x}(p) = 0$ outside of O. Denote by $\tilde{x}$ the preimage of $\tilde{x}(p)$ under the Gelfand isomorphism. Clearly, $\tilde{x} \neq 0$. Since $\mathfrak{D}_{\cap}$ is an ideal, we have $x \cap \tilde{x} = \tilde{x} \cap x \in \mathfrak{D}_{\cap}$ for any $x \in \mathfrak{D}_{\cap}$. Therefore,

$$\|x \cap \tilde{x}\|_H^2 = \int_Q |(x \cap \tilde{x})(p)|^2 dp = \int_Q |x(p)|^2 |\tilde{x}(p)|^2 dp = 0$$

because, by assumption, $m(O) = 0$. Thus, we have $x \cap \tilde{x} = 0$ for any $x \in \mathfrak{D}_{\cap}$. Since $\mathfrak{D}_{\cap}$ is dense in H, we get $L_{\cap}(\tilde{x}) = 0$, whence $\tilde{x} = 0$. We arrive at a contradiction.

Let us show that $F(H) = L_2(Q, m)$. It suffices to establish the denseness (in the sense of $L_2(Q, m)$) of the ideal $\mathfrak{D}_{\cap}$ in the space $C_0(Q)$ of finite continuous functions. Assume that $z(p) \in C_0(Q)$, $A = \operatorname{supp} z$, and U is a neighborhood of A with compact closure. Let us choose $y(p) \in C_\infty(Q)$ such that

$$y(p) = \begin{cases} 1 & \text{if} \quad p \in A, \\ 0 & \text{if} \quad p \in Q \setminus U. \end{cases}$$

Since $\mathfrak{D}_{\cap}$ is dense in $C_\infty(Q)$, one can find a sequence $y_n \in \mathfrak{D}_{\cap}$ that approximates y in the norm $\|\cdot\|_\infty$. Then $x_n(p) = y_n(p) z(p) \in \mathfrak{D}_{\cap}$ approximates z in the norm of $L_2(Q, m)$.

For any $x(p) \in L_1(Q, m)$, we have

$$\lim_{n \to \infty} \int_Q x(p) f_n(p)\, dp = \int_Q x(p)\, dp.$$

Indeed, the function $x(p)$ admits the representation $x(p) = y(p) z(p)$, where $y(p)$, $z(p) \in L_2(Q, m)$, e.g., $x(p) = |x(p)|^{1/2} (|x(p)|^{1/2} \exp[\arg x(p)])$. Therefore,

$$\int_Q x(p) f_n(p)\, dp$$

$$= (y, z^{\cap} \cap f_n)_H \xrightarrow[n \to \infty]{} (y, z^{\cap})_H = \int_Q y(p) z(p)\, dp = \int_Q x(p)\, dp.$$

Denote by I the image of $\mathfrak{D}$ under the considered isometry. Clearly, I is dense in $L_2(Q, m)$. Let us show that $m(A) = m(A^*)$ for $A \in \mathcal{B}_0(Q)$. Let $\kappa_A(t) \in L_2(Q, m)$ be the indicator of A. There is a sequence $x_n(t) \in I$ that converges to κ_A in the quadratic mean. Clearly, $x_n(t^*) \to \kappa_{A^*}(t)$ in $L_2(Q, m)$. This implies that

$$m(A) = \|\kappa_A\|_2^2 = \lim_{n\to\infty} \int_Q |x_n(t)|^2 dt$$

$$= \lim_{n\to\infty} \|x_n\|_H^2 = \lim_{n\to\infty} \int_Q |x_n^{\cup}(t)|^2 dt = \|\kappa_{A^*}\|_2^2 = m(A^*)$$

for any $A \in \mathcal{B}_0(Q)$.

The weak closure (in $L_2(Q, m)$) of the C^*-algebra of operators of multiplication by functions from $C_\infty(Q)$ coincides with the W^*-algebra of operators of multiplication by functions from $L_\infty(Q, m)$. As is known, $L_1(Q, m)$ is the predual of $L_\infty(Q, m)$. Therefore, the considered isometry can be continued to a mapping of $\mathfrak{L}_{\cap *}$ onto $L_1(Q, m)$. Let us show that this mapping is also isometric. Since $I \cdot I$ is dense in $L_2(Q, m)$, by using the decomposition $L_1(Q, m) \ni x(p) = y(p)z(p)$ $(y(p), z(p) \in L_2(Q, m))$, we establish that $I \cdot I$ is dense also in $L_1(Q, m)$.

Let $x = u \cap v \in \mathfrak{D} \cap \mathfrak{D}$ $(u, v \in \mathfrak{D})$. Then

$$\|x\|_{\mathfrak{L}_{\cap *}} = \|\omega_x\| = \sup_{A \in \mathfrak{L}_{\cup}} \frac{|(Au, v^{\cap})_H|}{\|A\|}$$

$$= \sup_{A(p) \in L_\infty(Q,m)} \frac{\left|\int_Q A(p)u(p)v(p)\,dp\right|}{\|A(p)\|_\infty} = \int_Q |u(p)v(p)|\,dp = \|x(p)\|_{L_1(Q,m)}$$

The inequality $\|u \cap v\|_{\mathfrak{L}_{\cap *}} \le \|W\| \, \|u\|_H \|v\|_H$ implies that $\mathfrak{D} \cap \mathfrak{D}$ is dense in $\mathfrak{L}_{\cap *}$. This and the fact that $I \cdot I$ is dense in $L_1(Q, m)$ yield that the mapping of $\mathfrak{L}_{\cap *}$ onto $L_1(Q, m)$ is an isometry.

■

By using the second assertion of the theorem, we can equip the space $L_1(Q, m)$ with convolution (generally speaking, noncommutative) and involution by transferring them from $\mathfrak{L}_{\cap *}$. The measure m is a Plancherel measure and, therefore, the Plancherel formula and the inversion formula (see Subsection 5.3) hold for this convolution.

If one of the operations is commutative, then, by Theorem 5.1, the quantized hypercomplex system is realized as an object similar to a hypercomplex system with locally compact basis. However, generally speaking, this object may not have the properties of the multiplicativity of the measure m, the existence of the basis unity, and the preservation of positivity under convolution. To guarantee that it possesses these properties we impose the following additional condition on the quantized hypercomplex system:

(A4) the mappings $\omega \mapsto \omega(I)$, where I is the identity operator and either $\omega \in \mathfrak{L}_{\cap *}$ or $\omega \in \mathfrak{L}_{\cup *}$, can be extended to characters of the C^*-algebras $E_\cap$ and $E_\cup$, respectively.

Lemma 5.2. *Suppose that the mappings* $\omega \mapsto \omega(I)$, *where* $\omega \in \mathfrak{L}_{\cup *}$ *or* $\omega \in \mathfrak{L}_{\cap *}$, *are multiplicative on* $\mathfrak{L}_{\cup *}$ *or* $\mathfrak{L}_{\cap *}$. *If there exist approximative units* $e_\alpha \in \mathfrak{L}_{\cup *}$ *and* $f_\beta \in \mathfrak{L}_{\cap *}$ *such that* $\| e_\alpha \|_{\mathfrak{L}_{\cup *}} \leq 1$, $\| f_\beta \|_{\mathfrak{L}_{\cap *}} \leq 1$, *and* $\rho_\cap(e_\alpha) \to I$ *and* $\rho_\cup(f_\beta) \to I$ $*$*-weakly in* $\mathfrak{L}_\cap$ *and* $\mathfrak{L}_\cup$, *respectively, then these mappings can be extended to characters of the* C^**-algebras* $E_\cap$ *and* $E_\cup$, *respectively.*

Proof. It suffices to check the inequalities

$$|\langle I, \omega_\cup \rangle| \leq \| \omega_\cup \|_\cap \quad \text{and} \quad |\langle I, \omega_\cap \rangle| \leq \| \omega_\cap \|_\cup \quad (\omega_\cup \in \mathfrak{L}_{\cup *}, \omega_\cap \in \mathfrak{L}_{\cap *}).$$

By virtue of (5.4), we have

$$\begin{aligned} |\langle I, \omega_\cup \rangle| &= \lim_\beta |\langle L_\cup(f_\beta), \omega_\cup \rangle| = \lim_\beta |\langle \rho_\cap(\omega_\cup^{\cap\cup}), f_\beta \rangle| \\ &\leq \sup_\beta \| \rho_\cap(\omega_\cup^{\cap\cup}) \| \, \| f_\beta \|_{\mathfrak{L}_{\cap *}} = \sup_\beta \| \omega_\cup \|_\cap \| f_\beta \|_{\mathfrak{L}_{\cap *}} \leq \| \omega_\cup \|_\cap, \end{aligned}$$

where $\omega_\cup \in \mathfrak{L}_{\cup *}$. The second inequality can be proved similarly.

■

Condition (A4) is satisfied for the normal hypercomplex system with basis unity $L_1(Q, m)$ satisfying the conditions of Theorem 3.5. Indeed, one can take the approximative unit e_n of the algebra $L_1(Q, m)$ constructed in Lemma 1.3 as f_α, and the sequence

$$x_n(t) = \frac{1}{m(A)} c(Q_n, A, t),$$

where $A \in \mathcal{B}_0(Q)$ is an arbitrary open set and $Q_n \subset Q_{n+1}$ is a sequence of compact sets that covers the entire space Q, as e_α.

Let a quantized hypercomplex system satisfy (A4) and let the operation $\cap$ *be commutative. Then there exists a basis unity, i.e., a point* $e = e^* \in Q$ *such that* $(f, g)_2 = (f * g^*)(e)$, *where* $f, g \in L_2(Q, m)$, $(\cdot, \cdot)_2$ *is the scalar product in* $L_2(Q, m)$, *and the measure* m *is multiplicative, i.e.,*

$$\langle f * g, m\rangle = \langle f, m\rangle\langle g, m\rangle \qquad (f, g \in L_2(Q, m)).$$

Indeed, since the functional $G(x)$ $(x \in E_\cup)$, which is an extension of $\langle I, \omega_\cup\rangle$ $(\omega_\cup \in \mathfrak{L}_{\cup *})$, is multiplicative on $E_\cap = C_\infty(Q)$, there exists a point $e \in Q$ such that $G(x) = x(e)$ $(x \in C_\infty(Q))$. It is easy to see that, on $\mathfrak{L}_{\cup *} = H \cup H$, the functional G has the form $\langle I, u \cup v\rangle = (u, v^\cup)_H$. Therefore, for any $f, g \in H$, we have $(f, g)_2 = G(f \cup g^\cup) = (f * g^*)(e)$. Since

$$G((u \cup v)^{\cup\cap}) = G(v^{\cup\cap} \cup u^{\cup\cap}) = (v^{\cup\cap}, u^\cap)_H = (u, v^\cup)_H = G(u \cup v),$$

we establish that $x(e^*) = x(e)$ for all $x \in \mathfrak{D} \cup \mathfrak{D}$, which implies that $e^* = e$ because $\mathfrak{D} \cup \mathfrak{D}$ is dense in $E_\cap$.

To prove the multiplicativity of m it suffices to note that any element $x(p) \in L_1(Q, m)$ can be represented as $x(p) = u(p)v(p)$ $(u(p), v(p) \in L_2(Q, m))$ and to use the multiplicativity of the functional

$$\langle I, \omega_x\rangle = (u, v^\cap)_H = \int_Q u(p)v(p)\,dp = \int_Q x(p)\,dp.$$

As a result, we get

$$\int_Q x * y(p)\,dp = \langle I, \omega_x \cup \omega_y\rangle = \langle I, \omega_x\rangle\langle I, \omega_y\rangle.$$

■

If we interchange the pairs of operations $\cup$ and $\cap$ in the quantized hypercomplex system $\mathfrak{D}$, we obtain a quantized hypercomplex system $\hat{\mathfrak{D}}$ (dual to $\mathfrak{D}$). Thus, if we pass from the pair $(*, \cdot)$ to the pair $(\cdot, *)$ in the hypercomplex system $L_1(Q, m)$ in Theorem 3.5, then the corresponding quantized hypercomplex system $L_1(Q, m) \cap L_\infty(Q, m)$ is realized with respect to the operation $*$ according to Theorem 5.1. As a result, we obtain the dual hypercomplex system $L_1(\hat{Q}, \hat{m})$. Clearly, $\left(\hat{\mathfrak{D}}\right)^\wedge = \mathfrak{D}$. The class of quantized hypercomplex systems satisfying (A4) is closed with respect to duality. All results presented above remain true if we require the commutativity of the operation $\cup$ instead of $\cap$.

In a quantized hypercomplex system with commutative operation $\bigcap$ satisfying (A4) and realized, according to Theorem 5.1 in the space $L_2(Q, m)$, we introduce generalized translation operators: $(L_p f, g)_2 = (f * g^*)(p)$, where $f, g \in L_2(Q, m)$ and $p \in Q$. This definition is correct because $f * g^* \in C_\infty(Q)$. Let us show that *these operators have the following properties:*

(i) $L_e = I$;

(ii) $(L_p f)(q) \in L_2(Q, m)$ *m-almost everywhere in q;*

(iii) $\big(L_p^q(L_q f)\big)(r) = \big(L_q^r(L_p f)\big)(r)$ *$m \times m \times m$-almost everywhere in p, r, and q;*

(iv) $(L_p f)(q) = \overline{(L_{q^*} f^*)(p^*)}$ *$m \times m$-almost everywhere;*

(v) $L_p^* = L_{p^*}$;

(vi) $\| L_p \| \le \| W \|$;

(vii) the operator function $Q \ni p \mapsto L_p$ *is weakly continuous;*

(viii) $(L_q \bar{f})(p) = \overline{(L_q f)(p)}$ *$m \times m$-almost everywhere in p and q;*

(ix) the measure m is invariant, i.e.,

$$\int_Q (L_p f)(q)\,dq = \int_Q f(q)\,dq$$

for m-almost all p.

Indeed, (i) follows from the definition of basis unity, (vi) follows from inequality (5.1), and (v) is implied by the axioms of Hilbert algebras. Let $g \in L_1 \bigcap L_2$, $h \in L_2$, $f_n \in I$, be a sequence which approximates $f \in L_2$. It is easy to show that $f_n * g^*$ converges to $f * g^*$ in mean square. It follows from inequality (5.1) that $f_n^* * h$ uniformly converges to $f^* * g$. In view of the Lebesgue theorem, this yields

$$\int_Q (L_p f)(q)\overline{g(q)}\,dq\,\overline{h(p)}\,dp = (f * g^*, h)_2$$

$$= \lim_{n \to \infty} (f_n * g^*, h)_2 = \lim_{n \to \infty} \big(g^*, f_n^* * h\big)_2$$

$$= \int_Q \overline{g(q^*)} \overline{\int_Q (L_q f)^*(p) h(p^*) dp dq}$$

$$= \int_Q \int_Q \overline{(L_{q^*} f^*)(p^*)}\, \overline{g(q)}\, \overline{h(p)}\, dp\, dq;$$

since $g \in L_1 \cap L_2$ and $h \in L_2$ are arbitrary, we arrive at (iv).

Let us check the associativity relation (iii). For all $g, h \in L_1 \cap L_2$ and $f \in L_2$, we have

$$\int_Q \int_Q \left(L_p^q (L_q f)\right)(r) \overline{g(r)}\, \overline{h(q)}\, dr\, dq = \int_Q \overline{g(r)} \int_Q (L_q f)(r) \overline{L_{p^*} h(q)}\, dq\, dr$$

$$= \int_Q (f * g^*)(q) \overline{L_{p^*} h(q)}\, dq$$

$$= ((f * g^*) * h^*)(p) = (f * (h * g)^*)(p)$$

$$= \int_Q (L_p f)(r) \int_Q \overline{(L_r h)(p)} g^*(q)\, dq\, dr$$

$$= \int_Q \int_Q (L_p f)(r) (L_{r^*} g^*)(q) \overline{h(q)}\, dr\, dq$$

$$= \int_Q \int_Q \left(L_q^r (L_p f)\right)(r) \overline{g(r)}\, dr\, \overline{h(q)}\, dq,$$

whence we get (iii) because g and h are arbitrary and $L_1 \cap L_2$ is dense in L_2. We note that (ii) immediately follows from (iv). Property (vii) is a consequence of the fact that $f * g \in C_\infty(Q)$ $(f, g \in L_2(Q, m))$ and property (viii) is obvious.

Finally, let us prove (ix). For $f, g \in L_1 \cap L_2$, we have

$$\int_Q \int_Q f(p)\, dp\, g(q)\, dq = \int_Q (f * g)(p)\, dp$$

$$= \int_Q \int_Q (L_p f)(q) g(q^*)\, dq\, dp = \int_Q \int_Q \overline{\left(L_{q^*} f^*\right)(p^*)} g(q^*)\, dq\, dp$$

$$= \int_Q \int_Q \overline{\left(L_q f^*\right)(p^*)} g(q)\, dq\, dp = \int_Q \int_Q \overline{\left(L_q f^*\right)(p)}\, dq\, g(q)\, dp.$$

Hence, for almost all q,

$$\int_Q \overline{(L_q f^*)(p)}\, dq = \int_Q f(p)\, dp = \int_Q \overline{f^*(p)}\, dp,$$

whence

$$\int_Q (L_q f)(p)\, dp = \int_Q f(p)\, dp.$$

■

The element $A \in \mathfrak{L}_{\cup}$ is called $\cap$-positive $\left(A \overset{\cap}{\geq} 0\right)$ if $\langle A, \omega \cap \omega^{\cap} \rangle \geq 0$ for any $\omega \in \mathfrak{L}_{\cup *}$. The inequality $A \overset{\cup}{\geq} 0$ is understood similarly.

We introduce the axioms

(A5) $A, B \overset{\cap}{\geq} 0 \;\Rightarrow\; AB \overset{\cap}{\geq} 0 \quad (A, B \in \mathfrak{L}_{\cup})$;

(A5′) $A, B \overset{\cup}{\geq} 0 \;\Rightarrow\; AB \overset{\cup}{\geq} 0 \quad (A, B \in \mathfrak{L}_{\cap})$.

Theorem 5.2. *Let $\mathfrak{D}$ be a quantized hypercomplex system which satisfies (A4) and (A5) and let the operation $\cap$ be commutative. Then $L_1(Q, m)$ is a Banach $*$-algebra and the convolution of nonnegative functions is nonnegative.*

Proof. We first show that the equality

$$\langle \rho_{\cap}(x), y \rangle = \int_Q x(p) y(p^*)\, dp \tag{5.7}$$

holds for any $x \in \mathfrak{L}_{\cap *}$ and $y \in \mathfrak{L}_{\cup *}$.

In fact, for any $a \in \mathfrak{D}$ and $y = u \cup v^{\cup} \in \mathfrak{L}_{\cup *}$, we have

$$\langle L_{\cup}(a), y \rangle = (a \cup u, v)_H = (a, v \cup u^{\cup})_H$$

$$= (a, (u \cup v^{\cup})^{\cup})_H = (a,\ y^{\cup})_H = \int_Q a(p) y(p^*)\, dp. \tag{5.8}$$

Since I is dense in L_1, one can approximate an arbitrary function $x \in L_1(Q, m)$ by functions $a \in \mathfrak{D}$. Then $L_{\cup}(a) \to L_{\cup}(x)$ in $\mathfrak{L}_{\cup}$. Hence, by passing to the limit on both sides of (5.8), we obtain (5.7).

It follows from (5.7) that if $x(p) \in L_1(Q, m)$ is almost everywhere nonnegative, then $L_{\cup}(x) \overset{\cap}{\geq} 0$. Indeed, for any $a \in \mathfrak{D} \cup \mathfrak{D}$, we have

$$\langle L_{\cup}(x), y \cap y^{\cup} \rangle = \int_Q x(p) |y(p^*)|^2 \, dp \geq 0$$

by virtue of (5.7). Since $\mathfrak{D} \cup \mathfrak{D}$ is dense in $C_\infty(Q)$, the converse assertion is also true.

Let $x, y \in L_1(Q, m)$ are real-valued functions. The inequality

$$|(x * y)(p)| \leq |x| * |y|(p) \tag{5.9}$$

holds for almost all p.

In fact, we have $x = x_+ - x_-$ and $y = y_+ - y_-$, where $x_\pm, y_\pm \in L_1(Q, m)$ and are nonnegative almost everywhere. Therefore, $L_{\cup}(x_\pm) \geq 0$ and $L_{\cup}(y_\pm) \geq 0$. Since $x_\pm * y_\pm$ can be represented as a linear combination of the functions $|x| * |y|$, $|x| * y$, $x * |y|$, and $x * y$ from $L_1(Q, m)$, we have $x_\pm * y_\pm \in L_1(Q, m)$. By virtue of (A5),

$$L_{\cup}(x_\pm * y_\pm) = L_{\cup}(x_\pm) L_{\cup}(y_\pm) \overset{\cap}{\geq} 0.$$

Therefore, $(x_\pm * y_\pm)(p) \geq 0$ almost everywhere. Hence, for almost all p, we have

$$\begin{aligned} |(x * y)(p)| &= |(x_+ * y_+)(p) - (x_+ * y_-)(p) - (x_- * y_+)(p) + (x_- * y_-)(p)| \\ &\leq (x_+ * y_+)(p) + (x_+ * y_-)(p) + (x_- * y_+)(p) + (x_- * y_-)(p) \\ &= ((x_+ + x_-) * (y_+ + y_-))(p) = (|x| * |y|)(p). \end{aligned}$$

By integrating (5.9) and using the fact that the measure m is multiplicative, we obtain

$$\|x * y\|_1 \leq \|x\|_1 * \|y\|_1.$$

If the functions $x, y \in L_1(Q, m)$ are complex-valued, then we represent them in the form $x = x_1 + ix_2$ and $y = y_1 + iy_2$, where x_k and y_k are real-valued functions, and easily conclude that

$$\|x * y\|_1 \leq 2\|x\|_1 * \|y\|_1.$$

■

Thus, for quantized hypercomplex systems which satisfy the conditions of Theorem 5.2, the Banach $*$-algebra $L_1(Q, m)$ possesses all properties of a hypercomplex system with locally compact basis except, maybe, the fact that the convolution of compactly supported functions may be not compactly supported. In order that a class of quantized hypercomplex systems of this sort be closed with respect to the operation of transition to the dual quantized hypercomplex system, one must require that both (A5) and (A5′) be satisfied. A class of finite-dimensional quantized hypercomplex systems satisfying (A4), (A5), and (A5′) is contained in the class of bialgebras considered by Vainerman [Vai4], [Vai5]. He also presented an example of a finite-dimensional quantized hypercomplex system noncommutative with respect to both operations.

5.3. Harmonic Analysis in Quantized Hypercomplex Systems with One Commutative Operation. In this subsection, we briefly describe harmonic analysis in quantized hypercomplex systems with commutative operation $\bigcap$. For these systems, we prove analogs of the Plancherel theorem and inversion formula and establish duality in the sense of Pontryagin.

Let $\mathfrak{D}$ be a quantized hypercomplex system satisfying (A4) and such that the operation $\bigcap$ is commutative. By using the results established in Subsection 5.2, we construct a locally compact space Q, involution $Q \ni p \mapsto p^* \in Q$, multiplicative measure m, and a basis unity $e = e^* \in Q$ such that the space $L_1 = L_1(Q, m)$ is a $*$-algebra with respect to the (noncommutative) generalized convolution $f * g$.

Let us show that $\tilde{\mathfrak{D}} = L_1 \bigcap L_\infty$ *is a Hilbert algebra in* L_2 *with respect to the convolution* $f * g$ *and involution* $f^*(t) = \overline{f(t^*)}$ *(the first pair of operations) and with respect to the pointwise multiplication of functions and complex conjugation (the second pair of operations).*

It is easy to see that $\tilde{\mathfrak{D}}$ is a Hilbert algebra with respect to ordinary multiplication and complex conjugation. Let us show that $(\tilde{\mathfrak{D}}, *)$ is a Hilbert algebra. Denote the linear span of $L_2 * L_2$ by $A(Q)$. Obviously, $A(Q)$ is an algebra dense in $C_\infty(Q)$ because $A(Q)$ is isomorphic to $\mathfrak{L}_{\bigcup *}$. Let $a, b \in \tilde{\mathfrak{D}}$. By virtue of (A3), $a * b \in L_1$, but since $\tilde{\mathfrak{D}} \subset L_2$, we have $a * b \in A(Q)$, and, hence $a * b \in L_\infty$. Thus, $\tilde{\mathfrak{D}}$ is a $*$-algebra with respect to convolution and involution $a^*(p) = \overline{a(p^*)}$. Let us verify that $\tilde{\mathfrak{D}}$ satisfies conditions (i)–(iv) in the definition of Hilbert algebra (see Subsection 1.7). Indeed, condition (i) follows from the unimodularity of m $(m(A^*) = m(A))$. Condition (ii) follows from the equality $L^*_{\bigcup}(x) = L_{\bigcup}(x^*)$ $(x \in \mathfrak{L}_{\bigcap *})$, namely,

$$(a * b, c)_2 = (L_{\bigcup}(a)b, c)_2 = (b, L_{\bigcup}(a^*)c)_2 = (b, a^* * c)_2 \quad (a, b \in \tilde{\mathfrak{D}}).$$

Condition (iii) follows from (5.5), namely,

$$\| a * f\|_2 \leq \| L_{\bigcup}(a)\| \, \|f\|_2$$

$$\leq \|W\| \, \|a\|_{\mathfrak{L}_{\cap *}} \|f\|_2 = \|W\| \, \|a\|_1 \|f\|_2 \quad (a \in \tilde{\mathfrak{D}}, f \in L_2).$$

Let us show that condition (iv) is also satisfied. Let I be the image of the original Hilbert bialgebra $\mathfrak{D}$ under the isomorphism $H \leftrightarrow L_2(Q, m)$. Then $I \subset L_\infty$ and $I \cdot I \subset L_1$. This implies that $I \cdot I \subset \tilde{\mathfrak{D}}$. Hence, to show that $\tilde{\mathfrak{D}} * \tilde{\mathfrak{D}}$ is dense in L_2 it suffices to establish the denseness of $(I \cdot I) * (I \cdot I)$ in $L_2(Q, m)$ or, which is the same, the denseness of $(\mathfrak{D} \cap \mathfrak{D}) \cup (\mathfrak{D} \cap \mathfrak{D})$ in H. The last assertion is already proved in Lemma 5.1. Thus, $(\tilde{\mathfrak{D}}, *)$ is a Hilbert algebra and it remains to prove the continuity of the binding operator $\|\tilde{W}\|$. Let $a_1, a_2, b_1, b_2 \in \tilde{\mathfrak{D}}$. By virtue of relation (5.5) and the Schwartz inequality, we have

$$\left| \left(a_1 \otimes \bar{b}_2, \tilde{W}(a_2^* \otimes b_1) \right)_{L_2 \otimes L_2} \right|$$

$$= |(a_1 * a_2, b_1 b_2)_2| = |(a_1, (b_1 b_2) * a_2^*)_2| \leq \|a_1\|_2 \|L_{\cup}(b_1 b_2) a_2^*\|_2$$

$$\leq \|W\| \, \|a_1\|_2 \|a_2\|_2 \|b_1 b_2\|_1 \leq \|W\| \, \|a_1\|_2 \|a_2\|_2 \|b_1\|_2 \|b_2\|_2,$$

where W is the binding operator of the original quantized hypercomplex system.

■

Let $\tilde{\mathfrak{D}} = L_1 \cap L_\infty$ be the Hilbert bialgebra introduced above. Consider the Hilbert algebra $(\tilde{\mathfrak{D}}, *)$. The set of elements $f \in L_2$ such that the operator $\tilde{\mathfrak{D}} \ni a \mapsto L_{\cup}(a) f = f * a$ is continuous in L_2 forms a Hilbert algebra $\mathfrak{D}'$ that contains $\tilde{\mathfrak{D}}$. The algebra $\mathfrak{D}'$ is called *perfect*. Let $\mathfrak{L}_{\cup}$ be the left W^*-algebra of the Hilbert algebra $(\tilde{\mathfrak{D}}, *)$ (or, which is the same, $(\mathfrak{D}', *)$) and let $\mathfrak{L}_{\cup}^+$ be the set of its positive elements. For $A \in \mathfrak{L}_{\cup}^+$, we set $\varphi(A) = (a, a)_{L_2}$ if $A^{1/2} = L_{\cup}(a)$, where $a \in \mathfrak{D}'$, and $\varphi(A) = +\infty$, otherwise. Then φ *is a semifinite faithful normal trace on* $\mathfrak{L}_{\cup}^+$, i.e., the function $\varphi : \mathfrak{L}_{\cup}^+ \to [0, \infty]$ has the following properties:

(i) $\varphi(A + B) = \varphi(A) + \varphi(B)$ $(A, B \in \mathfrak{L}_{\cup}^+)$;

(ii) $\varphi(\lambda A) = \lambda \varphi(A)$, $(\lambda \geq 0, A \in \mathfrak{L}_{\cup}^+$; we suppose that $0 \cdot \infty = 0)$;

(iii) $\varphi(U A U^{-1}) = \varphi(A)$ $(A \in \mathfrak{L}_{\cup}^+$; U is a unitary operator in $\mathfrak{L}_{\cup})$;

(iv) $\varphi(A) = \sup \{ \varphi(T) \mid T \leq A$ and $\varphi(T) < +\infty \}$ $(A \in \mathfrak{L}_{\cup}^+)$;

(v) $\varphi(A) = 0 \Leftrightarrow A = 0$;

(vi) $\varphi(A) = \sup\{\varphi(B) \mid B \in \mathfrak{M}\}$ for any increasing family $\mathfrak{M} \subset \mathfrak{L}_{\cup}^{+}$ with the upper bound $A \in \mathfrak{L}_{\cup}^{+}$.

The trace φ on $\mathfrak{L}_{\cup}^{+}$ is called *canonical*. We extend φ to the entire $\mathfrak{L}_{\cup}$ by linearity. Clearly, $\varphi(L_{\cup}(b)^* L_{\cup}(a)) = (a, b)_2$ $(a, b \in \mathfrak{D})$.

By applying the central decomposition (see, [Dix1]), we expand L_2 into the direct integral

$$L_2 = \int_Z H_\lambda \, d\rho(\lambda)$$

of Hilbert spaces H_λ so that $\mathfrak{L}_{\cup}$ can be represented as a direct integral of W^*-algebras:

$$\mathfrak{L}_{\cup} = \int_Z \mathfrak{L}_{\cup}(\lambda) \, d\rho(\lambda),$$

where Z is the *quasispectrum* of the W^*-algebra $\mathfrak{L}_{\cup}$ (see, e.g., [Dix1]), i.e., a specially topologized set of representations of $\mathfrak{L}_{\cup}$ such that the W^*-algebra spanned by the operators of each representation is a factor.

The Fourier transform of the function $a \in \mathfrak{D}$ is defined as an operator function

$$\hat{a}(\lambda) = \int_Q a(s) L_s(\lambda) \, dm(s) \quad (\lambda \in Z), \tag{5.10}$$

where $L_s(\lambda)$ are the components of the central decomposition of generalized translation operators $L_s \in \mathfrak{L}_{\cup}$ constructed for the quantized hypercomplex system $\mathfrak{D}$ in Subsection 5.2. As follows from the results obtained in Subsection 5.2, Theorem 1 in [VaL]) is applicable to the generalized translation operators L_s. As a result of its application, we obtain the following assertion:

Theorem 5.3. *The following Plancherel formula and the inversion formula hold for the Fourier transform (5.5):*

$$(a, b)_{L_2} = \int_Z \varphi_\lambda(L_\lambda(b)^* L_\lambda(a)) \, d\rho(\lambda), \tag{5.11}$$

$$a(s) = \int_Z \varphi_\lambda(L_s(\lambda) \hat{a}(\lambda)) \, d\rho(\lambda), \tag{5.12}$$

where φ_λ are the components of the canonical trace φ on $\mathfrak{L}_{\cup}$ and $L_\lambda(a)$ are the components of the operator $\mathfrak{L}_{\cup}(a)$ in the central decomposition of $\mathfrak{L}_{\cup}$.

Note that the measure ρ in (5.11) and (5.12) is defined uniquely up to equivalence. If the generalized translation operators L_s are commutative, then the measure ρ is unique. The Fourier transform (5.10) and the inversion formula (5.12) establish a unitary isomorphism between $L_2(Q, m)$ and the space H_φ constructed from the trace φ on $\mathfrak{L}_{\cup}$ by using the GNS-construction.

Remark 1. In the case of ordinary commutative hypercomplex system, Theorem 5.3 turns into Theorems 3.2 and 3.3. Thus, Theorems 3.2 and 3.3 can also be proved by using Theorem 1 in [VaL]. For simplicity, in Section 3, we gave more elementary proofs of these theorems.

Remark 2. The Plancherel theorem and inversion formula are also true for nonunimodular generalized translation operators (see [VaL]) and unbounded Hilbert algebras generated by unbounded generalized translation operators (see [Vai6]–[Vai8]).

It is convenient to express the Pontryagin duality in terms of quantized hypercomplex systems. Let the operations $\cup$ and $\cap$ be commutative. Realizing a quantized hypercomplex system with respect to the operation $\cup$, we obtain a locally compact space $\hat{Q}$ dual to Q. Denote the operation of convolution in $L_1(\hat{Q}, \hat{m})$ by $\hat{*}$ ($x \hat{*} y(\hat{p})$ is the image of $x \cap y \in \mathfrak{L}_{\cup *}$). The operator $A \in \mathfrak{L}_{\cap}$ is called a $\cup$-character (or the character of the hypercomplex system $L_1(Q, m)$) if $\langle A, x \cup y\rangle = \langle A, x\rangle\langle A, y\rangle$ $(x, y \in \mathfrak{L}_{\cap *})$ and $A \neq 0$. Similarly, we define the notion of $\cap$-character. By X and X_h we denote the sets of $\cup$-characters and Hermitian $\cup$-characters, respectively (a $\cup$-character A is called Hermitian if $A = A^{\cup\cap}$). Similarly, we define $\hat{X}$ and $\hat{X}_h$. Clearly, $\hat{Q} \subset X_h$.

If we realize a quantized hypercomplex system with respect to the operation $\cap$, then the W^*-algebra $\mathfrak{L}_{\cap}$ turns into $L_\infty(Q, m)$. Hence, the $\cup$-characters have the form $\chi(p)$ $(p \in Q)$; similarly, the $\cap$-characters have the form $\chi(\hat{p})$ $(\hat{p} \in \hat{Q})$. Since $\hat{Q} \subset X$, we conclude that $p(\chi) = \chi(p)$ $(\chi \in \hat{Q})$ is a $\cap$-character. If a quantized hypercomplex system satisfies (A4), (A5), and (A5′), then, clearly, there is a one-to-one correspondence between the set of $\cup$-characters ($\cap$-characters) and the set of maximal ideals of the Banach algebra $L_1(Q, m)$ $(L_1(\hat{Q}, \hat{m}))$. One can now easily formulate an analog of Theorem 3.5 for the Fourier transformation defined in the standard way (we do not present it here).

If $\mathfrak{L}_{\cap *}$ has a weak approximative unit, i.e., a net $e_\alpha \in \mathfrak{L}_{\cap *}$, $\|e_\alpha\|_{\mathfrak{L}_{\cap *}} \leq 1$, such that $e_\alpha \cup x \to x$ weakly in $\mathfrak{L}_{\cap *}$, then an analog of the Bochner theorem is true for the quantized hypercomplex system with the commutative operation $\cap$. The proof of this statement, in fact, coincides with the proof of Theorem 3.1 and is thus omitted.

5.4. Real Hypercomplex Systems with Compact and Discrete Bases. In this subsection, we briefly discuss hypercomplex systems with compact and discrete bases introduced by Vainerman (see [Vai2], [Vai3], and [Vai5]). Generally speaking, these hypercomplex systems do not belong to the class of quantized hypercomplex systems. Nevertheless, they also admit rich harmonic analysis. The hypercomplex systems introduced by Vainerman are defined by analogy with hypergroups as follows: a hypercomplex system is a set of Radon measures on Q (Q is the basis of this hypercomplex system) with the structure of a $*$-algebra with identity I (here, it is not assumed that the identity belongs to the basis, i.e., it is not required that $I = \delta_e$, where $e \in Q$). Rigorous definitions are independently formulated for the cases of discrete and compact bases.

Let Q be a compact space satisfying the second axiom of countability and equipped with involutory homeomorphism $Q \ni p \mapsto p^* \in Q$, let $M = M_b(Q)$ be the space of bounded Radon measures equipped, in addition to the standard norm, with the topology of ordinary convergence on a subspace T which is dense in $C(Q)$ and invariant under the mappings $f(r) \mapsto \overline{f(r)}$ and $f(r) \mapsto f^*(r)$ (this topology is called T-topology; for $T = C(Q)$, it coincides with weak topology). Assume that the set of Dirac measures $\mathfrak{D}(Q) = \{\delta_t \mid t \in Q\}$ is total in the T-topology (for $T = C(Q)$, this condition is clearly satisfied).

The space M is called a *real hypercomplex system with compact basis* if the following conditions are satisfied:

(i) the space M equipped with T-topology is a topological associative $*$-algebra with involution $*$ and identity 1;

(ii) the function $Q \times Q \ni (t, s) \mapsto \langle \delta_t * \delta_s, f \rangle$ is separately continuous (this condition implies that the form $\langle \overline{\mu}^* * \eta, f \rangle$ with fixed $f \in T$ and $\mu \in M$ is T-continuous on M and, hence, is given by a certain function $\mu * f \in T$);

(iii) there exists a unique (to within a positive factor) positive *Plancherel measure,* i.e., a measure $m \in M^+$, $\operatorname{supp} m = Q$, such that $\overline{m}^* = m$ and $\eta * fm = (\eta * f)m$ $(f \in T, \eta \in M)$;

(iv) $\|\eta * f\|_2 \leq C\|\eta\| \, \|f\|_2$, where $f \in T$, $C \geq 0$, $\eta \in M$, and $\|\cdot\|_2$ is a norm in $L_2(Q, m)$.

If $1 = \delta_e$ for some $e \in Q$, $\delta_x * \delta_y \in M^1(Q)$, and $1(x) \in T$ $(1(x) \equiv 1)$, we obtain an object similar to a compact hypergroup; such a hypercomplex system is called a *generalized hypergroup* and the Plancherel measure turns into an invariant measure (which is unique). If Q is connected, the uniqueness of the Plancherel measure follows from axioms (i), (ii), and (iv). If Q is disconnected, then the Plancherel measure can be nonunique; for example, in the case $M = \mathbb{C}^n$, $Q = \{1, \dots, n\}$, $\delta_i * \delta_j = \delta^i_j \delta_i$ (δ^i_j is the Kronecker symbol), the Plancherel measure is nonunique.

The operations $f * g = fm * g$ $(f, g \in T)$ and $f^*(x) = \overline{f(t^*)}$ define the structure of a $*$-algebra on T. Moreover, axiom (iv) implies that $\|f * g\|_\infty \le K\|f\|_2 \|g\|_2$ $(K \ge 0)$. This estimate allows us to extend the convolution to arbitrary functions from $L_2(Q, m)$. In this case, $L_2(Q, m)$ is a complete Hilbert algebra (i.e., $\mathfrak{D} = L_2(Q, m)$). Therefore, it can be decomposed into a direct sum of at most countably many minimal closed ideals H_k [Dix1]. One can prove that each H_k is finite-dimensional. This enables us to establish the following analog of the Peter–Weyl theorem for the hypercomplex system M: All irreducible $*$-representations of M are finite-dimensional, the set of their equivalence classes is at most countable, the matrix elements of all irreducible representations of M form a complete orthogonal system in $L_2(Q, m)$, and the linear span of this system is uniformly dense in $C(Q)$.

For a real hypercomplex system, by analogy with Theorem 5.3, one can prove the Plancherel theorem and inversion formula for the corresponding Fourier transformation; moreover, one can prove an analog of the Bochner theorem and study the Fourier algebra $A(Q)$ following [Eym]. Note that, in this case, $A(Q)$ is an algebra with respect to convolution (however, it is not an algebra with respect to pointwise multiplication). This and the fact that, generally speaking, $L_1(Q, m)$ is not a $*$-algebra explain that a real hypercomplex system with compact basis is not necessarily a quantized hypercomplex system.

Let us define a real hypercomplex system with discrete basis. Assume that Q is a countable set with discrete topology and M is a C^*-algebra with identity 1 which consists of Borel measures on Q and is a completion of $l_1(Q)$ in some topology generated by the C^*-norm $\|\cdot\|$ which is weaker than the topology of $l_1(Q)$ but stronger than the topology of pointwise convergence of measures on Q. The form $\langle f, \bar{\mu}^* * \nu\rangle$, where f has a compact support, is continuous in ν and, hence, is given by a bounded function $\mu * f$ with compact support. A Borel measure m is called a *Plancherel measure* on Q if $m^* = \bar{m}$ and $\mu * fm = (\mu * f)m$ $(\mu \in M, f \in C_0(Q))$. We say that M is a *real hypercomplex system with discrete basis* if the following conditions are satisfied:

(i) $\|\nu\| = \|\bar{\nu}\|$ and $|\nu(Q)| \le C\|\nu\|$ $(\nu \in l_1(Q),\ C \ge 0)$;

(ii) there exists a unique positive Plancherel measure $m = (m_k)_{k \in Q}$ whose support is the entire set Q;

(iii) $\|f\|_2 \le C\|f\mu\|$ $(C \ge 0,\ f \in C_0(Q)$, and $\|\cdot\|_2$ is the norm in $l_2(Q, m_k))$.

Ordinary hypercomplex systems with discrete bases satisfy the axioms of real hypercomplex systems with discrete bases (in this case, M is the enveloping C^*-algebra for the Banach $*$-algebra $l_1(m)$, i.e., the completion of $l_1(Q)$ in the norm

$$\| x \|' = \sup_{\pi \in \mathfrak{R}} \| \pi(x) \|,$$

where $x \in l_1(m)$ and $\mathfrak{R}$ is the set of representations of $l_1(m)$; it is clear that, in the case of commutative hypercomplex systems, we have $\| x \|' = \| \hat{x}(\chi) \|_{\infty}$).

Note that one can use the following equivalent definition of real hypercomplex systems with discrete bases in terms of C^*-algebras: A real hypercomplex system is a C^*-algebra M with identity 1 such that there exists a countable total subset $Q = \{ \delta_k \}$ of M invariant under the involution $*$ and satisfying the following conditions:

(i) there exists a unique (to within a constant factor) finite faithful trace μ on M such that $\mu(\delta_k^* * \delta_i) = \delta_j^k \mu_k^{-1}$ $(\mu_k > 0,\ \delta_i, \delta_k \in Q)$);

(ii) the numbers $\mu(\delta_{k^*} * \delta_i * \delta_j)$ are real and there exists a continuous Hermitian form τ on M such that $\| \delta_k \| \leq C\tau(\delta_k)$ for any $\delta_k \in Q$.

If M is commutative, then, applying the Gelfand theorem to the C^*-algebra M, we establish that a commutative real hypercomplex system with discrete basis is isomorphic to a C^*-algebra $C(\hat{Q})$ with identity, where $\hat{Q}$ is a Hausdorff compact set that satisfies the second axiom of countability. There exists a finite Borel measure μ $(\operatorname{supp} \mu = \hat{Q})$ and an at most countable set $Q = \{\varphi_i(t)\}_{i=1}^{\infty}$ of functions from $C(\hat{Q})$ mutually orthogonal with respect to the scalar product in $L_2(\hat{Q}, m)$. These functions form a total set in $C(\hat{Q})$ and satisfy the following conditions:

(i) if $\varphi_i(t) \in Q$, then $\overline{\varphi_i(t)} \in Q$;

(ii) the numbers $\int \varphi_i(t)\varphi_j(t)\overline{\varphi_k(t)}\, d\mu(t)$ are real;

(iii) there exists a finite real Borel measure τ on $\hat{Q}$ such that

$$\max_{t \in \hat{Q}} |\varphi_i(t)| \leq C \int \varphi_i(t)\, d\tau(t)$$

for all i.

One can show that the form $s(\mu) = (\mu, 1)$, where $\mu \in M$ and $(\cdot, \cdot)$ is the scalar product in $H = l_2(Q, 1/m_k)$, defines a finite faithful trace on the C^*-algebra M. Denote by $L(\mu)f = \mu * f$ the left regular representation of the C^*-algebra M in H. By applying the construction suggested in [VaL] to the W^*-algebra generated by the operators of this representation and following Subsection 5.3, one can prove the Plancherel theorem and inversion formula for the corresponding Fourier transformation. For such hypercomplex systems, one can also prove an analog of the Bochner theorem and the following analog of the Pontryagin duality principle: Commutative real hypercomplex systems with compact and discrete bases are in duality.

2. EXAMPLES OF HYPERCOMPLEX SYSTEMS

As indicated in the introduction to Section 2 of Chapter 1, the appearance of generalized translation operators is, as a rule, connected with the existence of a Fourier-type transformation satisfying the Plancherel theorem and the inversion formula. These generalized translation operators often possess additional properties which enable one to construct a hypercomplex system. In view of the existence of developed harmonic analysis for hypercomplex systems, it is possible to consider, from the general point of view, numerous results of harmonic analysis obtained in various special cases. Note that the application of duality theory to these cases sometimes gives new results (see, e.g., Vainerman's version of the inverse problem for the Sturm–Liouville equation [Vai3] in Subsection 4.5 of this chapter).

In Chapter 2, we consider four situations of this kind. In Section 1, we investigate a hypercomplex system isomorphic to the center of the group algebra of compact group and the dual hypercomplex system. In Section 2, we study hypercomplex systems associated with Gelfand pairs, i.e., with pairs of locally compact groups G and their compact subgroups H such that the subalgebra of the group algebra of G which consists of H-biinvariant functions is commutative. In particular, these hypercomplex systems contain hypercomplex systems studied in the Section 1. Section 3 is devoted to the analysis of hypercomplex system associated with orthogonal polynomials defined on bounded subsets of the real axis. In Section 4, we consider hypercomplex systems constructed by using the Sturm–Liouville equation. Here, we do not consider hypercomplex systems connected with association schemes, hypergroups $K // H$, where K is a locally compact hypergroup and H is a compact subhypergroup of K ([Jew]), and some other examples of hypercomplex systems.

1. Centers of Group Algebras of Compact Groups

We can consider the center of the group algebra of a finite group G as a commutative finite-dimensional hypercomplex system whose basis coincides with the set of conjugacy classes of the group, i.e., the collection of sets of the type $\{ aga^{-1} \mid a \in G\}$. In this section, we show that, for a compact topological group, the center of its group algebra is a commutative hypercomplex system whose basis coincides with the compact set of conjugacy classes of the group. By applying the results obtained in Chapter, 1 we obtain fundamental principles of the theory of representations of compact groups. In the second part of this section, we present some results from [BeG] devoted, in particular, to the investigation of hypercomplex systems whose basis is the compact set of conjugacy classes of compact semisimple Lie groups and dual hypercomplex systems.

The hypercomplex systems studied in this and subsequent sections are subalgebras of the group algebra.

1.1. General Construction of Hypercomplex Systems Corresponding to Locally Compact Groups. First, we describe a general method aimed at the construction of new examples of generalized translation operators based on the reduction of the operation of ordinary translation on a locally compact group G.

Let G be a locally compact Hausdorff group satisfying the second axiom of countability. Consider its group algebra, i.e., a set $L_1(G, dg)$ of functions defined on the group and summable with respect to the left Haar measure dg with norm

$$\|x\| = \int_G |x(g)|\,dg$$

and convolution

$$(x * y)(g) = \int_G x(a)y(a^{-1}g)\,da \quad (x, y \in L_1(G, dg);\; g \in G).$$

Let K be a closed subalgebra of the algebra $L_1(G, dg)$. Assume that there exists a decomposition of G into disjoint compact classes of elements $p, q, r, \ldots$ such that every function from K can be obtained as the limit in the norm of the space $L_1(G, dg)$ of a sequence of continuous finite functions defined on the group which take constant values on every class $p, q, r, \ldots$ and, vice versa, every limit of this sort belongs to K. We equip the set Q with the factor topology, namely, a set $A \subset Q$ is open in this topology if and only if $P^{-1}(A) \subset G$ is open (here, P is the natural projection which maps every $g \in G$ into the class p which contains g). Assume that P is a closed mapping. Then Q is a locally compact Hausdorff space with the second axiom of countability. This implies that every continuous function on G taking constant values on classes from Q can be regarded as a continuous function on Q, and vice versa.

In the space $C_0(Q)$, we consider a positive linear functional

$$I(f) = \int_G f(g)\,dg \quad (f \in C_0(Q)).$$

By virtue of the Riesz–Markov theorem, this equality defines a unique nonnegative regular Borel measure $m(E)$ $(E \in \mathcal{B}(Q))$ such that

$$\int_G f(g)dg = \int_Q f(p)dp \quad (f \in C_0(Q)) \tag{1.1}$$

(integration with respect to the measure m is denoted by dp). It follows from (1.1) that there exists an isometric mapping $x(g) \mapsto x(p)$ of the entire K onto the entire space $L_1(Q, dp)$ such that

$$\int_G x(g)dg = \int_Q x(p)dp \quad (x(g) \in K).$$

Note that $m(A) = \lambda(\mathfrak{A})$ ($A \in \mathcal{B}_0(Q)$, $\mathfrak{A} \in \mathcal{B}_0(G)$, $\mathfrak{A}$ is a preimage of the set A, and λ is the Haar measure). It is not difficult to show that the convolution of two functions $x, y \in K$ can be written in the form

$$(x * y)(r) = \int_Q x(p)d_p \int_Q y(q)d_q c(E_p, E_q, r) \quad (r, p, q \in Q),$$

where $c(A, B, r)$ is the structure measure given by the formula

$$c(A, B, r) = \lambda(g^{-1}\mathfrak{A} \cap \mathfrak{B}^{-1}) = \chi_{\mathfrak{A}} * \chi_{\mathfrak{B}}(g) \tag{1.2}$$

($A, B \in \mathcal{B}_0(Q)$, $g \in r \in Q$, and $\mathfrak{A}$ and $\mathfrak{B}$ are, respectively, the preimages of the sets A and B), and m is the multiplicative measure.

Thus, the subalgebra $K \subset L_1(G, dg)$ can be regarded as a hypercomplex system with locally compact basis Q.

1.2. Centers of Group Algebras of Compact Groups. Let us show that the center of the group algebra of a compact group is a compact hypercomplex system.

We apply the general constructions of Subsection 1.1 to the case where K is the *center* $ZL_1(G)$ *of the group algebra of a compact group* G, i.e., $K = ZL_1(G) = \{x \in L_1(G, dg) \mid x * y = y * x$ for all $y \in L_1(G, dg)\}$. It is well known that $ZL_1(G)$ *consists of functions* $x(g) \in L_1(G, dg)$ *such that* $x(gh) = x(hg)$ *for almost every* $(g, h) \in G \times G$ (these functions are called *central*). Indeed, for all $y \in L_1(G, dg)$ and almost all $g \in G$, by virtue of unimodularity of the group G and commutativity of $ZL_1(G)$, we have

$$\begin{aligned} 0 = x * y(g) - y * x(g) &= \int x(h)y(h^{-1}g)dh - \int y(h)x(h^{-1}g)dh \\ &= \int x(gh)y(h^{-1})dh - \int x(hg)y(h^{-1})dh \\ &= \int [x(gh) - x(hg)]y(h^{-1})dh. \end{aligned}$$

Since the function y is arbitrary, we conclude that $x \in ZL_1(G)$ if and only if $x(gh) = x(hg)$ for almost all $(g, h) \in G \times G$.

■

Let Q be the set of conjugacy classes of the group G. Denote by $p, q, r, \ldots$ elements of Q (recall that $p = \{aga^{-1} \mid a \in G\}$). Note that the equality $x(gh) = x(hg)$ $(g, h \in G)$ is equivalent to the equality $x(h^{-1}gh) = x(g)$. Thus, continuous functions from $ZL_1(G)$ take constant values on the conjugacy classes of the group G and, vice versa, every continuous function which takes constant values on these classes belongs to $ZL_1(G)$. Since $ZL_1(G)$ is a closed subalgebra, the limit of these functions in the norm of the space $L_1(G, dg)$ also belongs to $ZL_1(G)$. Conversely, if $x(g) \in ZL_1(G)$, then it is possible to construct a sequence of functions which are continuous on G such that $\| x - f_n \|_1 \to 0$ as $n \to \infty$. This means that continuous functions

$$f_n^{\vee}(g) = \int_G f_n(a^{-1}ga)da \in ZL_1(G)$$

approximate $x(g)$. If we equip the set Q with the factor topology and apply the general construction of Subsection 1.1, then we conclude that the center $ZL_1(G)$ of the group algebra of a compact group is a commutative hypercomplex system with compact basis Q of conjugacy classes. This hypercomplex system becomes a normal hypercomplex system if we introduce the involution $*$ as follows:

$$Q \ni p = \{a^{-1}ga \mid a \in G\} \mapsto \{a^{-1}g^{-1}a \mid a \in G\} = p^* \in G.$$

The identity e of the group is a basis unity of the obtained hypercomplex system. Consequently, $ZL_1(G)$ is semisimple. It easy to see that the hypercomplex system $ZL_1(G)$ satisfies the condition of separate continuity. It contains the identity if and only if G is finite.

1.3. Elements of the Theory of Representations of Compact Groups. Prior to the description of the space of characters of the hypercomplex system $ZL_1(G)$, we recall some principal notions and facts from the theory of representations of groups. A strongly continuous operator-valued function $G \ni g \mapsto \pi(g)$, where $\pi(g)$ is a bounded (unitary) operator in a separable Hilbert space H, is called a *representation (unitary representation) of a group* G if $\pi(gh) = \pi(g)\pi(h)$ $(g, h \in G)$. The space H is called the *space of the representation* and its dimension is denoted by $\dim \pi$. If this space is finite-dimensional, then the representation also is called *finite-dimensional.* By choosing an orthonormal basis $f_1, \ldots, f_n$ in H and expanding $\pi(g)f_j$ in this basis, we obtain

$$\pi(g)f_j = \sum_{i=1}^{\dim \pi} \pi_{ij}(g)f_i.$$

Thus, the matrix $(\pi_{ij}(g))_{i,j=1}^{\dim\pi}$ is associated with the unitary representation $\pi(g)$. The elements of this matrix are continuous functions $\pi_{ij}(g) = (\pi(g)f_j, f_i)_H$ satisfying the equations

$$\pi_{ij}(gh) = \sum_{k=1}^{\dim\pi} \pi_{ik}(g)\pi_{kj}(h).$$

Since $\pi(g)$ is unitary, the matrix $(\pi_{ij}(g))_{i,j=1}^{\dim\pi}$ is also unitary, i.e.,

$$\left((\pi_{ij}(g))_{i,j=1}^{\dim\pi}\right)^{-1} = \overline{(\pi_{ji}(g))}_{j,i=1}^{\dim\pi}.$$

Further, it easy to see that

$$\sum_{j=1}^{\dim\pi} \pi_{ij}(g)\overline{\pi_{kj}(g)} = \sum_{j=1}^{\dim\pi} \pi_{ij}(g)\pi_{jk}(g^{-1}) = \delta_{i,k}.$$

The trace of the matrix of a finite-dimensional representation is called the *character of this representation* and is denoted by

$$\chi_\pi(g) = \sum_{i=1}^{\dim\pi} \pi_{ii}(g).$$

Since

$$\chi_\pi(a^{-1}ga) = \sum_i \pi_{ii}(a^{-1}ga) = \sum_i \sum_{k,p} \pi_{ik}(a^{-1})\pi_{kp}(g)\pi_{pi}(a)$$

$$= \sum_{k,p} \sum_i \overline{\pi_{ki}(a)}\pi_{pi}(a)\pi_{kp}(g) = \sum_{k,p} \delta_{k,p}\pi_{kp}(g) = \chi_\pi(g),$$

we conclude that the character of a unitary representation is a continuous function on G which is constant on conjugacy classes.

We say that representations $\pi^{(1)}(g)$ and $\pi^{(2)}(g)$ of a group G in the spaces H_1 and H_2, respectively, are *equivalent (unitary equivalent)* if there exists a linear (unitary) operator $A\colon H_1 \to H_2$ with a bounded inverse A^{-1} such that $\pi^{(2)} = A\pi^{(1)}A^{-1}$. If representations $\pi^{(1)}(g)$ and $\pi^{(2)}(g)$ are equivalent ($\pi^{(1)} \sim \pi^{(2)}$), then, for properly chosen bases, their matrices coincide. Indeed, if we choose a basis $\{f_k\}$ in the space H_1 of the representation $\pi^{(1)}$, then it suffices to choose the basis $\{Af_k\}$ in the space H_2 of the representation $\pi^{(2)}$.

It follows from the properties of the trace of a matrix that the character of a representation does not depend on the choice of a basis in the space of this representation. This implies that the characters of equivalent representations coincide.

A subspace H_1 of the space of a representation $\pi(g)$ is called *invariant* if, for any $\varphi \in H_1$, we have $\pi(g)\varphi \in H_1$ for any $g \in G$. Representations that have no nontrivial invariant subspaces (i.e., other than the null space or the entire space of a representation) are called *irreducible.* Otherwise, representations are called *reducible.*

The representation $\pi(g)$ of a group G in the Hilbert space $H = \oplus_{k=1}^{N} H_k$ $(N \le \infty)$ given by the equality

$$\pi(g)\varphi = \sum_{k=1}^{N} \pi^{(k)}(g)\varphi_k \quad \left(\varphi = \sum_{k=1}^{N} \varphi_k, \ \varphi_k \in H_k \right)$$

(this representation is obviously reducible) is called the direct sum of unitary representations $\pi^{(k)}(g)$ of the group G in the spaces H_k and is denoted by $\pi(g) = \oplus_{k=1}^{N} \pi^{(k)}(g)$. A representation $\pi(g)$ is called *completely reducible* if it can be decomposed into a direct sum of irreducible representations.

All finite-dimensional unitary representations are completely reducible. Indeed, if H_1 is an invariant subspace of a unitary representation $\pi(g)$, then, by virtue of the equality $(\varphi, \pi(g)\psi)_H = (\pi(g^{-1})\varphi, \psi)_H = 0$ $(\varphi \in H_1, \ \psi \in H_1^{\perp}, \ g \in G)$, the subspace $H_2 = H_1^{\perp}$ is also invariant. Since the restrictions $\pi^{(1)}$ and $\pi^{(2)}$ of the representation $\pi(g)$ onto the subspaces H_1 and H_2, respectively, are also unitary, we get $\pi(g) = \pi^{(1)}(g) \oplus \pi^{(2)}(g)$. By repeating this procedure, after finitely many steps, we arrive at irreducible unitary representations.

■

This statement remains true for infinite-dimensional unitary representations of compact groups, i.e., every unitary representation can be decomposed into a direct sum of finite-dimensional irreducible representations (see, e.g., [Nai2]). In particular, this implies that irreducible representations of compact groups are finite-dimensional. Obviously, in the space of any finite-dimensional unitary representation, one can choose a basis in which the matrix of this representation takes a block-diagonal form, namely,

$$\begin{bmatrix} \pi^{(\sigma_1)}(g) & & & & & & 0 \\ & \ddots & & & & & \\ & & \pi^{(\sigma_1)}(g) & & & & \\ & & & \ddots & & & \\ & & & & \pi^{(\sigma_n)}(g) & & \\ & & & & & \ddots & \\ 0 & & & & & & \pi^{(\sigma_n)}(g) \end{bmatrix}.$$

Here, the matrices of irreducible representations $\pi^{(\sigma_k)}(g)$ of the group G are repeated on the main diagonal M_k times, where M_k is the multiplicity of $\pi^{(\sigma_k)}(g)$ in the decomposition of the representation $\pi(g)$ into irreducible representations. This decomposition can be written as follows:

$$\pi(g) \sim \bigoplus_{k=1}^{n} M_k \pi^{(\sigma_k)}(g), \tag{1.3}$$

where all σ_k are different, i.e., the representations $\pi^{(\sigma_k)}$ are pairwise nonequivalent.

Denote by $\hat{G}$ the set of equivalence classes of irreducible unitary representations of a group G. Let χ_σ be the character of a representation $\pi^{(\sigma)} \in \hat{G}$ and denote $d_\sigma = \dim \pi^{(\sigma)}$. Below, for simplicity, we write $\sigma \in \hat{G}$ instead of $\pi^{(\sigma)} \in \hat{G}$. It follows from (1.3) that

$$\chi_\pi(g) = M_1 \chi_{\sigma_1}(g) + \ldots + M_n \chi_{\sigma_n}(g). \tag{1.4}$$

If a group G is compact, then decomposition (1.3) is unique in the following sense: If, in addition to (1.3), the representation $\pi(g)$ admits a decomposition

$$\pi(g) \sim \bigoplus_{k=1}^{m} N_k \pi^{(\tau_k)}(g),$$

where $\pi^{(\tau_k)}$ are pairwise nonequivalent, then there exists a permutation ω of the set $\{1, \ldots, n\}$ such that $n = m$, $\tau_k = \sigma_{\omega(k)}$, and $N_k = M_{\omega(k)}$. Before proving this statement, we establish several properties of unitary representations of compact groups.

A unitary representation π is irreducible if and only if each bounded operator A commuting with all $\pi(g)$ $(g \in G)$ is a multiple of the identity operator. Indeed, let $A\pi(g) = \pi(g)A$ for any $g \in G$ and let the representation $\pi(g)$ be irreducible. Then $\pi(g)A^* = (A\pi(g^{-1}))^* = (\pi(g^{-1})A)^* = A^*\pi(g)$ and, since any bounded operator A can be represented in the form $A = A_1 + iA_2$, where $A_1 = (A_1 + A^*)/2$ and $A_2 = (A - A^*)/2$, it suffices to prove this statement for self-adjoint operators. Let A be a self-adjoint operator. Consider its spectral decomposition

$$A = \int_{-\infty}^{\infty} \lambda \, dE(\lambda).$$

Then $E(\lambda)\pi(g) = \pi(g)E(\lambda)$, whence, by virtue of the irreducibility of $\pi(g)$, we get either $E(\lambda) = 0$ or $E(\lambda) = I$. Since the function $(E(\lambda)\varphi, \varphi)_H$ is nondecreasing, one can find a real number λ_0 such that $E(\lambda) = 0$ for $\lambda < \lambda_0$ and $E(\lambda) = I$ for $\lambda \geq \lambda_0$. Consequently, $A = \lambda_0 I$. Let us prove the converse statement by contradiction. Let H_1 be an invariant subspace of the space of the representation $\pi(g)$. Denote by P the operator of projection onto H_1. Since $H_1^{\perp}$ is also invariant, we have $\pi(g)P = P\pi(g)$, i.e., the operator P commutes with $\pi(g)$. This implies that either $P = 0$ or $P = I$, and, hence, the representation $\pi(g)$ is irreducible.

■

The well-known *Schur lemma* (see, e.g., [Nai2]) appears to be quite useful in the theory of representations. We present this lemma without proof:

Let $\pi^{(1)}(g)$ and $\pi^{(2)}(g)$ be finite-dimensional irreducible representations of a group G in the spaces H_1 and H_2, respectively, and let an operator $A: H_1 \to H_2$ be permutable with these representations, i.e., $A\pi^{(1)}(g) = \pi^{(2)}(g)A$ $(g \in G)$. Then either A is the zero operator or it has the inverse operator (and, consequently, the representations $\pi^{(1)}$ and $\pi^{(2)}(g)$ are equivalent). In this case, A is uniquely defined (up to multiplication by a scalar).

A representation $\pi(g) = \pi^{(1)}(g) \otimes \pi^{(2)}(g)$ in the space $H_1 \otimes H_2$ is called the *tensor product of representations* $\pi^{(1)}(g)$ and $\pi^{(2)}(g)$ in the spaces H_1 and H_2, respectively. If $\{f_i^1\}$ and $\{f_j^2\}$ are bases in the spaces H_1 and H_2, respectively, then the matrix of the representation $\pi^{(1)}(g) \otimes \pi^{(2)}(g)$ in the basis $f_i^{(1)} \otimes f_j^{(2)}$ has the form $\pi_{ij,km} = \pi_{ik}^{(1)}(g)\,\pi_{jm}^{(2)}(g)$. It easy to see that the character of the tensor product of two finite-dimensional representations is equal to the product of their characters

$$\chi_{\pi^{(1)}\otimes\pi^{(2)}}(g) = \chi_{\pi^{(1)}}(g)\chi_{\pi^{(2)}}(g).$$

Let us prove the so-called orthogonality relations for matrix elements of irreducible representations of a compact group G, i.e.,

$$\int_G \pi_{ij}^{(\sigma)}(g)\overline{\pi_{kl}^{(\mu)}(g)}\,dg = 0, \tag{1.5}$$

if $\pi^{(\sigma)}$ and $\pi^{(\mu)}$ $(\sigma, \mu \in \hat{G})$ are not equivalent and

$$\int_G \pi_{ij}^{(\sigma)}(g)\overline{\pi_{kl}^{(\sigma)}(g)}\,dg = \frac{1}{d_\sigma}\delta_{i,k}\delta_{j,l}. \tag{1.6}$$

To prove (1.5), we consider the matrix

$$A_{ij} = \int_G \pi^{(\sigma)}(g)E_{ij}\pi^{(\mu)}(g^{-1})\,dg$$

where E_{ij} is a matrixunit, i.e., all elements of this matrix are equal to zero except the element e_{ij} which is equal to 1. Obviously, $\pi^{(\sigma)}(g)A_{ij} = A_{ij}\pi^{(\mu)}(g)$ $(g \in G)$. By virtue of the Schur lemma, this means that the matrix A_{ij} is equal to the zero matrix. The elements of this matrix are integrals of the form

$$\int_G \pi_{ki}^{(\sigma)}(g)\pi_{jp}^{(\mu)}(g^{-1})dg = \int \pi_{ki}^{(\sigma)}(g)\overline{\pi_{pj}^{(\mu)}(g)}dg,$$

whence we arrive at (15). To prove (1.6), we consider the matrix

$$A_{ij} = \int \pi^{(\sigma)}(g)E_{ij}\pi^{(\sigma)}(g^{-1})\,dg.$$

Since A_{ij} commutes with the matrix of the representation $\pi^{(\sigma)}(g)$, by using the Schur lemma, we conclude that A_{ij} is a scalar matrix, i.e., $A_{ij} = \lambda I$. Consequently,

$$\int_G \pi_{ki}^{(\sigma)}(g)\overline{\pi_{pj}^{(\sigma)}(g)}dg = \lambda\delta_{kp},$$

where λ does not depend on k and p. In order to find λ, we note that $\operatorname{Tr} A_{ij} = d_\sigma\lambda$. On the other hand, this trace is equal to the trace of matrix E_{ij}, i.e., to δ_{ij}.

■

Let us prove the uniqueness of decomposition (1.3) stated above. It follows from the orthogonality relations (1.5) and (1.6) that

$$\int_G \chi_\sigma(g)\chi_\mu(g)dg = \begin{cases} 1, & \text{if } \overline{\pi^{(\sigma)}} \sim \pi^{(\mu)}, \\ 0, & \text{otherwise.} \end{cases} \tag{1.7}$$

Here, the representation $\overline{\pi^{(\sigma)}}$ is written in the matrix form by the formula $G \ni g \mapsto \overline{(\pi_{ij}(g))}_{i,j=1}^{\dim\pi}$. Obviously, if $\pi^{(\sigma)}$ is irreducible, then $\overline{\pi^{(\sigma)}}$ is also irreducible. This implies that if the representation $\pi(g)$ can be decomposed into a sum of irreducible representations by formula (1.3), then it follows from (1.4) that

$$\int_G \chi_\pi(g)\chi_\sigma(g)dg = \begin{cases} M_j, & \text{if } \sigma \text{ coincides with } \sigma_j, \\ 0, & \text{otherwise.} \end{cases} \tag{1.8}$$

Since the character of the representation χ_π depends only on the equivalence class of the representation π, it follows from relations (1.8) that $\sigma_1, \dots, \sigma_n$ and $M_1, \dots, M_n$

are determined by the equivalence class which contains the representation π. This proves uniqueness of decomposition (1.3).

■

1.4. Peter–Weyl Theorem. In this subsection, we show that, up to a constant, the characters of the hypercomplex system $ZL_1(G)$ coincide with the characters of irreducible representations of the group G and establish fundamental statements of the theory of Peter and Weyl.

Theorem 1.1. *The set of characters $\chi(p)$ of the hypercomplex system $ZL_1(G)$ is in one-to-one correspondence with the set of characters $\chi_\sigma(g)$ of irreducible unitary representations of the group G. Moreover, these characters satisfy the relation*

$$\chi(p) = \frac{1}{d_\sigma}\chi_\sigma(g) \quad (g \in p,\ \chi(p) \in X,\ \sigma \in \hat{G}). \tag{1.9}$$

Proof. Let $\sigma \in \hat{G}$. Since the function $\chi_\sigma(g)$ is constant on conjugacy classes of the group G, we can introduce the function

$$\chi(p) = \frac{1}{d_\sigma}\chi_\sigma(g) \quad (g \in p \in Q)$$

defined on the set of conjugacy classes. Let us show that $\chi(p)$ is a character of the hypercomplex system $ZL_1(G)$. It is obvious that $|\chi(p)| \le 1$ $(p \in Q)$. Prior to checking equality (1.6) from Chapter 1, we show that, for all $x \in ZL_1(G)$,

$$t_{ik} = \int_G x(g)\pi_{ik}^{(\sigma)}(g)dg = \delta_{ik}\frac{1}{d_\sigma}\int_G x(g)\chi_\sigma(g)dg \tag{1.10}$$

$$(i, k = 1, \dots, d_\sigma),$$

where $\pi_{ik}^{(\sigma)}(g)$ are matrix elements of the irreducible unitary representation $\pi^{(\sigma)}(g)$. Indeed, consider the matrix $T = (t_{ik})_{i,k=1}^{d_\sigma}$. Since $x(gh) = x(hg)$ $(x \in ZL_1(G))$ almost everywhere on $G \times G$, by virtue of unimodularity of the group G, we obtain

$$\sum_k t_{ik}\pi_{kj}^{(\sigma)}(g) = \int_G x(a)\sum_k \pi_{ik}^{(\sigma)}(a)\,\pi_{kj}^{(\sigma)}(g)da = \int_G x(a)\pi_{ij}^{(\sigma)}(ag)da$$

$$= \int_G x(ag^{-1})\pi_{ij}^{(\sigma)}(a)da = \int_G x(g^{-1}a)\pi_{ij}^{(\sigma)}(a)da$$

$$= \int_G x(a)\pi_{ij}^{(\sigma)}(ga)da = \int_G x(a)\sum_k \pi_{ik}^{(\sigma)}(g)\,\pi_{kj}^{(\sigma)}(a)da$$

$$= \sum_k \pi_{ik}^{(\sigma)}(g)t_{kj},$$

whence $T\pi^{(\sigma)}(g) = \pi^{(\sigma)}(g)T$. Since the representation $\pi^{(\sigma)}(g)$ is irreducible, we have $T = \lambda I$. This yields (1.10). By virtue of (1.10), for all $A, B \in \mathcal{B}_0(Q)$, we obtain

$$\int_Q c(A, B, r)\chi(r)dr = \frac{1}{d_\sigma}\sum_i \int_G \kappa_{\mathfrak{A}} * \kappa_{\mathfrak{B}}(g)\pi_{ii}^{(\sigma)}(g)dg$$

$$= \frac{1}{d_\sigma}\sum_i \int_G \int_G \kappa_{\mathfrak{A}}(a)\kappa_{\mathfrak{B}}(a^{-1}g)da\pi_{ii}^{(\sigma)}(g)dg$$

$$= \frac{1}{d_\sigma}\sum_{i,j} \int_G \kappa_{\mathfrak{A}}(a)\int_G \kappa_{\mathfrak{B}}(g)\pi_{ij}^{(\sigma)}(a)\,\pi_{ji}^{(\sigma)}(g)dg\,da$$

$$= \frac{1}{d_\sigma}\int_G \kappa_{\mathfrak{A}}(a)\chi_\sigma(a)da\frac{1}{d_\sigma}\int_G \kappa_{\mathfrak{B}}(g)\chi_\sigma(g)dg = \chi(A)\chi(B),$$

where $\mathfrak{A}$ and $\mathfrak{B}$ are preimages of the sets A and B, respectively, under the natural mapping $G \ni g \mapsto p \in Q$ $(g \in p)$.

Conversely, let $\chi(p)$ be a character of the hypercomplex system $ZL_1(G)$. We regard it as a function on G (note that $\chi(gh) = \chi(hg)$ and $\chi(g^{-1}) = \overline{\chi(g)}$) and consider an integral operator

$$L_2(G, dg) \ni \varphi(g) \mapsto (K\varphi)(g) = \int_G \chi(gh^{-1})\varphi(h)dh.$$

Since

$$(K\varphi, \psi)_{L_2} = \int_G \int_G \chi(gh^{-1})\varphi(h)\overline{\psi(g)}dh\,dg$$

$$= \int_G \varphi(h)\overline{\int_G \chi(hg^{-1})\psi(g)dg}dh = (\varphi, K\psi)_{L_2},$$

the operator K is self-adjoint. Since $|\chi(g)| \le 1$ and the group G is compact, the operator K is compact. Let $\mu \ne 0$ be one of its eigenvalues and let $\varphi_1(g), \dots, \varphi_n(g)$ be an orthonormal basis of the corresponding eigenspace H_μ. Note that the functions

$\varphi_j(a)$ are continuous because $\varphi_j = (1/\mu)\varphi_j * \chi$. Since $\varphi(g) \in H_\mu$ and $\varphi(ga) \in H_\mu$ $(a \in G)$, we have

$$\varphi_j(ga) = \sum_k \pi_{jk}(a)\varphi_k(g) \quad (a, g \in G;\ j = 1, \dots, n). \tag{1.11}$$

It is easy to see that the matrices $\pi(g) = (\pi_{ik}(g))_{j,k=1}^n$ form a unitary representation of the group. Indeed, in view of the invariance of the Haar measure,

$$\pi_{jk}(a^{-1}) = \int_G \varphi_j(ga^{-1})\overline{\varphi_k(g)}dg = \overline{\int_G \varphi_k(ga)\overline{\varphi_j(g)}dg} = \overline{\pi_{kj}(a)},$$

$$\pi_{jk}(gh) = \int_G \varphi_j(agh)\overline{\varphi_k(a)}da = \int_G \varphi_j(ah)\overline{\varphi_k(ag^{-1})}da$$

$$= \sum_{p,q=1}^n \pi_{jp}(h)\overline{\pi_{kq}(g^{-1})}\int_G \varphi_p(a)\overline{\varphi_q(a)}da = \sum_{l=1}^n \pi_{jl}(h)\pi_{lk}(g).$$

It follows from the continuity of the convolution $\varphi * \psi(g)$ $(\varphi, \psi \in L_2(G, dg))$ with respect to g and the equality $\pi_{jk}(a) = \varphi_k^* * \varphi_j(a)$ $\left(\varphi_k^*(g) = \overline{\varphi_k(g^{-1})}\right)$ that the functions $G \ni a \mapsto \pi_{jk}(a)$ are continuous. The irreducibility of π is proved somewhat later. Let us show that $\chi(p) = d_\pi^{-1}\chi_\pi(g)$, where $\chi_\pi(g)$ is the character of the representation π. Let $\theta \neq \chi$ be any other character of the hypercomplex system $ZL_1(G)$. For any $\varphi(g) \in H_\mu$, in view of the equality $\chi * \theta = 0$ and the fact that the characters of the hypercomplex system $ZL_1(G)$ are Hermitian, we obtain

$$\int_G \varphi(g)\overline{\theta(g)}dg = \frac{1}{\mu}\int_G\int_G \chi(gh^{-1})\varphi(h)\,dh\overline{\theta(g)}dg$$

$$= \frac{1}{\mu}\int_G \varphi(h)\int_G \chi(h^{-1}g)\theta(g^{-1})\,dg\,dh$$

$$= \frac{1}{\mu}\int_G \varphi(h)(\chi * \theta)(h^{-1})\,dh = 0. \tag{1.12}$$

By virtue of $\chi(gh) = \chi(hg)$ $(h, g \in G)$, it is not difficult to show that every $\varphi_j(ag)$ regarded as a function of g belongs to H_μ. This, (1.11), and (1.12), enable us to conclude that

$$\int_G \chi_\pi(g)\overline{\theta(g)}dg = \sum_j \int_G \int_G \varphi_j(hg)\varphi_j(h)dh\overline{\theta(g)}dg$$

$$= \sum_j \int_G \int_G \varphi_j(hg)\overline{\theta(g)}dg\varphi_j(h)dh = 0. \tag{1.13}$$

Since the characters of the hypercomplex system $ZL_1(G)$ form a complete system in $L_2(G, dg)$, it follows from (1.13) that $\chi_\pi(g) = C\chi(g)$. At the same time, $\chi_\pi(e) = \dim \pi$ and $\chi(e) = 1$. Consequently, $\chi_\pi(g) = \dim \pi \chi(g)$.

Let us show that the representation π is irreducible. Indeed, we assume the opposite and decompose π into a direct sum of irreducible representations as follows:

$$\pi(g) = \bigoplus_{k=1}^{\nu} M_k \pi^{(\sigma_\kappa)}(g).$$

This gives

$$\chi(g) = \frac{1}{\dim \pi}\chi_\pi(g) = \frac{1}{\dim \pi}\sum_{k=1}^{\nu} M_k \chi_{\sigma_\kappa}(g).$$

By virtue of the first part of the theorem, the functions

$$\theta_k(g) = \frac{1}{d_\sigma}\chi_{\sigma_\kappa}(g)$$

are characters of the hypercomplex system $ZL_1(G)$ but this leads to a contradiction since the characters of the hypercomplex system $ZL_1(G)$ are orthogonal.

The established correspondence is one-to-one in view of the orthogonality of the characters of the group G and the same property of the characters of the hypercomplex system $ZL_1(G)$.

■

As a consequence of this theorem, we establish fundamental results of the theory of Peter and Weyl (see, e.g., [Nai1]).

Since characters of an arbitrary hypercomplex system with compact basis form an A-system of functions, by virtue of Theorem 3.8 in Chapter 1 and relations (1.8) and (1.9), the Plancherel measure on the set of classes of equivalence of irreducible representations is given by the formula

$$m(\sigma) = \left(\int_Q |\chi(p)|^2 dp\right)^{-1} = \left(\int_G \frac{1}{d_\sigma^2}|\chi_\sigma(g)|^2 dg\right)^{-1} = d_\sigma^2 \tag{1.14}$$

(here and in what follows, instead of $m(\pi^{(\sigma)})$, $c(\pi^{(\sigma)}, \pi^{(\mu)}, \pi^{(\tau)})$, etc., we write $m(\sigma)$, $c(\sigma, \mu, \tau), \ldots$, respectively).

The system of irreducible unitary representations $\pi^{(\sigma)}(g)$ $(\sigma \in \hat{G};\ g \in G)$ *is complete, i.e., for any* $g_1 \neq g_2$, *there exists a representation* $\pi^{(\sigma)}$ *such that* $\pi^{(\sigma)}(g_1) \neq \pi^{(\sigma)}(g_2)$ (the Gelfand–Raikov theorem for compact groups). This statement follows from Theorem 3.1 in Chapter 1.

Denote by Trig $\hat{G}$ the linear span of $\{\pi_{ij}^{(\sigma)}(g) \mid \sigma \in \hat{G},\ i, j = 1, \ldots, d_\sigma\}$. By decomposing the tensor product $\pi^{(\sigma)}(g) \otimes \pi^{(\mu)}(g)$ $(\sigma, \mu \in \hat{G})$ according to (1.3), we conclude, in view of the fact that $(\pi^{(\sigma)}(g) \otimes \pi^{(\mu)}(g))_{ij,kl} = \pi_{ij}^{(\sigma)}(g)\, \pi_{kl}^{(\mu)}(g)$, the lineal Trig $\hat{G}$ is an algebra with respect to the ordinary operations of summation and pointwise multiplication of functions. It is called the *representing algebra* of the group G.

In view of Theorem 3.13 from Chapter 1, Trig $\hat{G}$ *is dense in* $C(G)$ *in the uniform norm.*

We can now prove the main Peter–Weyl theorem which completely describes the structure of $L_2(G, dg)$.

Theorem 1.2. *Let* $\pi^{(\sigma)}(g)$ $(\sigma \in \hat{G})$ *be a complete system of pairwise inequivalent irreducible unitary representations of a compact group* G. *Then the functions* $\sqrt{d_\sigma}\, \pi_{ij}^{(\sigma)}(g)$ $(i, j = 1, \ldots, d_\sigma;\ \sigma \in \hat{G})$ *form an orthonormal basis of the space* $L_2(G, dg)$. *Every function* $f \in L_2(G, dg)$ *can be expanded into a convergent (in mean square) Fourier series with respect to the functions* $\pi_{ij}^{(\sigma)}(g)$, *namely,*

$$f(g) = \sum_{\sigma \in \hat{G}} \sum_{i,j=1}^{d_\sigma} a_{ij}^{(\sigma)} \pi_{ij}^{(\sigma)}(g). \tag{1.15}$$

Its coefficients are given by the formulas

$$a_{ij}^{(\sigma)} = d_\sigma \int_G f(g) \overline{\pi_{ij}^{(\sigma)}(g)}\, dg \tag{1.16}$$

and the Parseval equality is true

$$\int_G |f(g)|^2 dg = \sum_{\sigma \in \hat{G}} \frac{1}{d_\sigma} \sum_{i,j=1}^{d_\sigma} |a_{ij}^{(\sigma)}|^2. \tag{1.17}$$

Proof. The orthonormality of the system of functions $\sqrt{d_\sigma}\, \pi_{ij}^{(\sigma)}(g)$ follows from the orthogonality relations (1.5) and (1.6). The completeness of this system of functions follows from the denseness of Trig $\hat{G}$ in $C(G)$ in the uniform norm, the fact that $C(G)$ is dense in $L_2(G, dg)$, and the fact that the Haar measure is finite. The other statements follow from the elementary theory of Hilbert spaces.

■

Let us now present several properties of characters which are used in what follows. Assume that $x \in L_1(G, dg)$ and denote

$$x^{\natural}(g) = \int_G x(aga^{-1})\,da.$$

Clearly, $x^{\natural} \in ZL_1(G)$. It is not difficult to show that

$$\int_G \chi(g)x(g)\,dg = \int_G \chi(g)x^{\natural}(g)\,dg = \int_Q \chi(p)x^{\natural}(p)\,dp, \tag{1.18}$$

$$\chi(h)\chi(g) = \int_G \chi(haga^{-1})\,da \qquad (g, h \in G) \tag{1.19}$$

for any character χ *of the hypercomplex system* $ZL_1(G)$. Indeed, it suffices to prove (1.19). For any $x, y \in L_1(G, dg)$, we have

$$\begin{aligned}
\int_G x(g)\chi(g)\,dg \int_G y(h)\chi(h)\,dh &= \int_Q x^{\natural}(p)\chi(p)\,dp \int_Q y^{\natural}(q)\chi(q)\,dq \\
&= \int_Q (x^{\natural} * y^{\natural})(p)\chi(p)\,dp \\
&= \int_G \chi(g) \int_G x^{\natural}(h)y^{\natural}(h^{-1}g)\,dh\,dg \\
&= \int_G \int_G \int_G \chi(g) \int_G x(aha^{-1})y(bh^{-1}gb^{-1})\,dh\,da\,db\,dg \\
&= \int_G \int_G \chi(g)x(h) \int_G \int_G y(ba^{-1}h^{-1}gab^{-1})\,da\,db\,dg\,dh \\
&= \int_G \int_G \chi(g)x(h) \int_G y(a^{-1}h^{-1}ga)\,da\,dh\,dg \\
&= \int_G \int_G \int_G \chi(haga^{-1})\,da\,x(h)y(g)\,dh\,dg.
\end{aligned}$$

Relation (1.19) holds for almost all x and y owing to the arbitrariness of $g, h \in G$. By virtue of the continuity of $\chi(g)$, this implies (1.19) for all $g, h \in G$.

∎

It is not difficult to understand what modifications should be made in this reasoning to obtain the converse assertion: *If a function $f \in L_1(G, dg)$ satisfies the functional equation (1.19) for all $g, h \in G$, then either $f = 0$ or $f(g) = \chi(g)$ for some character $\chi(g)$ of the hypercomplex system $ZL_1(G)$.*

Let us now describe the dual hypercomplex system. By virtue of Theorem 1.1 and the fact that every character of the hypercomplex system with compact basis is essential, it is possible to consider the set $\hat{G}$ of equivalence classes of unitary representations of the group G as the basis $\hat{Q} = X_h$ of the dual hypercomplex system. Since Q is a compact set, $\hat{G}$ is discrete. We can find the structure measure $\hat{c}(\sigma, \mu, \tau)$ $(\sigma, \mu, \tau \in \hat{G})$ of the dual hypercomplex system by using relation (3.6) from Chapter 1. Indeed,

$$\hat{c}(\sigma, \mu, \tau) = \frac{d_\sigma d_\mu}{d_\tau} \int_G \chi_\sigma(g)\chi_\mu(g)\overline{\chi_\tau(g)}dg.$$

By virtue of (1.4),

$$\chi_\sigma(g)\chi_\mu(g) = \chi_{\pi(\sigma)\otimes\pi(\mu)}(g) = \sum_\tau M(\sigma, \mu, \tau)\chi_\tau(g). \tag{1.20}$$

Taking (1.8) into account, we conclude that

$$\hat{c}(\sigma, \mu, \tau) = \frac{M(\sigma,\mu,\tau)d_\sigma d_\mu}{d_\tau}$$

if the representation $\pi^{(\tau)}(g)$ belongs to expansion (1.3) of $\pi^{(\sigma)}(g) \oplus \pi^{(\mu)}(g)$ into irreducible representations; otherwise, $\hat{c}(\sigma, \mu, \tau)$ is equal to zero. Thus, $\hat{c}(\sigma, \mu, \tau) \geq 0$ and its finiteness with respect to τ is a consequence of (1.3). It follows from (1.20) and Corollary 3.1 in Chapter 1 that the product of any two positive definite continuous functions on the hypercomplex system $ZL_1(G)$ is positive definite. As already mentioned (see (1.14)), the multiplicative measure of the dual hypercomplex system is given by the equality $\hat{m}(\sigma) = d_\sigma^2$ $(\sigma \in \hat{G})$. By virtue of Theorem 3.5 in Chapter 1, *duality takes place for the hypercomplex system $ZL_1(G)$, i.e., the support of the Plancherel measure of the dual hypercomplex system is homeomorphic to the compact set of conjugacy classes of the group G.*

1.5. Tannaka–M. Krein Duality Theorem. As is known, the set $\hat{G}$ of irreducible unitary representations of an Abelian locally compact group G coincides with the group of characters of the group G and, therefore, it is the object dual to the group G. If G

is a compact, generally speaking, noncommutative group, then $\hat{G}$ is not a group but a hypergroup. Nevertheless, according to the Tannaka–M. Krein theorem, it is possible to reconstruct a group if the tensor product of any pair of irreducible representations and the adjoint of each irreducible representation are known (in a certain sense). In the present subsection, we prove the Tannaka–M. Krein theorem by using the duality theorem for hypercomplex systems. To do this, we apply the results of Section 3 in Chapter 1, to the hypercomplex system $ZL_1(G)$ and establish the Tannaka–M. Krein duality principle for compact groups ([Tan], [Kre1], and [Kre2]).

If $G \ni g \mapsto \left(\pi_{ij}(g)\right)_{i,j=1}^{\dim \pi}$ is a unitary representation of G in the matrix form, then the mapping $G \ni g \mapsto \overline{\left(\pi_{ij}(g)\right)_{i,j=1}^{\dim \pi}}$ is also a unitary representation of G. Consequently, together with any function, the representing algebra $\operatorname{Trig} \hat{G}$ contains its complex conjugate.

Since every finite-dimensional unitary representation is equivalent to a direct sum of finitely many irreducible representations, the matrix elements of irreducible representations $\pi_{ij}^{(\sigma)}(g)$ $(\sigma \in \hat{G};\ i, j = 1, \dots, d_\sigma)$ form a basis in $\operatorname{Trig} \hat{G}$. In particular, by expanding the tensor product $\pi^{(\sigma)} \otimes \pi^{(\mu)}$ of any two irreducible representations according to (1.3), we obtain the matrix equality

$$\pi^{(\sigma)}(g) \otimes \pi^{(\mu)}(g) = S_{\sigma\mu}^{-1} \bigoplus_{i=1}^{k} M(\sigma, \mu, \tau_i) \pi^{(\tau_i)}(g) S_{\sigma\mu}, \tag{1.21}$$

where $S_{\sigma\mu}$ is a unitary numerical matrix which does not depend on g. The character of representation $\pi^{(\sigma)} \otimes \pi^{(\mu)}$ is computed according to (1.20). By integrating (1.20), we get

$$\int_H \chi_\sigma(g) \chi_\mu(g)\, dg = \sum_{i=1}^{k} M(\sigma, \mu, \tau_i) \int_G \chi_{\tau_i}(g)\, dg. \tag{1.22}$$

By virtue of (1.7), the left-hand side of (1.22) is equal to 1 or 0 if the representation $\pi^{(\mu)}$ is equivalent or not equivalent to the representation $\overline{\pi^{(\sigma)}}$, respectively. The integral $\int_G \chi_{\tau_i}(g)\, dg$ is equal to 1 or 0 if the representation $\pi^{(\tau_i)}(g)$ is trivial (one-dimensional and identically equal to one) or nontrivial, respectively.

Thus, *the multiplicity of the trivial representation in decomposition (1.21) is equal to zero if* $\overline{\pi^{(\sigma)}}$ *is not equivalent to* $\pi^{(\mu)}$ *and is equal to one whenever* $\overline{\pi^{(\sigma)}} \sim \pi^{(\mu)}$.

A commutative algebra $\mathfrak{A}$ with identity e and involution $*$ is called a *Krein algebra* (or a *square block algebra* according to M. Krein) if there exists an at most countable basis which can be split into disjoint sets (blocks) U_σ with d_σ^2 elements (d_σ is called the *rank of a block)* which have the form of square matrices $U_\sigma = (u_{ij}^{(\sigma)})_{i,j=1}^{d_\sigma}$ such that following conditions are satisfied:

1. The collection of sets U_σ contains a set with a single element e.

2. Each block U_σ is associated with a numerical matrix S_σ of order d_σ such that the set $S_\sigma^{-1} U_\sigma^* S_\sigma$, where $U_\sigma^* = (u_{ij}^{(\sigma)*})_{i,j=1}^{d_\sigma}$, is also a block from the indicated collection of sets.

3. For any two blocks U_σ and U_μ of orders d_σ and d_μ, respectively, there exists a unitary numerical matrix $S_{\sigma\mu}$ of order $d_\sigma \times d_\mu$ such that

$$U_\sigma \otimes U_\mu = S_{\sigma\mu}^{-1} \bigoplus_{i=1}^{p} U_{\sigma_i} S_{\sigma\mu} \tag{1.23}$$

for some $\sigma_1, \ldots, \sigma_p$ ($U_\sigma \otimes U_\mu$ denotes the matrix $\left(u_{ij}^{(\sigma)} u_{ki}^{(\mu)}\right)$).

4. The block $\{e\}$ either does not belong to the right-hand side of (1.23) or appears there exactly once; the latter takes place if and only if $U_\mu = S_\sigma^{-1} U_\sigma^* S_\sigma$, where S_σ is a unitary numerical matrix.

5. Each block $U_\sigma = (u_{ij}^{(\sigma)})_{i,j=1}^{d_\sigma}$ satisfies the condition

$$U_\sigma U_\sigma^{*T} = \left(\delta_{i,k} e\right)_{i,k=1}^{d_\sigma}, \tag{1.24}$$

where $\delta_{i,k}$ is the Kronecker symbol and U_σ^T is the relevant transposed matrix. Note that relation (1.23) defines, in fact, the operation of multiplication of basis elements of the algebra $\mathfrak{A}$ (and, hence, of its arbitrary elements).

The reasoning presented above demonstrates that *the representing algebra* Trig $\hat{G}$ *of a compact group* G *is a Krein algebra.*

Every element φ of the Krein algebra $\mathfrak{A}$ can be represented in the form

$$\varphi = \sum_\sigma \operatorname{Tr}(F_\sigma U_\sigma), \tag{1.25}$$

where each F_σ is a square matrix of order d_σ and d_σ is the order of the block U_σ (the sum in (1.25) is finite). This fact enables us to introduce a norm in $\mathfrak{A}$ by the formula

$$\|\varphi\| = \sum_\sigma \operatorname{Tr}|F_\sigma| \tag{1.26}$$

(recall that $|F_\sigma| = \sqrt{F_\sigma F_\sigma^*}$). It is not difficult to show that $\mathfrak{A}$ with norm (1.26) is a normed space (not necessary complete). Let us prove that $\mathfrak{A}$ *with this norm is an invo-*

lutive normed algebra, i.e., $\|\varphi\psi\| \le \|\varphi\|\|\psi\|$ *and* $\|\varphi^*\| = \|\varphi\|$ *for all* $\varphi, \psi \in \mathfrak{A}$.

It suffices to establish these relations for $\varphi = \mathrm{Tr}(AU_\sigma)$ and $\psi = \mathrm{Tr}(BU_\mu)$. By using (1.23), we obtain

$$\begin{aligned}\varphi\cdot\psi = \mathrm{Tr}(AU_\sigma)\mathrm{Tr}(BU_\mu) &= \mathrm{Tr}((A\otimes B)(U_\sigma\otimes U_\mu))\\ &= \mathrm{Tr}\left((A\otimes B)S_{\sigma\mu}^{-1}\bigoplus_{k=1}^{p} U_{\sigma_k}S_{\sigma\mu}\right)\\ &= \mathrm{Tr}\left(S_{\sigma\mu}(A\otimes B)S_{\sigma\mu}^{-1}\bigoplus_{k=1}^{p} U_{\sigma_k}\right) = \sum_{k=1}^{p}\mathrm{Tr}(C_kU_{\sigma_\kappa}),\end{aligned}$$

where C_k $(k = 1, \dots, p)$ are matrices composed of elements of the matrix $C = S_{\sigma\mu}(A\otimes B)S_{\sigma\mu}^{-1}$ and located in C at the same places as the matrices $U\sigma_\kappa$ in the sum $\bigoplus_{k=1}^{p} U_{\sigma_\kappa}$. Taking into account the fact that the set of matrices U_{σ_κ} may contain equal matrices, we obtain

$$\begin{aligned}\|\varphi\psi\| \le \sum_{k=1}^{p}\mathrm{Tr}|C_k| &= \mathrm{Tr}\left|\bigoplus_{k=1}^{p} C_k\right| \le \mathrm{Tr}|C|\\ &= \mathrm{Tr}|A\otimes B| = \mathrm{Tr}|A|\mathrm{Tr}|B| = \|\varphi\|\|\psi\|\end{aligned}$$

(the proof is based on the use of the well-known properties of traces).

In order to check the equality $\|\varphi^*\| = \|\varphi\|$, we note that $\varphi^* = \sum_\sigma \mathrm{Tr}\left(\overline{A}_\sigma U_\sigma^*\right)$. According to condition 2 in the definition of the Krein algebra, every block U_σ is associated with a block U_μ such that $U_\sigma^* = S_\mu U_\mu S_\mu^{-1}$. As a consequence φ^* can be represented in the form

$$\varphi^* = \sum_\sigma \mathrm{Tr}\left(S_\mu^{-1}\overline{A}_\sigma S_\mu U_\mu\right).$$

Consequently,

$$\|\varphi^*\| = \sum_\sigma \mathrm{Tr}|S_\mu^{-1}A_\sigma S_\mu| = \|\varphi\|.$$

■

By completing the Krein algebra $\mathfrak{A}$ with respect to norm (1.26), we arrive at a commutative Banach $*$-algebra $\mathfrak{R}(\mathfrak{A})$ which is called a *complete Krein algebra.* The algebra $\mathfrak{R}(\mathfrak{A})$ can be regarded as a set of infinite formal sums (1.25) for which series (1.26) is convergent. The Krein algebra $\mathfrak{A}$ is a dense part of $\mathfrak{R}(\mathfrak{A})$ and consists of sums (1.25) with finitely many terms.

Let us show that the *involutive Banach algebra* $\mathfrak{R}(\mathfrak{A})$ *is symmetric*. To do this, it suffices to prove the equality $g(U_\sigma^*) = \overline{g(U_\sigma)}$, where U_σ is an arbitrary block and

$$\overline{g(U_\sigma)} = \overline{\left(g\left(u_{ij}^{(\sigma)}\right)\right)}_{i,j=1}^{d_\sigma},$$

for every maximal ideal g of the algebra $\mathfrak{R}(\mathfrak{A})$. Indeed, the ideal g satisfies the equalities $g(e)=1$ and $|g(\varphi)| \le \|\varphi\|$ $(\varphi \in \mathfrak{R}(\mathfrak{A}))$. Applying the last inequality to the element $\varphi = \operatorname{Tr}(AU_\sigma)$ and using the known fact that $\operatorname{Tr}(AB) \le \operatorname{Tr}|A| \cdot \|B\|$ (see, e.g., [ReS]), we get $\|g(U_\sigma)\| \le 1$. Further, by virtue of condition 5 in the definition of a Krein algebra, we have $U_\sigma U_\sigma^{*T} = (\delta_{i,j}e)_{i,j=1}^{d_\sigma}$, whence $g(U_\sigma)^{-1} = g(U_\sigma^*)^T$. Consequently, by virtue of condition 2,

$$\|g(U_\sigma)^{-1}\| = \|g(U_\sigma^*)\| = \|S_\sigma^{-1} g(U_\sigma) S_\sigma\| = \|g(U_\sigma)\| = 1.$$

This implies that the matrix $g(U_\sigma)$ is unitary and, hence, $g(U_\sigma^*) = g((U_\sigma)^T)^{-1} = \overline{g(U_\sigma)}$.

■

A complete Krein algebra generated by the representing algebra Trig $\hat{G}$ of a compact group G coincides with the Fourier algebra $A(g)$ of the group G (see, e.g., [Eym] and [HeR]); moreover, it also coincides with the linear span of the set of continuous positive definite functions on G.

Let us introduce the notion of a *representing group* of a Krein algebra $\mathfrak{A}$. A functional $g(a)$ $(a \in \mathfrak{A})$ is called *elementary* if it is a symmetric homomorphism of the algebra $\mathfrak{A}$ in the field of complex numbers, i.e., if $g(ab) = g(a)g(b)$ and $g(a^*) = \overline{g(a)}$ $(a, b \in \mathfrak{A})$. Every square block $U_\sigma = (u_{ij}^{(\sigma)})_{i,j=1}^{d_\sigma}$ with elements from $\mathfrak{A}$ can be associated with a numerical matrix

$$g(U_\sigma) = \left(g\left(u_{ij}^{(\sigma)}\right)\right)_{i,j=1}^{d_\sigma},$$

where g is an elementary functional. The matrices $g(U_\sigma)$ are unitary by virtue of (1.24) and satisfy the same conditions (1.23) as the blocks themselves. Conversely, it is obvious that every mapping of the blocks U_σ into a set of unitary matrices of the same orders d_σ that satisfy conditions (1.23) defines an elementary functional. Therefore, we can introduce a group operation in the set $G_{\mathfrak{A}}$ of all elementary functionals of the Krein algebra $\mathfrak{A}$. Namely, we set $(gh)(U_\sigma) = g(U_\sigma)h(U_\sigma)$. It is obvious that relations (1.23) are invariant under the mapping $U_\sigma \mapsto gh(U_\sigma)$ and, therefore, it generates an elementary functional. One can now easily check that $G_{\mathfrak{A}}$ is a group. The group $G_{\mathfrak{A}}$ is called the *representing group* of a Krein algebra $\mathfrak{A}$.

There is a one-to-one correspondence between the set of elementary functionals $G_{\mathfrak{A}}$ *and the set of maximal ideals of the complete Krein algebra* $\mathfrak{R}(\mathfrak{A})$. Indeed, since the algebra $\mathfrak{R}(\mathfrak{A})$ is symmetric, it is clear that any maximal ideal is an elementary functional. Conversely, by virtue of the inequality $|\operatorname{Tr} AB| \leq \|A\| \cdot \operatorname{Tr}|B|$, we have

$$|g(\varphi)| = \left|\sum_{\sigma} \operatorname{Tr}[F_\sigma g(U_\sigma)]\right| \leq \left|\sum_{\sigma} \operatorname{Tr}\left[g(U_\sigma)^{-1} F_\sigma\right]\right| \leq \sum_{\sigma} \operatorname{Tr}|F_\sigma| = \|\varphi\|$$

for any $\varphi \in \mathfrak{A}$ and $g \in G_{\mathfrak{A}}$. Since $\mathfrak{A}$ is dense in $\mathfrak{R}(\mathfrak{A})$, we conclude that the functional g is bounded and, furthermore, $\|g\| \leq 1$. Thus, it is a maximal ideal of the complete Krein algebra $\mathfrak{R}(\mathfrak{A})$.

■

Note that the group $G_{\mathfrak{A}}$ is compact in the topology of the space of maximal ideals.

As mentioned above, the representing algebra $\operatorname{Trig} \hat{G}$ of a compact group G is a Krein algebra. Let $\hat{\hat{G}}$ denote the representing group of the Krein algebra $\operatorname{Trig} \hat{G}$. Since the mapping $\operatorname{Trig} \hat{G} \ni f \mapsto f(g)$ (here, g is a fixed element of the group G) is an elementary functional, we conclude that $G \subset \hat{\hat{G}}$. It is clear that G is a closed subset of $\hat{\hat{G}}$ by the definition of topology in the group $\hat{\hat{G}}$.

Let $G \ni g \mapsto \pi(g)$ be a finite-dimensional unitary representation of the group G. The correspondence $\hat{\hat{G}} \ni \mathfrak{g} \mapsto \pi(\mathfrak{g}) = \mathfrak{g}(\pi) = (\mathfrak{g}(\pi_{ij}))_{i,j=1}^{\dim \pi} = (\pi_{ij}(\mathfrak{g}))_{i,j=1}^{\dim \pi}$ associates every $\mathfrak{g} \in \hat{\hat{G}}$ with the unitary matrix $(\pi_{ij}(\mathfrak{g}))_{i,j=1}^{\dim \pi}$. Obviously, this mapping is a unitary representation of the group $\hat{\hat{G}}$. If the representation $\pi(g)$ of the group G is irreducible, then the representation $\pi(\mathfrak{g})$ of the group $\hat{\hat{G}}$ generated by the representation $\pi(g)$ is also irreducible. Note that *representations of the group* $\hat{\hat{G}}$ *generated by all irreducible unitary representations* $\pi^{(\sigma)}(g)$ $(\sigma \in \hat{G})$ *of the group* G *form a complete system of representations of* $\hat{\hat{G}}$. Indeed, let $\mathfrak{g}_1 \neq \mathfrak{g}_2$; this means that there exists a unitary representation $\pi(g)$ of the group G such that $\mathfrak{g}_1(\pi) \neq \mathfrak{g}_2(\pi)$. By decomposing $\pi(g)$ into irreducible representations according to (1.3), we establish that $\mathfrak{g}_1(\pi^{(\sigma)}) \neq \mathfrak{g}_2(\pi^{(\sigma)})$ at least for one $\sigma \in \hat{G}$.

■

Below, we present the well-known Tannaka–M. Krein duality theorem (see [Tan] and [Kre1], [Kre2]).

Theorem 1.3.

(i) If G is a compact group and $\mathrm{Trig}\,\hat{G}$ *is its representing algebra, then the representing group* $\hat{\hat{G}}$ *of the algebra* $\mathrm{Trig}\,\hat{G}$ *is isomorphic to the group* G.

(ii) If $\mathfrak{A}$ *is a Krein algebra and* $G_{\mathfrak{A}}$ *is its representing group, then there exists an isomorphism between the algebra* $\mathfrak{A}$ *and the representing algebra* $\mathrm{Trig}\,\hat{G}_{\mathfrak{A}}$ *of the group* $G_{\mathfrak{A}}$. *This isomorphism associates the block-basis* U_σ *of the algebra* $\mathfrak{A}$ *with the complete system of irreducible unitary representations of the group* $G_{\mathfrak{A}}$.

Proof. Assertion (i) follows from the fact that the set $\hat{\hat{G}}$ of elementary functionals of the algebra $\mathrm{Trig}\,\hat{G}$ is homeomorphic to the space of maximal ideals of the Fourier algebra $A(g)$, which is homeomorphic to the group G (see, e.g., [Eym] and [HeR]).

We prove assertion (ii) in several steps.

1. Let $\mathfrak{A}$ be a Krein algebra and let Σ denote the set of indices σ of the blocks U_σ equipped with discrete topology.

Each block U_σ $(\sigma \in \Sigma)$ generates an irreducible unitary representation of the group $G_{\mathfrak{A}}$ according to the formula $\pi^{(\sigma)}(g) = g(U_\sigma)$. To prove this it suffices to show that the representation $\pi^{(\sigma)}$ is irreducible. Assume the contrary. Then one can find a bounded operator A such that $A \neq \lambda I$ and $A\pi^{(\sigma)}(g) = \pi^{(\sigma)}(g)A$. For any $g \in G_{\mathfrak{A}}$, $i, j = 1, \ldots, d_\sigma$, we have

$$0 = \sum_k \left(a_{ik}\pi_{kj}^{(\sigma)}(g) - a_{kj}\pi_{ik}^{(\sigma)}(g)\right) = g\left[\sum_k \left(a_{ik}u_{kj}^{(\sigma)} - a_{kj}u_{ik}^{(\sigma)}\right)\right].$$

Since $G_{\mathfrak{A}}$ coincides with the space of maximal ideals of the Banach algebra $\mathfrak{R}_{\mathfrak{A}}$, the last equality yields

$$\sum_k \left(a_{ik}u_{kj}^{(\sigma)} - a_{kj}u_{ik}^{(\sigma)}\right) = 0$$

for all $i, j = 1, \ldots, d_\sigma$. Since $A \neq \lambda I$, it follows from these equalities that the vectors $u_{ij}^{(\sigma)} \in U_\sigma$ $(i, j = 1, \ldots, d_\sigma)$ are linearly dependent. On the other hand, as elements of a basis of the algebra $\mathfrak{A}$, they must be linearly independent.

2. The system of irreducible unitary representations $\pi^{(\sigma)}(g) = g(U_\sigma)$ $(\sigma \in \Sigma$, $g \in G_{\mathfrak{A}})$ of the group $G_{\mathfrak{A}}$ is complete, i.e., for every $g_1 \neq g_2$, one can find $\sigma \in \Sigma$ such that $\pi^{(\sigma)}(g_1) \neq \pi^{(\sigma)}(g_2)$. This statement obviously follows from the facts that

$G_{\mathfrak{A}}$ coincides with the space of maximal ideals of the Banach algebra $\mathfrak{R}(\mathfrak{A})$, and the elements $u_{ij}^{(\sigma)} \in U_\sigma$ ($\sigma \in \Sigma$, $i, j = 1, \dots, d_\sigma$) form a basis in $\mathfrak{R}(\mathfrak{A})$.

By using the results of steps 1 and 2, one can establish that there is an isomorphism between the Krein algebra $\mathfrak{A}$ and a certain subalgebra $\mathfrak{A}'$ of the representing algebra Trig $\hat{G}_{\mathfrak{A}}$ of the group $G_{\mathfrak{A}}$ and, furthermore, this isomorphism associates the block-basis U_σ of the algebra $\mathfrak{A}$ with the complete system of irreducible unitary representations of the group $G_{\mathfrak{A}}$. To complete the proof it suffices to verify that $\mathfrak{A}'$ coincides with Trig $G_{\mathfrak{A}}$. We prove this fact by using the duality theory for hypercomplex systems.

3. We set $c(\sigma, \mu, \tau) = M(\sigma, \mu, \tau) d_\sigma d_\mu (d_\tau)^{-1}$, where $M(\sigma, \mu, \tau)$ is the multiplicity of the block U_τ in expansion (1.23) of the product $U_\sigma \otimes U_\mu$ in the "block-basis" ($M(\sigma, \mu, \tau) = 0$ if the right-hand side of (1.23) does not contain U_τ). Also denote $m(\sigma) = d_\sigma^2$. One can easily verify that $c(\sigma, \mu, \tau)$ is a structure measure and $m(\sigma)$ is a multiplicative measure. Denote the obtained hypercomplex system by $Zl_1(\Sigma)$. We introduce the involution $\Sigma \ni \sigma \mapsto \sigma^* \in \Sigma$ in the hypercomplex system $Zl_1(\Sigma)$, where σ^* is the index of the block $S_\sigma^{-1} U_\sigma^* S_\sigma$ in condition 2 in the definition of Krein algebras. Obviously, the hypercomplex system $Zl_1(\Sigma)$ is normal and possesses the basis unity (the index of the block $U_e = \{e\}$). It is easy to see that the hypercomplex system $Zl_1(\Sigma)$ is isometrically isomorphic to a closed subalgebra of the Banach $*$-algebra $\mathfrak{R}(\mathfrak{A})$ which consists of vectors of the form

$$\varphi = \sum_\sigma \mathrm{Tr}(F_\sigma U_\sigma),$$

where

$$F_\sigma = x(\sigma) d_\sigma I_\sigma, \quad \sum_\sigma |x(\sigma)| d_\sigma^2 < \infty.$$

According to the results obtained in [FeT], the hypercomplex system $Zl_1(\Sigma)$ is symmetric because $\mathfrak{R}(\mathfrak{A})$ is symmetric and $Zl_1(\Sigma)$ contains the identity.

4. Denote by $\overline{Q}$ the space of characters of the hypercomplex system $Zl_1(\Sigma)$. Let us show that $\overline{Q}$ is homeomorphic to the compact set $Q_{\mathfrak{A}}$ of conjugacy classes of the group $G_{\mathfrak{A}}$. Indeed, assume that $\mathfrak{g}(\sigma)$ ($\sigma \in \Sigma$) is a character of the hypercomplex system $Zl_1(\Sigma)$. As indicated above, $Zl_1(\Sigma)$ is isometrically isomorphic to a closed subalgebra of the Banach algebra $\mathfrak{R}(\mathfrak{A})$. Therefore, according to the Shilov theorem on the extensions of maximal ideals [Shi], the character $\mathfrak{g}(\sigma)$ can be extended to a maximal ideal $g \in G_{\mathfrak{A}}$ of the algebra $\mathfrak{R}(\mathfrak{A})$ and, obviously,

$$\mathfrak{g}(\sigma) = \frac{1}{d_\sigma} \mathrm{Tr}\, g(U_\sigma) = \frac{1}{d_\sigma} \chi_\sigma(g), \tag{1.27}$$

where χ_σ is the character of the irreducible representation $\pi^{(\sigma)}$ of the group $G_{\mathfrak{A}}$ corresponding to the block U_σ. We associate the character $\mathfrak{g}(\sigma)$ with the conjugacy class p of the group $G_{\mathfrak{A}}$ that contains the element g. It follows from (1.27) that p does not depend on the choice of g.

Conversely, let $p \in Q_{\mathfrak{A}}$. We choose an arbitrary element $g \in p$ and define a character $\mathfrak{g}(\sigma)$ of the hypercomplex system $Zl_1(\Sigma)$ by relation (1.27); the fact that $\mathfrak{g}(\sigma)$ thus defined is a character follows from (1.27), the definition of $c(\sigma, \mu, \tau)$, and condition 3 in the definition of Krein algebras. Obviously, $\mathfrak{g}(\sigma)$ does not depend on the choice of $g \in p$. Thus, there is a correspondence between the space of characters $\overline{Q}$ of the hypercomplex system $Zl_1(\Sigma)$ and the compact set $Q_{\mathfrak{A}}$ of conjugacy classes of the group $G_{\mathfrak{A}}$. This correspondence is bijective by virtue of (1.27) and the completeness of the system of irreducible representations $\pi^{(\sigma)}$ of the group $G_{\mathfrak{A}}$ generated by the blocks U_σ $(\sigma \in \Sigma)$. Since $\overline{Q}$ and $Q_{\mathfrak{A}}$ are equipped with the same topology as the spaces of maximal ideals of the algebra $\mathfrak{R}(\mathfrak{A})$, we conclude that this correspondence is a homeomorphism.

5. Let us show that duality takes place for the hypercomplex system $Zl_1(\Sigma)$. First, note that the measure $\hat{m}$ on the space $Q_{\mathfrak{A}}$ of conjugacy classes of the group $G_{\mathfrak{A}}$ generated by the Haar measure of $G_{\mathfrak{A}}$ according to (1.1) coincides with the multiplicative measure of the hypercomplex system dual to $Zl_1(\Sigma)$. Since $\operatorname{supp} \hat{m} = Q_{\mathfrak{A}}$, by virtue of Theorem 3.7 in Chapter 1, it suffices to check that, for any $\mathfrak{g}_1, \mathfrak{g}_2 \in \overline{Q}$, the product $\mathfrak{g}_1(\sigma)\mathfrak{g}_2(\sigma)$ is positive definite. Let p_1 and p_2 be the conjugacy classes of the group $G_{\mathfrak{A}}$ that correspond to the characters $\mathfrak{g}_1$ and $\mathfrak{g}_2$. Let $g_1 \in p_1$ and $g_2 \in p_2$ be arbitrary elements of these classes. Taking into account (1.27), (1.19), and the invariance of the Haar measure of the group $G_{\mathfrak{A}}$, we get

$$\mathfrak{g}_1(\sigma)\mathfrak{g}_2(\sigma) = \chi(g_1)\chi(g_2) = \int_{G_{\mathfrak{A}}} \chi(g_1 a g_2 a^{-1})\, da$$

$$= \int_{G_{\mathfrak{A}}} \chi(b g_1 a g_2 a^{-1} b^{-1})\, da = \int_{G_{\mathfrak{A}}} \int_{G_{\mathfrak{A}}} \chi(b g_1 b^{-1} a g_2 a^{-1})\, da\, db,$$

where $\chi(g) = \chi_\sigma(g)/d_\sigma$ is the character of the hypercomplex system $Zl_1(G_{\mathfrak{A}})$ corresponding to the character χ_σ of the irreducible representation $\pi^{(\sigma)}$ generated by the block U_σ $(\sigma \in \Sigma)$. Let us define a measure $d\mu_{p_1, p_2}(p)$ on $\overline{Q}$ by the equality

$$\int_{\overline{Q}} f(p)\, d\mu_{p_1, p_2}(p) = \int_{G_{\mathfrak{A}}} \int_{G_{\mathfrak{A}}} f(a g_1 a^{-1} b g_2 b^{-1})\, da\, db \quad (f \in C(\overline{Q})).$$

Then, by virtue of Theorem 3.1 in Chapter 1, the product $\mathfrak{g}_1(\sigma)\mathfrak{g}_2(\sigma)$ of characters

$\mathfrak{g}_1$ and $\mathfrak{g}_2$ of the hypercomplex system $Zl_1(\Sigma)$ is positive definite.

It is obvious that the hypercomplex system $Zl_1(G_{\mathfrak{A}})$ is dual to the hypercomplex system $Zl_1(\Sigma)$.

6. Let us complete the proof of Theorem 1.3. Assume that π is an arbitrary irreducible unitary representation of the group $G_{\mathfrak{A}}$ and

$$\chi(p) = \frac{1}{\dim \pi}\chi_\pi(g) \quad (g \in p;\ p \in Q_{\mathfrak{A}})$$

is the character of the hypercomplex system $Zl_1(G_{\mathfrak{A}})$ that corresponds to the character χ_π of the representation π. Since duality takes place for the hypercomplex system $Zl_1(\Sigma)$, there exists $\sigma \in \Sigma$ such that $\chi(p) = \mathfrak{g}(\sigma)$ (here, $\mathfrak{g}$ is the character of the hypercomplex system $Zl_1(\Sigma)$ that corresponds to the class $p \in Q_{\mathfrak{A}}$). By virtue of (1.27), we have $\chi_\pi(g) = \chi_\sigma(g)$, where χ_σ is the character of the representation $\pi^{(\sigma)}$ of the group $G_{\mathfrak{A}}$ generated by the block U_σ. This means that the representations π and $\pi^{(\sigma)}$ are equivalent. Since the matrix elements of all irreducible representations of the group $G_{\mathfrak{A}}$ form a basis of the representing algebra Trig $\hat{G}_{\mathfrak{A}}$ of the group $G_{\mathfrak{A}}$, we conclude that $\mathfrak{A}'$ coincides with Trig $\hat{G}_{\mathfrak{A}}$.

■

1.6. Elements of the Theory of Semisimple Groups and Lie Algebras. Below, we describe the hypercomplex system $Zl_1(G)$ in the case where G is a compact semisimple Lie group. These results were obtained by Berezin and Gelfand [BeG] and Gelfand [Gel1]. First, recall basic definitions and results concerning semisimple Lie groups and Lie algebras.

A set G is called a *Lie group* if it is a topological group and a smooth manifold such that the mapping $\varphi\colon G \times G \to G$ defined by the equality $\varphi(x, y) = xy^{-1}$ is smooth. A linear space $\mathfrak{g}$ with a bilinear operation $\mathfrak{g} \times \mathfrak{g} \ni (X, Y) \mapsto [X, Y] \in \mathfrak{g}$ is called a *Lie algebra* if the following conditions are satisfied:

(i) $[X, Y] = -[Y, X]$ for all $X, Y \in \mathfrak{g}$;

(ii) the Jacobi identity

$$[[X, Y], Z] + [[Y, Z], X] + [[Z, X], Y] = 0$$

holds for all $X, Y, Z \in \mathfrak{g}$.

With every Lie group, one can associate a finite Lie algebra. Namely, denote by $A(g)$ $(g \in G)$ an inner automorphism of the group G; every inner automorphism is a diffeomorphism of the group G. By $\mathfrak{g} = T_e G$ we denote the tangent space at the neutral element e of the group G and consider the tangent mapping $Ad_g = dA(g)(e)$: $\mathfrak{g} \to \mathfrak{g}$. The mapping $G \ni g \mapsto Ad_g$ is a homomorphism of the group G into the group $\operatorname{Aut}\mathfrak{g}$ of linear invertible transformations of the space $\mathfrak{g}$ and is called the *adjoint representation of the group* G. Since the adjoint representation is a smooth mapping of G into $\operatorname{Aut}\mathfrak{g}$, one can define the tangent mapping $\operatorname{ad} = dAd_e$ of the space $\mathfrak{g}$ into the space $\operatorname{End}\mathfrak{g}$ of all linear transformations of $\mathfrak{g}$ (which is obviously a tangent space to $\operatorname{Aut}\mathfrak{g}$). For every $X, Y \in \mathfrak{g}$, we set $[X, Y] = \operatorname{ad}X(Y)$. One can easily verify that the tangent space $\mathfrak{g}$ with such multiplication is a Lie algebra and the mapping $\mathfrak{g} \ni X \mapsto \operatorname{ad}X \in \operatorname{End}\mathfrak{g}$ is a representation of the Lie algebra $\mathfrak{g}$ (it is called the *regular representation of the Lie algebra).* Note that we can define a regular representation of an arbitrary Lie algebra $\mathfrak{g}$ by setting $\operatorname{ad}X(Y) = [X, Y]$ $(X, Y \in \mathfrak{g})$. For each $X \in \mathfrak{g}$, the mapping $\operatorname{ad}X$ is a differentiation of the Lie algebra $\mathfrak{g}$, i.e., $\operatorname{ad}X([Y, Z]) = [\operatorname{ad}X(Y), Z] + [Y, \operatorname{ad}X(Z)]$ for every $X, Y, Z \in \mathfrak{g}$. Note that, with every homomorphism of Lie groups, one can associate a homomorphism of their Lie algebras and with every Lie subgroup H of the Lie group G, one can associate a Lie subalgebra $\mathfrak{h} \subset \mathfrak{g}$. In this case, the sum of the Lie subalgebras corresponds to the product of Lie subgroups, the intersection of Lie subalgebras corresponds to the intersection of Lie subgroups, and ideals correspond to normal subgroups.

Every finite-dimensional Lie algebra uniquely (to within a local isomorphism) determines a local Lie group (roughly speaking, it determines a rule of multiplication in a certain vicinity of the neutral element of a Lie group). Moreover, every finite-dimensional Lie algebra $\mathfrak{g}$ can be associated with a unique connected simply connected Lie group G for which $\mathfrak{g}$ is a Lie algebra. All connected Lie groups with this property have the form G/D, where D is a discrete normal subgroup which belongs to the center of the group G.

Let $\mathfrak{g}$ be a Lie algebra over the field of the characteristic 0 and let $B(X, Y) = \operatorname{Tr}(\operatorname{ad}X \operatorname{ad}Y)$ $(X, Y \in \mathfrak{g})$ be the *Killing form* on $\mathfrak{g}$. A Lie algebra $\mathfrak{g}$ is called *semisimple* if the Killing form B of the algebra $\mathfrak{g}$ is nondegenerate. A Lie group is called *semisimlple* if its Lie algebra is semisimple.

Let $\mathfrak{g}$ be a complex Lie algebra. A maximal Abelian subalgebra $\mathfrak{h}$ is called a *Cartan subalgebra* if its adjoint representation $\mathfrak{h} \ni H \mapsto \operatorname{ad}H$ is completely reducible in $\mathfrak{g}$. Every semisimple complex Lie algebra contains a Cartan subalgebra. By $\mathfrak{h}'$ we denote the space dual to the complex space $\mathfrak{h}$ of a given Cartan subalgebra in $\mathfrak{g}$. For every $\alpha \in \mathfrak{h}'$, we set

$$\mathfrak{g}^{\alpha} = \{X \in \mathfrak{g} \mid [H, X] = \alpha(H)X \text{ for all } H \in \mathfrak{h}\}.$$

If $\mathfrak{g}^{\alpha} \neq (0)$, then the functional α is called a *root* (of the algebra $\mathfrak{g}$ with respect to $\mathfrak{h}$) and $\mathfrak{g}^{\alpha}$ is called the *root subspace.* Obviously, $\mathfrak{h} = \mathfrak{g}^0$. By Δ we denote the set of nonzero roots. Any complex semisimple Lie algebra $\mathfrak{g}$ can be represented as a direct sum

$$\mathfrak{g} = \mathfrak{h} + \sum_{\alpha \in \Delta} \mathfrak{g}^{\alpha}$$

such that

(i) $\dim \mathfrak{g}^{\alpha} = 1 \ (\alpha \in \Delta)$;

(ii) if $\alpha, \beta \in \Delta$, $\alpha + \beta \neq 0$, then $\mathfrak{g}^{\alpha}$ and $\mathfrak{g}^{\beta}$ are orthogonal with respect to the Killing form;

(iii) the restriction of B to $\mathfrak{h} \times \mathfrak{h}$ is nondegenerate and, for every $\alpha \in \Delta$, there exists a unique element $H_{\alpha} \in \mathfrak{h}$ such that $B(H, H_{\alpha}) = \alpha(H) \ (H \in \mathfrak{h})$;

(iv) if $\alpha \in \Delta$, then $-\alpha \in \Delta$, $[\mathfrak{g}^{\alpha}, \mathfrak{g}^{-\alpha}] = \mathbb{C} H_{\alpha}$, and $\alpha(H_{\alpha}) \neq 0$.

Let

$$\mathfrak{h}_0 = \sum_{\alpha \in \Delta} \mathbb{R} H_{\alpha}.$$

The Killing form is real and positive definite on $\mathfrak{h}_0 \times \mathfrak{h}_0$ and the Cartan subalgebra can be written as a direct sum, i.e., $\mathfrak{h} = \mathfrak{h}_0 + i\mathfrak{h}_0$.

Since the restriction of the Killing form B to $\mathfrak{h} \times \mathfrak{h}$ is nondegenerate, for every $\lambda \in \mathfrak{h}'$, there exists a unique element $H_{\lambda} \in \mathfrak{h}$ such that $\lambda(H) = B(H, H_{\lambda})$. Therefore, on $\mathfrak{h}'$, we can define a nondegenerate scalar product by the formula

$$\langle \lambda, \mu \rangle = B(H_{\lambda}, H_{\mu}) = \lambda(H_{\mu}) = \mu(H_{\lambda}) \quad (\lambda, \mu \in \mathfrak{h}').$$

For any $\lambda, \mu \in \mathfrak{h}'$, we have

$$\langle \lambda, \mu \rangle = \sum_{\alpha \in \Delta} \langle \lambda, \alpha \rangle \langle \mu, \alpha \rangle$$

and $\langle \alpha, \alpha \rangle > 0$ for $\alpha \in \Delta$ (since $\alpha \in \mathfrak{h}_0'$ and the restriction of the Killing form to $\mathfrak{h}_0 \times \mathfrak{h}_0$ is positive definite).

Let $\alpha, \beta \in \Delta$. Then the set of all roots of the form $\beta + n\alpha \ (n \in \mathbb{Z})$ is called the α-series which contains the root β. If $\alpha, \beta \in \Delta$, then the α-series containing the root β has the form $\beta + n\alpha \ (p \leq n \leq q)$ and

$$-2\frac{\langle \alpha, \beta \rangle}{\langle \alpha, \alpha \rangle} = p + q$$

(obviously, $p \le p+q \le q$). The only roots proportional to the root α are $-\alpha$, 0, α.

For any $\alpha \in \Delta$, there exists an element $E_\alpha \in \mathfrak{g}^\alpha$ such that

$$B(E_\alpha, E_{-\alpha}) = -1,$$

$$\mathfrak{g} = \mathfrak{h} + \sum_{\alpha\in\Delta} \mathbb{C} E_\alpha,$$

$$[H, E_\alpha] = \lambda(H) E_\alpha \quad (H \in \mathfrak{h}),$$

$$[E_\alpha, E_{-\alpha}] = -H_\alpha,$$

$$[E_\alpha, E_\beta] = \begin{cases} N_{\alpha,\beta} E_\alpha, & \text{if } \alpha+\beta \text{ is a nonzero root,} \\ 0, & \text{if } \alpha+\beta \neq 0 \text{ is not a root,} \end{cases}$$

where $N_{\alpha,\beta} = -N_{-\alpha,-\beta} \in \mathbb{R}$. The set $\{E_\alpha \mid \alpha \in \Delta\}$ is called the *Weyl basis* (modulo $\mathfrak{h}$).

Let $H_1, \dots, H_r$ be a basis of the space $\mathfrak{h}_0$ and let $F_1, \dots, F_r$ be a basis of the space $\mathfrak{h}_0'$ dual to $\mathfrak{h}_0$ and biorthogonal to $H_1, \dots, H_r$ (i.e., $\langle F_i, H_j\rangle = \delta_{i,j}$ for all $i, j = 1, \dots, r$). In the space $\mathfrak{h}_0'$, we introduce lexicographic ordering with respect to the basis $F_1, \dots, F_r$, namely, we say that $F > E$ $(E, F \in \mathfrak{h}_0')$ if $F(H_i) = E(H_i)$ for $i = 1, \dots, k$ and $F(H_{k+1}) > E(H_{k+1})$. It is clear that the space $\mathfrak{h}_0'$ is completely ordered. A root $\alpha \in \Delta$ is called *positive* if it is a positive element of $\mathfrak{h}_0'$. Obviously, $\alpha > 0$ if and only if $0 > -\alpha$ and, therefore, $\Delta = \Delta_+ \cup (-\Delta_+)$, where Δ_+ is the set of positive roots. A root $\alpha \in \Delta_+$ is called *simple* if it cannot be represented as a sum of positive roots. Let Π be a space of all simple roots. Then $\Pi = \{\alpha_1, \dots, \alpha_r\}$ forms a basis in the space $\mathfrak{h}_0'$ and every root $\beta \in \Delta$ admits a representation in the form $\beta = n_1\alpha_1 + \dots + n_r\alpha_r$, where $n_i \in \mathbb{Z}$ and either all n are nonnegative or all n_i are nonpositive. The number

$$n = \sum_{i=1}^{r} n_i$$

is called the *order* of the root β.

Let $c_{ij} = -2\langle\alpha_i, \alpha_j\rangle / \langle\alpha_j, \alpha_j\rangle$ $(\alpha_i, \alpha_j \in \Pi;\ i, j = 1, \dots, r)$. Then $c_{ij} \in \mathbb{Z}$ and $c_{ij} \ge 0$ for $i \neq j$. The numbers c_{ij} are called *Cartan numbers* and $(c_{ij})_{i,j=1}^r$ is called the *Cartan matrix*. The set $E_{\alpha_1}, \dots, E_{\alpha_r}$, $E_{-\alpha_1}, \dots, E_{-\alpha_r}$, where $\alpha_k \in \Pi$ and E_{α_k} are elements of the Weyl basis corresponding to simple roots, generates $\mathfrak{g}$.

For every $\lambda \in \mathfrak{h}_0'$, we define a mapping $S_\lambda \in GL(\mathfrak{h}_0')$ by the formula

$$S_\lambda \xi = \xi - 2\frac{\langle \xi, \lambda \rangle}{\langle \lambda, \lambda \rangle}\lambda \quad (\xi \in \mathfrak{h}_0');$$

S_λ is the operator of reflection with respect to the hyperplane $P_\lambda = \{\xi \in \mathfrak{h}_0' \mid \langle \lambda, \xi \rangle = 0\}$. Obviously, $S_\alpha \beta \in \Delta$ for $\beta \in \Delta$. The subgroup of $O(\mathfrak{h}_0')$ generated by the set $\{S_\alpha \mid \alpha \in \Delta\}$ is called the *Weyl group* of the system Δ and is denoted by $\operatorname{Ad}\Delta$. Since every element $\omega \in \operatorname{Ad}\Delta$ is an orthogonal transformation, its determinant $\det \omega = \pm 1$. Obviously, $\det S_\alpha = -1$. We set $(\mathfrak{h}_0')^{\sim} = \mathfrak{h}_0' \setminus \bigcup_{\alpha \in \Delta} P_\alpha$. The connected components of the set $(\mathfrak{h}_0')^{\sim}$ are called *Weyl chambers.* Let $\Pi = \{\alpha_1, \dots, \alpha_r\}$ be the set of simple roots. Then $W_0 = \{\xi \in \mathfrak{h}_0' \mid \langle \alpha_1, \xi \rangle > 0, \dots, \langle \alpha_r, \xi \rangle > 0\}$ is a Weyl chamber. The Weyl chamber W_0 is called *dominant* for the system Π. The Weyl group acts transitively on the set of Weyl chambers and the elements $S_{\alpha_1}, \dots, S_{\alpha_r}$ are generators of the group $\operatorname{Ad}\Delta$. Every root $\beta \in \Delta$ admits a representation

$$\beta = S_{\alpha_{j_1}} S_{\alpha_{j_2}} \cdots S_{\alpha_{j_k}} \alpha_i,$$

where $\alpha_{j_p}, \alpha_i \in \Pi$ $(p = 1, \dots, k)$. Let $\operatorname{Aut}\Delta = \{S \in O(\mathfrak{h}_0') \mid S\Delta = \Delta\}$ be the group of automorphisms of the system Δ and let $\operatorname{Aut}\Pi = \{S \in O(\mathfrak{h}_0') \mid S\Pi = \Pi\}$ be the group of automorphisms of the system Π. Then $\operatorname{Aut}\Delta$ is a finite group and $\operatorname{Aut}\Delta = \operatorname{Ad}\Delta \rtimes \operatorname{Aut}\Delta$. The groups $\operatorname{Aut}\mathfrak{g}/\operatorname{Ad}\mathfrak{g}$, $\operatorname{Aut}\Delta/\operatorname{Ad}\Delta$, and $\operatorname{Aut}\Pi$ are isomorphic.

Let $\pi(g)$ be a finite-dimensional representation of the Lie group G in a real or complex space V. Then the operator-valued function $G \ni g \mapsto \pi(g)$ is differentiable and its derivative $d\pi$ at the point e determines a representation of the Lie algebra $\mathfrak{g}$ of the group G in the space V. If G is a connected simply connected Lie group, then every representation of $\mathfrak{g}$ can obtained from a certain representation of G in this way (i.e., the problem of description of finite-dimensional representations of a connected simply connected Lie group reduces to a similar problem for its Lie algebra). Every finite-dimensional representation of a connected semisimple Lie group is completely reducible.

Let π be a linear representation of a semisimple complex Lie algebra $\mathfrak{g}$ in a space V. For every $\lambda \in \mathfrak{h}'$, we denote $V_\lambda = \{v \in V \mid \pi(H)v = \lambda(H)v \text{ for all } H \in \mathfrak{h}\}$. If $V_\lambda \neq (0)$, then λ is called a *weight* of the representation π, the subspace V_λ is called a *weight subspace,* and the number $n_\lambda = \dim V_\lambda$ is called the *multiplicity of the weight* λ. Let

$$\mathfrak{n}_+ = \bigoplus_{\alpha \in \Delta_+} L^\alpha \quad \text{and} \quad \mathfrak{n}_- = \bigoplus_{\alpha \in -\Delta_+} L^\alpha$$

be subalgebras of $\mathfrak{g}$. A vector $v \in V$ is called a *dominant vector* of the representation π if $v \in V_\lambda$ for some weight λ, $\pi(\mathfrak{n}_+)v = 0$, and the least subspace of V which is invariant under the action of all operators $\pi(X)$ $(X \in \mathfrak{g})$ and contains the vector v coincides with V. The relevant weight λ is called the *dominant weight* of the representation π. If λ is the dominant weight of the representation π, then the weight subspace V_λ is one-dimensional and any weight μ of the representation π admits a representation

$$\mu = \lambda - \sum_{i=1}^{r} m_i \alpha_i$$

where $\alpha_i \in \Pi$ and $m_i \geq 0$ (whence, in particular, it follows that λ is the maximal weight in the collection of all weights of the representation π). Moreover, $V = \oplus V_\mu = \pi(U_-)v$, where U_- is a universal enveloping algebra of $\mathfrak{n}_-$. Every finite-dimensional representation π of the algebra $\mathfrak{g}$ possesses a dominant vector which is determined uniquely (to within multiplication by a scalar). The set of all weights of the representation π is finite and invariant under the action of the Weyl group $\mathrm{Ad}\,\Delta$. An irreducible representation is uniquely determined by its dominant weight. A linear functional $\lambda \in \mathfrak{h}'$ is the dominant weight of a certain finite-dimensional representation of the Lie algebra $\mathfrak{g}$ if and only if all numbers $2\langle\lambda, \alpha_i\rangle/\langle\alpha_i, \alpha_i\rangle$ $(\alpha_i \in \Pi)$ are integer and nonnegative.

Let $\mathfrak{g}$ be a complex Lie algebra. Obviously, we can consider $\mathfrak{g}$ as a real Lie algebra $\mathfrak{g}_{\mathbb{R}}$ whose dimensionality is twice as large. Moreover, there exists a *complex structure* on $\mathfrak{g}_{\mathbb{R}}$, i.e., a linear mapping $J: \mathfrak{g}_{\mathbb{R}} \to \mathfrak{g}_{\mathbb{R}}$ such that $J^2 = -id$ and $[X, JY] = J[X, Y]$ $(X, Y \in \mathfrak{g}_{\mathbb{R}})$. We define a *real form* of a complex Lie algebra $\mathfrak{g}$ as a subalgebra $\mathfrak{g}_0 \subset \mathfrak{g}_{\mathbb{R}}$ such that $\mathfrak{g}_{\mathbb{R}} = \mathfrak{g}_0 + J\mathfrak{g}_0$. A mapping $\sigma: X + iY \mapsto X - iY$ $(X, Y \in \mathfrak{g}_0)$ is called the *conjugation* of a complex Lie algebra $\mathfrak{g}$ with respect to its real form $\mathfrak{g}_0$. On the other hand, for every real Lie algebra $\mathfrak{g}_0$, it is possible to construct a complex Lie algebra $\mathfrak{g} = (\mathfrak{g}_0)_{\mathbb{C}}$. Indeed, we set $\mathfrak{g} = \{X + iY \mid X, Y \in \mathfrak{g}_0\}$ with the operation of commutation defined by the formula

$$[X+iY, Z+iT] = [X, Z] - [Y, T] + i([Y, Z] + [X, T]) \quad (X, Y, Z, T \in \mathfrak{g}_0).$$

The algebra $(\mathfrak{g}_0)_{\mathbb{C}}$ is called the *complexification* of $\mathfrak{g}_0$.

A real Lie algebra is called *compact* if it possesses an invariant (positive definite) scalar product. A subalgebra $\mathfrak{k}$ of the real Lie algebra $\mathfrak{g}$ is called *compactly embedded* if there exists a $\mathfrak{k}$-invariant scalar product on $\mathfrak{g}$. Obviously, a compactly embedded subalgebra is compact. In order that a real semisimple Lie algebra be compact, it is necessary and sufficient that its Killing form be negative definite, i.e., $B(X, X) < 0$ for $X \neq 0$. Every semisimple complex Lie algebra possesses a compact real form. Let $\mathfrak{g}_0$ be a real semisimple Lie algebra and let $\mathfrak{g}$ be its complexification. The decomposition $\mathfrak{g}_0 = \mathfrak{k}_0 + \mathfrak{p}_0$ of a Lie algebra $\mathfrak{g}_0$ into the direct sum of a subalgebra $\mathfrak{k}_0$ and a vector subspace $\mathfrak{p}_0$ is called the *Cartan decomposition* if there exists a compact real form $\mathfrak{u}$ of the algebra $\mathfrak{g}$ such that $\mathfrak{k}_0 = \mathfrak{g} \cap \mathfrak{u}$ and $\mathfrak{p}_0 = \mathfrak{g} \cap i\mathfrak{u}$. The mapping $\tau: X + Y \mapsto X - Y$ $(X \in \mathfrak{k}_0,$

$Y \in \mathfrak{p}_0$) is an involutive automorphism of the algebra $\mathfrak{g}_0$; $[\mathfrak{k}_0, \mathfrak{k}_0] \subset \mathfrak{k}_0$, $[\mathfrak{k}_0, \mathfrak{p}_0] \subset \mathfrak{p}_0$, and $[\mathfrak{p}_0, \mathfrak{p}_0] \subset \mathfrak{k}_0$. For any $X \in \mathfrak{k}_0$ and $Y \in \mathfrak{p}_0$, we have $B(X, X) < 0$ $(X \neq 0)$, $B(Y, Y) > 0$ $(Y \neq 0)$, $B(X, Y) = 0$, and $\mathfrak{k}_0 + i\mathfrak{p}_0$ is a compact real form of the algebra $\mathfrak{g}$. The subalgebra $\mathfrak{k}_0$ is a maximal compactly embedded subalgebra of $\mathfrak{g}_0$.

Let $\mathfrak{g}_0 = \mathfrak{k}_0 + \mathfrak{p}_0$ be a Cartan decomposition of the real semisimple Lie algebra $\mathfrak{g}_0$ and let $\mathfrak{g}$ be the complexification of $\mathfrak{g}_0$. By $\mathfrak{a}$ we denote the maximal Abelian subspace of $\mathfrak{p}_0$. The dimension of $\mathfrak{a}$ is called the *real rank* of the algebra $\mathfrak{g}_0$. Let $\mathfrak{h}_{\mathfrak{a}}$ be a maximal Abelian subalgebra of the algebra $\mathfrak{g}_0$ that contains $\mathfrak{a}$ and let $\mathfrak{h}$ be the subalgebra of $\mathfrak{g}$ generated by $\mathfrak{h}_{\mathfrak{a}}$. Then $\mathfrak{h}$ is a Cartan subalgebra of $\mathfrak{g}$, and $\mathfrak{a}$ is a subalgebra of

$$\mathfrak{h}_0 = \sum_{\alpha \in \Delta} \mathbb{R} H_\alpha.$$

We choose a basis $X_1, \dots, X_r$ in $\mathfrak{h}_0$ so that $X_1, \dots, X_{\dim \mathfrak{a}}$ forms a basis in $\mathfrak{a}$, and order $\mathfrak{h}_0'$ lexicographically with respect to this basis. As a result, we obtain an order on the set of roots Δ. Let $P_- = \{\alpha \in \Delta_+ \mid \alpha \equiv 0 \text{ on } \mathfrak{a}\}$ and let $P_+ = \Delta_+ \backslash P_-$. We construct the subalgebra

$$\mathfrak{n} = \sum_{\alpha \in P_+} \mathfrak{g}^\alpha$$

of the algebra $\mathfrak{g}$ and denote $\mathfrak{n}_0 = \mathfrak{g}_0 \cap \mathfrak{n}$. Then $\mathfrak{n}$ and $\mathfrak{n}_0$ are nilpotent Lie algebras, and $\mathfrak{g}_0$ can be decomposed into the direct sum of subspaces: $\mathfrak{g}_0 = \mathfrak{k}_0 + \mathfrak{a} + \mathfrak{n}_0$. This decomposition is called the *Iwasava decomposition.*

A global version of Cartan and Iwasava decompositions is also true. Let G be a connected semisimple Lie group with finite center and Lie algebra $\mathfrak{g}_0$. Let K be an analytic subgroup (i.e., connected Lie subgroup) of G with Lie algebra $\mathfrak{k}_0$ and let P be the image of the vector space $\mathfrak{p}_0$ under the exponential mapping. Then the set PK is everywhere dense in G. If G is a connected semisimple Lie group with Lie algebra $\mathfrak{g}_0$ and if K, A, and N are the analytic subgroups of G that correspond to the subalgebras $\mathfrak{k}_0$, $\mathfrak{a}$, and $\mathfrak{n}_0$ respectively, then the mapping $\varphi: (k, a, n) \mapsto kan$ $(k \in K$, $a \in A$, $n \in N)$ is an analytic diffeomorphism of the manifold $K \times A \times N$ onto G, and the groups A and N are simple connected. The group G also admits the Cartan decomposition $G = KA^+K$, where $A^+ = \exp \mathfrak{a}^+$ and $\mathfrak{a}^+$ is the closure of the dominant Weyl chamber in the algebra $\mathfrak{a}$. The Cartan decomposition becomes unique if the left factor K is replaced by K/M, where M is the centralizer of A in K.

Let G be a noncompact semisimple Lie group with finite center and let K be its maximal compact subgroup (the group K corresponds to the Lie algebra $\mathfrak{k}_0$ in the Iwasava decomposition $\mathfrak{g}_0 = \mathfrak{k}_0 + \mathfrak{a} + \mathfrak{k}_0$). Let M be the centralizer of the subalgebra $\mathfrak{a}$ in K, let $\mathfrak{m}_0$ be the Lie algebra of the group M, and let $\mathfrak{l}_0$ be the orthogonal complement to $\mathfrak{m}_0$ in $\mathfrak{k}_0$ with respect to the Killing form B of the algebra $\mathfrak{g}_0$. We identify

$\mathfrak{l}_0$ with the tangent space to the manifold K/M at the point e. Let dk be the Haar measure on K normalized by the condition

$$\int_K dk = 1$$

and let dg be the invariant measure on G such that

$$\int_G f(g)dg = \int_{G/K}\int_K f(gk)dk\,dg_K$$

for every $f \in C_0(G)$ (here, dg_K is the Riemannian measure on the manifold G/K). The following integral equality holds for any $f \in C_0(G)$:

$$\int_G f(g)dg = \frac{\mathrm{vol}(K/M)}{|Ad\Delta|}\int_{\mathfrak{a}}\left|\prod_{\alpha\in P_+}\sinh\alpha(H)\right|\int_K\int_K f(k_1\exp Hk_2)\,dk_1dk_2dH, \quad (1.28)$$

where $\mathrm{vol}(K/M)$ is the volume of the manifold K/M in the K-invariant metric induced by the restriction of the form B to $\mathfrak{l}_0$.

Let U be a connected simply connected compact semisimple Lie group, let $\mathfrak{u}$ be the corresponding Lie algebra, and let $\mathfrak{t}$ be a maximal Abelian subalgebra of $\mathfrak{u}$. Then

$$\mathfrak{u} = \bigcup_{g\in U} Ad(g)\mathfrak{t}.$$

The connected Abelian subgroup T that corresponds to the algebra $\mathfrak{t}$ is called the *maximal torus*. The set $\mathfrak{t}_e = \{H \in \mathfrak{t} \mid \exp H = e\}$ is called the *unit lattice* for U. Denote by $\mathfrak{g}$ the complexification of the Lie algebra $\mathfrak{u}$ of the group U. Then the subalgebra $\mathfrak{h} = \mathfrak{t} + i\mathfrak{t}$ is the Cartan subalgebra of $\mathfrak{g}$, and

$$\mathfrak{t}_e = \left\{4\pi i\sum_{\alpha\in\Delta}\frac{n_\alpha H_\alpha}{\langle\alpha,\alpha\rangle}\,\middle|\, n_\alpha \in \mathbb{Z}\right\}.$$

For any $f \in C(U)$, the following equality is true:

$$\int_U f(u)du = \frac{1}{|\mathrm{Ad}\Delta|}\int_U\int_T f(utu^{-1})\delta(t)\,dt\,du, \quad (1.29)$$

where du and dt are the Haar measures on U and T normalized by the conditions

$$\int_U du = \int_T dt = 1,$$

$$\delta(t) = \left|\prod_{\alpha\in\Delta} \sin\alpha\left(\frac{iH}{2}\right)\right| \qquad (t = \exp H,\ \ H\in \mathfrak{t}).$$

Denote by Λ the set of all weights of finite-dimensional representations of the group U. For every $\mu \in \Lambda$, we define the function e^{μ} on T by the equality $e^{\mu}(\exp H) = e^{\mu(H)}$ $(H \in \mathfrak{t})$ (this definition is correct because $\Lambda = \{\lambda \in \mathfrak{h}' \mid \lambda(\mathfrak{t}_e) \subset 2\pi i\mathbb{Z}\}$). Let π_λ be an irreducible representation of the group U with dominant weight λ. Then the character χ_λ and dimension $\dim \pi_\lambda$ of this representation are given by the equalities

$$\chi_\lambda(\exp H) = \frac{\sum\limits_{\omega\in \operatorname{Ad}\Delta} (\det\omega)e^{\omega(\lambda+\rho)(H)}}{\sum\limits_{\omega\in \operatorname{Ad}\Delta} (\det\omega)e^{\omega\rho(H)}} \qquad (H \in \mathfrak{t}), \tag{1.30}$$

$$\dim \pi_\lambda = \prod_{\alpha\in\Delta_+} \frac{\langle\lambda+\rho,\alpha\rangle}{\langle\rho,\alpha\rangle}, \tag{1.31}$$

where

$$\rho = \frac{1}{2}\sum_{\alpha\in\Delta_+}\alpha.$$

Equality (1.30) is called the *Weyl formula* for the characters of representations of compact groups.

Taking into account (1.29), (1.30), and the denseness of linear combinations of characters in $C(Q)$ (here, Q is the set of conjugacy classes of the group U), we conclude that the algebra $ZL_1(U)$ is isomorphic to some algebra of functions on $\mathfrak{t}$ which are symmetric with respect to the Weyl group $\operatorname{Ad}\Delta$ and periodic with respect to the unit lattice $\mathfrak{t}_e$.

Let $n_\mu = n(\mu,\lambda)$ be the multiplicity of the weight μ in the representation π_λ with dominant weight λ. Then, obviously,

$$\chi_\lambda(\exp H) = \operatorname{tr}\pi_\lambda(H) = \sum_{\mu} n_\mu e^{\mu(H)}. \tag{1.32}$$

The numbers n_μ can be determined, e.g., by the reccurent *Freudenthal formula*

$$n(\mu, \lambda) = \frac{2}{\langle \lambda+\rho, \lambda+\rho \rangle - \langle \mu+\rho, \mu+\rho \rangle} \sum_{\alpha \in \Delta_+} \sum_{k=1}^{\infty} n(\mu + k\alpha, \lambda) \langle \mu + k\alpha, \alpha \rangle$$

or the following simpler reccurent formula:

$$n(\mu, \lambda) = -\sum_{\substack{\omega \in \mathrm{Ad}\,\Delta \\ \omega \neq e}} (\det \omega) n(\mu + \rho - \omega\rho, \lambda).$$

Let us illustrate the notions introduced above, taking the Lie algebra $sl(n, \mathbb{C})$ $(n \geq 2)$ as an example ($sl(n, \mathbb{C})$ is the algebra of complex matrices with trace zero). Let E_{ij} be the matrix units. The complex dimension of $sl(n, \mathbb{C})$ is equal to $n^2 - 1$, and the elements E_{ij} $(i \neq j)$ and $H_{ij} = E_{ii} - E_{jj}$ form a basis in $sl(n, \mathbb{C})$. The Killing form on the algebra $sl(n, \mathbb{C})$ is given by the formula $B(X, Y) = 2n \operatorname{tr} XY$. The algebra $\mathfrak{h} = \{ H \mid H = \operatorname{diag}(x_1, \dots, x_n),\ x_1 + \dots + x_n = 0 \}$ is the Cartan subalgebra of $sl(n, \mathbb{C})$. Let $\varepsilon_i \in \mathfrak{h}'$ $(i = 1, \dots, n)$ be defined by the equality $\varepsilon_i(\operatorname{diag}(x_1, \dots, x_n)) = x_i$. Then $\Delta = \{ \alpha_{ij} = \varepsilon_i - \varepsilon_j \mid i \neq j \}$, $\Delta_+ = \{ \alpha_{ij} \mid i < j \}$, $\Pi = \{ \alpha_i = \varepsilon_i - \varepsilon_{i+1} \mid i = 1, \dots, n-1 \}$, and $H_{\alpha_i} = H_{i\,i+1}/2n$. Since $\langle \alpha_i, \alpha_j \rangle = B(H_{\alpha_i}, H_{\alpha_j}) = 2n \operatorname{tr} H_{\alpha_i} H_{\alpha_j}$, it is obvious that

$$\langle \alpha_i, \alpha_j \rangle = \begin{cases} \dfrac{1}{n}, & i = j, \\ -\dfrac{1}{2n}, & |i-j| = 1, \\ 0, & |i-j| \geq 2 \end{cases} \qquad (i, j = 1, \dots, n-1).$$

The Weyl group $\mathrm{Ad}\,\Delta$ *of the algebra* $sl(n, \mathbb{C})$ *is isomorphic to the symmetric group* S_n. Indeed, it is easy to find the form of the generators S_{α_i} of the group $\mathrm{Ad}\,\Delta$. We have

$$S_{\alpha_i} \alpha_j = \begin{cases} \alpha_j, & |i-j| > 2, \\ \alpha_i + \alpha_j, & |i-j| = 1, \\ -\alpha_i, & i = j \end{cases} \qquad (i, j = 1, \dots, n-1).$$

By using the equality $\varepsilon_1 + \dots + \varepsilon_n = 0$, we express ε_i $(i = 1, \dots, n)$ in terms of α_i $(i = 1, \dots, n-1)$. As a result, we get

$$\varepsilon_i = -\frac{1}{n}[\alpha_1 + 2\alpha_2 + \dots + (i-1)\alpha_{i-1} - ((n-i)\alpha_i + (n-i-1)\alpha_{i+1} + \dots + \alpha_{n-1})]$$

$$(i = 1, \dots, n-1),$$

$$\varepsilon_n = -\frac{1}{n}(\alpha_1 + 2\alpha_2 + \ldots + (n-1)\alpha_{n-1}).$$

It is easy to see that the transformations S_{α_i} transpose the vectors ε_i and ε_{i+1}.

■

Since $H_{ii+1} = H_{\alpha_i}/\langle\alpha_i, \alpha_i\rangle$ is a basis in $\mathfrak{h}$, the dominant weight λ of the irreducible representation π is uniquely determined by the set of nonnegative integers $(k_1, \ldots, k_{n-1})$, where $k_i = \lambda(H_{ii+1})$. For what follows, it is convenient to define the weight λ in terms of the set of nonnegative integers $m_1' \geq m_2' \geq \ldots \geq m_n' \geq 0$ such that $k_i = m_i' - m_{i+1}'$ $(i = 1, \ldots, n-1)$. It is obvious that the set $(m_1', \ldots, m_n')$ is uniquely determined by λ up to a summand constant for all m_k' $(k = 1, \ldots, n)$.

We can take $su(n) = \{X \in sl(n, \mathbb{C}) \mid X^* = -X\}$ as a compact real form of the algebra $sl(n, \mathbb{C})$. Let us now present the Cartan and Iwasava decompositions for the Lie algebra $sl(n, \mathbb{R})$ of real matrices with zero trace. Clearly, $sl(n, \mathbb{R})$ is also a real form of the algebra $sl(n, \mathbb{C})$. We set $\mathfrak{k}_0 = so(n) = \{X \in sl(n, \mathbb{R}) \mid X^T = -X\}$ and $\mathfrak{p}_0 = \{Y \in sl(n, \mathbb{R}) \mid Y^T = Y\}$. It is easy to see that $so(n) = sl(n, \mathbb{R}) \cap su(n)$ and $\mathfrak{p}_0 = sl(n, \mathbb{R}) \cap isu(n)$. Hence, $sl(n, \mathbb{R}) = so(n) + \mathfrak{p}_0$ is the Cartan decomposition for $sl(n, \mathbb{R})$. The decomposition $sl(n, \mathbb{R}) = so(n) + \mathfrak{a} + \mathfrak{n}$, where $\mathfrak{a} = \{X \in sl(n, \mathbb{R}) \mid x_{ij} = 0,\ i \neq j\}$ and $\mathfrak{n} = \{X \in sl(n, \mathbb{R}) \mid x_{ij} = 0,\ i \geq j\}$ is the Iwasava decomposition for the algebra $sl(n, \mathbb{R})$.

Let us determine the explicit form of relation (1.29) for the compact semisimple Lie group $SU(n)$—the group of unitary unimodular matrices. The Lie algebra $su(n)$ is the Lie algebra of the group $SU(n)$ and T is the set of diagonal matrices of the form $\operatorname{diag}\left(e^{it_1}, \ldots, e^{it_n}\right)$, where $t_1 + \ldots + t_n = 0$. Obviously, $it = \{(t_1, \ldots, t_n) \mid t_1 + \ldots + t_n = 0\}$ and $it_e = \{(t_1, \ldots, t_n) \in it \mid t_k = 2\pi n_k,\ n_k \in \mathbb{Z},\ k = 1, \ldots, n\}$. Consider a manifold $D = \{(t_1, \ldots, t_n) \mid t_1 + \ldots + t_n = 0,\ -\pi \leq t_k \leq \pi,\ k = \overline{1,n}\}$. It is clear that $T = \exp iD$. We introduce coordinates in D as follows: Let

$$D \ni t = (t_1, \ldots, t_n) \xrightarrow{\varphi_n} (t_1, \ldots, t_{n-1}) \in M_n = \{t \in \mathbb{R}^{n-1} \mid$$

$$-\pi \leq t_1 + \ldots + t_{n-1} \leq \pi,\ -\pi \leq t_i \leq \pi,\ i = 1, \ldots, n-1\}.$$

The mapping φ_n is a diffeomorphism of D onto a closed convex bounded region $M \subset \mathbb{R}^{n-1}$ symmetric with respect to $t_1, \ldots, t_{n-1}$.

For every $z(t_1, \ldots, t_n) \in C(D)$, we set

$$Z_n(t_1, \ldots, t_{n-1}) = z(t_1, \ldots, t_{n-1}, -t_1 - \ldots - t_{n-1}) = z(t_1, \ldots, t_n) \quad ((t_1, \ldots, t_n) \in D).$$

Obviously, $Z_n(t_1, \dots, t_{n-1}) \in C(M_n)$. If the function $z(t_1, \dots, t_n)$ is symmetric (skew-symmetric), then $Z_n(t_1, \dots, t_{n-1})$ is also symmetric (skew-symmetric) and Z_n satisfies the additional condition of symmetry (S):

$$Z_n(t_1, \dots, t_i, \dots, t_{n-1}) = Z_n(t_1, \dots, -t_1 - \dots -t_{n-1}, \dots, t_{n-1}) \quad (i = 1, \dots, n-1)$$

(condition of antisymmetry (A):

$$Z_n(t_1, \dots, t_i, \dots, t_{n-1}) = -Z_n(t_1, \dots, -t_1 - \dots -t_{n-1}, \dots, t_{n-1}) \quad (i = 1, \dots, n-1)).$$

Let $dt_1 \dots dt_{n-1}$ be the Lebesgue measure on $\mathbb{R}^{n-1}$. We introduce a measure dt on D by the formula

$$\int_D z(t_1, \dots, t_n)\,dt = \int_{M_n} Z_n(t_1, \dots, t_{n-1})\,dt_1 \dots dt_{n-1} \quad (z \in C(D)).$$

Clearly, dt is the Haar measure on T. By virtue of (1.29), for any function $x(t) \in ZL_1(SU(n))$, we can write

$$\begin{aligned} \int_{SU(n)} x(g)\,dg &= \frac{c}{n!}\int_D x(t_1, \dots, t_n)\left|\prod_{p\neq q} 2\sin\frac{t_p - t_q}{2}\right| dt \\ &= \frac{c}{n!}\int_D x(t_1, \dots, t_n)\prod_{p<q} 2\big(1-\cos(t_p - t_q)\big)\,dt \\ &= \frac{c}{n!}\int_D x(t_1, \dots, t_n)\prod_{p<q} 2\big(e^{it_p} - e^{it_q}\big)^2\,dt, \end{aligned} \tag{1.33}$$

where

$$c = \left(\int_{M_n} dt_1 \dots dt_{n-1}\right)^{-1}.$$

Let us now determine the explicit form of relations (1.30) and (1.31). Since $\alpha_{ij} = \varepsilon_i - \varepsilon_j = \alpha_i + \alpha_{i+1} + \dots + \alpha_{j-1}$ $(i<j)$, we have

$$\rho = \frac{1}{2}[(n-1)\alpha_1 + 2(n-2)\alpha_2 + \dots + k(n-k)\alpha_k + \dots + (n-1)\alpha_{n-1}],$$

whence

$$(\omega\rho)(H) = \frac{i}{2}[(n-1)(t_{\omega(1)} - t_{\omega(2)}) + \ldots + k(n-k)(t_{\omega(k)} - t_{\omega(k+1)}) + \ldots$$
$$+ (n-1)(t_{\omega(n-1)} - t_{\omega(n)})]$$
$$(\omega \in S_n,\ \ H = i(t_1, \ldots, t_n) \in \mathfrak{t}).$$

By using the equality $t_1 + \ldots + t_n = 0$, we obtain

$$(\omega\rho)(H) = \frac{i}{2}[(n-1)(t_{\omega(1)} - t_{\omega(2)}) + \ldots + k(n-k)(t_{\omega(k)} - t_{\omega(k+1)}) + \ldots$$
$$+ (n-1)(t_{\omega(n-1)} - t_{\omega(n)})] - \frac{i}{2}(n-1)(t_{\omega(1)} + \ldots + t_{\omega(n)})$$
$$= -i\left(0 \cdot t_{\omega(1)} + t_{\omega(2)} + 2t_{\omega(3)} + \ldots + (n-1)t_{\omega(n)}\right).$$

This yields

$$\sum_{\omega \in \operatorname{Ad}\Delta} (\det \omega) e^{\omega\rho(H)} = \sum_{\omega \in S_n} (-1)^{\sigma(\omega)}\, e^{-i\left(t_{\omega(2)} + \ldots + (n-1)t_{\omega(n)}\right)},$$

where $\sigma(\omega)$ is the parity of the permutation ω. It is easy to see that the right-hand side of this formula is the Vandermonde determinant. Therefore,

$$\sum_{\omega \in \operatorname{Ad}\Delta} (\det \omega) e^{\omega\rho(H)} = \prod_{p>q} \left(e^{-it_p} - e^{-it_q}\right) = \prod_{p<q} \left(e^{it_p} - e^{it_q}\right).$$

Similarly, we can find the explicit form of

$$\sum_{\omega \in \operatorname{Ad}\Delta} (\det \omega) e^{\omega(\rho+\lambda)(H)}.$$

One should only note that

$$\omega(\rho + \lambda)(H) = i\left((m_1' + n - 1)\, t_{\omega(1)} + \ldots + (m_k' + n - k)\, t_{\omega(k)} + \ldots + m_n' t_{\omega(n)}\right)$$
$$(H = i(t_1, \ldots, t_n) \in \mathfrak{t},\ \omega \in S_n)$$

Hence, denoting $m_k = m_k' + n - k$, we conclude that the irreducible representations $SU(n)$ are enumerated by the collections $m_1 > m_2 > \ldots > m_n$ which are determined uniquely up to a constant summand for all m_k $(k = 1, \dots, n)$. The explicit form of relation (1.31) can easily be found and we leave the relevant calculations to the reader. Finally, for every character

$$\chi(g) = \frac{1}{\dim \pi_\lambda} \chi_\lambda(g) \quad (g \in SU(n)),$$

of the hypercomplex system $ZL_1(SU(n))$, we conclude that the following equality is true:

$$\chi(g) = \varphi_m(t) = \varphi(t, \mu) = \frac{1! \dots (n-1)!}{\prod\limits_{p<q}(m_p - m_q)} \frac{\begin{vmatrix} e^{it_1 m_1} & \dots & e^{it_n m_1} \\ \dots & \dots & \dots \\ e^{it_1 m_n} & \dots & e^{it_n m_n} \end{vmatrix}}{\prod\limits_{p<q}(e^{it_p} - e^{it_q})} \quad (t \in D), \tag{1.34}$$

where e^{it_k} $(k = 1, \dots, n)$ are eigenvalues of the matrix $g \in SU(n)$ and $m_1 > \dots > m_n$ are integers which uniquely determine the weight λ up to a constant summand for all m_k $(k = 1, \dots, n)$.

1.7. Center of the Group Algebra of Compact Semisimple Lie Groups. In this subsection, we study the hypercomplex system $ZL_1(G)$, where G is a compact connected simply connected Lie group. First, we consider the case where $G = SU(n)$ and then the general case.

By virtue of (1.33), the hypercomplex system $ZL_1(SU(n))$ consists of symmetric functions $x(t)$ on D summable with respect to the measure

$$\frac{c}{n!} \prod_{p<q} (e^{it_p} - e^{it_q})^2 \, dt.$$

The functions $\varphi_m(t)$ from equality (1.34) are characters of the hypercomplex system $ZL_1(SU(n))$. We set

$$\Delta(t) = \prod_{p<q} (e^{it_p} - e^{it_q}) \quad (t \in D).$$

Clearly, $\Delta(t)$ is a skew-symmetric function. Consider an isometric operator

$$ZL_1(SU(n)) \ni x(t) \mapsto (Ax)(t) = \Delta(t) x(t) \in AL_1(D, |\Delta(t)| \, dt),$$

where $AL_1(D, |\Delta(t)|dt)$ is the space of skew-symmetric functions defined on D and summable with respect to the measure

$$\frac{c}{n!}|\Delta(t)|dt.$$

The operation $\star$ of multiplication in the space $AL_1(D, |\Delta(t)|dt)$ is defined by setting $(Ax \star Ay)(t) = A(x * y)(t)$, $x, y \in ZL_1(SU(n))$. It is obvious that the Banach algebras $ZL_1(SU(n))$ and $AL_1(D, |\Delta(t)|dt)$ are isomorphic. The homomorphisms of the algebra $AL_1(D, |\Delta(t)|dt)$ in the field of complex numbers are given by the relation

$$AL_1(D, |\Delta(t)|dt) \ni z \mapsto \frac{c}{n!}\int_D z(t)\psi_m(t)dt,$$

where

$$\psi_m(t) = \varphi_m(t)\Delta(t) = \frac{1!\dots(n-1)!}{\prod_{p<q}(m_p - m_q)}\begin{vmatrix} e^{it_1m_1} & \dots & e^{it_nm_1} \\ \dots & \dots & \dots \\ e^{it_1m_n} & \dots & e^{it_nm_n}\end{vmatrix} \quad (t \in D). \tag{1.35}$$

Consider a differential operator L in $C^\infty(D)$ defined by the formula

$$(Lz)(t) = \frac{1}{1!2!\dots(n-1)!}\prod_{p<q}(L_{pq}^{(n)}Z_n(t_1, \dots, t_{n-1})) \quad (t \in D),$$

where

$$L_{pq}^{(n)} = \frac{1}{i}\left(\frac{\partial}{\partial t_p} - \frac{\partial}{\partial t_q}\right) \quad (1 \le p < q \le n-1),$$

$$L_{pn}^{(n)} = \frac{1}{i}\frac{\partial}{\partial t_{p_n}} \quad (p = 1, \dots, n-1).$$

It is clear that the operators $(L_{pq}z)(t) := (L_{pq}^{(n)}Z_n)(t_1, \dots, t_{n-1})$ commute with each other. Moreover, the operators L_{pq} are symmetric in $L_2(D, dt)$ (to prove this fact, it suffices to apply the Stokes theorem). We now introduce other coordinates in D and write L in these coordinates. Indeed,

$$D \ni (t_1, \dots, t_k, \dots, t_n) \xrightarrow{\varphi_k} (t_1, \dots, \hat{t}_k, \dots, t_n) \in M_k = \{t \in \mathbb{R}^{n-1} \mid$$

$$-\pi \le t_1 + \dots + \hat{t}_k + \dots + t_n \le \pi,\ -\pi \le t_i \le \pi,\ i = 1, \dots, k-1, k+1, \dots, n\}$$

(the sign $\wedge$ over a coordinate t_k means that this coordinate is omitted in the corresponding expressions). Denote $Z_k(t_1, \dots, \hat{t}_k, \dots, t_n) = z(t)$ $(t \in D)$. We have

$$\frac{\partial Z_n}{\partial t_p} = \lim_{h \to 0} \frac{1}{h}(z(t_1, \dots, t_p + h, \dots, -t_1 - \dots - t_{n-1} - h) - z(t_1, \dots, t_n))$$

$$= \lim_{h \to 0} \frac{1}{h}(z(t_1, \dots, t_p + h, \dots, t_n - h) - z(t_1, \dots, t_n))$$

$$= \lim_{h \to 0} \frac{1}{h}(Z_k(t_1, \dots, \hat{t}_k, \dots, t_p + h, \dots, t_n - h) - Z_k(t_1, \dots, \hat{t}_k, \dots, t_n))$$

$$= \frac{\partial Z_k}{\partial t_p} - \frac{\partial Z_k}{\partial t_n}$$

(for $p \neq k$). Similarly,

$$\frac{\partial Z_n}{\partial t_k} = -\frac{\partial Z_k}{\partial t_n}.$$

Consequently,

$$L_{pq}^{(k)} = \frac{1}{i}\left(\frac{\partial}{\partial t_p} - \frac{\partial}{\partial t_q}\right) \quad (1 \le p < q \le n,\ p \neq k,\ q \neq k),$$

$$L_{kq}^{(k)} = i\frac{\partial}{\partial t_q} \quad (q \ge k+1),$$

$$L_{pk}^{(k)} = \frac{1}{i}\frac{\partial}{\partial t_p} \quad (p \le k-1),$$

and

$$Lz(t) = \frac{1}{1! \dots (n-1)!} \prod_{p<q} \left(L_{pq}^{(k)} Z_k\right)(t) \quad (t \in D).$$

Lemma 1.1. *Every skew-symmetric solution of the equation* $Lz(t) = 0$ $(t \in D)$ *is identically equal to zero.*

Proof. We rewrite the equation under consideration in the form

$$\prod_{p<q} \left(L_{pq}^{(n)} Z_n\right)(t_1, \dots, t_{n-1}) = 0.$$

We set

$$Z_n^{(1)} = \prod_{2\le p<q\le n} L_{pq}^{(n)} \prod_{q=3}^{n} L_{1q}^{(n)} Z_n.$$

It is obvious that the function $Z_n^{(1)}$ is a solution of the equation

$$\frac{\partial Z_n^{(1)}}{\partial t_1} - \frac{\partial Z_n^{(1)}}{\partial t_2} = 0.$$

The general solution of this equation is given by the formula

$$Z_n^{(1)}(t_1, \dots, t_{n-1}) = f(t_1 + t_2, \dots, t_{n-1}). \tag{1.36}$$

On the other hand, the function Z_n is skew-symmetric with respect to the variables $t_1, \dots, t_{n-1}$. Let us show that $Z_n^{(1)}$ is skew-symmetric with respect to the variables t_1 and t_2.

Indeed, it follows from the equality

$$Z_n(t_1, \dots, t_{n-1}) = -Z_n(t_2, t_1, \dots, t_{n-1})$$

that

$$\left(L_{pq}^{(n)} Z_n\right)(t_1, t_2, \dots, t_{n-1}) = -\left(L_{pq}^{(n)} Z_n\right)(t_2, t_1, \dots, t_{n-1}) \quad (1 < p < q \le n),$$

$$\left(L_{1q}^{(n)} Z_n\right)(t_1, t_2, \dots, t_{n-1}) = -\left(L_{2q}^{(n)} Z_n\right)(t_2, t_1, \dots, t_{n-1}) \quad (q > 2)$$

and

$$\left(L_{2q}^{(n)} Z_n\right)(t_1, t_2, \dots, t_{n-1}) = -\left(L_{1q}^{(n)} Z_n\right)(t_2, t_1, \dots, t_{n-1}) \quad (q > 2).$$

This yields

$$Z_n^{(1)}(t_1, t_2, \dots, t_{n-1}) = -Z_n^{(1)}(t_2, t_1, \dots, t_{n-1})$$

which contradicts (1.36). Therefore,

$$\prod_{2\le p<q\le n} L_{pq} \prod_{q=3}^{n} L_{1q} z = \prod_{2\le p<q\le n} L_{pq}^{(n)} \prod_{q=3}^{n} L_{1q}^{(n)} Z_n = Z_n^{(1)} = 0.$$

By repeating this reasoning (for L_{pn}, it is necessary to use the coordinates $(t_1, \dots, \hat{t}_k, \dots, t_n)$, $k \neq n$), after finitely many steps, we conclude that $z \equiv 0$.

■

Clearly, the operator L transforms any symmetric function on D into a skew-symmetric function, and vice versa. Denote $\mathfrak{S} = \{z(t) \in C^\infty(D) \,|\, z \text{ is symmetric}\}$ and let P denote the hyperplane $t_1 + \dots + t_n = 0$.

Lemma 1.2. *Assume that a measurable function* $w \in L_\infty(D, dt)$ *is skew-symmetric. If*

$$\int_D (Lz)(t)w(t)dt = 0$$

for any function $z \in \mathfrak{S}\ell$, *then* $w(t) = 0$ *almost everywhere.*

Proof. Consider a fixed smooth function $\Phi(s)$ $(s \in [0, \infty))$ such that $\Phi(s) = 0$ $(s \geq 1)$, $1 \geq \Phi(s) \geq 0$ $(s < 1)$. For any function $z \in L_\infty(D, dt)$, we set

$$z^h(t_1, \dots, t_n) = \frac{\kappa}{h^{n-1}} \int_D z(\tau_1, \dots, \tau_n)\, \Phi\Big(\frac{r}{h}\Big) d\tau \quad (t \in P,\ h > 0), \tag{1.37}$$

where

$$r = \sqrt{\sum_{k=1}^{n} (t_k - \tau_k)^2} \quad \text{and} \quad \kappa = \Big(\int_D \Phi(\tau)d\tau\Big)^{-1}.$$

It is obvious that the functions z^h are smooth and equal to zero outside the set of points of the hyperplane P such that the distances from these points to a certain point of the set D do not exceed h. Relation (1.37) readily implies that the function z^h is symmetric if z is symmetric and w^h is skew-symmetric by virtue of the skew-symmetry of w. The functions w^h are standard regularizations of the distribution w. Therefore, w is the weak limit of w^h as $h \to 0$ in the sense of duality $\langle \mathfrak{D}, C_0^\infty(P) \rangle$.

If $z \,|\, \partial D = 0$, then, by using the Stokes formula, we obtain

$$(Lz^h)(t_1, \dots, t_n) = \frac{\kappa}{h^{n-1}} \int_D (Lz)(\tau_1, \dots, \tau_n)\, \Phi\Big(\frac{r}{h}\Big) d\tau = (Lz)^h(t_1, \dots, t_n).$$

Let $z \in \mathfrak{S}$, $z \,|\, \partial D = 0$. Then $z^h \in \mathfrak{S}$ and we conclude that

$$0 = \int_D (Lz^h)(t)w(t)dt = \int_D (Lz)^h(t)w(t)dt. \tag{1.38}$$

It follows from (1.37) that the transition from z to z^h is an integral operator with symmetric kernel. Therefore, we can continue equality (1.38) as follows:

$$0 = \int_D (Lz)^h(t)w(t)dt = \int_D (Lz)(t)w^h(t)dt = (-1)^{n(n-1)/2}\int_D z(t)(Lw^h)(t)dt.$$

The function Lw^h is symmetric because w^h is skew-symmetric and L transforms skew-symmetric functions into symmetric functions. Therefore, $(Lw^h)(t) = 0$ for all $t \in D$. At the same time, by virtue of Lemma 1.1, $w^h(t) = 0$ $(t \in D)$.

■

Lemma 1.3. *The set of all functions* Lz $(z \in \mathfrak{S})$ *is dense in* $AL_1(D, |\Delta(t)|dt)$.

Proof. Assume the opposite. Then one can find a nonzero functional F such that $F(Lz) = 0$ $(z \in \mathfrak{S})$. This means that there exists a nonzero function $w_1(t) \in L_\infty(D, |\Delta(t)|dt)$ such that

$$F(z) = \frac{c}{n!}\int z(t)w_1(t)\Delta(t)dt \quad (z \in AL_1(D, |\Delta(t)|dt)).$$

By assumption,

$$\int_D (Lz)(t)w_1(t)\Delta(t)dt = 0 \quad (z \in \mathfrak{S}).$$

We set $w(t) = w_1(t)|\Delta(t)|$ $(t \in D)$. Since the function $Lz(t)$ is skew-symmetric, we can assume that the function $w(t)$ is also skew-symmetric. Indeed, if this is not true, we replace $w(t)$ by the alternating expression

$$\hat{w}(t) = \frac{1}{n!}\sum_{\omega \in S_n} (-1)^{\sigma(\omega)} w(t_{\omega(1)}, \dots, t_{\omega(n)}),$$

where $\sigma(\omega)$ is the parity of the permutation $\omega \in S_n$. By virtue of the skew-symmetry of the function $z(t)$, we have

$$F(z) = \frac{c}{n!}\int z(t)w(t)dt = \frac{c}{n!}\int z(t)\hat{w}(t)dt.$$

Consequently,

$$F(Lz) = \frac{c}{n!}\int (Lz)(t)\hat{w}(t)\,dt = 0 \quad (z \in \mathfrak{S}).$$

It follows from Lemma 1.2 that $\hat{w}(t) = 0$ $(t \in D)$, whence $F = 0$.

■

Lemma 1.4. *Let* $\psi_m(t) = \varphi_m(t)\Delta(t)$ *be a function defined by equality (1.35). Then*

$$(L\psi_m)(t) = \sum_{\omega \in S_n} e^{i(m_{\omega(1)}t_1 + \ldots + m_{\omega(n)}t_n)} \quad (t \in D). \tag{1.39}$$

Proof. Let d denote the determinant on the right-hand side of (1.35). Clearly,

$$d = \sum_{\omega \in S_n} (-1)^{\sigma(\omega)} e^{i(m_{\omega(1)}t_1 + \ldots + m_{\omega(n)}t_n)}.$$

Since $t_n = -t_1 - \ldots - t_{n-1}$, we have

$$\mathcal{D}_n = \sum_{\omega \in S_n} (-1)^{\sigma(\omega)} e^{i[(m_{\omega(1)} - m_{\omega(n)})t_1 + \ldots + (m_{\omega(n-1)} - m_{\omega(n)})t_{n-1}]}$$

(recall that $\mathcal{D}_n(t_1, \ldots, t_{n-1}) = d(t_1, \ldots, t_n)$ $(t \in D)$). This implies that

$$L_{pn}^{(k)}\mathcal{D}_n = \sum_{\omega \in S_n} (-1)^{\sigma(\omega)} (m_{\omega(p)} - m_{\omega(n)}) e^{i[(m_{\omega(1)} - m_{\omega(n)})t_1 + \ldots + (m_{\omega(n-1)} - m_{\omega(n)})t_{n-1}]},$$

for $1 \le p < n$ and

$$L_{pq}^{(n)}\mathcal{D}_n = \sum_{\omega \in S_n} (-1)^{\sigma(\omega)} (m_{\omega(p)} - m_{\omega(q)}) e^{i[(m_{\omega(1)} - m_{\omega(n)})t_1 + \ldots + (m_{\omega(n-1)} - m_{\omega(n)})t_{n-1}]}.$$

for $1 \le p < q \le n-1$. Consequently,

$$Ld = \prod_{p<q} L_{pq}^{(n)}\mathcal{D}_n = \sum_{\omega \in S_n} (-1)^{\sigma(\omega)} D(m_{\omega(1)}, m_{\omega(2)}, \ldots, m_{\omega(n)})$$

$$\times \exp\{i[(m_{\omega(1)} - m_{\omega(n)})t_1 + \ldots + (m_{\omega(n-1)} - m_{\omega(n)})t_{n-1}]\},$$

where

$$D(m_1, \dots, m_n) = \prod_{1 \le p < q \le n} (m_p - m_q).$$

Since $D(m_1, \dots, m_n)$ is a skew-symmetric function, we have

$$D(m_{\omega(1)}, \dots, m_{\omega(n)}) = (-1)^{\sigma(\omega)} D(m_1, \dots, m_n).$$

This implies that

$$(Ld)(t) = D(m_1, \dots, m_n) \sum_{\omega \in S_n} e^{i[(m_{\omega(1)} - m_{\omega(n)})t_1 + \dots + (m_{\omega(n-1)} - m_{\omega(n)})t_{n-1}]}$$

$$= \prod_{p<q} (m_p - m_q) \sum_{\omega \in S_n} e^{i[m_{\omega(1)}t_1 + \dots + m_{\omega(n)}t_n]} \quad (t \in D).$$

■

We introduce a new operation over symmetric functions. For any $z_1, z_2 \in \mathfrak{S}$, we set

$$z_1 \circ z_2(t) = c \int_D z_1(t_1 - \tau_1, \dots, t_n - \tau_n) z_2(\tau_1, \dots, \tau_n) d\tau \quad (t \in D). \tag{1.40}$$

Note that addition in this formula is performed modulo 2π. Obviously, $\mathfrak{S}$ is an algebra with respect to the multiplication defined by (1.40). Let $\mathfrak{S}L_1(D, dt)$ denote the completion of $\mathfrak{S}$ in the norm

$$\| z \|_{\mathfrak{S}} = \frac{c}{n!} \int_D |(Lz)(t) \Delta(t)| dt.$$

Theorem 1.4. *The space $\mathfrak{S}L_1(D, dt)$ is a Banach algebra with multiplication (1.40) isomorphic to $AL_1(D, |\Delta(t)|, dt)$. The homomorphisms of $\mathfrak{S}L_1(D, dt)$ in the field of complex numbers are given by the formula*

$$\mathfrak{S}L_1(D, dt) \ni z \mapsto c \int_D z_1(t_1, \dots, t_n) e^{i[m_1 t_1 + \dots + m_n t_n]} dt \in \mathbb{C}, \tag{1.41}$$

where $m_1 > \dots > m_n$ are integers uniquely defined (to within a common summand).

Proof. With every function $z \in \mathfrak{S}$, we associate a function

$$\hat{z}(t) = (-1)^{n(n-1)/2} (Lz)(t)$$

and show that the correspondence $z \mapsto \hat{z}$ is an isomorphism of the algebra $\mathfrak{S}$ into $AL_1(D, |\Delta(t)|, dt)$. First, we establish the equality $z_1 \circ z_2(t) = \hat{z}_1 \star \hat{z}_2$. For every $z \in \mathfrak{S}$, in view of the symmetry of the operator L, equality (1.39), and symmetry of the function $z(t) \in \mathfrak{S}$, we obtain

$$\frac{c}{n!}\int_D \hat{z}(t)\psi_m(t)\,dt = \frac{(-1)^{n(n-1)/2}c}{n!}\int_D (Lz)(t)\psi_m(t)\,dt$$

$$= \frac{c}{n!}\int_D z(t)(L\psi_m)(t)\,dt$$

$$= \frac{c}{n!}\int_D z(t)\sum_{\omega \in S_n} e^{i(m_{\omega(1)}t_1 + \ldots + m_{\omega(n)}t_n)}dt$$

$$= c\int_D z(t)e^{i(m_1t_1 + \ldots + m_nt_n)}dt. \tag{1.42}$$

The mapping

$$AL_1(D, |\Delta(t)|, dt) \ni z \mapsto \frac{c}{n!}\int_D z(t)\psi_m(t)\,dt$$

is a homomorphism of $AL_1(D, |\Delta(t)|, dt)$ in $\mathbb{C}$ and multiplication (1.40) is nothing but ordinary convolution. Therefore, by using the fact that the right-hand side of (1.42) is an ordinary Fourier transform, we obtain

$$\frac{c}{n!}\int_D (z_1 \circ z_2)^\wedge(t)\psi_m(t)\,dt = c\int_D (z_1 \circ z_2)(t)e^{i(m_1t_1 + \ldots + m_nt_n)}dt$$

$$= c\int_D z_1(t)e^{i(m_1t_1 + \ldots + m_nt_n)}dt\, c\int_D z_2(s)e^{i(m_1s_1 + \ldots + m_ns_n)}ds$$

$$= \frac{c}{n!}\int_D \hat{z}_1(t)\psi_m(t)\,dt\,\frac{c}{n!}\int_D \hat{z}_2(s)\psi_m(s)\,ds$$

$$= \frac{c}{n!}\int_D (\hat{z}_1 \star \hat{z}_2)(t)\psi_m(t)\,dt.$$

Therefore,

$$\int_D [(z_1 \circ z_2)(t) - (\hat{z}_1 \star \hat{z}_2)(t)]\psi_m(t)dt = 0 \quad (m_1 > \ldots > m_n, \quad m_i \in \mathbb{Z}). \quad (1.43)$$

Note that $(z_1 \circ z_2)^\wedge - \hat{z}_1 \star \hat{z}_2 \in AL_1(D, |\Delta(t)|, dt)$. Therefore, $(z_1 \circ z_2)^\wedge = \hat{z}_1 \star \hat{z}_2$ because (1.43) means that $(z_1 \circ z_2)^\wedge - \hat{z}_1 \star \hat{z}_2$ belong to all maximal ideals of the Banach algebra $AL_1(D, |\Delta(t)|, dt)$.

Thus, the mapping $\mathfrak{S} \ni z \mapsto \hat{z}$ is a homomorphism. Let us show that this mapping is injective. Let $\hat{z} = 0$. Then, by virtue of (1.42), the Fourier transformation of the function z is equal to zero and, consequently, $z = 0$.

We now show that the product $\circ$ is continuous in the norm $\|\cdot\|_{\mathfrak{S}}$. For any $z_1, z_2 \in \mathfrak{S}$, we have

$$\|z_1 \circ z_2\|_{\mathfrak{S}} = \frac{c}{n!}\int_D |(z_1 \circ z_2)^\wedge(t)\Delta(t)|dt$$

$$\leq \left(\frac{c}{n!}\right)^2 \int_D |\hat{z}_1(t)\Delta(t)|dt \int_D |\hat{z}_2(s)\Delta(s)|ds = \|z_1\|_{\mathfrak{S}}\|z_2\|_{\mathfrak{S}}.$$

This implies that $\mathfrak{S}L_1(D, dt)$ is a Banach algebra. The remaining statements of the theorem are obvious.

■

Consider a differential operator

$$\tilde{L} = \frac{1}{1!\ldots(n-1)!}\prod_{p<q}\frac{1}{i}\left(\frac{\partial}{\partial t_p} - \frac{\partial}{\partial t_q}\right)$$

acting in $C_0^\infty(\mathbb{R}^n)$. It is obvious that

$$\left(\tilde{L}Z_k\right)(t_1, \ldots, \hat{t}_k, \ldots, t_n) = \left(L^{(k)}Z_k\right)(t_1, \ldots, \hat{t}_k, \ldots, t_n)$$

for any k and, hence, the restriction of $\tilde{L}$ to $C^\infty(P)$ coincides with the operator L. Consider the equation

$$\tilde{L}_\tau \tilde{G}(t, \tau) = \sum_{(k_1, \ldots, k_n)\in\mathbb{Z}^n}\sum_{\omega\in S_n}\delta(t - \tau_{\omega(1)} + 2k_1\pi, \ldots, t_n - \tau_{\omega(n)} + 2k_n\pi)$$

$$= \sum_{\omega\in S_n}\bar{\delta}(t - \omega(\tau)) \quad (t, \tau \in \mathbb{R}^n), \quad (1.44)$$

where $\bar{\delta}(x_1, \dots, x_n)$ is the n-dimensional δ-function on the torus $(e^{it_1}, \dots, e^{it_n})$

By repeating the reasoning used in the proof of Lemma 1.1, we conclude that every skew-symmetric solution of the equation $(\tilde{L}z)(t) = 0$ is identically equal to zero in any convex region $O \subset \mathbb{R}^n$. Consequently, equation (1.44) possesses a unique solution in the class of distributions skew-symmetric with respect to $\tau_1, \dots, \tau_n$. Since the right-hand side of (1.44) is symmetric with respect to $t_1, \dots, t_n$ and $\tilde{L}_\tau$ does not act upon these arguments, the function $\tilde{G}(t, \tau)$ is symmetric with respect to $t_1, \dots, t_n$. Finally, the function $\tilde{G}(t, \tau)$ is periodic in each argument t_i and τ_j $(i, j = 1, \dots, n)$ because the right-hand side of (1.44) is periodic. By using relation (1.35), we extend the function $\psi_m(t)$ from the hyperplane P to the entire space $\mathbb{R}^n$. By introducing obvious modifications in the proof of Lemma 1.4, we conclude that

$$\left(\tilde{L\psi_m}\right)(t) = \sum_{\omega \in S_n} e^{i(m_{\omega(1)}t_1 + \dots + m_{\omega(n)}t_n)} \quad (t \in \mathbb{R}^n). \tag{1.45}$$

Let us show that the *expansion of the function* $\tilde{G}$ *in the Fourier series has the form*

$$\tilde{G}(t, \tau) = \sum_{m \in \mathbb{Z}^n} e^{-i(m_1 t_1 + \dots + m_n t_n)} \psi_m(\tau). \tag{1.46}$$

Indeed, by applying the operator $\tilde{L}_\tau$ to both sides of (1.46), we see that the function $\tilde{G}(t, \tau)$ given by (1.46) satisfies equation (1.44). Moreover, the fact that $\tilde{G}(t, \tau)$ is skew-symmetric with respect to τ follows from the skew-symmetry of $\psi_m(\tau)$ and this implies that the solution of equation (1.44) is determined uniquely.

■

Lemma 1.5. *The function* $\tilde{G}(t, \tau)$ *satisfies the equation*

$$\left(\tilde{L}_t\tilde{G}\right)(t, \tau) = (-1)^{n(n-1)/2} \sum_{k \in \mathbb{Z}^n} \sum_{\omega \in S_n} (-1)^{\sigma(\omega)} \delta(t_1 - \tau_{\omega(1)} + 2k_1\pi, \dots, t_n - \tau_{\omega(n)} + 2k_n\pi)$$

$$= (-1)^{n(n-1)/2} \sum_{\omega \in S_n} (-1)^{\sigma(\omega)} \bar{\delta}(t - \omega(\tau)) \quad (t, \tau \in \mathbb{R}^n) \tag{1.47}$$

and can be represented in the form

$$\tilde{G}(t, \tau) = \delta(t_1 + \dots + t_n - \tau_1 - \dots - \tau_n) G(t, \tau). \tag{1.48}$$

Proof. Let us apply the operator $\tilde{L}_t$ to relation (1.46). This gives

$$\tilde{L}_t e^{-i(m_1t_1+\ldots+m_nt_n)}\psi_m(\tau) = (-1)^{n(n-1)/2}\sum_{\omega\in S_n}(-1)^{\sigma(\omega)}e^{-i[m_1(t_1-\tau_{\omega(1)})+\ldots+m_n(t_n-\tau_{\omega(n)})]}.$$

Carrying out summation over all $m \in \mathbb{R}^n$, we arrive at (1.47).

To obtain (1.48), we rewrite (1.46) in a somewhat different form: Let $t^* = t_1 + \ldots + t_n$, $\tau^* = \tau_1 + \ldots + \tau_n$, $t_i' = t_i - t_i^*$ and $\tau_i' = \tau_i - \tau^*$ $(i = 1, \ldots, n)$. Then $t_1' + \ldots + t_n' = \tau_1' + \ldots + \tau_n' = 0$ and

$$\tilde{G}(t,\tau) = \sum_{m\in\mathbb{Z}^n} e^{-i(m_1t_1'+\ldots+m_nt_n')}\psi_m(\tau')\,e^{-i(m_1+\ldots+m_n)(t^*-\tau^*)}.$$

Note that this formula easily follows from the explicit expression for $\psi_m(\tau)$. In the last equality, we first carry out summation over $m_1 + \ldots + m_n$ and then over the collections $(m_1, \ldots, m_n)$ such that one can never indicate two collections of this sort which differ from each other by a common summand. As a result, we obtain (1.48).

■

It follows from (1.48) that $\tilde{G}(t,\tau)|_{D\times D} = G(t,\tau)$ and it follows from the proof of Lemma 1.5 that

$$G(t,\tau) = \sum_m e^{-i(m_1t_1+\ldots+m_nt_n)}\psi_m(\tau) \quad (t,\tau\in D), \tag{1.49}$$

where the summation is carried out over all collections $(m_1, \ldots, m_n)$ such that one can never indicate two collections of this sort that differ from each other by a common summand.

It follows from (1.46) that

$$\psi_m(t) = \frac{1}{(2\pi)^n}\int_{\{s\,|\,-\pi\le s_k\le\pi;\,k=1,\ldots,n\}} e^{i(m_1s_1+\ldots+m_ns_n)}\tilde{G}(s,t)\,ds_1\ldots ds_n \quad (t\in\mathbb{R}^n).$$

Therefore,

$$\psi_m(t) = \frac{K}{(2\pi)^{n-1}}\int_D e^{i(m_1s_1+\ldots+m_ns_n)}G(s,t)\,ds \quad (t\in D), \tag{1.50}$$

where K is a constant. Since the function $\psi_m(t)$ is symmetric with respect to m and the function $G(s,t)$ is symmetric with respect to s, one can rewrite relations (1.49) and (1.50) in the following equivalent form:

$$G(t,\tau) = \sum_{m_1 > \ldots > m_n} \left(\sum_{\omega \in S_n} e^{i(m_{\omega(1)}t_1 + \ldots m_{\omega(n)}t_n)} \right) \psi_m(\tau) \quad (t, \tau \in D), \tag{1.51}$$

$$\psi_m(t) = \frac{K}{(2\pi)^{n-1}} \int_{D_+} \left(\sum_{\omega \in S_n} e^{i(m_{\omega(1)}s_1 + \ldots m_{\omega(n)}s_n)} \right) G(s,t)\,ds \quad (t \in D), \tag{1.52}$$

where $D_+ = D \cap \{s \in \mathbb{R}^n \mid s_1 > \ldots > s_n\}$ (note that, since $t \in D$, we can assume that all $m_j \geq 0$ in (1.51)). Since the set $m_1 > \ldots > m_n \geq 0$ determines an irreducible representation of $SU(n)$, we can assume that the outer sum in (1.51) is taken over all irreducible representations of $SU(n)$.

Let us determine the structure measure of the hypercomplex system $ZL_1(SU(n))$. By virtue of (1.47) and (1.48), the operator inverse to L exists and is determined by the Green function

$$\frac{1}{n!}(-1)^{n(n-1)/2} G(t,\tau).$$

By applying the operator L to both sides of (1.40), we get

$$Lz(t) = L(z_1 \circ z_2)(t) = c \int_D (Lz_1)(t-\tau) z_2(\tau)\,d\tau \qquad (t \in D).$$

The functions Lz and Lz_1 belong to $AL_1(D, |\Delta(t)|, dt)$. Therefore, we can rewrite the last equality as follows:

$$(y_1 \star y_2)(t) = (-1)^{n(n-1)/2} c \int_D y_1(t-\tau)(L^{-1} y_2)(\tau)\,d\tau$$

$$= \frac{c}{n!} \int_D \int_D G(\tau_1, \tau_2)\, y_1(t-\tau_1)\, y_2(\tau_2)\, d\tau_1\, d\tau_2$$

$$= \frac{c}{n!} \int_D \int_D G(t-\tau_1, \tau_2)\, y_1(\tau_1)\, y_2(\tau_2)\, d\tau_1\, d\tau_2 \tag{1.53}$$

$$\left(t \in D;\ y_i = (-1)^{n(n-1)/2} Lz_i;\ i = 1, 2\right).$$

Here, the last equality holds by virtue of the periodicity of the integrands. Let us alternate the function $G(t-\tau_1, \tau_2)$ with respect to each variable and denote the function obtained by $a(\tau_1, \tau_2, t)$. We substitute $a(\tau_1, \tau_2, t)$ in (1.53) for $G(t-\tau_1, \tau_2)$. Obviously, the integral on the right-hand side of (1.53) does not change, i.e.,

$$(y_1 \star y_2)(t) = \frac{c}{n!}\int_D\int_D y_1(\tau_1)y_2(\tau_2)a(\tau_1,\tau_2,t)d\tau_1 d\tau_2.$$

Recall that the operator of multiplication by

$$\Delta(t) = \prod_{p<q}(e^{it_p} - e^{it_q})$$

isomorphically maps $ZL_1(SU(n))$ onto $AL_1(D, |\Delta(t)|dt)$. Therefore, the structure measure of $ZL_1(SU(n))$ is determined by the function

$$c(\tau_1,\tau_2,t) = \frac{c}{n!}\frac{\Delta(\tau_1)\Delta(\tau_2)}{\Delta(t)}a(\tau_1,\tau_2,t) \qquad (t,\tau_1,\tau_2 \in D).$$

Thus, we have proved the following statement:

Theorem 1.5. *The function*

$$c(\tau_1,\tau_2,t) = \frac{c}{n!}\frac{\Delta(\tau_1)\Delta(\tau_2)}{\Delta(t)}a(\tau_1,\tau_2,t)$$

determines the convolution in $ZL_1(SU(n))$ *according to the formula*

$$(x_1 * x_2)(t) = \int_D\int_D x_1(\tau_1)x_2(\tau_2)c(\tau_1,\tau_2,t)d\tau_1 d\tau_2 \quad (t\in D). \tag{1.54}$$

Here, $c = 1/\int_D dt$, *the function* $a(\tau_1,\tau_2,t)$ *is obtained by alternating* $G(t-\tau_1,\tau_2)$ *with respect to all variables, and the function* $G(t-\tau_1,\tau_2)$ *is determined by the characters* $\varphi_m(t)$ *of the hypercomplex system* $ZL_1(SU(n))$, *namely,*

$$G(t,\tau) = \sum_m e^{-i(m_1t_1+\ldots+m_nt_n)}\varphi_m(\tau)\Delta(\tau)$$

$$= \sum_{m_1>\ldots>m_n}\left(\sum_{\omega\in S_n} e^{i(m_{\omega(1)}t_1+\ldots+m_{\omega(n)}t_n)}\right)\varphi_m(\tau)\Delta(\tau) \quad (t,\tau\in D), \tag{1.55}$$

where the summation is carried out over all collections $m_1 > \ldots > m_n$ *such that one can never indicate two collections of this sort that differ from each other by a common summand (or, which is the same, the summation is carried out over all irreducible representations of* $SU(n)$*).*

The results obtained in this subsection remain true in the case of an arbitrary compact semisimple connected simply connected Lie group G. We present here only main statements. The hypercomplex system $ZL_1(G)$ is isomorphic to the Banach algebra of functions on $\mathfrak{t}$ that are symmetric with respect to the Weyl group $\operatorname{Ad}\Delta$, periodic with respect to the lattice $\mathfrak{t}_e$, and integrable with respect to the measure $\delta(t)dt$ (here,

$$\delta(t) = \left|\prod_{\alpha\in\Delta} \sin\frac{\alpha(iH)}{2}\right| \qquad (t = \exp H,\ H \in \mathfrak{t})$$

and dt is the Haar measure on the maximal torus T normalized by the condition $\int dt = 1$). The algebra AL_1 is the Banach algebra of functions on $\mathfrak{t}$ that are skew-symmetric with respect to $\operatorname{Ad}\Delta$, periodic with respect to $\mathfrak{t}_e$, and integrable with respect to the measure $|\Delta(t)|dt$, where

$$\Delta(t) = \prod_{\alpha\in\Delta_-} \sin\frac{\alpha(iH)}{2}.$$

Finally, $\mathfrak{S}L_1$ is the completion of the set of functions having the same properties as functions from ZL_1 with respect to the norm

$$\|z\|_{\mathfrak{S}} = \frac{1}{|\operatorname{Ad}\Delta|}\int_T |Lz(t)||\Delta(t)|dt,$$

where the operator L is the product of operators of differentiation along positive roots α. The following statement is true:

Theorem 1.6. *The function*

$$c(\tau_1, \tau_2, t) = \frac{\Delta(\tau_1)\Delta(\tau_2)}{|\operatorname{Ad}\Delta|\Delta(t)} a(\tau_1, \tau_2, t) \qquad (t, \tau_1, \tau_2 \in T)$$

determines the convolution in $ZL_1(SU(n))$ according to the formula

$$(x_1 * x_2)(t) = \int_T\int_T x_1(\tau_1)x_2(\tau_2)c(\tau_1, \tau_2, t)d\tau_1 d\tau_2. \tag{1.56}$$

Here,

$$\Delta(t) = \prod_{\alpha\in\Delta_-} \sin\frac{\alpha(iH)}{2} \qquad (t = \exp H,\ H \in \mathfrak{t}),$$

the function $a(\tau_1, \tau_2, t)$ *is obtained by alternating the function* $G(t-\tau_1, \tau_2)$ *with respect to all variables, and* $G(t, \tau)$ *is determined by the relation*

$$G(t, \tau) = \sum_{\pi \in \hat{G}} \sum_{\omega \in \mathrm{Ad}\,\Delta} e^{i\langle \omega m, H \rangle} \chi_m(t) \Delta(t), \tag{1.57}$$

where

$$\chi_m(t) = \frac{1}{\dim \pi} \chi_\pi(t) \quad (t \in T,\ \pi \in \hat{G})$$

are the characters of the hypercomplex system $ZL_1(G)$, $m = \lambda + \rho$, λ *is the dominant weight of the representation, and*

$$\rho = \frac{1}{2} \sum_{\alpha \in \Delta_+} \alpha .$$

The outer sum in (1.57) is taken over all equivalence classes of irreducible representations of G. *The function* $G(t, \tau)$ *is skew-symmetric with respect to* τ *and is a solution of the equation*

$$L_\tau G(t, \tau) = \sum_{\omega \in \mathrm{Ad}\,\Delta} \delta(t - \omega\tau) \quad (\omega\tau = \exp \omega H,\ \tau = \exp H,\ H \in \mathfrak{t}).$$

1.8. Algebra $\hat{G}$ ***of Equivalence Classes of Irreducible Representations of a Compact Semisimple Lie Group*** G. In this subsection, we study the hypercomplex system $Zl_1(\hat{G})$, where $\hat{G}$ is the set of equivalence classes of irreducible representations of a compact semisimple connected simply connected Lie group G. This hypercomplex system is dual to the hypercomplex system considered in Subsection 1.7. As in Subsection 1.7, we carefully analyze the case of the group $SU(n)$; the results obtained can easily be extended to the general case.

We introduce the following equivalence relation on the integer-valued lattice $\mathbb{Z}^n$: vectors $m = (m_1, \ldots, m_n)$ and $m' = (m_1', \ldots, m_n')$ are regarded as equivalent if they differ from each other by a vector with equal coordinates, i.e., $m_i = m_i' + k$ for all $i = 1, \ldots, n$. Let $\hat{\mathbb{Z}}^n = \mathbb{Z}^n/\sim$. In what follows, we consider vectors $m \in \mathbb{Z}^n$ such that $m_1 > \ldots > m_n$. Let $\hat{Q} \subset \hat{\mathbb{Z}}^n$ be the set of equivalence classes of such vectors. Relation (1.34) enables us to identify $\widehat{SU(n)}$ with the set $\hat{Q}$. Let $\mu \in \hat{Q}$. The dimension d_μ of an irreducible representation π_μ is determined by the equality

$$d_\mu = \frac{\prod_{p>q} (m_p - m_q)}{1! \ldots (n-1)!}, \tag{1.58}$$

the multiplicative measure of the hypercomplex system $ZL_1(\widehat{SU(n)})$ is equal to $\hat{m}(\mu) = d_\mu^2$, and the characters $\chi(\mu) = \chi_t(\mu) = \varphi(t, \mu)$ $(t \in D,\ \mu \in \hat{Q})$ are defined by (1.34). Denote by $M(\mu_1, \mu_2; \mu)$ $(\mu_1, \mu_2, \mu \in \hat{Q})$ the multiplicity of the representation π_μ in the decomposition of the tensor product $\pi_{\mu_1} \otimes \pi_{\mu_2}$ into irreducible representations. Then the structure measure of $ZL_1(\widehat{SU(n)})$ is equal to

$$\hat{c}(\mu_1, \mu_2, \mu) = \frac{1}{d_\mu} M(\mu_1, \mu_2; \mu) d_{\mu_1} d_{\mu_2}.$$

Our aim is to find the explicit form of the function $M(\mu_1, \mu_2; \mu)$.

Let us extend a function $x \in ZL_1(\widehat{SU(n)})$ to a symmetric function on $\hat{\mathbb{Z}}^n$ according to the formula $x(\omega\mu) = x(\mu)$ $(\omega \in S_n,\ \mu \in \hat{Q})$. Further, we extend d_μ according to (1.58). For any function $x \in ZL_1(\widehat{SU(n)})$, we set $Ax(\mu) = x(\mu)d_\mu$ $(\mu \in \hat{\mathbb{Z}}^n)$. Obviously, the function $Ax(\mu)$ is skew-symmetric. One can easily verify that the operator A realizes an isometric isomorphism between the Banach algebras $Zl_1(\widehat{SU(n)})$ and $Al_1(\hat{\mathbb{Z}}^n)$, where $Al_1(\hat{\mathbb{Z}}^n)$ is the space of skew-symmetric functions on $\hat{\mathbb{Z}}^n$ summable with respect to the norm $|d_\mu|/n!$ $(\mu \in \hat{\mathbb{Z}}^n)$ with the multiplication

$$y_1 \star y_2 = \frac{1}{(n!)^2} \sum_{\mu_1, \mu_2 \in \hat{\mathbb{Z}}^n} y_1(\mu_1) y_2(\mu_2) \tilde{M}(\mu_1, \mu_2; \mu),$$

and $\tilde{M}(\mu_1, \mu_2; \mu)$ is a function skew-symmetric with respect to each variable and such that $\tilde{M}(\mu_1, \mu_2; \mu) = M(\mu_1, \mu_2; \mu)$ $(\mu_1, \mu_2, \mu \in \hat{Q})$. It is obvious that $\tilde{M}(\mu_1, \mu_2; \mu)$ is uniquely determined by this condition and, furthermore, $\tilde{M}(\omega_1\mu_1, \omega_2\mu_2; \omega_3\mu_3) = (-1)^{\sigma(\omega_1\omega_2\omega_3)} M(\mu_1, \mu_2; \mu_3)$ $(\mu_i \in \hat{Q},\ i = 1, 2, 3)$. The homomorphisms of the Banach algebra $Al_1(\hat{\mathbb{Z}}^n)$ in the field of complex numbers are determined by the relation

$$Al_1(\hat{\mathbb{Z}}^n) \ni y \mapsto \frac{1}{n!} \sum_{\mu \in \hat{\mathbb{Z}}^n} y(\mu)\theta(t, \mu) \qquad (t \in D),$$

where

$$\theta(t, \mu) = \frac{\begin{vmatrix} e^{it_1 m_1} & \dots & e^{it_n m_1} \\ \dots & \dots & \dots \\ e^{it_1 m_n} & \dots & e^{it_n m_n} \end{vmatrix}}{\prod\limits_{p<q} \left(e^{it_p} - e^{it_q}\right)} \qquad (t \in D,\ m \in \mu \in \hat{\mathbb{Z}}^n) \tag{1.59}$$

(the right-hand side of (1.59) is well defined because $t_1 + \ldots + t_n = 0$ for all $t \in D$).

Let us introduce difference operators $\Lambda^{\pm}$ in the space of functions on $\hat{\mathbb{Z}}^n$ as follows:

$$\Lambda^{\pm}_{pq} = \frac{\partial^{\pm}}{\partial m_p} - \frac{\partial^{\pm}}{\partial m_q} \qquad (1 \le p < q \le n),$$

where

$$\frac{\partial^{\pm} x}{\partial m_p}(\mu) = x(m_1, \ldots, m_{p-1}, m_p \pm 1, m_{p+1}, \ldots, m_n) - x(m_1, \ldots, m_p, \ldots, m_n)$$

$$(m \in \mu \in \hat{\mathbb{Z}}^n).$$

We set

$$(\Lambda^{\pm} x)(\mu) = \prod_{p<q} (\Lambda^{\pm}_{pq} x)(\mu) \qquad (\mu \in \hat{\mathbb{Z}}^n).$$

It is easy to see that the operators $\Lambda^{\pm}$ map the set $c_0(\hat{\mathbb{Z}}^n)$ of finite functions into itself. Moreover, $\Lambda^{\pm}$ map functions symmetric in $m_1, \ldots, m_n$ into skew-symmetric and vice versa. By direct calculation, one can easily verify that

$$\sum_{\mu \in \hat{\mathbb{Z}}^n} (\Lambda^{\pm} x)(\mu) \overline{y(\mu)} = \sum_{\mu \in \hat{\mathbb{Z}}^n} x(\mu) \overline{(\Lambda^{\mp} y)(\mu)} \tag{1.60}$$

for arbitrary functions x and y on $\hat{\mathbb{Z}}^n$ such that the series in (1.60) are convergent. Properly modifying the reasoning in the proofs of Lemmas 1.1–1.3, we conclude that every skew-symmetric solution of the equation $(\Lambda^{\pm} z)(\mu) = 0$ is identically equal to zero, and the set of functions $\Lambda^{\pm} z$ $(z \in \mathfrak{S} = \{ z \in c_0(\hat{\mathbb{Z}}^n) \mid z \text{ is symmetric} \})$ is dense in $Al_1(\hat{\mathbb{Z}}^n)$ (in this case, the proofs are simpler, e.g., it is not necessary to regularize the functions $w(\mu)$ as in Lemma 1.2). One can easily check that

$$(\Lambda^{+} \theta)(t, \mu) = \sum_{\omega \in S_n} e^{i(m_1 t_{\omega(1)} + \ldots + m_n t_{\omega(n)})},$$

$$(\Lambda^{-} \theta)(t, \mu) = (-1)^{n(n-1)/2} \sum_{\omega \in S_n} e^{i(m_1 t_{\omega(1)} + \ldots + m_n t_{\omega(n)})} \quad (t \in D,\ \mu \in \hat{\mathbb{Z}}^n). \tag{1.61}$$

Let us introduce multiplication $\circ$ in $\mathfrak{S}$. We set

$$z_1 \circ z_2 = \sum_{\kappa \in \hat{\mathbb{Z}}^n} z_1(m_1 - k_1, \ldots, m_n - k_n) z_2(k_1, \ldots, k_n) \quad (k \in \kappa,\ m \in \mu \in \hat{\mathbb{Z}}^n)$$

for any $z_1, z_2 \in \mathfrak{S}$. Clearly, $\mathfrak{S}$ is an algebra with respect to the multiplication $\circ$. Denote by $\mathfrak{S}l_1(\hat{\mathbb{Z}}^n)$ the completion of $\mathfrak{S}$ with respect to the norm

$$\| z \| = \frac{1}{n!} \sum_{\mu \in \hat{\mathbb{Z}}^n} |(\Lambda^+ z)(\mu) d_\mu |.$$

Following the proof of Theorem 1.4, we conclude that $\mathfrak{S}l_1(\hat{\mathbb{Z}}^n)$ is a Banach algebra with respect to the multiplication $\circ$ and is isomorphic to $Al_1(\mathbb{Z}^n)$. The corresponding isomorphism has the form

$$\mathfrak{S} \ni z \mapsto \hat{z} = (-1)^{n(n-1)/2} \Lambda^+ z \in Al_1(\hat{\mathbb{Z}}^n).$$

The homomorphisms of $\mathfrak{S}l_1(\hat{\mathbb{Z}}^n)$ into the field of complex numbers are given by the relation

$$\mathfrak{S}l_1(\hat{\mathbb{Z}}^n) \ni z \mapsto \sum_{\mu \in \hat{\mathbb{Z}}^n} z(\mu) e^{i(m_1 t_1 + \ldots + m_n t_n)} \quad (t \in D).$$

Consider the equation

$$\left(\Lambda_\kappa^+ G\right)(\mu, \kappa) = \sum_{\omega \in S_n} \delta_0(m_1 - k_{\omega(1)}, \ldots, m_n - k_{\omega(n)})$$

$$(m \in \mu \in \hat{\mathbb{Z}}^n,\ k \in \kappa \in \hat{\mathbb{Z}}^n), \tag{1.62}$$

where $\delta_\mu(\kappa) = 1$ for $\kappa = \mu$, and $\delta_\mu(\kappa) = 0$ for $\kappa \neq \mu$. Equation (1.62) has a unique solution in the class of functions skew-symmetric in $k_1, \ldots, k_n$ Since the right-hand side of (1.62) is symmetric with respect to $m_1, \ldots, m_n$ and Λ_κ^+ does not act on these variables, the function $G(\mu, \kappa)$ is symmetric with respect to μ. Let us show that the *the function* $G(\mu, \kappa)$ *admits the following integral representation:*

$$G(\mu, \kappa) = \frac{1}{(2\pi)^{n-1}} \int_{D'} e^{-i(m_1 t_1 + \ldots + m_n t_n)} \theta(t, \kappa)\, dt \quad (\mu, \kappa \in \hat{\mathbb{Z}}^n), \tag{1.63}$$

where

$$D' = \{(t_1, \ldots, t_n) \mid -\pi \le t_i \le \pi \ (i = 1, \ldots, n-1),\ t_1 + \ldots + t_n = 0\}.$$

Indeed, by applying the operator Λ_κ^+ to both sides of (1.63) and using (1.61), we get

$$\left(\Lambda_\kappa^+ G\right)(\mu, \kappa) = \frac{1}{(2\pi)^{n-1}} \int_{D'} e^{-i(m_1 t_1 + \ldots + m_n t_n)} \Lambda_+ \theta(t, \kappa)\, dt$$

$$= \frac{1}{(2\pi)^{n-1}} \sum_{\omega \in S_n} \int_{-\pi}^{\pi} \ldots \int_{-\pi}^{\pi} \exp\left\{ -i\left[\left(m_1 - m_n - \left(k_{\omega(1)} - k_{\omega(n)}\right)\right) t_1 + \ldots\right.\right.$$

$$\left.\left. + \left(m_{n-1} - m_n - \left(k_{\omega(n-1)} - k_{\omega(n)}\right)\right) t_{n-1}\right]\right\} dt_1 \ldots dt_{n-1}$$

$$= \sum_{\omega \in S_n} \delta(\mu - \omega(\kappa)) \qquad (\mu, \kappa \in \hat{\mathbb{Z}}^n).$$

Consequently, the function $G(\mu, \kappa)$ defined by (1.63) satisfies equation (1.62). Moreover, the fact that $\theta(t, \kappa)$ is skew-symmetric with respect to κ implies that $G(\mu, \kappa)$ is also skew-symmetric with respect to κ. To complete the proof it remains to use the fact that, as proved above, (1.62) is uniquely solvable in the class of functions skew-symmetric with respect to κ.

■

It follows from (1.63) that

$$\theta(t, \kappa) = \sum_{\mu \in \hat{\mathbb{Z}}^n} G(\mu, \kappa) e^{i(m_1 t_1 + \ldots + m_n t_n)} \quad (\kappa \in \hat{\mathbb{Z}}^n,\ t \in D). \tag{1.64}$$

By comparing (1.64) with (1.32), we establish that the *function* $G(\mu, \kappa)$ *is equal to the multiplicity of the weight* $\mu + \rho$ *in the representation with dominant weight* $\kappa + \rho$.

By applying the operator Λ_μ^+ to both sides of (1.63), we obtain

$$\left(\Lambda_\mu^+ G\right)(\mu, \kappa) = (-1)^{n(n-1)/2} \sum_{\omega \in S_n} \sigma(\omega) \delta(\mu - \omega(\kappa)).$$

Hence, the operator Λ^+ has the inverse which acts in the class of symmetric and skew-symmetric functions and is determined by the "Green function" $\dfrac{(-1)^{n(n-1)/2}}{n!} G(\mu, \kappa)$ as follows:

$$(\Lambda^{-1} x)(\mu) = \frac{(-1)^{n(n-1)/2}}{n!} \sum_{\kappa} G(\mu, \kappa) x(\kappa)$$

if $x(\kappa)$ is a skew-symmetric function, and

$$\Lambda^{-1}x(\kappa) = \frac{(-1)^{n(n-1)/2}}{n!} \sum_{\mu} G(\mu, \kappa)x(\mu)$$

if x is a symmetric function. Following the reasoning of Subsection 1.7, we obtain

$$(y_1 \star y_2)(\mu) = \frac{1}{n!} \sum_{\kappa_1, \kappa_2} G(\mu - \kappa_1, \kappa_2) y_1(\kappa_1) y_2(\kappa_2).$$

Consequently,

$$M(\mu_1, \mu_2; \mu) = \frac{1}{n!} \sum_{\omega_1, \omega_2 \in S_n} (-1)^{\sigma(\omega_1 \omega_2)} G(\mu - \omega_1(\mu_1), \omega_2(\mu_2))$$

$$= \sum_{\omega \in S_n} (-1)^{\sigma(\omega)} G(\mu - \omega(\mu_1), \mu_2) \qquad (\mu_1, \mu_2, \mu \in \hat{Q}). \tag{1.65}$$

Relation (1.65) was obtained by Weyl. By using this relation, one can easily determine $\hat{c}(\mu_1, \mu_2; \mu)$.

The results of this subsection can also be used in the case of an arbitrary semisimple Lie group. The corresponding relation for determining the multiplicity $M(\lambda_1, \lambda_2; \lambda)$ of the representation π_λ with dominant weight λ in the decomposition of $\pi_{\lambda_1} \otimes \pi_{\lambda_2}$ into irreducible representations coincides with the following Weyl formula:

$$M(\lambda_1, \lambda_2; \lambda) = \sum_{\omega \in \mathrm{Ad}\,\Delta} (\det \omega) n(\lambda + \delta - \omega(\lambda_2 + \delta), \lambda_1),$$

where $n(\mu, \lambda_1)$ is the multiplicity of the weight μ in the representation π_{λ_1} with dominant weight λ_1.

2. Gelfand Pairs

In this section, we give an example of a hypercomplex system whose basis is the set of double cosets of a locally compact group with respect to a compact subgroup. Such a hypercomplex system is isomorphic to a subalgebra of the group algebra of the original group G which consists of functions invariant under the right and left actions of a compact subgroup H. In this section, we mainly consider the case where this hypercomplex system is commutative; in this case, the pair (G, H) is called a Gelfand pair. For example, the pair (G, H) that generates a Riemannian symmetric space is a Gelfand pair; the hypercomplex systems considered in Section 1 are also related to Gelfand pairs.

In this section, we present basic theorems of harmonic analysis on Gelfand pairs. We omit the well-known facts from the theory of functions on symmetric spaces and the theory of discrete hypercomplex systems associated with the group of motions of a connected nonoriented graph. At the end of the section, we give an important example of a hypercomplex system associated with the Delsarte generalized translation operators; every hypercomplex system of this sort is isomorphic to a hypercomplex system associated with a certain Gelfand pair.

2.1. Definition of Gelfand Pairs. Consider a hypercomplex system that consists of functions defined on a locally compact group G and biinvariant under the action of its compact subgroup H. If this hypercomplex system is commutative, then (G, H) is a called a Gelfand pair. Below, we give basic examples of Gelfand pairs and describe their properties.

Let G be a locally compact group, let H be its compact subgroup, let $d\lambda(g) = dg$ be a left Haar measure on G, and let $d\mu(h) = dh$ be the normalized $(\mu(H) = 1)$ Haar measure on H. The set G/H of left cosets of the group G with respect to the subgroup H is a homogeneous space with the group of motions G. Namely, an element $g \in G$ transforms a class aH into the class gaH $(a \in G)$. Let $\alpha, \beta, \gamma, \ldots$ denote elements of the space G/H and let e be the element corresponding to the class H (*the fixed point*). Let $\gamma \in G/H$. Then the set $\{ h\gamma | h \in H\}$ is called an H-*orbit* of the point γ (or a sphere centered at the fixed point), and G/H can be decomposed into the set Q of H-orbits. Denote elements of Q by $p, q, r, \ldots$. Obviously, Q is the set of double cosets $G//H$ of the group G with respect to the subgroup H (in other words, Q is the set of points of the form $HgH = \{h_1 g h_2 | h_1, h_2 \in H\}$ $(g \in G)$).

A function $x \in L_1(G, dg)$ is called *orbital* (or *biinvariant*) if it is constant on the double cosets of G with respect to H, i.e., if $x(h_1 g h_2) = x(g)$ for almost all $g \in G$ and all $h_1, h_2 \in H$. It easy to see that the *set* K *of all orbital functions is a closed subalgebra of the group algebra* $L_1(G, dg)$ *of the group* G. Indeed, K is obviously closed. If $x, y \in K$, $g \in G$, and $h_1, h_2 \in H$, then

$$(x * y)(h_1 g h_2) = \int_G x(h_1 g h_2 a) y(a^{-1}) da$$

$$= \int_G x(h_1 g a) y(a^{-1} h_2) da = \int_G x(ga) y(a^{-1}) da = (x * y)(g).$$

■

The algebra K can be regarded as a hypercomplex system whose basis is the space $Q = G//H$. Indeed, by virtue of the general construction presented in Subsection 1.1, it suffices to show that every function $x(g) \in K$ can be approximated in the norm of the space $L_1(G, dg)$ by finite continuous functions taking constant values on the double cosets. For this purpose, we consider a sequence $f_n(g) \in C_0(G)$ of functions such that $\| x - f_n \|_1 \xrightarrow[n \to \infty]{} 0$. Then, the functions

$$x_n(g) = \int_H \int_H f_n(h_1 g h_2) dh_1 dh_2$$

are constant on the classes HgH and approximate $x(g)$.

■

Denote the obtained hypercomplex system by $L_1(G, H)$. Note that the multiplicative measure m and the structure measure $c(A, B, r)$ satisfy the following relations: $m(A) = \lambda(\mathfrak{A})$ $(A \in \mathcal{B}(Q))$ and $c(A, B, r) = \lambda(g^{-1}\mathfrak{A} \cap \mathfrak{B}^{-1}) = (\kappa_{\mathfrak{A}} * \kappa_{\mathfrak{B}})(g)$ $(A, B \in \mathcal{B}_0(Q)$, $g \in r = HgH \in Q)$, where λ is the Haar measure of the group G, and $\mathfrak{A}$ and $\mathfrak{B}$ are complete preimages of the sets A and B, respectively. The hypercomplex system will be normal if we introduce involution $*$ in Q as follows: $Q \ni HgH = p \mapsto p^* = Hg^{-1}H \in Q$. Obviously, the class H is the basis unity of the hypercomplex system $L_1(G, H)$. Thus, by virtue of Corollary 1.2 in Chapter 1, the algebra K is semi-simple.

Generalized translation operators associated with $L_1(G, H)$ have the form

$$L_s f(t) = \int_H f(g_1 h g_2) dh \quad (g_1 \in s,\ g_2 \in t,\ f \in L_1(G, H)). \tag{2.1}$$

Indeed, for any biinvariant functions $x, y \in C_0(G)$, we have

$$\int_Q L_s x(t) y(t^*) dt = (x * y)(s) = \int_H (x * y)(g_1 h) dh$$

$$= \int_H \int_G x(g_1 h g_2) y(g_2^{-1})\, dg_2 dh = \int_G \int_H x(g_1 h g_2) dh\, y(g_2^{-1})\, dg_2$$

$$(g_1 \in s, \quad g_2 \in t, \quad t, s \in Q),$$

which implies (2.1).

■

In particular, it follows from (2.1) that the hypercomplex system $L_1(G, H)$ satisfies the condition of separate continuity. It is not difficult to show that *$Q = G//H$ is a hypergroup.*

The modular function $\Delta(g)$ $(g \in G)$ of the group G is biinvariant and its projection onto Q is the modular function of the hypercomplex system $L_1(G, H)$. Indeed, since H is a compact subgroup, the set $\{\Delta(h) \mid h \in H\}$ is a compact subgroup of the multiplicative group $(0, \infty)$. On the other hand, the unique compact subgroup of $(0, \infty)$ is $\{1\}$. Thus, $\Delta(h) = 1$ $(h \in H)$ and, hence, $\Delta(h_1 g h_2) = \Delta(h_1)\Delta(g)\Delta(h_2) = \Delta(g)$ $(g \in G,\ h_1, h_2 \in H)$. The last statement follows from the equality

$$\int_Q (R_s * f)(t) dt = \int_G \int_H f(g_1 h g_2^{-1})\, dh dg_1 = \int_H \int_G f(g_1 h g_2^{-1})\, dg_1 dh$$

$$= \int_H \Delta(g_2 h^{-1}) dh \int_G f(g_1) dg_1 = \Delta(s) \int_Q f(t) dt$$

$$(g_1 \in t, \quad g_2 \in s, \quad f \in C_0(G) \cap L_1(G, H)).$$

■

Below, we study the case of commutative hypercomplex systems. The pair of groups (G, H), where H is a compact subgroup of a locally compact group G, is called a *Gelfand pair* if the algebra K of orbital functions is commutative.

Note that *if (G, H) is a Gelfand pair, then the group G is unimodular.* Indeed, let $\Delta(g)$ be the modular function of the group G. As shown above, $\Delta(h) = 1$ $(h \in H)$. Hence, for any function $f \in L_1(G, dg)$, we have

$$\int_G \int_H \int_H f(h_1 g h_2) dh_1 dh_2 dg = \int_H \int_H \int_G f(g) dg \Delta(h_2) dh_1 dh_2 = \int_G f(g) dg.$$

Therefore, to prove the unimodularity of G it suffices to show that

$$\int_G x(g) dg = \int_G x(g^{-1}) dg$$

for any finite orbital function $x \in K$. Denote $A = \operatorname{supp} x \cup \operatorname{supp} x^*$, where $x^*(g) = \overline{x(g^{-1})}$. Obviuosly, $\kappa_A(g) \in K$. By virtue of commutativity of K, we obtain

$$\int_G x(g)\,dg = (x * \kappa_A)(e) = (\kappa_A * x)(e) = \int_G x(g^{-1})\,dg.$$

■

Theorem 2.1. *If there exists an involutive automorphism σ of the group G such that $\sigma(g) \in Hg^{-1}H$ for all $g \in G$, then (G, H) is a Gelfand pair.*

Proof. Let $x, y \in K$. Since $x(\sigma(g)) = x(g^{-1})$, we have

$$x * y(g) = \int_G x(ga)y(a^{-1})\,da = \int_G x(\sigma((ga)^{-1}))y(\sigma(a))\,da$$

$$= \int_G x((\sigma(a))^{-1}\sigma(g^{-1}))y(\sigma(a))\,da.$$

Since the automorphism σ is involutive, it preserves the Haar measure (see, e.g., [HeR]). Therefore, taking into account that $y * x \in K$, we can extend the last equality as follows:

$$\int_G x((\sigma(a^{-1}))\sigma(g^{-1}))y(\sigma(a))\,da = \int_G x(a^{-1}\sigma(g^{-1}))g(a)\,da$$

$$= (y * x)(\sigma(g^{-1})) = (y * x)(g).$$

■

Remark. It follows from the proof of Theorem 2.1 that it is not necessary that the automorphism σ be involutive; it is sufficient to require that it preserve the Haar measure, i.e., $\lambda(\sigma(A)) = \lambda(A)\ (A \in \mathcal{B}_0(G))$.

A pair (G, H) that generates a Riemannian symmetric space[1] *is a Gelfand pair.* Indeed, since every element $g \in G$ can be represented in the form $g = h \cdot k$, where $h \in H$ and $\sigma(k) = k^{-1}$ (see, e.g., [Hel1]), we have $\sigma(g) = hk^{-1} \in Hk^{-1}H = Hg^{-1}H$.

■

Riemannian symmetric spaces are thoroughly studied and the reader can find a rather

[1] This is, in fact, the space G/H, where G is a connected Lie group and H is a compact subgroup of G for which there exists an involutive automorphism σ of the group G such that $(G_\sigma)_0 \subset H \subset G_\sigma$, where G_σ is the set of fixed points of the automorphism σ and $(G_\sigma)_0$ is the connected component of G_σ.

complete presentation of the theory of such spaces in many monographs; therefore, we do not present here the corresponding results and refer the reader to the well-known monographs of Helgason [Hel1] and [Hel2]. Note only that, in this case, the algebra $\mathcal{D}(G/H)$ of G-invariant differential operators on G/H is commutative and finitely generated; the number of algebraically independent generators of the algebra $\mathcal{D}(G/H)$ is called the *rank* of the symmetric space G/H. Spherical functions (see Subsection 2.2) are determined by the following conditions:

(i) $\varphi(e) = 1$,

(ii) $\varphi \in C^\infty(G)$,

(iii) $\varphi(hgh') = \varphi(g)$ for all $g \in G$ and $h, h' \in H$,

(iv) $D\varphi = \lambda_D \varphi$ $(\lambda_D \in \mathbb{C})$ for every $D \in \mathcal{D}(G/H)$ (note that, by virtue of condition (iii), the function φ can be regarded as a function on G/H).

The converse statement is also true: If G is a Lie group, H is its compact subgroup, and the algebra $\mathcal{D}(G/H)$ is commutative, then (G, H) is a Gelfand pair (see [Tho2]).

Another example of a Gelfand pair is the pair (G, H) such that, for any two points $\alpha, \beta \in G/H$, there exists a motion $g \in G$ which permutes these points, i..e., $g\alpha = \beta$ and $g\beta = \alpha$.[2] Indeed, one can easily show that the classes HgH and $Hg^{-1}H$ coincide. Let $a \in G$ be such that $a\alpha = \beta$ and $a\beta = \alpha$, where $\alpha = gH$ and $\beta = H$. In other words, there exist $h_1, h_2 \in H$ such that $ag = h_1$ and $a = gh_2$. Therefore, $g^{-1} = g^{-1}a^{-1}a = (ag)^{-1}a = h_1^{-1}gh_2$, i.e., $g^{-1} \in HgH$. Obviously, we can take the identical automorphism as the automorphism σ required in Theorem 2.1.

2.2. Spherical Functions. In this subsection, we show that characters of the hypercomplex system introduced in Subsection 1.1 coincide with bounded spherical functions and study general properties of functions of this sort.

A complex-valued function $\varphi(g)$ $(g \in G)$ which is not identically equal to zero is called *spherical* if

$$\int_H \varphi(g_1 h g_2)\, dh = \varphi(g_1)\varphi(g_1) \tag{2.2}$$

for any $g_1, g_2 \in G$. Note that *spherical functions are constant on double cosets.* Indeed, let $\varphi(g_1) \neq 0$. Since

[2] Gelfand pairs of this sort are called *symmetric.* As we know, they were considered for the first time by M. Krein [Kre1].

$$\varphi(g_1)\varphi(g) = \int_H \varphi(g_1 hg)dh = \int_H \varphi(g_1 hh'g)dh = \varphi(g_1)\varphi(h'g),$$

we see that $\varphi(g)=\varphi(h'g)$ $(g \in G, h' \in H)$. Similarly, we conclude that $\varphi(gh')= \varphi(g)$ $(h' \in H, g \in G)$.

■

Theorem 2.2. *A biinvariant continuous function* $\varphi(g)$ $(g \in G)$ *is spherical if and only if the following conditions are satisfied:*

(*i*) *the equality* $\varphi * f = \lambda(\varphi, f)\varphi$ $(\lambda(\varphi, f) \in \mathbb{C}^1)$ *holds for any continuous finite biinvariant function* $f \in K$;

(*ii*) $\varphi(e)=1$.

Proof. *Necessity.* Since

$$\varphi(e)\varphi(g) = \int_H \varphi(hg)dh = \int_H \varphi(g)dh = \varphi(g) \quad (g \in G)$$

and $\varphi(g) \neq 0$, we have $\varphi(e)=1$. Let $f \in K$ be an arbitrary continuous biinvariant function. Since

$$\begin{aligned}(\varphi * f)(g) &= \int_G \varphi(ga^{-1})f(a)da = \int_H \int_G \varphi(ga^{-1})f(a)dadh \\ &= \int_H \int_G \varphi(gha^{-1})f(ah^{-1})dadh = \int_G \int_H \varphi(gha^{-1})dhf(a)da \\ &= \int_G \varphi(g)\varphi(a^{-1})f(a)da = (\varphi * f)(e)\varphi(g),\end{aligned}$$

condition (i) is satisfied with $\lambda(\varphi, f) = (\varphi * f)(e)$.

Sufficiency. Since $\varphi(e)=1$, we have $\lambda(\varphi, f) = \lambda(\varphi, f)\varphi(e) = (\varphi * f)(e)$. Since $\varphi(g) \neq 0$, there exists a finite biinvariant continuous function f such that

$$\lambda(\varphi, f) = (\varphi * f)(e) = \int_G \varphi(a^{-1})f(a)da \neq 0.$$

In view of the unimodularity of the group G, we have

$$(\varphi * f)(e)\int_H \varphi(g_1hg_2)\,dh = \int_H (\varphi * f)(g_1hg_2)\,dh$$

$$= \int_H \int_G \varphi(g_1hg_2a^{-1})f(a)\,da\,dh$$

$$= \int_H \int_G \varphi(g_1a^{-1})f(ahg_2)\,da\,dh$$

$$= (\varphi * f'_{g_2})(g_1) = \lambda(\varphi, f'_{g_2})\varphi(g_1),$$

where

$$f'_{g_2}(a) = \int_H f(ahg_2)\,dh$$

is obviously a finite biinvariant continuous function. We compute $\lambda(\varphi, f'_{g_2})$. Indeed,

$$\lambda(\varphi, f'_{g_2}) = (\varphi * f'_{g_2})(e) = \int_G \varphi(a^{-1})f'_{g_2}(a)\,da$$

$$= \int_H \int_G \varphi(a^{-1})f(ahg_2)\,da\,dh = \int_H \int_G \varphi(ha^{-1})f(ag_2)\,da\,dh$$

$$= \int_G \varphi(a^{-1})f(ag_2)\,da = (\varphi * f)(g_2) = \lambda(\varphi, f)\varphi(g_2)$$

$$= (\varphi * f)(e)\varphi(g_2).$$

Thus,

$$(\varphi * f)(e)\int_H \varphi(g_1hg_2)\,dh = (\varphi * f)(e)\varphi(g_1)\varphi(g_2)$$

for any $g_1, g_2 \in G$. Consequently, in view of the fact that $\varphi * f(e) \neq 0$, we conclude that the function $\varphi(g)$ is spherical.

■

Let us now present an important example of spherical functions. Assume that $G = P \cdot H$, where P is a closed subgroup and $P \times H \to PH = G$ is a homeomorphism. Let $\rho : P \to \mathbb{C}^1 \setminus \{0\}$ be a continuous homomorphism of the group P. By assuming that $\rho(ph) = \rho(p)$ $(p \in P, h \in H)$, we extend it to a function on G. Obviuosly, $\rho(gh) = \rho(g)$ $(g \in G, h \in H)$. Moreover,

$$\rho(pg) = \rho(pp'h) = \rho(pp') = \rho(p)\rho(p') = \rho(p)\rho(g) \quad (g = p'h \in G,\ p \in P).$$

The function

$$\varphi(g) = \int_H \rho(hg)\,dh \tag{2.3}$$

is spherical with respect to the subgroup H. Indeed, since $P \times H \to PH = G$ is a homeomorphism, for any $h \in H$ and $g \in G$, there are elements $p_g(h) \in P$ and $u_g(h) \in H$ such that $hg = p_g(h)u_g(h)$ and the functions $H \ni h \mapsto p_g(h) \in P$ and $H \ni h \mapsto u_g(h) \in H$ are continuous in h for any fixed $g \in G$. In view of the invariance of the Haar measure and the properties of ρ established above, we obtain

$$\begin{aligned}
\int_H \varphi(g_1 h g_2)\,dh &= \int_H \int_H \rho(hg_1h'g_2)\,dh'dh \\
&= \int_H \int_H \rho(p_{g_1}(h)u_{g_1}(h h'g_2))\,dh'dh \\
&= \int_H \int_H \rho(p_{g_1}(h)h'g_2)\,dh'dh \\
&= \int_H \int_H \rho(p_{g_1}(h))\rho(h'g_2)\,dh'dh \\
&= \int_H \rho(p_{g_1}(h)u_{g_1}(h))\,dh \int_H \rho(h'g_2)\,dh' \\
&= \int_H \rho(hg_1)\,dh \int_H \rho(h'g_2)\,dh' = \varphi(g_1)\varphi(g_2).
\end{aligned}$$

■

Example. Let $G = SL_2(\mathbb{R})$. Then $G = PH$, where $H = SO(2)$ and the group P consists of upper triangular unimodular matrices, i.e., matrices of the type

$$P = \begin{pmatrix} a & b \\ 0 & a^{-1} \end{pmatrix} \quad (a > 0),$$

Let $s \in \mathbb{C}^1$. Obviuosly, the mapping $\rho^s : P \to \mathbb{C}^1 \setminus \{0\}$ defined by the formula

$$\rho^s(p) = \rho^s\left(\begin{pmatrix} a & b \\ 0 & a^{-1} \end{pmatrix}\right) = a^s,$$

is a continuous homeomorphism of the group P in $\mathbb{C}^1\backslash\{0\}$ and, hence, the functions

$$\varphi_s(g) = \int_H \rho^{s+1}(hg)dh \quad (s \in \mathbb{C}^1\backslash\{0\}) \tag{2.4}$$

are spherical. Note that the last formula exhausts the set of spherical functions (see, e.g., [Lan]).

Let us now describe the space of characters of the hypercomplex system $L_1(G, H)$.

Theorem 2.3. *The set of characters of the hypercomplex system $L_1(G, H)$ is in a one-to-one correspondence with the set of bounded spherical functions. Moreover, a character $\chi(p)$ and the relevant bounded spherical function $\varphi(g)$ satisfy the equality $\chi(p) = \varphi(g)$ $(g \in p \in G//H)$.*

Proof. Let $\varphi(g)$ $(g \in G)$ be a bounded spherical function. We set $\chi(p) = \varphi(g)$ $(g \in p \in Q)$ and show that $\chi(p)$ is a character of the hypercomplex system $L_1(G, H)$. It is obvious that the function $\chi(p)$ is bounded and continuous. Let $A, B \in \mathcal{B}_0(Q)$ and let $\mathfrak{A}, \mathfrak{B} \in \mathcal{B}_0(G)$ be their complete preimages. Since the indicator $\kappa_{\mathfrak{B}}$ is biinvariant, we obtain

$$\begin{aligned}\int_Q c(A, B, r)\chi(r)dr &= \int_G (\kappa_{\mathfrak{A}} * \kappa_{\mathfrak{B}})(g_1)\varphi(g_1)dg_1 \\ &= \int_G \kappa_{\mathfrak{A}}(g_2)\int_H \int_G \kappa_{\mathfrak{B}}(g_1)\varphi(g_2 h g_1)dg_1 dh\, dg_2 \\ &= \int_G \int_G \kappa_{\mathfrak{A}}(g_2)\varphi(g_2)\kappa_{\mathfrak{B}}(g_1)\varphi(g_1)dg_1 dg_2 \\ &= \int_A \chi(p)dp\int_B \chi(q)dq = \chi(A)\chi(B).\end{aligned}$$

Conversely, let $\chi(p)$ be a character of the hypercomplex system $L_1(G, H)$. We set $\varphi(g) = \chi(p)$ for all $g \in p$ and show that (2.2) is satisfied. Consider a function

$$\Phi(g_1, g_2) = \int_H \varphi(g_1 h g_2)dh.$$

In view of the unimodularity of the group H, the function $\Phi(g_1, g_2)$ is biinvariant with respect H in each variable, i.e., for any $h_1, h_2, h_3, h_4 \in H$, we have $\Phi(h_2 g_1 h_2,$

$h_3 g_2 h_4) = \Phi(g_1, g_2)$. Consequently, we can introduce the function $\Phi(p_1, p_2) = \Phi(g_1, g_2)$ $(g_i \in p_i;\ g_i = 1, 2)$ defined on $Q \times Q$. The function $\Phi(g_1, g_2)$ is continuous by the Lebesgue theorem (the function $\varphi(g)$ is continuous and bounded since $|\varphi(g)| = |\chi(p)| \le 1$). We fix $g_1, g_2 \in G$ and set $p_1 = Hg_1H, p_2 = Hg_2H$. Let $E_n^{(1)}$ and $E_n^{(2)}$ be sequences of neighborhoods of the points p_1 and p_2, respectively, contracting to these points and let $\mathfrak{E}_n^{(1)}$ and $\mathfrak{E}_n^{(2)}$ be their complete preimages. Then

$$\varphi(g_1)\varphi(g_2) = \chi(p_1)\chi(p_2)$$

$$= \lim_{k,n\to\infty} \frac{1}{m(E_n^{(1)})} \int_{E_n^{(1)}} \chi(q_1)\,dq_1 \frac{1}{m(E_k^{(2)})} \int_{E_k^{(2)}} \chi(q_2)\,dq_2$$

$$= \lim_{k,n\to\infty} \frac{1}{m(E_n^{(1)})m(E_k^{(2)})} \int_Q c\big(E_n^{(1)}, E_k^{(2)}, q\big)\chi(q)\,dq$$

$$= \lim_{k,n\to\infty} \frac{1}{m(E_n^{(1)})m(E_k^{(2)})} \int_G \int_G \kappa_{\mathfrak{E}_n^{(1)}}(g_1)\kappa_{\mathfrak{E}_k^{(2)}}\big(g_1^{-1}g_2\big)\,dg_1\varphi(g_2)\,dg_2$$

$$= \lim_{k,n\to\infty} \frac{1}{m(E_n^{(1)})m(E_k^{(2)})} \int_G \kappa_{\mathfrak{E}_n^{(1)}}(g_1) \int_H \int_G \kappa_{\mathfrak{E}_k^{(2)}}\big(h^{-1}g_1^{-1}g_2\big)\varphi(g_2)\,dg_2\,dh\,dg_1$$

$$= \lim_{k,n\to\infty} \frac{1}{m(E_n^{(1)})m(E_k^{(2)})} \int_G \int_G \kappa_{\mathfrak{E}_n^{(1)}}(g_1)\kappa_{\mathfrak{E}_k^{(2)}}(g) \int_H \varphi(g_1 h g_2)\,dh\,dg_1\,dg_2$$

$$= \lim_{k,n\to\infty} \frac{1}{m(E_n^{(1)})m(E_k^{(2)})} \int_Q \int_Q \kappa_{E_n^{(1)}}(q_1)\kappa_{E_k^{(2)}}(q_2)\Phi(q_1, q_2)\,dq_1\,dq_2$$

$$= \Phi(p_1, p_2) = \Phi(g_1, g_2) = \int_H \varphi(g_1 h g_2)\,dh.$$

■

By virtue of the last theorem, it is important to know conditions of boundedness of spherical functions. Thus, functions of the type (2.4) are bounded if and only if $-1 \le \operatorname{Re} s \le 1$ (see [Lan]).

2.3. *Representations of Class I.* In this subsection, we establish a one-to-one correspondence between the set of spherical positive definite functions on the group G and the set of irreducible unitary representations of G with H-invariant vector.

A unitary representation π of the group G in a Hilbert space $\mathcal{H}$ is called *repre-*

sentation of Class I with respect to a subgroup H if it is irreducible and there exists a vector $\xi \neq 0$ in $\mathcal{H}$ such that $\pi(h)\xi = \xi$ for all $h \in H$.

Theorem 2.4. *Let (G, H) be a Gelfand pair. Then, for any irreducible unitary representation of the group G, the subspace of H-invariant vectors is at most one-dimensional. The set of equivalence classes of representations of Class I is in a one-to-one correspondence with the set of positive definite spherical functions. This correspondence is given by the formula $\varphi(g) = (\pi(g)\xi, \xi)_{\mathcal{H}}$ (π is a representation of Class I, ξ is a unit vector invariant with respect to all $\pi(h)$, $h \in H$, and φ is a positive definite spherical function).*

Proof. Let us prove the first part of the theorem. Let $G \ni g \mapsto \pi(g)$ be an irreducible unitary representation of G in the Hilbert space $\mathcal{H}$. We consider a representation of the subalgebra $K \subset L_1(G, dg)$ of biinvariant functions

$$K \ni x \mapsto \pi(x) = \int_G x(g)\pi(g)dg$$

and set $\mathfrak{M} = \{\xi \in \mathcal{H} \mid \pi(h)\xi = \xi \text{ for all } h \in H\}$. The subspace $\mathfrak{M}$ is invariant with respect to $\pi(x)$ $(x \in K)$ because, for all $h \in H$,

$$\pi(h)\pi(x)\xi = \int_G \pi(h)\pi(a)x(a)da\xi = \int_G \pi(ha)x(a)da\xi$$

$$= \int_G \pi(a)x(h^{-1}a)da\xi = \int_G \pi(a)x(a)da\xi = \pi(x)\xi.$$

We set $x^*(g) = \overline{x(g^{-1})}$ $(x \in K)$. Then $x^* \in K$ and, since G is unimodular,

$$(\pi(x)\varphi, \psi)_H = \int_G x(g)(\pi(g)\varphi, \psi)_H dg$$

$$= \int_G x(g^{-1})(\varphi, \pi(g)\psi)_H dg = (\varphi, \pi(x^*)\psi)_{\mathcal{H}}. \tag{2.5}$$

By A_x $(x \in K)$ we denote the restriction of the operator $\pi(x)$ to $\mathfrak{M}$. By virtue of (2.5) and the fact that K is commutative, the operators A_x form a commutative family of bounded normal operators in the Hilbert space $\mathfrak{M}$. By virtue of the spectral theorem (see, e.g., [BeKo]), every operator A_x can be represented in the form

$$A_x = \int_{\mathbb{C}^K} \lambda(x)dE(\lambda(\cdot)),$$

where $dE(\lambda(\cdot))$ is a joint resolution of the identity for the operators A_x. Since $E(\Delta)$ $(\Delta \subset \mathbb{C}^K)$ commute with all operators A_x $(x \in K)$, the subspace $E(\Delta)\mathfrak{M}$ is invariant with respect to all A_x $(x \in K)$.

Assume that $\dim \mathfrak{M} > 1$. Then either all operators A_x are multiples of the identity operator or $E(\Delta)\mathfrak{M}$ differs from $\{0\}$ and $\mathfrak{M}$ for some Δ. In both cases, $\mathfrak{M}$ can be represented in the form $\mathfrak{M} = \mathfrak{N}_1 \oplus \mathfrak{N}_2$, where $\mathfrak{N}_1$ and $\mathfrak{N}_2$ are mutually orthogonal nonzero closed subspaces of $\mathfrak{M}$ invariant under the action of all operators A_x and, hence, $\pi(x)$ $(x \in K)$. We take $\xi \in \mathfrak{N}_1$ and denote by Ω the closure of the set $\{\pi(x)\xi \mid x \in L_1(G, dg)\}$. Since

$$\pi(g)\pi(x)\xi_1 = \int_G x(a)\pi(ga)\xi_1 da = \int_G x(g^{-1}a)\pi(a)\xi_2 da = \pi({}_{g^{-1}}x)\xi_1$$

$$(g \in G, x \in L_1(G, dg);\ {}_g x(h) = x(gh), g, h \in G),$$

the space Ω is invariant under the action of the group G. Moreover, $\Omega \neq (0)$ because $\xi_1 \in \Omega$. Let us show that Ω does not coincide with the entire $\mathcal{H}$ (this contradicts the irreducibility of π). To do this, it suffices to show that $\Omega \perp \mathfrak{N}_2$. Indeed, by virtue of the equality $\Delta(h) = 1$ $(x \in K)$, we have

$$
\begin{aligned}
(\pi(x)\xi_1, \xi_2)_{\mathcal{H}} &= \int_G x(g)(\pi(g)\xi_1, \xi_2)_{\mathcal{H}} dg \\
&= \int_G \int_H \int_H x(g)(\pi(h_1 g h_2)\xi_1, \xi_2)_{\mathcal{H}} dg\, dh_1 dh_2 \\
&= \int_H \int_G \int_H x(h_1 g h_2) dh_1 dh_2 (\pi(g)\xi_1, \xi_2)_{\mathcal{H}} dg \\
&= (\pi(x^{\natural})\xi_1, \xi_2)_{\mathcal{H}} = 0,
\end{aligned}
$$

for all $x \in L_1(G, dg)$ and $\xi_2 \in \mathfrak{N}_2$ since

$$x^{\natural}(g) = \int_H \int_H x(h_1 g h_2) dh_1 dh_2 \in K.$$

The first part of Theorem 2.4 is proved.

Let us prove the second part of the theorem. Let $\varphi(g)$ be a positive definite spherical function. By applying the GNS construction to the function φ, we construct a unitary representation of the group $\pi(g)$ in the Hilbert space $\mathcal{H}_\varphi$ with a cyclic vector ξ_0 such that $\varphi(g) = (\pi(g)\xi_0, \xi_0)_{\mathcal{H}_\varphi}$. Let us show that $\pi(g)$ is a representation of Class I. By virtue of the invariance of the function φ with respect to H, we have

$$(\pi(g)\xi_0, \xi_0)_{\mathcal{H}_\varphi} = \varphi(g) = \varphi(h^{-1}g) = (\pi(g)\xi_0, \pi(h)\xi_0)_{\mathcal{H}_\varphi} \quad (h \in H).$$

Since the vector ξ_0 is cyclic, we have $\pi(h)\xi_0 = \xi_0$ $(h \in H)$.

In order to prove the irreducibility of the representation $\pi(g)$, we consider a bilinear form in the space $\mathcal{H}_\varphi$

$$B(\xi, \eta) = \int_H (\pi(h)\xi, \eta)_{\mathcal{H}_\varphi} dh.$$

Since $|B(\xi, \eta)| \le \|\xi\|_{\mathcal{H}_\varphi} \|\eta\|_{\mathcal{H}_\varphi}$, there exists a bounded operator P in $\mathcal{H}_\varphi$ such that $(P\xi, \eta)_{\mathcal{H}_\varphi} = B(\xi, \eta)$ $\quad \xi, \eta \in \mathcal{H}_\varphi)$. By virtue of (2.2), for all $\eta = \pi(a)\xi_0$ $(a \in G)$, we have

$$\begin{aligned}(P\pi(g)\xi_0, \eta)_{\mathcal{H}_\varphi} &= \int_H (\pi(h)\pi(g)\xi_0, \pi(a)\xi_0)_{\mathcal{H}_\varphi} dh \\ &= \int_H (\pi(a^{-1}hg)\xi_0, \xi_0)_{\mathcal{H}_\varphi} dh = \int_H \varphi(a^{-1}hg)dh = \varphi(a^{-1})\varphi(g) \\ &= \varphi(g)(\pi(a^{-1})\xi_0, \xi_0)_{\mathcal{H}_\varphi} = \varphi(g)(\xi_0, \eta)_{\mathcal{H}_\varphi} \quad (g \in G).\end{aligned}$$

Since ξ_0 is cyclic, we conclude that $P\pi(g)\xi_0 = \varphi(g)\xi_0$ and P is a projector onto the one-dimensional space generated by the vector ξ_0. Let $\mathcal{H}'$ be the closure of the sum of all closed subspaces $U' \subset \mathcal{H}_\varphi$ invariant under π and satisfying the condition $PU' = \{0\}$. Then $\mathcal{H}'$ and its orthogonal complement $\mathcal{H}''$ in $\mathcal{H}_\varphi$ are invariant under π. Let U be a closed subspace of $\mathcal{H}''$ invariant under π $(U \ne \{0\})$. Then $PU \ne \{0\}$ and, in view of the fact that $P\mathcal{H}_\varphi = \{\lambda\xi_0 \mid \lambda \in \mathbb{C}\}$, we have $\xi_0 \in U$. But then $\pi(g)\xi_0 \in U$ for all $g \in G$ which gives $U = \mathcal{H}_\varphi$ because ξ_0 is cyclic. Consequently, $\mathcal{H}_\varphi = \mathcal{H}''$ and the representation $\pi(g)$ is irreducible.

Conversely, let $\pi(g)$ be a representation of Class I of the group G in a Hilbert space $\mathcal{H}$, let $\mathfrak{M}$ be a (one-dimensional) subspace of vectors invariant under the action of transformations from H, and let $\xi_0 \in \mathfrak{M}$ be a unit vector $(\|\xi_0\|_{\mathcal{H}} = 1)$. Let us show that the positive definite function $\varphi(g) = (\pi(g)\xi_0, \xi_0)_{\mathcal{H}}$ is spherical. Since $\mathfrak{M}$ is invariant under the action of $\pi(x)$ $(x \in K)$ and is one-dimensional, the vector $\pi(x)\xi_0$ is a scalar multiple of the vector ξ_0. Since

$$(\pi(x)\xi_0, \xi_0)_{\mathcal{H}} = \int_G x(g)\varphi(g)dg,$$

we have

$$\pi(x)\xi_0 = \left(\int_G x(g)\varphi(g)dg\right)\xi_0. \tag{2.6}$$

Since the mapping $x \mapsto \pi(x)$ is a representation of the subalgebra K of the group algebra $L_1(G, dg)$ formed by orbital functions and the latter is isometrically isomorphic to the hypercomplex system $L_1(G, H)$, the mapping

$$L_1(G, H) \ni x \mapsto \int_G x(g)\varphi(g)dg$$

is a homomorphism of $L_1(G, H)$ in $\mathbb{C}^1$. By virtue of Theorem 2.3, this implies that φ is a spherical function on G.

Since $(\pi(g)\xi, \xi)_{\mathcal{H}}$ does not depend on the choice of the unit vector $\xi \in \mathfrak{M}$ and on the choice of the representation π in the corresponding equivalence class of irreducible representations, we conclude that the formula $\varphi(g) = (\pi(g)\xi, \xi)_{\mathcal{H}}$ defines a mapping of the set Ω of equivalence classes of irreducible representations of Class I into the set S of positive definite spherical functions. This mapping is surjective since $\varphi(g) = (\pi_\varphi(x)\xi_0, \xi_0)_\varphi$, where $\varphi \in S$ and $\pi_\varphi(x)$ is the representation associated with φ by the GNS construction. To prove that this mapping is bijective, it suffices to show that if $\omega \in \Omega$, $\pi \in \omega$, and $\varphi(g) = (\pi(g)\xi, \xi)$, where ξ is a unit vector from $\mathfrak{M}$, then $\pi_\varphi \in \omega$. The required mapping of the Hilbert space $\mathcal{H}$ (the space of the representation π) onto $\mathcal{H}_\varphi$ is given by the formula

$$\sum_{i=1}^{r} a_i\pi(g_i)\xi \mapsto \sum_{i=1}^{r} a_i\pi_\varphi(g_i)\xi_0,$$

where $g_1, \ldots, g_r \in G$ and $a_1, \ldots, a_r \in \mathbb{C}^1$.

■

It is well known (see, e.g., [Hel2]) that bounded spherical functions are not necessarily positive definite. In [Hul1] (Theorem 4.2), it was proved that, for every bounded spherical function to be positive definite, it suffices that the group G be such that, for every compact set $A \subset G$ and every $c > 0$,

$$\lambda(A^n) = \lambda(\underbrace{A \cdot A \cdot \ldots \cdot A}_{n}) = o(c^n)$$

as $n \to \infty$ (here, λ is the Haar measure of the unimodular group G). Clearly, compact groups satisfy this condition. Moreover, in [Hul1], it was shown that locally compact nilpotent groups (in particular, Abelian groups) and some other classes of groups also satisfy this condition. By using the fact that positive definite functions satisfy the equality $\varphi(g^{-1}) = \overline{\varphi(g)}$ $(g \in G)$, Theorem 4.2 in [Hul1], and Theorem 2.3, we conclude that *if* (G, H) *is a Gelfand pair, and the group* G *is compact or nilpotent, then the hypercomplex system* $L_1(G, H)$ *is symmetric, i.e., every its character is Hermitian.*

For real noncompact semisimple Lie groups, the description of positive definite spherical functions (with respect to the maximal compact subgroup) was given by Kostant [Kos] (see also [FlK2]).

2.4. Harmonic Analysis on Gelfand Pairs. In the present subsection, on the basis of the results of Section 3 in Chapter 1, we establish some theorems of harmonic analysis for the hypercomplex system $L_1(G, H)$ and give three examples of hypercomplex systems associated with Gelfand pairs. These hypercomplex systems are related to simple examples of symmetric spaces, namely, to spaces of compact, noncompact, or Euclidean type.

Lemma 2.1. *The support of the Plancherel measure* $dm(\varphi) = d\varphi$ *lies in the set* S *of positive definite spherical functions.*

Proof. Note that if a function $y \in L_\alpha(G, dg)$ is orbital, then the following equality is true for any function $x \in L_{\alpha'}(G, dg)$ $(1/\alpha + 1/\alpha' = 1)$:

$$\begin{aligned}\int_G x(g)y(g)\,dg &= \int_G x(g)\int_H\int_H y(h^{-1}gh'^{-1})\,dh\,dh'dg \\ &= \int_G\int_H\int_H x(hgh')\,dh\,dh'y(g)\,dg \\ &= \int_G x^{\natural}(g)y(g)\,dg = \int_Q x^{\natural}(p)y(p)\,dp,\end{aligned} \tag{2.7}$$

where

$$x^{\natural}(g) = \int_H\int_H x(hgh')\,dh\,dh'.$$

The function $z(p) = (x^* * x)^{\natural}(p)$ $(p \in Q)$ is positive definite in the hypercomplex system $L_1(G, H)$ for any function $x(g) \in C_0(G)$. Indeed, for any function $y \in L_1(G, H)$, we have

$$\int_Q z(p)(y^{*} * y)(p)\,dp = \int_G (x^{*} * x)(g)(y^{*} * y)(g)\,dg$$

$$= (x^{*} * x, \overline{y^{*} * y})_2 = (x * \bar{y}^{*}, x * \bar{y}^{*})_2 = \|x * \bar{y}^{*}\|_2^2 \geq 0.$$

It is obvious that the function $z(p)$ is bounded and continuous. Moreover, since $x^{*} * x \in L_1(G, dg)$, we have $z(p) \in L_1(Q, dp)$.

By virtue of Corollary 3.4 in Chapter 1, $\hat{z}(\chi) \geq 0$ for all $\chi \in \operatorname{supp} \hat{m} \subset X_h$. At the same time, according to Theorem 2.3, X_h belongs to the set of bounded spherical functions. Consequently, if $\varphi \in \operatorname{supp} \hat{m}$, then, for any function $x(g) \in C_0(G)$, we get

$$\int_G \varphi(g)(x^{*} * x)(g)\,dg = \int_G z(p)\varphi(p)\,dp = \hat{z}(\varphi) \geq 0,$$

i.e., the function $\varphi(g)$ is positive definite.

■

The following equalities take place for all functions $x \in K$ *and* $y \in L_1(G, dg)$:

$$(x * y)^{\natural} = x * y^{\natural} \quad \text{and} \quad (y * x)^{\natural} = y^{\natural} * x. \tag{2.8}$$

Indeed,

$$(x * y)^{\natural}(g) = \int_H \int_H (x * y)(hgh')\,dh\,dh'$$

$$= \int_H \int_H \int_G x(hgh'a)y(a^{-1})\,da\,dh\,dh'$$

$$= \int_H \int_H \int_G x(hgh'a^{-1})y(a)\,da\,dh\,dh'$$

$$= \int_H \int_H \int_G x(gh'a^{-1}h)y(a)\,da\,dh\,dh'$$

$$= \int_H \int_H x(ga^{-1})y(h'ah)\,da\,dh\,dh'$$

$$= \int_G x(ga^{-1})y^{\natural}(a)\,da = (x * y^{\natural})(g).$$

The second equality in (2.8) is proved similarly.

∎

By applying the results of Section 3 in Chapter 1 to the hypercomplex system $L_1(G, H)$, we obtain the following theorem which actually contains the main results of harmonic analysis on the hypercomplex system $L_1(G, H)$:

Theorem 2.5. *On the set S of positive definite spherical functions equipped with the topology of the space of maximal ideals of the hypercomplex system $L_1(G, H)$, there exists a regular Borel measure $d\varphi$ such that the following statements are true:*

(i) (Plancherel–Godement theorem, see [God]) the space $L_2(Q, dp)$ of biinvariant functions square summable with respect to the Haar measure on G is isometrically isomorphic to the space $L_2(S, d\varphi)$ and the Parseval equality takes place

$$\int_G x(g)\overline{y(g)}dg = \int_S \hat{x}(\varphi)\overline{\hat{y}(\varphi)}d\varphi \quad (x, y \in L_2(Q, dp)), \tag{2.9}$$

where $\hat{x}(\varphi)$ is the spherical Fourier transform of the function x given by

$$\hat{x}(\varphi) = \int_G x(g)\overline{\varphi(g)}dg \tag{2.10}$$

for $x \in L_1(G, dg)$.

(ii) (inverse theorem, see [Ber]). Let $\mathfrak{N}$ be the linear span of $P(Q) \cap L_1(G, dg)$, where $P(Q)$ is the set of continuous biinvariant functions on G positive definite with respect to the hypercomplex system $L_1(G, H)$ (i.e., the set of functions satisfying relation (3.1) of Chapter 1 for all $x(g) \in L_1(G, H)$). For any function $x \in \mathfrak{N}$, its spherical Fourier transform $\hat{x}(\varphi)$ belongs to $L_1(S, d\varphi)$ and the following inversion formula holds:

$$x(g) = \int_S \hat{x}(\varphi)\varphi(g)d\varphi. \tag{2.11}$$

(iii) (an analog of the Bochner theorem, see [God]). For every biinvariant function $f(g)$ $(g \in G)$ to be representable in the form

$$f(g) = \int_{X_h} \varphi(g)d\mu(\varphi) \quad (g \in G) \tag{2.12}$$

where μ is a nonnegative finite regular measure on the space X_h of bounded spherical functions, it is necessary and sufficient that it be continuous and positive definite with respect to the hypercomplex system $L_1(G, H)$. If $f(g)$ is positive definite on the group G, then the measure $d\mu$ in (2.12) is concentrated on the set S of positive definite spherical functions.

(iv) *(an analog of the Pontryagin duality theorem). The product of any two positive definite spherical functions is positive definite; therefore, if G is amenable, then duality takes place for the hypercomplex system $L_1(G, H)$.*

(v) *every continuous biinvariant function on G can be uniformly approximated by linear combinations of spherical functions on every compact set.*

Proof. The statements (i), (ii), (v), and the first part of (iii) follow from Theorems 3.3, 3.2, 3.13, and 3.1 in Chapter 1 and the fact that $\operatorname{supp}\hat{m} \subset S$ (Lemma 2.1). It is well known that the product of any two continuous positive definite functions on a group is positive definite (see, e.g., [Dix2], Subsection 13.14.9). Moreover, Voit proved that $1 \in \operatorname{supp}\hat{m}$ if and only if G is amenable [Voi7]. In this case, $\operatorname{supp}\hat{m} = S$ and (iv) follows from Theorem 3.5. Therefore, it remains to prove the second part of (iii).

Let $f(g)$ be a continuous biinvariant positive definite function. It can be represented in the form (2.12). Assume that there exists $\varphi(g) \in X_h$ such that $\varphi \in (X_h \backslash S) \cap \operatorname{supp}\mu$. It is well known ([Dix2], Section 13.5) that S is closed in X_h and the topology of S coincides with the topology of convergence on compact sets. Therefore, there exists a neighborhood U of the point φ such that $U \subset X_h \backslash S$ and $\mu(U) > 0$. Since the function $\varphi(g)$ is not positive definite on the group G, we can find a function $x(g) \in L_1(G, dg)$ such that

$$(x^* * x)^{\#\wedge}(\varphi) = \int_G (x^* * x)^{\natural}(g)\varphi(g)\,dg$$

$$= \int_G (x^* * x)(g)\varphi(g)\,dg = C_\varphi < 0$$

for all $\varphi \in U$ (here, we have used equality (2.7)). Therefore,

$$\int_U (x^* * x)^\wedge(\varphi)\,d\mu(\varphi) = \int_{X_h} (x^* * x)^{\#\wedge}(\varphi)\kappa_U(\varphi)\,d\mu(\varphi) < 0.$$

On the other hand, since the set $\hat{L}_1(Q, dp)$ is dense in $C_\infty(X_h)$ in the uniform norm, $C_\infty(X_h)$ is dense in $L_2(X_h, d\mu)$, and the measure μ is finite, we can find a sequence of functions $x_n \in L_1(G, H)$ such that $\| \hat{x}_n - \kappa_U \|_{L_2(X_h, d\mu)} \to 0$ as $n \to \infty$. By using the fact that the function $f(g)$ is positive definite, equalities (2.7) and (2.8), representation (2.12), and the Lebesgue theorem, we obtain

$$0 \le \int_G f(g)[(x*x_n)^* * (x*x_n)](g)\,dg$$

$$= \int_G f(g)(x_n^* * (x^* * x)^{\natural} * x_n)(g)\,dg$$

$$= \int_G \int_{X_h} \varphi(g)\,d\mu(\varphi)(x_n^* * (x^* * x)^{\natural} * x_n)(g)\,dg$$

$$= \int_{X_h} (x_n^* * (x^* * x)^{\natural} * x_n)^{\frown}(\varphi)\,d\lambda(\varphi)$$

$$= \int_{X_h} |\hat{x}_n(\varphi)|^2 (x^* * x)^{\#\wedge}(\varphi)\,d\mu(\varphi)$$

$$\xrightarrow[n\to\infty]{} \int_{X_h} \kappa_U(\varphi)(x^* * x)^{\#\wedge}(\varphi)\,d\varphi.$$

At the same time, somewhat earlier it was shown that the last integral is negative.

■

Remark. Let (G, H) be a Gelfand pair and let G be a noncompact group. Then the conditions of Theorem 3.5 in Chapter 1 are, generally speaking, not satisfied. Thus, let $G = SL_2(\mathbb{R})$ and $H = SO(2)$. The spherical functions $\varphi_s(g)$ can be found from (2.4), $X = \{s \mid -1 \le \operatorname{Re} s \le 1\}$, and the Plancherel measure

$$d\hat{m}(s) = \frac{s}{i}\tanh\left(\frac{\pi s}{i}\right)\frac{ds}{2\pi i} \qquad (\operatorname{Re} s = 0)$$

(see, e.g., [Lan]). Thus, $\hat{Q} = \operatorname{supp} \hat{m} = \{s \in \mathbb{C} \mid \operatorname{Re} s = 0\}$ and $1(g) = \varphi_{-1}(g) \notin Q$. Hence, in the general case where the group G is not amenable, we fail to prove the duality theorem (iv). Nevertheless, we can construct a hypercomplex system on X_h (generally speaking without basis unity). Hypercomplex systems of this sort (in terms of the operation of generalized convolution of measures on X_h) were studied in [May] (see also [Miz1] and [Miz2]).

In conclusion, we present several examples of hypercomplex systems associated with Gelfand pairs.

Example 1. Let $G = SO(3)$ and $H = SO(2)$. For definiteness, we assume that $SO(2)$ is realized in $SO(3)$ as the group of rotations about the axis OZ. As is khown, every element g of the group $SO(3)$ is parametrized by the *Euler angles* $g = g_{\varphi_1} g^{\theta} g_{\varphi_2}$ $(0 \le \varphi_1, \varphi_2 \le 2\pi, 0 \le \theta \le \pi)$, where

$$H \ni g_{\varphi} = \begin{pmatrix} \cos\varphi & \sin\varphi & 0 \\ -\sin\varphi & \cos\varphi & 0 \\ 0 & 0 & 1 \end{pmatrix}, \quad g^{\theta} = \begin{pmatrix} 1 & 0 & 0 \\ 0 & \cos\varphi & \sin\varphi \\ 0 & -\sin\varphi & \cos\varphi \end{pmatrix};$$

this parametrization is unique if $0 < \theta < \pi$. The homogeneous space $SO(3)/SO(2)$ is homeomorphic to the unit sphere in $\mathbb{R}^3$, and $SO(2)$ is the stationary subgroup of the point $z_0 = (0, 0, 1)$. Obviously, $SO(3)//SO(2)$ is homeomorphic to $[0, \pi]$. Since $dg = (8\pi^2)^{-1} \sin\theta \, d\varphi_1 d\varphi_2 d\theta$ is the Haar measure on $SO(3)$ (see, e.g., [Vil]), the measure $(1/2)\sin\varphi \, d\varphi$ is the multiplicative measure of the hypercomplex system $L_1(G, H)$. For convenience, we consider a hypercomplex system isomorphic to $L_1(G, H)$. Consider the homeomorphism $[0, \pi] \ni \theta \mapsto t = \cos\theta \in [-1, 1] = Q$. The measure $(1/2)\sin\varphi \, d\varphi$ turns into the Lebesgue measure $(1/2)dt$ on Q. By virtue of (2.1), we have

$$(L_{g^{\theta_1}} f)(g^{\theta_2}) = \frac{1}{2\pi} \int_0^{2\pi} f(g^{\theta_1} g_{\varphi} g^{\theta_2}) \, d\varphi \qquad (f \in L_1(G, H)).$$

By representing $g = g^{\theta_1} g_{\varphi} g^{\theta_2}$ in the form $g_{\tilde{\varphi}_1} g^{\tilde{\theta}} g_{\tilde{\varphi}_2}$, we get $\cos\tilde{\theta} = \cos\theta_1 \cos\theta_2 - \sin\theta_1 \sin\theta_2 \cos\varphi$ and, hence, the generalized translation operators in $L_1(Q, (1/2)dt)$ have the following form:

$$(L_s f)(t) = \frac{1}{2\pi} \int_0^{2\pi} f(\cos\theta_1 \cos\theta_2 - \sin\theta_1 \sin\theta_2 \cos\varphi) \, d\varphi$$

$$= \frac{1}{2\pi} \int_0^{2\pi} f\left(ts - \sqrt{1-t^2}\sqrt{1-s^2}\cos\varphi\right) d\varphi \quad (f \in C(Q)).$$

This enables us to establish the multiplication law in $L_1(Q, (1/2)dt)$:

$$(x * y)(t) = \frac{1}{2\pi} \int_{-1}^{1} \int_{ts-\sqrt{(1-t^2)(1-s^2)}}^{ts+\sqrt{(1-t^2)(1-s^2)}} x(r) y(s) [(1-t^2)(1-s^2) - (ts-r)^2]^{-1/2} \, dr \, ds$$

$$(x, y \in L_1(Q, (1/2)dt)).$$

As is khown (see, e.g., [Hel1]), in this case, the spherical functions have the form $\varphi_n(g^\theta) = P_n(\cos\theta)$ $(n = 0, 1, \dots ;\ \theta \in [0, \pi])$, where P_n are the Legendre polynomials normalized by the condition $P_n(1) = 1$. Therefore, the set of Legendre polynomials $P_n(t)$ $(t \in [-1, 1],\ n = 0, 1, \dots)$ forms the system of characters of the hypercomplex system $L_1(Q, (1/2)dt)$, and relation (2.2) turns into the multiplication formula for the Legendre polynomials, namely,

$$P_n(t)P_n(s) = \frac{1}{2\pi}\int_0^{2\pi} P_n\left(ts - \sqrt{1-t^2}\sqrt{1-s^2}\cos\varphi\right)d\varphi.$$

The set $\hat{Q} = \{0, 1, \dots\}$ is the basis of the dual hypercomplex system, and the Plancherel measure is given by the relation

$$\hat{m}(k) = \left(\frac{1}{2}\int_1^1 |P_k(t)|^2\, dt\right)^{-1} = 2k + 1$$

(see, e.g., [Sze]). Thus, harmonic analysis on the hypercomplex system $L_1(Q, (1/2)dt)$ reduces to the Fourier analysis on the Legendre polynomials.

Remark. Similarly, for the Gelfand pair $(SO(n), SO(n-1))$, we establish that the characters of the hypercomplex system $L_1(SO(n), SO(n-1))$ have the form

$$\chi_k(t) = \frac{U_k^{(n-2)/2}(t)}{U_k^{(n-2)/2}(1)} \qquad (k = 0, 1, \dots ;\ t \in [-1, 1]),$$

where $U_k^{(n-2)/2}(t)$ are ultraspherical polynomials (see, e.g., [Vil]).

Example 2. Let $G = SL_2(\mathbb{R})$, let $H = SO(2)$, and let A be the group of diagonal unimodular matrices

$$a = \begin{pmatrix} \alpha & 0 \\ 0 & \alpha^{-1} \end{pmatrix} \qquad (\alpha > 0).$$

Denote by A^+ the set of matrices $a \in A$ such that $\alpha \geq 0$. By using the polar decomposition, one can uniquely represent any matrix $g \in SL_2(\mathbb{R})$ in the form $g = sh$, where $h \in SO(2)$ and s is a positive symmetric matrix. Moreover, by orthogonal transformation, the matrix s can be diagonalized. If the obtained diagonal matrix a belongs to $A \backslash A^+$, then the adjoint matrix $\omega a \omega^{-1}$ belongs to A^+ (here,

$$\omega = \begin{pmatrix} 0 & 1 \\ -1 & 0 \end{pmatrix} \in SO(2)$$

is the Weyl matrix). If $a \neq 1$, then the matrix h_1 in the expression $s = h_1^{-1} a h_1$ $(a \in A^+$, $h_1 \in H)$ is uniquely defined to within multiplication by ± 1. Thus, we obtain the decomposition $G = HA^+H$ (the Cartan decomposition), which is uniquely defined for $a \neq 1$ to within the multiplication of the matrices h_1, h_2 by ± 1. Consequently, $Q = G // H = A^+$. Below, we represent matrices from A^+ in the form

$$a_\lambda = \begin{pmatrix} e^{\lambda/2} & 0 \\ 0 & e^{-\lambda/2} \end{pmatrix} \qquad (\lambda \geq 0).$$

Thus, Q can be identified with the semiaxis $[0, \infty)$. Denote the elements of $H = SO(2)$ by $h = g_\varphi$ $(0 \leq \varphi \leq 2\pi)$. Relation (1.31) takes the form

$$\int_G f(g)\,dg = \frac{1}{8\pi} \int_0^{2\pi} \int_0^{2\pi} \int_0^{\infty} f(g_{\varphi_1} a_\lambda g_{\varphi_2}) \sinh \lambda \, d\lambda \, d\varphi_1 \, d\varphi_2$$

and, hence, the measure $(\pi/2) \sinh \lambda \, d\lambda$ is the multiplicative measure on Q. It is convenient to consider a hypercomplex system isomorphic to $L_1(G, H)$. Namely, consider the homeomorphism $[0, \infty) \ni \lambda \mapsto t = \cosh \lambda \in [1, \infty)$. The measure $(\pi/2) \sinh \lambda \, d\lambda$ turns into the Lebesgue measure $(\pi/2)\,dt$ on $[1, \infty)$. By virtue of (2.1),

$$L_{t_1} f(t_2) = \frac{1}{2\pi} \int_0^{2\pi} f(a_{\lambda_1} g_\varphi a_{\lambda_2})\, d\varphi.$$

Rewriting $a_{\lambda_1} g_\varphi a_{\lambda_2}$ in the form $g_{\theta_1} a_\lambda g_{\theta_2}$, we obtain $\cosh \lambda = \cosh \lambda_1 \cosh \lambda_2 + \sinh \lambda_1 \sinh \lambda_2 \cos 2\varphi$. Consequently,

$$(L_{t_1} f)(t_2) = \frac{1}{2\pi} \int_0^{2\pi} f(\cosh \lambda_1 \cosh \lambda_2 + \sinh \lambda_1 \sinh \lambda_2 \cos 2\varphi)\, d\varphi$$

$$= \frac{1}{2\pi} \int_0^{2\pi} f\left(t_1 t_2 + \sqrt{1 + t_1^2}\sqrt{1 + t_2^2} \cos \varphi\right) d\varphi.$$

By using the explicit form of the generalized translation operators, one can easily establish the multiplication law in $L_1(Q, (\pi/2)\,dt)$. The spherical functions in this case are expressed via the Legendre functions (see, e.g., [Vil]):

$$\varphi_s(a_\lambda) = P_\rho(\cosh\lambda) \quad \left(\rho = \frac{s+1}{2},\ \lambda \in [0,\infty)\right).$$

The integral representation of the Legendre functions

$$P_\rho(\cosh\lambda) = \frac{1}{2\pi}\int_0^{2\pi} (\cosh\lambda + \sinh\lambda\cos u)^\rho\, du$$

and the multiplication formula

$$P_\rho(t)P_\rho(s) = \frac{1}{2\pi}\int_0^{2\pi} P_\rho\left(ts + \sqrt{1+t^2}\sqrt{1+s^2}\cos\varphi\right)d\varphi$$

follow from (2.4) and (2.2), respectively.

The relations obtained in Example 2 are similar to the corresponding relations in Example 1. This is explained by the fact that the corresponding Riemann spaces are in Cartan duality (see, e.g., [Hel1]).

Example 3. Let $G = M_0(2)$, where $M_0(2)$ is the identity component of the group $M(2)$ of all motions of plane, and let $H = SO(2)$. The group $M_0(2) = \mathbb{R}^2 \rtimes SO(2)$ is isomorphic to the group of matrices of the form

$$g(a, b, \alpha) = \begin{pmatrix} \cos\alpha & \sin\alpha & a \\ -\sin\alpha & \cos\alpha & b \\ 0 & 0 & 1 \end{pmatrix} \quad (a, b \in \mathbb{R}^1,\ \alpha \in [0, 2\pi]).$$

It is easy to see that $G/H = \mathbb{R}^2$, and $Q = G//H$ is parametrized by the semiaxis $[0, \infty)$ (every orbit $Hg(a, b, \varphi)H$ is a circle of radius $r = \sqrt{a^2+b^2} \in [0, \infty)$ in $\mathbb{R}^2$). Since $(1/4\pi^2)\,da\,db\,d\alpha$ is the Haar measure of the group $M_0(\mathbb{R}^2)$, by passing to the polar coordinates, we establish that $r\,dr$ is the multiplicative measure of the hypercomplex system $L_1(G, H)$. Let us find the explicit form of the generalized translation operators

$$(L_{r_1}f)(r_2) = \frac{1}{2\pi}\int_0^{2\pi} f(g(a_1, b_1, 0)g(0, 0, \alpha)g(a_2, b_2, 0))\,d\alpha$$

associated with $L_1(G, H)$. Rewriting $g(a_1, b_1, 0)g(0, 0, \alpha)g(a_2, b_2, 0)$ in the form $g(\overline{a}, \overline{b}, \overline{\alpha})$, we obtain

$$r = \sqrt{\overline{a}^2 + \overline{b}^2} = \sqrt{r_1^2 + r_2^2 + 2r_1r_2\cos(\varphi_2 - \varphi_1 - \alpha)}$$

and, therefore, .

$$(L_{r_1}f)(r_2) = \frac{1}{2\pi}\int_0^{2\pi} f\left(\sqrt{r_1^2+r_2^2+2r_1r_2\cos\alpha}\right)d\alpha.$$

It is khown (see, e.g., [Vil]) that, in this case, spherical functions are expressed in terms of the Bessel functions

$$\varphi_{i\rho}(g(a,b,\alpha)) = J_0(\rho r) = \frac{1}{2\pi}\int_0^{2\pi} e^{i\rho r\sin\varphi}\,d\varphi \quad (r = \sqrt{a^2+b^2},\ \rho \geq 0).$$

It is easy to derive the multiplication formula for Bessel functions from the explicit form of the generalized translation operators L_s. The Fourier transformation associated with the hypercomplex system $L_1(G,H)$ reduces to the Hankel transformation. As is khown, in this case, the Plancherel measure is concentrated on $\hat{Q} = [0,\infty)$ and has the form $\rho\,d\rho$.

Remark. If G is the group $M_0(n) = \mathbb{R}^n \rtimes SO(n)$ of motions of the Euclidean space $\mathbb{R}^n$ preserving orientation and the subgroup H is equal to $SO(n)$, then $Q = G//H = [0,\infty)$ and the characters of the hypercomplex system $L_1(G,H)$ are expressed in terms of Bessel functions $J_{(n-2)/2}$ as follows:

$$\chi_{i\rho}(r) = \Gamma\left(\frac{n}{2}\right)\frac{J_{(n-2)/2}(\rho r)}{(\frac{1}{2}\rho r)^{(n-2)/2}} \qquad (r = [0,\infty),\ \rho \geq 0)$$

(see [Vil]).

2.5. Hypercomplex Systems Associated with the Delsarte Generalized Translation Operators. Let G be a locally compact group and let B be a compact subgroup of the group of automorphisms of G. Then the set of B-orbits equipped with the quotient topology is a hypergroup with respect to natural convolution. This important class of hypergroups was introduced by Delsarte [Del4]. In the present subsection, we show that hypergroups of this sort are generated by Gelfand pairs.

Let G be a locally compact unimodular group and let B be a subgroup of the group $\operatorname{Aut} G$ containing the group of inner automorphisms $I(G)$. The group G is called a $[FIA]_B^-$-group if the closure B^- of B in $\operatorname{Aut} G$ is compact. The set $B^-(g) = \{\beta(g) \mid \beta \in B^-\}$ is called a B^--orbit of the element $g \in G$. Consider the space G_B of B-orbits equipped with the quotient topology.

The subalgebra

$$Z_B L_1(G) = \{f \in L_1(G, dg) | f \circ \beta = f \text{ for all } \beta \in B\}$$

of the group algebra of the group G is commutative. Moreover, it is a hypercomplex system and its basis G_B consists of B^--orbits of the group G.

Obviously, $f \in Z_B L_1(G)$ if and only if it is constant on B^--orbits. For all $g \in G$ and $x, y \in Z_B L_1(G)$, we have

$$(x * y)(g) = \int_G x(a) y(a^{-1} g) da = \int_G x(a) y(a^{-1} g a^{-1} a) da$$

$$= \int_G x(a) y(g a^{-1}) da = (y * x)(g).$$

By the general construction presented in Subsection 1.1 of Chapter 2, it suffices to show that any function $x(g) \in Z_B L_1(G)$ can be approximated in the norm of the space $L_1(G, dg)$ by continuous finite functions taking constant values on B^--orbits. This can be done in exactly the same way as in Subsections 1.2 or 2.1 of Chapter 2.

■

If B does not contain $I(G)$, then $Z_B L_1(G)$ is also a hypercomplex system (generally speaking, noncommutative).

Generalized translation operators associated with the hypercomplex system $Z_B L_1(G)$ are called Delsarte generalized translation operators and have the form

$$(L_s)(t) = \int_{B^-} f(\gamma(g_1) g_2) d\gamma = \int_{B^-} f(g_1 \gamma(g_2)) d\gamma \tag{2.13}$$

$$(g_1 \in s;\ g_2 \in t;\ t, s \in G_B;\ f \in Z_B L_1(G)),$$

where γ is the Haar measure on the group B^- normalized by the condition $\gamma(B^-) = 1$.

Indeed, by introducing necessary (obvious) modifications in the proof of relation (2.1), we arrive at the first formula in (2.13). The second formula follows from the invariance of the function f and the fact that the compact group B^- is unimodular.

■

In particular, it follows from (2.13) that the hypercomplex system $Z_B L_1(G)$ satisfies the condition of separate continuity. Clearly, the orbit $\{e\}$ is the basis unity of this hypercomplex system. It is also easy to check that the convolution of measures on G_B

generated by generalized translation operators (2.13) satisfies all axioms of hypergroup.

Let us show that the space G_B *can be identified with the space* $Q = \tilde{G}//\tilde{H}$ *of double cosets of the group* $\tilde{G}$ *with respect to* $\tilde{H}$ *for a properly chosen Gelfand pair* $(\tilde{G},\tilde{H})$.

Indeed, we set $\tilde{G} = G \rtimes B^-$ and $\tilde{H} = \{e\} \otimes B^- = B^-$. Let $(g, \beta) \in \tilde{G}$, then

$$\tilde{H}(g, \beta)\tilde{H} = \{(\gamma_1(g), \gamma_1 \circ \beta \circ \gamma_2) \,|\, \gamma_1, \gamma_2 \in B^-\} = \tilde{H}(g, 1)\tilde{H}.$$

This implies that $\tilde{G}//\tilde{H}$ is homeomorphic to G_B. In order to show that $(\tilde{G},\tilde{H})$ is a Gelfand pair, we define an involutive automorphism σ as follows: $\sigma((g, \beta)) = (g^{-1}, p_g \circ \beta)$, where p_g is the inner automorphism of the group G generated by an element g. Let us show that σ is indeed an automorphism. Since $\beta \circ p_g = p_{\beta(g)} \circ \beta$, we have

$$\begin{aligned}
\sigma((g_1, \beta_1))\sigma((g_2, \beta_2)) &= \left(g_1^{-1}, p_{g_1} \circ \beta_1\right)\left(g_2^{-1}, p_{g_2} \circ \beta_2\right) \\
&= \left(g_1^{-1}(p_{g_1} \circ \beta_1)(g_2^{-1}), p_{g_1} \circ \beta_1 \circ p_{g_2} \circ \beta_2\right) \\
&= \left(\beta_1(g_2^{-1})g_1^{-1}, p_{g_1\beta_1(g_2)} \circ \beta_1 \circ \beta_2\right) = \sigma((g_1\beta_1(g_2), \beta_1 \circ \beta_2)) \\
&= \sigma((g_1, \beta_1)(g_2, \beta_2)).
\end{aligned}$$

Obviously, σ is an involutive automorphism. Since

$$\begin{aligned}
(g^{-1}, p_g \circ \beta) &= (e, \beta)(\beta^{-1}(g^{-1}), \beta^{-1})(e, p_g \circ \beta) \\
&= (e, \beta)[(g, \beta)]^{-1}(e, p_g \circ \beta) \in \tilde{H}[(g, \beta)]^{-1}\tilde{H},
\end{aligned}$$

by virtue of Theorem 2.1, we conclude that $(\tilde{G},\tilde{H})$ is a Gelfand pair.

■

Note that if G is commutative, then we can take a simpler automorphism $\sigma((g, \beta)) = (g^{-1}, \beta)$ instead of σ.

Thus, the subalgebra $Z_B L_1(G)$ of the group algebra $L_1(G, dg)$ coincides with the hypercomplex system $L_1(\tilde{G},\tilde{H})$. It is easy to see that characters of the hypercomplex system $Z_B L_1(G)$ are continuous functions not identically equal to zero on G taking constant values on B^{-}-orbits, and such that

$$\varphi(g_1)\varphi(g_2) = \int_{B^-} \varphi(g_1\beta(g_2))\,d\beta. \tag{2.14}$$

As shown in [Mos], the space of characters of the hypercomplex system $Z_B L_1(G)$ can be identified with the set of extreme points of the space of B^--invariant positive definite functions φ on G normalized by the condition $\varphi(e)=1$ and equipped with the topology of uniform convergence on compact sets. Harmonic analysis on the hypercomplex system $Z_B L_1(G)$ was constructed in [Mos] and [LiM]. Obviously, these results can also be derived from Theorem 2.5. In [HHL], it was also shown that the space of Hermitian characters $E(G,B)$ of the hypergroup G_B is a commutative hypergroup with respect to the convolution of measures on $E(G,B)$ which is defined as follows: Since the product of positive definite functions $\varphi, \psi \in E(G,B)$ is positive definite, by virtue of Theorem 3.1 in Chapter 1, we have

$$(\varphi\psi)(p) = \int_{E(G,B)} \theta(p)\,d\mu_{\varphi,\psi}(\theta) \quad (p \in G_B),$$

where $\mu_{\varphi,\psi}$ is a probability measure on $E(G,B)$. Further, for any $\varphi, \psi \in E(G,B)$, we set $\delta_\varphi * \delta_\psi = \mu_{\varphi,\psi}$. According to [Hart], on the hypergroup $E(G,B)$, there exists the left-invariant measure $\hat{m}$ whose support coincides with $E(G,B)$. The measure m normalized in a proper way turns into the Plancherel measure. By applying Theorem 3.5 to the hypercomplex system $Z_B L_1(G)$, we conclude that *duality is true for the hypercomplex system* $Z_B L_1(G)$ (this result was first established in [Hart]). It is worth noting that the subclass of $[FIA]_B^-$-groups composed by so-called Z-groups, i.e., locally compact groups G with center Z such that the group G/Z is compact, is fairly well studied (see [GrM], [GMM], [Ros1], [Ros2], and [HKK]). In this case, $B^- = I(G)$.

2.6. Center of Group Algebra as a Gelfand Pair. If G is a compact group and B is the group of all inner automorphisms of the group G, then the reasoning used above can be significantly simplified. In this case, the conjugacy classes of the group G are B^--orbits and the hypercomplex system $Z_B L_1(G)$ coincides with the center $Z L_1(G)$ of the group algebra of the group G. This hypercomplex system was studied in detail in the previous subsection. In this case, we take the direct product $G \otimes G$ as $\tilde{G}$ and the subgroup of all pairs (g,g) $(g \in G)$ as $\tilde{H}$. Clearly, relation (2.12) for characters of the hypercomplex system $ZL_1(G)$ can be reduced to the Weyl formula (1.20).

3. Orthogonal Polynomials

As shown in Section 2, some known systems of orthogonal polynomials form sets of spherical functions of a Gelfand pair (G, H), where G is a compact group and H is its subgroup. In particular, spherical functions of Riemannian symmetric spaces of compact type are orthogonal polynomials of, generally speaking, several variables (the number of variables is equal to the rank of the symmetric space, [Vret]). Thus, these families of orthogonal polynomials form the set of characters of a normal hypercomplex system with basis unity and compact basis G/H. It is natural to construct examples of hypercomplex systems with compact basis $E \subset \mathbb{R}^n$ $(n \geq 1)$ whose sets of characters coincide with given systems of orthogonal polynomials on E. This is done in Subsection 3.5. The main attention in this section is focused on the analysis of the dual case, namely, for fairly general families of orthonormal polynomials $P_n(t)$ $(t \in \mathbb{R}^1;\ n \in \mathbb{N}_0 = \{0, 1, \dots\})$, we construct discrete hypercomplex systems with basis $\mathbb{N}_0$ whose characters are $\chi(n) = A_n P_n(z)$ $(A_n \in \mathbb{R},\ z \in \mathbb{C})$ and study harmonic analysis for these systems.

3.1. Discrete Hypercomplex Systems Associated with Orthogonal Systems of Polynomials. In this subsection, we construct a discrete Hermitian hypercomplex system with the basis $Q = \{0, 1, \dots\}$, the nonnegative structure measure, and the basis unity $e = 0$ associated with a given family of orthogonal polynomials defined on a bounded closed subset $E \subset \mathbb{R}^1$. We describe the set of characters and give a sufficient condition for this hypercomplex system to be symmetric.

Let E be a closed bounded subset of the real axis and let $d\sigma$ be a nonnegative measure on E with infinitely many points of growth satisfying the conditions $\sigma(E) = 1$, $\int |t|^n d\sigma(t) < \infty$ $(n = 1, 2, \dots)$, and $\operatorname{supp} \sigma = E$. Let $P_0(t), P_1(t), \dots$ be a sequence of polynomials orthonormal with respect to the measure $d\sigma(t)$ with positive leading coefficients and let $a = \inf E$ and $b = \sup E$. As is known (see [Sze]), all zeros of the family of orthogonal polynomials $P_n(t)$ are located inside the interval (a, b). Hence, $P_n(b) > 0$ and $(-1)^n P_n(a) > 0$ $(n \in \mathbb{N})$. We set

$$R_p(t) = \frac{P_p(t)}{P_p(b)}, \quad m(p) = [P_p(b)]^2,$$

$$c(p, q, r) = \frac{P_p(b) P_q(b)}{P_r(b)} \int P_p(t) P_q(t) P_r(t) d\sigma(t) \quad (p, q, r = 0, 1, \dots). \tag{3.1}$$

Theorem 3.1. *If* $\int P_p(t) P_q(t) P_r(t) d\sigma(t) \geq 0$ *for all* $p, q, r = 0, 1, \dots$, *then* $c(p, q, r)$ *are structure constants and* $m(p)$ *is the multiplicative weight of a certain*

Hermitian hypercomplex system with the discrete basis $\mathbb{N}_0 = \{0, 1, 2, \ldots\}$ *and basis unity* 0. *We denote this hypercomplex system by* $l_1(d\sigma, m)$. *The space* X *of characters of the hypercomplex system* $l_1(d\sigma, m)$ *consists of the set of points* z *of the complex plane such that the sequence* $R_n(z)$ *is bounded. The space of Hermitian characters is* $X_h = X \cap \mathbb{R}$.

Proof. Orthogonal polynomials possess the property that $\int Q(t)P_n(t)d\sigma(t) = 0$ for any polynomial $Q(t)$ of degree less than n. Hence, $\int P_p(t)P_q(t)P_r(t)d\sigma(t) = 0$ for all $r > p+q$ and $r < |p-q|$. This obviously implies that the functions $R_p(t)$ $(p \in \mathbb{N}_0)$ form an A-system. By applying Theorem 3.8 from Chapter 1, we establish the validity of the first part of the theorem.

Note that, for all $z \in \mathbb{C}^1$, orthogonal polynomials satisfy the equality

$$P_p(z)P_q(z) = \sum_{r=0}^{\infty} P_r(z) \int P_p(t)P_q(t)P_r(t)d\sigma(t) \quad (p, q \in \mathbb{N}_0). \tag{3.2}$$

By dividing both sides of (3.2) by $P_p(b)P_q(b)$, we obtain relation (1.27) from Chapter 1, whence it follows that if the sequence $R_p(z)$ is bounded, then $\chi(p) = R_p(z)$ is a character of the hypercomplex system $l_1(d\sigma, m)$.

Conversely, assume that $\chi(p)$ $(p \in \mathbb{N}_0)$ is a character of the hypercomplex system $l_1(d\sigma, m)$. Then it follows from (3.2) that

$$\chi(p)\chi(q)P_p(b)P_q(b)$$

$$= \sum_{r=|p-q|}^{p+q} \chi(r)P_r(b) \int P_p(t)P_q(t)P_r(t)d\sigma(t) \quad (p, q \in \mathbb{N}_0).$$

This implies that all values of the character $\chi(p)$ are uniquely determined by the values $\chi(0) = 1$ and $\chi(1)$. We set $z = \chi(1)(b - \sigma_1) + \sigma_1$, where $\sigma_1 = \int t d\sigma(t)$ is the first moment of the measure $d\sigma$. Since

$$P_1(z) = \frac{1}{\alpha_0}(z - \sigma_1),$$

we have $\chi(0) = P_0(z)$ and $\chi(1)P_1(b) = P_1(z)$. This and (3.2) imply that $\chi(r)P_r(b) = P_r(z)$ or $\chi(r) = R_r(z)$ $(r \in \mathbb{N}_0)$. Since the character $\chi(p)$ is bounded, the sequence $R_p(z)$ is also bounded. The equality $X_h = X \cap \mathbb{R}$ follows from the fact that Hermitian characters of an Hermitian hypercomplex system are real-valued.

■

Remark 1. It may happen that $(-1)^{p+q+r}\int P_p(t)P_q(t)P_r(t)d\sigma(t) \geq 0$ for all $p, q, r \in \mathbb{N}_0$. In this case, it is necessary to set

$$S_p(t) = \frac{P_p(t)}{P_p(a)}, \qquad m(p) = [P_p(a)]^2,$$

$$c(p, q, r) = m(p)m(q)\int S_p(t)S_q(t)S_r(t)d\sigma(t).$$

For the numbers $c(p, q, r)$ and $m(r)$ thus defined, all statements of Theorem 3.1 are true.

Remark 2. Note that $\mathrm{X} \subseteq \{z \in \mathbb{C}^1 \mid |z - \sigma_1| \leq b - \sigma_1\}$ and, in particular, $\mathrm{X}_h \subseteq \{2\sigma_1 - b, b\}$.

Remark 3. By virtue of Theorems 3.2 and 3.3 in Chapter 1, there exists a regular Borel measure $d\pi$ on X_h such that the polynomials $P_r(t)$ are orthogonal with respect to it. For a system of polynomials orthonormal on a bounded subset of the real axis, such a measure is unique (see, e.g., [Ber10]). Consequently, we have $d\sigma = d\pi$. Since $\hat{Q} = E = \operatorname{supp}\sigma = \operatorname{supp}\pi \subseteq \mathrm{X}_h$, we conclude that $|R_r(t)| \leq 1$ for all $t \in Q$ or $|P_r(t)| \leq P_r(b)$ for all $t \in E$.

Remark 4. One can easily check that if conditions of Theorem 3.1 are satisfied, the functions $R'_p(t) = P_p(t)/P_p(c)$ form an A-system for any $c > b$. Consequently, they also can be used for the construction of an Hermitian hypercomplex system $l'_1(d\sigma, m')$ with basis unity. This hypercomplex system has the set of characters larger than E because, on the one hand, $E \subseteq \mathrm{X}_h$ and, on the other hand, the point $c \notin E$ generates a character of the hypercomplex system identically equal to one.

By applying Theorem 3.1 from Chapter 1 to the constructed hypercomplex system, we establish the following statement: *In order that a bounded sequence $s(p)$ of real numbers be representable as*

$$s(p) = \int_{\mathrm{X}_h} R_p(t)d\mu(t),$$

where μ is a nonnegative finite measure on X_h, it is necessary and sufficient that this sequence be positive definite in the sense of (3.1′) in Chapter 1.

Theorem 3.2. *Suppose that the conditions of Theorem 3.1 (or the conditions of Remark 1) are satisfied. If a sequence $P_r(b)$ is bounded, then the hypercomplex system $l_1(d\sigma, m)$ is symmetric, i.e., $\mathrm{X} = \mathrm{X}_h$.*

Proof. It is necessary to verify that $\chi(r) = \overline{\chi(r)}$ or $P_r(z) = \overline{P_r(z)}$ $(z \in \mathrm{X})$. If $z \in \mathrm{X}_h$, then this equality is obvious. Therefore, assume that $z \notin \mathbb{R}^1$. Let us fix r. For sufficiently large N, we have

$$\frac{1}{\sum_{p=0}^{N} |P_p(z)|^2} \sum_{p,q=0}^{N} \int P_p(t)P_q(t)P_r(t)d\sigma(t)\,\overline{P_p(z)}\,P_q(z)$$

$$= \frac{1}{\sum_{p=0}^{N} |P_p(z)|^2} \left[\sum_{p=0}^{N} \sum_{q=1}^{\infty} \int P_r(t)P_p(t)P_q(t)d\sigma(t)\,\overline{P_p(z)}\,P_q(z) \right.$$

$$\left. - \sum_{p=0}^{N} \sum_{q=N+1}^{\infty} \int P_r(t)P_p(t)P_q(t)d\sigma(t)\,\overline{P_p(z)}\,P_q(z) \right]$$

$$= P_r(z) - \frac{1}{\sum_{p=0}^{N} |P_p(z)|^2} \sum_{p=N-r+1}^{N} \sum_{q=N+1}^{N+r} \int P_r(t)P_p(t)P_q(t)d\sigma(t)\,\overline{P_p(z)}\,P_q(z) \tag{3.3}$$

(here, we have used the fact that $\int P_p(t)P_q(t)P_r(t)d\sigma(t) = 0$ for all $r > p+q$ and $r < |p-q|$). It follows from Remark 3 that $|P_r(z)| \le P_r(b)$ for all $z \in \mathrm{X}$ and, hence, the numbers $P_r(z)$ are uniformly bounded. But then the sum

$$\sum_{p=N-r+1}^{N} \sum_{q=N+1}^{N+r} \int P_r(t)P_p(t)P_q(t)d\sigma(t)\,\overline{P_p(z)}\,P_q(z)$$

is bounded in N. It is known (see, e.g., [Ber10]) that the series $\sum_{p=0}^{\infty} |P_p(z)|^2$ is divergent for real z. Therefore, it follows from (3.3) that

$$P_r(z) = \lim_{N\to\infty} \frac{1}{\sum_{p=0}^{N} |P_p(z)|^2} \sum_{p,q=0}^{N} \int P_r(t)P_p(t)P_q(t)d\sigma(t)\,\overline{P_p(z)}\,P_q(z).$$

On the other hand, it is easy to see that the numbers

$$\sum_{p,q=0}^{N} \int P_p(t)P_q(t)P_r(t)d\sigma(t)\,\overline{P_p(z)}\,P_p(z)$$

are real.

■

3.2. Jacobian Matrices and Generalized Translation Operators. Here, we outline another approach to the construction of hypercomplex systems in terms of orthonormal polynomials $P_r(t)$. First, we briefly recall several facts from the spectral theory of difference expressions (Jacobi matrices) (see [Ber10]).

For the difference expression with coefficients α_j and β_j

$$(Lu)_j = \alpha_{j-1}u_{j-1} + \beta_j u_j + \alpha_j u_{j+1} \quad (\alpha_j > 0,\ \beta_j \in \mathbb{R}^1,\ \alpha_{-1} = 0,\ j \in \mathbb{N}_0), \tag{3.4}$$

in the space l_2 of sequences $u = (u_j)_{j=0}^{\infty}$, we construct an operator Λ as the closure of the mapping $l_0 \ni u \mapsto Lu$ (l_0 is the linear subspace of l_2 consisting of finite sequences). To calculate $(Lu)_0$ we assume that the boundary condition $u_{-1} = 0$ is satisfied. The operator Λ is Hermitian. We assume that α_j and β_j are such that this operator is self-adjoint (we can even take bounded coefficients α_j and β_j; in this case, the operator Λ is bounded and, hence, self-adjoint). The Parseval equality for the decomposition of the operator Λ in the eigenfunctions has the form

$$(u, v)_{l_2} = \int_{\mathbb{R}^1} \tilde{u}(t)\overline{\tilde{v}(t)}\,d\sigma(t) \qquad (u, v \in l_0). \tag{3.5}$$

Here, $d\sigma(t)$ is the spectral measure (equal to $d(E(t)\delta_0, \delta_0)$, E is the resolution of the identity for the operator Λ, $\delta_0 = (1, 0, \dots) \in l_2$, and

$$\tilde{u}(t) = \sum_{p=0}^{\infty} u_p P_p(t)$$

is the Fourier transform associated with the sequence $(P_p(t))_{p=0}^{\infty}$ of so-called polynomials of the first kind, which can be recursively calculated as solutions of the Cauchy problem for the difference equation

$$(LP.(z))_p = zP_p(z) \quad (p \in \mathbb{N}_0,\ z \in \mathbb{C}^1,\ P_{-1}(z) \equiv 0,\ P_0(z) \equiv 1). \tag{3.6}$$

As follows from (3.5), the polynomials $P_p(t)$ form an orthonormal basis in $L_2(\mathbb{R}^1, d\sigma)$.

The inverse spectral problem now takes a fairly simple form, namely, for an arbitrary nonnegative probability measure $d\sigma(t)$ defined on the axis $(-\infty, \infty)$, having infinitely many growth points, and such that all powers t^p $(p \in \mathbb{N}_0)$ are integrable, by applying the orthogonalization procedure to the powers $1, t, t^2, \dots,$ we construct polynomials $P_p(t)$ with a positive leading coefficient. These polynomials are polynomials of the first kind for the difference expression L whose coefficients are uniquely determined by the relations

$$\alpha_p = \int_{-\infty}^{\infty} tP_p(t)P_{p+1}(t)d\sigma(t), \qquad \beta_p = \int_{-\infty}^{\infty} tP_p^2(t)\,d\sigma(t).$$

Consider the partial difference equation

$$\alpha_p u_{p+1,q} + \beta_p u_{p,q} + \alpha_{p-1} u_{p-1,q} = \alpha_q u_{p,q+1} + \beta_q u_{p,q} + \alpha_{q-1} u_{p,q-1}, \tag{3.7}$$

where $(u_{p,q})_{p,q=0}^{\infty}$ is an unknown infinite matrix. It follows from (3.7) that the elements of the $(q+1)$th column of the matrix $(u_{p,q})_{p,q=0}^{\infty}$ are uniquely determined by the elements of the qth and $(q-1)$th columns. This implies that the matrix $u_{p,q}$ is completely determined by its "initial" values—the column with index q equal to zero. On the linear span of the sequences $x = (x(p))_{p=0}^{\infty}$, we define a family of operators $(T_q'x)(p) = u_{p,q}$, where the matrix $(u_{p,q})_{p,q=0}^{\infty}$ is determined by the "initial" value $u_{p,0} = x(p)$ $(p \in Q)$. Let us show that $T_q'x$ can be directly calculated by using the relation

$$(T_q'x)(p) = u_{p,q} = \sum_r x(r) \int P_p(t)P_q(t)P_r(t)d\sigma(t) \quad (p, q \in Q) \tag{3.8}$$

(clearly, the summation is carried out over finitely many indices $|p-q| \le r \le p+q$). Indeed, since an element $u_{p,q}$ is completely determined by finitely many "initial" values $u_{r,0}$ $(|p-q| \le r \le p+q)$, it is sufficient to prove equality (3.8) for sequences x with finitely many nonzero terms. We set

$$x(t) = \sum_p x(p)P_p(t).$$

Then, by the Parseval equality,

$$\sum_r x(r) \int P_p(t)P_q(t)P_r(t)d\sigma(t) = \int P_p(t)P_q(t)x(t)d\sigma(t) = v_{p,q}.$$

The matrix $(v_{p,q})_{p,q=0}^{\infty}$ is a solution of equation (3.7). Indeed, by using (3.6), we get

$$\begin{aligned}
&\alpha_p v_{p+1,q} + \beta_p v_{p,q} + \alpha_{p-1} v_{p-1,q} \\
&\qquad = \int [\alpha_p P_{p+1}(t) + \beta_p P_p(t) + \alpha_{p-1} P_{p-1}(t)]\, P_q(t)x(t)d\sigma(t) \\
&\qquad = \int tP_p(t)P_q(t)x(t)d\sigma(t)
\end{aligned}$$

$$= \int P_p(t)[\alpha_q P_{q+1}(t) + \beta_q P_q(t) + \alpha_{q-1} P_{q-1}(t)]x(t)d\sigma(t)$$

$$= \alpha_q v_{p,q+1} + \beta_q v_{p,q} + \alpha_{q-1} v_{p,q-1}.$$

Since $v_{p,0} = \int P_p(t)x(t)d\sigma(t) = x(p) = u_{p,0}$ $(p \in \mathbb{Z}_+)$, we have $v_{p,q} = u_{p,q}$, and relation (3.8) is thus proved. Also note that $(T_q' P.(z))(p) = P_p(z)P_q(z)$ for any $p, q \in Q$, $z \in \mathbb{C}^1$.

We set

$$(T_q x)(p) = \frac{[T_q' x P.(b)](p)}{[T_q' P.(b)](p)} = \frac{[T_q' x P.(b)](q)}{P_p(b)P_q(b)}, \tag{3.9}$$

$$m(p) = P_p^2(b) \quad (p, q \in \mathbb{N}_0).$$

We introduce a convolution in $l_1(m)$ as follows:

$$(\xi * \eta)(r) = \sum_p (T_p \xi)(r)\eta(p)m(p) \quad (\xi, \eta \in l_1(m),\ r \in \mathbb{N}_0).$$

It is easy to see that if the conditions of Theorem 3.1 are satisfied, then $l_1(m)$ endowed with such a convolution is a commutative hypercomplex system with the structure measure

$$c(p, q, r) = \sum_l T_r \kappa_p(l)\kappa_q(l)m(l) = T_r \kappa_p(q)m(q)$$

$$= \frac{P_p(b)P_q(b)}{P_r(b)} \int P_q(t)P_p(t)P_r(t)d\sigma(t).$$

The operator T_p defined by (3.9) is a generalized translation operator in the hypercomplex system $l_1(d\sigma, m)$. Conversely, since $u_{p,q}$ are calculated in terms of the coefficients α_p, β_p of the difference expression L by the recurrence relation (3.7), formula (3.8) allows us to calculate the integrals $\int P_p(t)P_q(t)P_r(t)d\sigma(t)$ (and, hence, the structure constants $c(p, q, r)$ of the hypercomplex system $l_1(d\sigma, m)$ for a given difference expression.

3.3. Characterization of Hypercomplex Systems Associated with Orthogonal Polynomials. Below, we give a criterion for a particular hypercomplex system to be generated by a system of orthogonal polynomials.

Theorem 3.3. *For an Hermitian hypercomplex system* $l_1(m)$ *with basis* $Q = \mathbb{N}_0$ *to be generated by a system of orthogonal polynomials, it is necessary and sufficient that the following conditions be satisfied:*

(i) $m(0) = 1;$

(ii) $\operatorname{supp}_r c(p, 1, r) \subset [|p-1|, p+1]$ *and* $c(p, 1, p+1) > 0$. *This condition implies that* $\operatorname{supp}_r c(p, q, r) \subset [|p-q|, p+q]$ *for any* $p, q \in Q$.

Proof. The necessity obviously follows from Theorem 3.1. Let us establish the sufficiency. Consider the mapping $\theta : \mathrm{X}_h \to [-1, 1]$ given by the relation $\mathrm{X}_h \ni \chi \mapsto \theta(\chi) = \chi(1) \in [-1, 1]$ (since the hypercomplex system is Hermitian, we have $\chi(1) \in \mathbb{R}^1$). Since

$$\int_{X_h} \chi(1)\chi(0) d\chi = 0,$$

we get $\chi(1) \not\equiv \chi(0) = 1$. Consequently, $\theta(\mathrm{X}_h) \neq \{1\}$. Let us show that θ is injective. Let $\chi(1) = \psi(1)$ for some $\chi, \psi \in \mathrm{X}_h$. By using equality (1.27) from Chapter 1 with $q = 1$ and condition (ii), we obtain

$$\chi(p)\chi(1)m(p)m(1) = \sum_{r=p-1}^{p+1} c(p, 1, r)\chi(r)m(r) \quad (p \geq 1) \tag{3.10}$$

or

$$c(p, 1, p+1)\chi(p+1)m(p+1)$$

$$= \chi(p)\chi(1)m(1)m(p) - c(p, 1, p)\chi(p)m(p) - c(p, 1, p-1)\chi(p-1)m(p-1).$$

Since $c(p, 1, p+1) > 0$, any value of the character $\chi(p)$ is uniquely determined by the values $\chi(0) = 1$ and $\chi(1)$. Consequently, $\chi(p) = \psi(p)$ for all $p \in Q$, i.e., $\chi = \psi$ and θ is injective. Let $E = \theta(\mathrm{X}_h) \subseteq [1, 1]$. It is clear that θ is continuous and, since X_h is compact, the set E is closed.

By writing (3.10) for the character $\chi_t \in \mathrm{X}_h$, setting $P_p(t) = \chi_t(p)\sqrt{m(p)}$, where $\chi_t = \theta^{-1}(t)$ $(t \in E)$, and using the equality $c(p, q, r)m(r) = c(r, q, p)m(p)$, we obtain

$$P_p(t)t\sqrt{m(p)}\,m(1)$$

$$= \frac{c(p-1,1,p)m(p)P_{p-1}(t)}{\sqrt{m(p-1)}} + c(p,1,p)\sqrt{m(p)}\,P_p(t)$$

$$+ c(p,1,p+1)\sqrt{m(p+1)}\,P_{p+1}(t)$$

or

$$tP_p(t) = \frac{c(p-1,1,p)}{m(1)}\sqrt{\frac{m(p)}{m(p-1)}}\,P_{p-1}(t)$$

$$+ \frac{c(p,1,p)}{m(1)}P_p(t) + \frac{c(p,1,p+1)}{m(1)}\sqrt{\frac{m(p+1)}{m(p)}}\,P_{p+1}(t).$$

If we set

$$\alpha_p = \frac{c(p,1,p+1)}{m(1)}\sqrt{\frac{m(p+1)}{m(p)}}, \qquad \beta_p = \frac{c(p,1,p)}{m(1)},$$

the last equality can be rewritten in the form

$$(LP.(t))_p = \alpha_{p-1}P_{p-1}(t) + \beta_p P_p(t) + \alpha_p P_{p+1}(t) = tP_p(t) \quad (p \in Q,\ t \in E) \tag{3.11}$$

(by virtue of (i), $P_0(t) = \chi_t(0)\sqrt{m(0)} = 1$; it is also easy to see that (3.11) holds for $p = 0$). By virtue of condition (ii) and the fact that the structure constants are nonnegative, we have $\alpha_p > 0$ and $\beta_p \geq 0$.

Let us show that α_p and β_p are such that the operator Λ is essentially self-adjoint. It is sufficient to show that α_p and β_p are bounded. Since the hypercomplex system is normal,

$$\alpha_p = \frac{c(p,p+1,1)}{\sqrt{m(p)m(p+1)}} \quad (p \in Q).$$

But by virtue of (1.13) in Chapter 1, $c(p,p+1,1) \leq \min\{m(p), m(p+1)\}$ and

$$0 \leq \alpha_p \leq \frac{\min\{m(p), m(p+1)\}}{\sqrt{m(p)m(p+1)}} \leq 1.$$

Similarly, we get $0 \leq \beta_p \leq 1$. Consequently, the polynomials of the first kind $P_p(t)$

form an orthonormal basis in $L_2(E, d\sigma)$, where $d\sigma(t)$ is the spectral measure of the operator Λ.

Let us show that the relationship between $c(p, q, r)$, $m(r)$ and the polynomials $P_p(t)$ is described by formulas (3.1). It is clear that $P_p^2(1) = \chi_1^2(p)m(p) = m(p)$ because $\chi_1(p) = \theta^{-1}(1) \equiv 1$. The measure $d\chi$ is the image of the Plancherel measure $d\sigma$ under the mapping θ. Indeed, denoting by $d\rho$ the image of $d\chi$ under the mapping θ, by virtue of (3.14) in Chapter 1, we get

$$\int_E P_p(t)P_q(t)d\rho(t) = \sqrt{m(p)m(q)} \int_{X_h} \chi(p)\chi(q)d\chi = \delta_{p,q},$$

whence it follows that the polynomials $P_p(t)$ are orthogonal with respect to the measure $d\rho$. Since such a measure is unique for a system of polynomials orthonormal on a bounded subset of the real axis,, we get $d\sigma = d\rho$. By (3.15) in Chapter 1, we have

$$c(p, q, r) = m(p)m(q) \int_{X_h} \chi(p)\chi(q)\chi(r)d\chi$$

$$= \frac{P_p(1)P_q(1)}{P_r(1)} \int P_p(t)P_q(t)P_r(t)d\sigma(t).$$

■

Remark. As follows from the example of the join of countably many hypergroups constructed in Subsection 4.4 in Chapter 1, the condition $c(p, 1, p+1)$ is essential.

3.4. Another Method for the Construction of a Hypercomplex System Associated with Orthogonal Polynomials. Examples. In this subsection, we give another method, which is due to Lasser [Las2], for the construction of a discrete hypercomplex system associated with a system of orthogonal polynomials. First, let us prove an auxiliary result, which, nevertheless, is of independent interest.

Theorem 3.4. *If a hypercomplex system $l_1(m)$ is associated with a system of orthogonal polynomials, then the structure measure $c(p, q, r)$ is completely determined by its values $c(p, p, 0) = m(p)$, $c(p, 1, p)$, and $c(p, 1, p+1)$ $(p \in \mathbb{N}_0)$.*

Proof. By Theorem 3.3, $\operatorname{supp}_r c(p, q, r) \subseteq [|p-q|, p+q]$. Let us write the associativity relation (1.25) in Chapter 1 for $p = 1$, $q = n-1$, $l = k$, $s = n+1$ $(n, k \in \mathbb{N}_0)$. Taking into account that $\operatorname{supp}_r c(1, n-1, r) \subseteq [|n-2|, n]$, $\operatorname{supp}_r c(r, k, n+k) = \operatorname{supp}_r c(k, n+k, r) \subseteq [n, n+2k]$, $\operatorname{supp}_r c(n-1, k, r) \subseteq [|n-k-1|, n+k-1]$, and

$\operatorname{supp}_r c(r, 1, n+k) = \operatorname{supp}_r c(n+k, 1, r) \subseteq [n+k-1, n+k+1]$, we see that relation (1.25) in Chapter 1 takes the form

$$c(n-1, 1, n)c(n, k, n+k) = c(n-1, k, n+k-1)c(n+k-1, 1, n+k),$$

whence

$$c(n, k, n+k) = \frac{c(n+k-1, 1, n+k)c(n-1, k, n+k-1)}{c(n-1, 1, n)} \tag{3.12}$$

because $c(p, 1, p+1) > 0$ $(p \in \mathbb{N}_0)$. Since all numbers $c(k, 1, k+1)$ are known, relation (3.12) enables us to recursively determine from $c(p, 1, p+1)$ all numbers $c(p, q, p+q)$. By writing the associativity relation (1.25) from Chapter 1 for $p = 1$, $q = n-1$, $l = k$, $s = n+k-1$ and repeating the same argument, we obtain the equality

$$\begin{aligned} c(n, k, n+k-1) \\ &= \frac{c(n-1, k, n+k-1)[c(n+k-1, 1, n+k-1) - c(n-1, 1, n-1)]}{c(n-1, 1, n)} \\ &\quad + \frac{c(n-1, k, n+k-2)c(n+k-2, 1, n+k-1)}{c(n-1, 1, n)}, \end{aligned} \tag{3.13}$$

which allows us to determine all numbers $c(p, q, p+q-1)$ by recursion in n. Since the hypercomplex system is commutative, without loss of generality, we can assume that $p \le q$. Thus, by setting $p = 1$, $q = n-1$, $l = k$, $s = n+k-r$ $(r = 2, 3, \ldots, 2n-2)$ in relation (1.25) in Chapter 1, we obtain the following equality (recall that $n \le k$):

$$\begin{aligned} c(n, k, n+k-r) \\ &= \frac{c(n-1, k, n+k-r-1)c(n+k-r-1, 1, n+k-r)}{c(n-1, 1, n)} \\ &\quad + \frac{c(n-1, k, n+k-r)[c(n+k-r, 1, n+k-r) - c(n-1, k, n+k-r)]}{c(n-1, 1, n)} \\ &\quad + \frac{c(n-1, k, n+k-r+1)c(n+k-r+1, k, n+k-r)}{c(n-1, 1, n)} \\ &\quad - \frac{c(n-1, 1, n-2)c(n-2, k, n+k-r)}{c(n-1, 1, n)}. \end{aligned} \tag{3.14}$$

Note that

$$c(n-1,\,1,\,n-2) \;=\; \frac{c(n-2,1,n-1)m(n-1)}{m(n-2)}$$

and

$$c(n+k-r+1,\,k,\,n+k-r) \;=\; \frac{c(n+k-r,k,n+k-r+1)m(n+k-r+1)}{m(n+k-r)}\,.$$

By using (3.14), we determine by recursion (first, in n and then in r) the structure constants $c(p,q,p+q-r)$ $(r=2,3,\dots,2p-2)$.

Finally, the remaining structure constants $c(p,q,q-p+1)$ and $c(p,q,q-p)$ can be found as follows:

$$c(p,q,q-p) \;=\; c(q,p,q-p) \;=\; c(q-p,p,q)\,\frac{m(q)}{m(q-p)} \qquad (p\le q),$$

$$c(p,q,q-p+1) \;=\; c(q,p,q-p+1) \;=\; c(q-p+1,p,q)\,\frac{m(q)}{m(q-p+1)} \tag{3.15}$$

(note that $c(q-p,p,q)$ and $c(q-p+1,p,q)$ are numbers of the form $c(n,m,n+m)$ and $c(n,m,n+m-1)$, respectively).

■

Theorem 3.4 allows one to introduce the following method for the construction of a discrete hypercomplex system: Let three sequences $(a_n)_{n=0}^{\infty}$, $(b_n)_{n=0}^{\infty}$, and $(c_n)_{n=0}^{\infty}$ $(a_n>0,\ c_n>0,\ b_n\ge 0,\ n\ge 1,\ c_0=0)$, $a_n+b_n+c_n=1$, be given. We set

$$c(p,0,p) \;=\; 1, \quad c(p,1,p+1) \;=\; \frac{c_{p+1}}{c_1},$$

$$c(p,1,p) \;=\; \frac{b_p}{c_1}, \quad c(p,1,p-1) \;=\; \frac{a_{p-1}}{c_1} \quad (p=1,2,\dots)$$

define $c(p,q,r) = c(q,p,r)$ for $r\in[\,|p-q|,p+q]$ by using formulas (3.12)–(3.15), and set $c(p,q,r) = 0$ for other r.

Theorem 3.5. *If* $c(p,q,r)\ge 0$ $(p,q,r\in \mathbb{N}_0)$, *then* $c(p,q,r)$ *is the structure measure for the hypercomplex system* $l_1(m)$ *with basis unity* 0 *associated with a certain system of orthogonal polynomials.*

Proof. Consider the difference expression (3.4) with the coefficients determined by the formulas $\alpha_0 = a_0\sqrt{c_1}$, $\beta_0 = b_0$, $\alpha_n = a_0\sqrt{c_{n+1}a_n}$, and $\beta_n = a_0 b_n + b_0$ $(n \geq 1)$. Let $P_n(t)$ be the corresponding polynomials of the first kind. Since

$$0 < \alpha_n \leq a_0 \frac{c_{n+1} + a_n}{2} < \frac{a_0}{2}(a_n + b_n + c_n + a_{n+1} + b_{n+1} + c_{n+1}) = a_0$$

and $0 \leq \beta_n = a_0 b_n + b_0 < a_0(a_n + b_n + c_n) + b_0 = 1$, the operator Λ, that corresponds to the difference expression L is bounded and, hence, self-adjoint. Thus, $P_n(t)$ $(n \in \mathbb{N}_0)$ is a sequence of polynomials orthogonal on a bounded subset of the real axis with respect to a certain measure $d\sigma$. It is not difficult to show by induction that

$$P_n(1) = \frac{\prod_{k=1}^{n-1}\sqrt{a_k}}{\prod_{k=1}^{n}\sqrt{c_k}} > 0.$$

Consider the numbers

$$c'(p, q, r) = \frac{P_p(1)P_q(1)}{P_r(1)} \int P_p(t)P_q(t)P_r(t)d\sigma(t).$$

By repeating the argument used in the proof of Theorem 3.4, we see that $c'(p, q, r)$ satisfy the associativity relation and, hence, are completely determined, according to relations (3.12)–(3.15), by their values $c'(p, 0, p)$, $c'(p, 1, p+1)$, $c'(p, 1, p)$, and $c'(p, 1, p-1)$. But it is not difficult to check that

$$c'(p, 0, p) = 1 = c(p, 0, p), \quad c'(p, 1, p+1) = \frac{c_{p+1}}{c_1} = c(p, 1, p+1),$$

$$c'(p, 1, p) = \frac{b_p}{c_1} = c(p, 1, p), \quad \text{and} \quad c'(p, 1, p-1) = \frac{a_{p-1}}{c_1} = c(p, 1, p-1).$$

Thus, $c'(p, q, r) = c(p, q, r)$ and, since $c(p, q, r) \geq 0$ by the condition of the theorem, it remains to use Theorem 3.1 to complete the proof.

■

As was noticed by Bloom and Selvanathan [BlS], this method for the construction of a hypercomplex system is universal in the sense that, *for any hypercomplex system associated with a system of orthogonal polynomials, it is possible to find sequences* $(a_n)_{n=0}^{\infty}$, $(b_n)_{n=0}^{\infty}$, *and* $(c_n)_{n=0}^{\infty}$ *which generate this hypercomplex system as indicated above.* More precisely, let $l_1(d\sigma, m)$ be associated with a system of orthogonal polynomials $P_n(t)$, which are polynomials of the first kind for the difference expression $(Lu)_p = \alpha_{p-1}u_{p-1} + \beta_p u_p + \alpha_p u_{p+1}$. We set

$$a_p = \frac{P_{p+1}(1)}{P_1(1)P_p(1)} \int P_1(t)P_p(t)P_{p+1}(t)d\sigma(t) = \frac{P_{p+1}(1)}{P_1(1)P_p(1)} \frac{\alpha_p}{\alpha_0} > 0,$$

$$b_p = \frac{1}{P_1(1)} \int P_1(t)P_p^2(t)d\sigma(t) \geq 0,$$

$$c_0 = 0, \quad c_p = \frac{P_{p-1}(1)}{P_1(1)P_p(1)} \int P_1(t)P_{p-1}(t)P_p(t)d\sigma(t) = \frac{P_{p-1}(1)}{P_1(1)P_p(1)} \frac{\alpha_{p-1}}{\alpha_p} > 0.$$

Since

$$P_1(t) = \frac{1}{\alpha_0}(t - \beta_0),$$

we have

$$a_p + b_p + c_p = [\alpha_0 P_1(1)P_p(1)]^{-1}[\alpha_p P_{p+1}(1) + \beta_p P_p(1)$$

$$- \beta_0 P_p(1) + \alpha_{p-1}P_{p-1}(1)] = \frac{1-\beta_0}{\alpha_0 P_1(1)} = 1.$$

The reader can easily verify that $c(p, q, r)$ are related to a_p, b_p, and c_p as required.

■

Below, we give sufficient conditions for the integrals $\int P_p(t)P_q(t)P_r(t)d\sigma(t)$ to be nonnegative and present several examples of hypercomplex systems associated with orthogonal polynomials and having nonnegative structure measure.

Theorem 3.6. *Let* α_p *and* β_p *be coefficients of the difference expression (3.4) associated with a system of orthogonal polynomials* $P_p(t)$. *If the sequences* α_p *and* β_p *are not decreasing, then* $\int P_p(t)P_q(t)P_r(t)d\sigma(t) \geq 0$ $(p, q, r \in \mathbb{N}_0)$.

Proof. The integrals $\int P_p(t)P_q(t)P_r(t)d\sigma(t)$ are, in fact, the Fourier coefficients in the decomposition of the product $P_p(t)P_q(t)$ in the polynomials $P_r(t)$. Without loss of generality, we can assume that $p \leq q$. To prove the theorem, we use induction on p. Since $P_1(t) = (t - \beta_0)/\alpha$, we have

$$P_1(t)P_q(t) = \frac{1}{\alpha_0}\,[tP_q(t) - \beta_0 P_q(t)]$$

$$= \frac{1}{\alpha_0}\,[\alpha_q P_{q+1}(t) + (\beta_q - \beta_0)P_q(t) + \alpha_{q-1}P_{q-1}(t)]$$

for all $q \geq 1$. Since $\alpha_p > 0$ and $\beta_q - \beta_0 \geq 0$, the theorem is true for $p = 1,\ q \geq 1$. Let us assume that the statement of theorem is true for $p = 1, \dots, l$ and prove it for $p = l+1,\ q > l$. We have

$$P_{l+1}(t)P_q(t)$$

$$= \frac{1}{\alpha_l}\,[(t - \beta_l)P_l(t)P_q(t) - \alpha_{l-1}P_{l-1}(t)P_q(t)]$$

$$= \frac{1}{\alpha_l}\,[(\alpha_q P_{q+1}(t) + \beta_q P_q(t) + \alpha_{q-1}P_{q-1}(t) - \beta_l P_q(t))P_l(t) - \alpha_{l-1}P_{l-1}(t)P_q(t)]$$

$$= \frac{1}{\alpha_l}\,[\alpha_q P_l(t)P_{q+1}(t) + (\beta_q - \beta_l)P_q(t)P_l(t)$$

$$+ (\alpha_{q-1} - \alpha_{l-1})P_{q-1}(t)P_l(t) + \alpha_{l-1}(P_l(t)P_{q-1}(t) - P_{l-1}(t)P_q(t))\,].$$

By using the induction hypothesis and the inequalities $\alpha_{q-1} - \alpha_{l-1} \geq 0,\ \beta_q - \beta_l \geq 0$, and $\alpha_q > 0\ (q > l)$, we see that the theorem will be proved if we show that the last term is nonnegative. We have

$$\alpha_{l-1}(P_l(t)P_{q-1}(t) - P_{l-1}(t)P_q(t))$$

$$= P_{q-1}(t)[(t - \beta_{l-1})P_{l-1}(t) - \alpha_{l-2}P_{l-2}(t)] - \alpha_{l-1}P_{l-1}(t)P_q(t)$$

$$= (\alpha_{q-1}P_q(t) + \beta_{q-1}P_{q-1}(t) + \alpha_{q-1}P_{q-2}(t))P_{l-1}(t)$$

$$- \alpha_{l-2}P_{l-2}(t)P_{q-1}(t) - \alpha_{l-1}P_{l-1}(t)P_q(t)$$

$$= (\alpha_{q-1} - \alpha_{l-1})P_q(t)P_{l-1}(t) + (\beta_{q-1} - \beta_{l-1})P_{l-1}(t)P_{q-1}(t)$$

$$+ (\alpha_{q-2} - \alpha_{l-2})P_{l-1}(t)P_{q-2}(t) + \alpha_{l-2}(P_{l-1}(t)P_{q-2}(t) - P_{l-2}(t)P_{q-1}(t)).$$

Continuing in this manner, we get terms which have nonnegative Fourier coefficients in decomposition (3.2) except, possibly, the last term $\alpha_0(P_1(t)P_{q-l}(t) - P_{q-l+1}(t))$. But for this term, we have $\alpha_0(P_1(t)P_{q-l}(t) - P_{q-l+1}(t)) = (t-\beta_0)P_{q-l}(t) - \alpha_0 P_{q-l+1}(t) = (\alpha_{q-1}-\alpha_0)P_{q-l+1}(t) + (\beta_{q-1}-\beta_0)P_{q-l}(t) + \alpha_{q-l-1}P_{q-l-1}(t)$, which completes the proof.

■

The second condition for a structure measure to be nonnegative was established by Koornwinder [Koo2] and involves an additional formula. It was formulated in Section 2 in Chapter 1 for an arbitrary A-system of functions.

Examples of hypercomplex systems associated with orthogonal polynomials are described in detail in [Las2]–[Las4]. Here, we just give a list of systems of orthogonal polynomials which lead to hypercomplex systems with nonnegative structure constants. These systems include the Jacobi polynomials $P_n^{(\alpha,\beta)}(t)$ $(\alpha \geq \beta > -1,\ a(a+5)(a+3)^2 \geq (a^2-7a-24)b^2)$, where $a = \alpha+\beta+1$ and $b = \alpha - \beta$, more general (see [Rah1], [Rah2]) q-Jacobi polynomials $P_n^{(\alpha,\beta)}(t;q)$ $(\alpha \geq \beta > -1,\ \alpha+\beta+1 \geq 0,\ 0 < q < 1)$, which turn into the Jacobi polynomials as $q \to 1$, q-ultraspherical polynomials $C_n(t;\beta\,|\,q)$ $(-1 < \beta < 1,\ 0 < q < 1)$ (see [Bre]), associated Legendre polynomials $C_n(t,\nu)$ $(\nu \geq 0)$ (see [BaD]), associated q-ultraspherical polynomials $C_n^{\alpha}(t;\beta\,|\,q)$ $(0 < \beta \leq q < 1,\ 0 < \alpha < 1)$ (see [BuI]), polynomials related to homogeneous trees $P_n(t,a)$ $(a \geq 2)$ (see [Let1]), Pollacheck polynomials $P_n(t;a,c,\alpha)$ $(a \geq 0,\ c \geq 0,\ \alpha \geq -1/2,\ 2\alpha+1+c \geq 0,\ 1 \leq 4a\alpha + 4\alpha^2 + 4ac + 6a)$ (see [Sze]), generalized Chebyshev polynomials $T_n^{\alpha,\beta}(t)$ $(\alpha \geq \beta+1,\ \beta > -1)$ (see [Lai]), and polynomials orthogonal with respect to the weight functions $\sqrt{1-t^2}\,/\,(1-\mu t^2)$ $(\mu = a-(a/2)^2,\ a \geq 2)$ and $1/[1-\nu t^2]\sqrt{1-t^2}$ $(\nu = 2a-a^2,\ a \geq 2)$ (see [Ger]).

As follows from Theorem 3.7 and Corollary 3.1 in Chapter 1, for an object dual to a hypercomplex system associated with orthogonal polynomials to be a hypercomplex system, it is necessary and sufficient that, for any $t, s \in E$, the following integral representation hold:

$$R_p(t)R_p(s) = \int_E R_p(r)\,d\mu_{t,s}(t) \quad (p \in \mathbb{N}_0), \tag{3.16}$$

where $\mu_{t,s}(t)$ is a nonnegative Radon measure on the space X_h. In this case, we have duality. Necessary and sufficient conditions for representation (3.16) to exist are established in [Gas3] for the case of Jacobi polynomials, and a sufficient condition for generalized Chebyshev polynomials is found in [Lai].

3.5. Compact Hypercomplex Systems Associated with Orthogonal Polynomials. In this subsection, we consider hypercomplex systems with compact basis associated with a system of orthogonal polynomials. We preserve the notation of the Subsection 3.4. Let $\rho_n = \max_{t \in E} |P_n(t)|$ and set $R_n(t) = P_n(t)/\rho_n$. Denote by $P(F)$ the set of all polynomials on E. By the Weierstrass theorem, $P(E)$ is dense in $C(E)$ and, hence, in $L_2(E, \sigma)$.

Suppose that the following conditions are satisfied:

(i) there exists a point $e \in E$ such that $\rho_n = P_n(e)$ for all $e \in \mathbb{N}_0$;

(ii) for any $t, s \in E$, there exists a measure $d\mu(t, s; z) \in M_b^+(E)$ such that

$$R_n(t)R_n(s) = \int R_n(z)d\mu(t, s; z) \tag{3.17}$$

for all $n \in \mathbb{N}_0$.

Remark 1. It follows from Remark 3 to Theorem 3.1 that if it is possible to construct a hypercomplex system with discrete basis associated with polynomials $P_n(t)$, then condition (i) is automatically satisfied.

Remark 2. As a rule, $d\mu(t, s; z)$ should be sought in the form $d\mu(t, s; z) = K(t, s, z)\,d\sigma(z)$, where

$$K(t, s, z) = \sum_{n=0}^{\infty} P_n(t)P_n(s)R_n(z)$$

(we are not concerned with the convergence of the series). If $K(t, s, z) \geq 0$, then, after formal calculations, we obtain

$$\int R_n(z)K(t, s, z)d\sigma(z)$$

$$= \sum_{k=0}^{\infty} P_k(t)P_k(s)\,\frac{1}{\rho_k\rho_n}\int P_n(z)P_k(z)\,d\sigma(z) = R_n(t)R_n(s).$$

We do not prove these formulas here and refer the reader to [Gas2], [Gas3], and [Osi1]. In the theory of special functions, relations of the form (3.17) are called *multiplication formulas*. If a multiplication formula exists for a certain system of functions, then, as a rule, it is possible to construct a hypercomplex system associated with these functions. The necessity to have a multiplication formula for the construction of a hypercomplex system associated with a certain system of functions and having a nonnegative structure measure follows from relation (3.10) in Chapter 1.

Let us define operators L_s $(s \in E)$ on $P(E)$ by the equality

$$(L_s f)(t) = \int f(z)d\mu(t, s; z) \quad (f \in P(E),\ t, s \in E). \tag{3.18}$$

Let

$$f(t) = \sum_{k=1}^{N} f_k P_k(t).$$

It is easy to see that

$$\|(L_s f)(t)\|_2^2 = \sum_{k=1}^{N} |f_k|^2 |R_k(s)|^2 \leq \sum_{k=1}^{N} |f_k|^2 = \|f\|_2^2.$$

Consequently, the operators $L_s\ (s \in E)$ can be continuously extended to the entire $L_2(E, \sigma)$ with $\|L_s\| \leq 1\ \ (s \in E)$. Approximating an arbitrary function $f \in C(E)$ uniformly on E by polynomials and passing to the limit in (3.18), we conclude that relation (3.18) is true for all functions continuous on E.

By setting $p = 0$ in (3.16), we establish that, for all $t, s \in E$, the measure $d\mu(t, s; z)$ belongs to $M_1(E)$ and $L_s 1 = 1$. Since $d\mu(t, s; z) \in M_1(\mathbb{R})$, it follows from (3.18) that inequality (2.7) in Chapter 1 holds for the operators $L_s\ (s \in E)$.

Let us show that *the operators* $L_s\ (s \in E)$ *form an Hermitian commutative family of generalized translation operators in* $L_2(E, \sigma)$ *which preserves positiveness, is weakly continuous with respect to the invariant measure* σ, *and satisfies the condition of separate continuity*. Indeed, let us show that L_s are, in fact, generalized translation operators. It is clear that $(L_s f)(t) = (L_s f)(s)\ \ (t, s \in E;\ f \in C(E))$. This implies that $(L_\cdot f)(t) = (L_t f)(\cdot) \in L_2(E, \sigma)$ for almost all $t \in E$ with respect to the measure σ. Let us show that $L_e = I$. We have $R_n(t) = \int R_n(z)d\mu(t, e; z)$ for every $n \in \mathbb{N}_0$. Since the polynomials $R_n(t)$ form a basis in $P(E)$, we have $t^n = \int z^n d\mu(t, e; z)$. Thus, we have obtained the sequence of moments $s_n = t^n$. Clearly, this moment problem is solvable (i.e., the measure $d\mu(t, e; z)$ can be uniquely constructed from its moments t^n; see, e.g., [Ber10]). Consequently, $d\mu(t, e; z) = \delta_t$ and $(L_e f)(t) = f(t)\ \ (f \in C(E))$. Hence, $L_e = I$. Since $P(E)$ is dense in $L_2(E, \sigma)$, it is sufficient to verify the associativity axiom only for the polynomials $R_n(t)$. But for $R_n(t)$, the corresponding equality is obvious. Thus, L_s are generalized translation operators. It follows from (3.18) that L_s preserve positiveness; the commutativity of the generalized translation operators L_s is obvious. Since involution on E is trivial, the generalized translation operators are involutory if and only if they are real, which again follows from (3.18): $(L_s \bar{f})(t) = \overline{L_s f(t)}\ \ (f \in L_2(E, \sigma))$ for almost all $(t, s) \in E \times E$. Let us show that the measure σ is invariant under the action of $L_s\ (s \in E)$. For every $n, m \in \mathbb{N}_0$, we have

$$\int (L_s R_n)(t)\overline{R_m(t)}\,d\sigma(t) = \delta_{n,m} R_n(s) = \int R_n(t)\overline{(L_s R_m)(t)}\,d\sigma(t)$$

and it remains to recall that the family of the functions $R_n(t)$ is total in $L_2(E, \sigma)$. Let us show that the condition of separate continuity is satisfied. Let $s_n \to s$, $n \to \infty$ (s, $s_n \in E$). Since, for every $n \in \mathbb{N}_0$, $(L_{s_k} R_n)(t) = R_n(t)R_n(s_k) \xrightarrow[n\to\infty]{} R_n(t)R_n(s) = (L_s R_n)(t)$, the function $L_s Q(t)$ is separately continuous for an arbitrary polynomial $Q(t)$. Let $f \in C(E)$ be an arbitrary function. We fix $\varepsilon > 0$. By the Weierstrass theorem, there exists a polynomial $Q(t)$ such that $\|f - Q\|_\infty < \varepsilon/3$. Then

$$|(L_{s_n} f)(t) - (L_s f)(t)| \le \int |f(z) - Q(z)|\,d\mu(t, s_n; z)$$

$$+ |(L_{s_n} Q)(t) - (L_s Q)(t)| + \int |f(z) - Q(z)|\,d\mu(t, s; z)$$

$$\le 2\|f - Q\|_\infty + |(L_{s_n} Q)(t) - (L_s Q)(t)| < \varepsilon$$

for sufficiently large n. Finally, let us prove that the generalized translation operators L_s ($s \in E$) are weakly continuous. Let $f \in C(E)$ and $g \in L_2(E, \sigma)$. By using separate continuity, the inequality $\|L_s\| \le 1$, and the Lebesgue theorem, we establish that the function $(L_s f, g)_2$ is continuous. Now let $f, g \in L_2(E, \sigma)$. Approximating the function f by continuous functions f_n and using the inequality $|(L_s(f - f_n), g)_2| \le \|f_n - f\|_2 \|g\|_2$, we find that the sequence of functions $(L_s f_n, g)_2$ uniformly converges to $(L_s f, g)$. This implies that the generalized translation operators L_s ($s \in E$) are weakly continuous.

■

If we apply Theorem 2.1 in Chapter 1 to the generalized translation operators L_s ($s \in E$), we obtain the Hermitian hypercomplex system $L_1(E, \sigma)$ with basis unity satisfying the condition of separate continuity. We summarize this discussion by the following statement:

Theorem 3.7. *Let $P_n(t)$ ($n \in \mathbb{N}_0$, $t \in E$) be polynomials orthonormal with respect to a measure σ with compact support E. Suppose that the following conditions are satisfied:*

(i) There exists a point $e \in E$ such that $P_n(e) = \max_{t \in E} |P_n(t)|$ ($n \in \mathbb{N}_0$);

(ii) for any $t, s \in E$, there exists a measure $d\mu(t, s; z) \in M_b^+(E)$ such that $R_n(t)R_n(s) = \int R_n(z)\,d\mu(t, s; z)$ ($n \in \mathbb{N}_0$), where $R_n(t) = P_n(t)/P_n(e)$.

Then the operators $(L_s f)(t) = \int f(z)d\mu(t, s; z)$ $(f \in C(E);\ t, s \in E)$ *can be extended to Hermitian generalized translation operators in* $L_2(E, \sigma)$ *which preserve the identity and positiveness, are weakly continuous with respect to the invariant measure, and satisfy the condition of separate continuity. The space* $L_1(E, \sigma)$ *with convolution* $f * g(t) = \int (L_t f)(r)g(r)d\sigma(r)$ $(f, g \in L_2(E, \sigma);\ t \in E)$ *generated by these generalized translation operators becomes an Hermitian hypercomplex system with compact basis* E *and basis unity* e *satisfying the condition of separate continuity. The set of polynomials* $R_n(t)$ $(n \in \mathbb{N}_0)$ *forms the set of characters of the hypercomplex system* $L_1(E, \sigma)$.

Proof. It is necessary to prove only the last statement of the theorem. By virtue of (3.17), $(L_s R_n)(t) = R_n(t)R_n(s)$ and, consequently, by using the statement proved at the end of Section 2.1 in Chapter 1, we conclude that $R_n(t)$ is a generalized character. But, by condition (i) of the theorem, $|R_n(t)| \le 1$ $(t \in E)$ and, hence, $R_n(t)$ is a character. Let us show that the hypercomplex system $L_1(E, \sigma)$ does not have other characters. Indeed, let $\chi(t)$ $(t \in E)$ be a character of the hypercomplex system $L_1(E, \sigma)$ different form $R_n(t)$ $(n \in \mathbb{N}_0)$. By Theorem 1.4 in Chapter 1, $\chi(t)$ is orthogonal to all $R_n(t)$ with respect to the measure σ. But $R_n(t)$ $(n \in \mathbb{N}_0)$ is an orthonormal basis in $L_2(E, \sigma)$, whence $\chi(t) \equiv 0$.

∎

Let $P_n^{(\alpha,\beta)}(t)$ $(t \in [-1, 1];\ \alpha, \beta > -1)$ be Jacobi polynomials orthonormal with respect to the measure $\sigma = (1-t)^{\alpha}(1-t)^{\beta}dt$. Since $P_n^{(\alpha,\beta)}(-t) = (-1)^n P_n^{(\alpha,\beta)}(t)$, we can assume that $\alpha \ge \beta$. But then we have (see, e.g., [Sze]) $|P_n^{(\alpha,\beta)}(t)| \le P_n^{(\alpha,\beta)}(1)$ if $\max\{\alpha, \beta\} > -1/2$ (if $\beta > \alpha$, we have $|P_n^{(\alpha,\beta)}(t)| \le (-1)^n P_n^{(\alpha,\beta)}(-1)$). For $\alpha \ge \beta$ and $\max\{\alpha, \beta\} > -1/2$, denote $R_n(t) = P_n^{(\alpha,\beta)}(t) / P_n^{(\alpha,\beta)}(1)$. It was shown in [Gas3] that the integral representation

$$R_n(t)R_n(s) = \int_{-1}^{1} R_n(z)d\mu(t, s; z),$$

where the measure $\mu(t, s; \cdot) \in M_b([-1, 1])$ is uniquely determined, holds if and only if $\alpha \ge \beta$, $\alpha + \beta + 1 \ge 0$. The measure $d\mu(t, s; z)$ is positive if and only if $\beta \ge -1/2$ or $\alpha + \beta \ge 0$. Hence, if $\alpha \ge \beta$ and $\alpha + \beta \ge 0$ or $\beta \ge -1/2$, then, by Theorem 3.7, one can construct the Hermitian hypercomplex system $L_2([-1, 1],\ (1-t)^{\alpha}(1+t)^{\beta}dt)$ associated with Jacobi polynomials and having a nonnegative structure measure. Clearly, this hypercomplex system is dual to the discrete hypercomplex system associated with

Jacobi polynomials constructed in Subsection 3.4. However, it should be noted that the set of values (α, β) for which a discrete hypergroup with nonnegative structure measure can be constructed is larger than the corresponding set for a compact hypercomplex system with nonnegative structure measure. More precise, let us introduce the following sets: $U = \{(\alpha, \beta) \mid \alpha \geq \beta,\ \alpha + \beta \geq 0\} \cup \{(\alpha, \beta) \mid \alpha \geq \beta,\ \beta \geq -1/2\}$, $V = \{(\alpha, \beta) \mid \alpha \geq \beta,\ \alpha + \beta + 1 \geq 0\}$, and $W = \{(\alpha, \beta) \mid \alpha \geq \beta,\ a(a+5)(a+5)^2 \geq (a^2 - 7a - 24)b^2$, where $a = \alpha + \beta + 1,\ b = \alpha - \beta\}$. It is clear that $U \subset V$. It is also easy to show that $V \subset W$ and $V \setminus U \neq \varnothing$, $W \setminus V \neq \varnothing$. If $(\alpha, \beta) \in U$, the structure measures of the compact and discrete hypercomplex systems are nonnegative and the duality theorem (Theorem 3.7 in Chapter 1) is true. If $(\alpha, \beta) \in V \setminus U$, then the structure measure of the discrete hypercomplex system is nonnegative, and the structure measure of the hypercomplex system $L_1([-1, 1], (1-t)^{\alpha}(1+t)^{\beta} dt)$ is real. Nevertheless, in this case, duality holds in the sense of quantized hypercomplex systems (see Section 5 in Chapter 1). If $(\alpha, \beta) \in W \setminus V$, the structure measure of the discrete hypercomplex system is still nonnegative, but equality (3.17) does not hold. This is why the generalized translation operators L_s $(s \in E)$ should be introduced in a different way, namely, if $f(t) = \Sigma f_k R_k(t)$ is the expansion of a function $f \in L_2(E, \sigma)$ in the basis $R_k(t)$, then $L_s f(t) = \Sigma f_k R_k(s) R_k(t)$. Generally speaking, the object generated by these generalized translation operators is not a quantized hypercomplex system but, since $|R_k(t)| \leq 1$ for $(\alpha, \beta) \in W$ (see Remark 3 to Theorem 3.1), it is a real hypercomplex system with compact basis in the sense of Vainerman (see Subsection 5.4 in Chapter 1). In this case, duality also takes place. As shown in [Vai9], for an arbitrary system of orthogonal polynomials satisfying the estimate

$$\max_{t \in E} |P_n(t)| \leq K \int_E P_n(t) d\mu(t)$$

with some $K > 0$ and $\mu \in M_b(E)$, it is possible to construct compact and discrete hypercomplex systems (in the sense of Vainerman) dual to each other. In the case where $(\alpha, \beta) \notin W$ $(\alpha, \beta > -1)$, we did not succeed to verify this estimate. And although we succeeded in constructing a discrete hypercomplex system in the sense of Subsection 2.4 in Chapter 1 (see below), the case $(\alpha, \beta) \notin W$ remains unclear from the duality point of view.

Formulas of type (3.17) are also established for generalized Chebyshev polynomials [Lai] and q-Jacobi polynomials [Rah2]. Moreover, in [CoS2], the following result was announced: Suppose that, for some family of orthogonal polynomials on an interval $I \subset \mathbb{R}^1$, the multiplication formula (3.17) holds and the measure $d\mu(t, s; z)$ possesses the following properties:

(i) There exists a point $e \in I$ such that $d\mu(t, e; \cdot) = \delta_t$;

(ii) $\lim_{s \to e} \operatorname{diam} \operatorname{supp} \mu(t, s; \cdot) = 0$.

Then, to within a linear change of variables, these polynomials coincide with the Jacobi polynomials $R_n^{(\alpha,\beta)}(t)$ with $(\alpha,\beta) \subset U$. This theorem is converse to the result of Gasper [Gas3]. The existence of a multiplication formula for generalized Chebyshev polynomials and q-Jacobi polynomials does not contradict this theorem because, in the first case, condition (ii) is not satisfied and, in the second case, condition (i) is not satisfied. Thus, generalized Chebyshev polynomials produce an example of a hypercomplex system which is not a hypergroup.

In [Osi1], a multiplication formula of type (3.17) was obtained for a wider class of orthogonal polynomials with measure $\mu(t, s; \cdot)$ not necessarily positive, and the corresponding generalized translation operators were studied.

It should also be noted that similar constructions of compact and discrete hypercomplex systems can be carried out for polynomials in several variables. However, we do not consider this case in detail but give an example (see, e.g., [AnT] and [Koo2]). Let $D = \{z = x + iy \mid x^2 + y^2 \le 1\}$ be the unit disk. For arbitrary $\alpha > 0$, we define a probability measure on D as follows:

$$dm_\alpha(z) = \frac{\alpha+1}{\pi}(1 - x^2 - y^2)^\alpha dx dy.$$

Consider the following family of polynomials in two variables x and y: $R_{m,n}^{(\alpha)}(z) = R_{m,n}^{(\alpha)}(re^{i\varphi}) = r^{|m-n|} e^{i(m-n)\varphi} R_{\min(m,n)}^{(\alpha,|m-n|)}(2r^2 - 1)$, where $m, n \in \mathbb{N}_0$ and $R_k^{(\alpha,\beta)}(t)$ are Jacobi polynomials normalized by the condition $R_k^{(\alpha,\beta)}(t) = 1$. These polynomials form an orthogonal basis in $L_2(D, m_\alpha)$ and satisfy the relation

$$(L_w R_{m,n}^{(\alpha)})(v) = \frac{\alpha}{\alpha+1} \int_D R_{m,n}^{(\alpha)}\left(vw + \sqrt{1-|v|^2}\sqrt{1-|w|^2}\, z\right) dm_\alpha(z)$$

$$= \int_D R_{m,n}^{(\alpha)}(z) E_\alpha(v, w, z) dm_\alpha(z) = R_{m,n}^{(\alpha)}(v) R_{m,n}^{(\alpha)}(w) \quad (v, w \in D),$$

where

$$E_\alpha(v, w, z) = \frac{\alpha}{\alpha+1} \frac{(1 - |v|^2 - |w|^2 - |z|^2 + 2\operatorname{Re} vwz)^{\alpha-1}}{(1-|v|^2)^\alpha (1-|w|^2)^\alpha (1-|z|^2)^\alpha}.$$

We set

$$(L_v f)(w) = \int_D f(z) E_\alpha(v, w, z) dm_\alpha(z).$$

Annabi and Trimeche [AnT] established that L_v are commutative generalized translation operators in the space $L_2(D, m_\alpha)$ which preserve the identity and positiveness and satisfy the condition of separate continuity; furthermore, $L_v^* = L_{\bar{v}}$ and $(L_v f)(w) = \overline{L_{\bar{w}} f^*(\bar{v})}$, where $f^*(v) = \overline{f(\bar{v})}$. Thus, the space $L_1(D, m_\alpha)$ with convolution induced by these generalized translation operators and with involution $D \ni z \to \bar{z} \in D$ is a normal hypercomplex system with the basis unity $e = (1, 0)$ satisfying the condition of separate continuity. It is obvious that he characters of this hypercomplex system are the polynomials $R_{m,n}^{(\alpha)}(z)$. By virtue of the equality $R_{m,n}^{(\alpha)}(\bar{z}) = \overline{R_{m,n}^{(\alpha)}(z)}$, all characters of the hypercomplex system $L_1(D, m_\alpha)$ are Hermitian. Hypercomplex systems dual to $L_1(D, m_\alpha)$ $(\alpha > 0)$ were, in fact, studied by Kanjin [Kan]. It should be noted that, in the discrete case, the nonnegativity of the structure measure was proved by Koornwinder [Koo2]. The polynomials $R_{m,n}^{(\alpha)}(z)$ with $\alpha \in \mathbb{N}_0$ are the characters of the hypergroup $SU(n)/SU(n-1)$ and were first studied by Shapiro [Sha1], [Sha2].

3.6. The Case of Not Necessarily Nonnegative Structure Constants. In this subsection, we consider discrete hypercomplex systems associated with orthogonal polynomials with not necessarily positive structure constants. Let $P_p(t)$ be a sequence of polynomials defined on a bounded subset E of the real line and orthogonal with respect to the measure $d\sigma$. Suppose that there exists a sequence $m(p)$ $(p = 0, 1, \ldots)$ of positive numbers such that

$$\sum_r \sqrt{m(r)} \left| \int P_p(t) P_p(t) P_p(t) d\sigma(t) \right| \leq C\sqrt{m(p)m(q)} \quad (p, q \in \mathbb{N}_0). \tag{3.19}$$

We set

$$R_p(t) = \frac{P_p(t)}{\sqrt{m(p)}}, \quad c(p, q, r) = \int R_p(t) R_q(t) R_r(t) d\sigma(t) \quad (p, q, r \in Q = \mathbb{N}_0). \tag{3.20}$$

By replacing $P_p(b)$ with $\sqrt{m(p)}$ and repeating the argument of Theorem 3.1, we can show that $c(p, q, r)$ are structure constants (generally speaking, nonpositive) and $m(p)$ is the multiplicative weight of some Hermitian hypercomplex system $l_1(d\sigma, m)$ with basis Q. The space of characters of this hypercomplex system consists of the set of points $z \in \mathbb{C}^1$ such that the sequence $R_p(z)$ is bounded; $X_h = X \cap \mathbb{R}^1$. Note that these statements remain valid if we set $R_p(t) = (-1)^{\varepsilon_p} P_p(t)/\sqrt{m(p)}$ in (3.20), where ε_p is an arbitrary sequence of natural numbers.

Let us describe a method for choosing the numbers $m(p)$. Suppose that $\left| \int P_p(t) P_q(t) P_r(t) d\sigma(t) \right| \leq C$ for all $p, q, r \in Q$. Note that if the numbers $\nu_p > 0$ satisfy the inequality

$$\sum_{r=|p-q|}^{p+q} \nu_r \le \nu_p \nu_q,$$

we can set $\sqrt{m(p)} = C\nu_p$ in (3.19). Indeed,

$$\sum_r \left|\int P_p(t)P_p(t)P_p(t)d\sigma(t)\right|\sqrt{m(r)} = \sum_{r=|p-q|}^{p+q} \left|\int P_p(t)P_p(t)P_p(t)d\sigma(t)\right| C\nu_r$$

$$\le C^2 \sum_{r=|p-q|}^{p+q} \nu_p \le C^2 \nu_p \nu_q = \sqrt{m(p)m(q)}.$$

The numbers ν_p can be chosen as $\nu_p = (2p+1)^A$ or $\nu_p = \sinh(2p+1)\lambda / \sinh\lambda$. Indeed, by using the formulas for the sum of sines, one can easily check that

$$\sum_{r=|p-q|}^{p+q} \frac{\sin(2r+1)z}{\sin z} = \frac{\sin(2p+1)z}{\sin z}\,\frac{\sin(2q+1)z}{\sin z}.$$

By setting here $z = 0$ and $z = i\lambda$ $(\lambda > 0)$, we get

$$\sum_{r=|p-q|}^{p+q} (2r+1) = (2p+1)(2q+1),$$

$$\sum_{r=|p-q|}^{p+q} \frac{\sinh(2r+1)\lambda}{\sinh\lambda} = \frac{\sinh(2p+1)\lambda}{\sinh\lambda}\,\frac{\sinh(2q+1)\lambda}{\sinh\lambda}.$$

Since

$$\sum_i x_i^A \le \left(\sum_i x_i\right)^A \quad (x_i \ge 0,\ A \ge 1),$$

it follows from the first relation that

$$\sum_{r=|p-q|}^{p+q} (2r+1)^A \le \left[\sum_{r=|p-q|}^{p+q} (2r+1)\right]^A = (2p+1)^A(2q+1)^A \quad (A \ge 1).$$

In particular, *if the polynomials $P_r(t)$ are uniformly bounded in $r \in Q$ and $t \in E$, we can set $\sqrt{m(r)} = K\,(2r+1)^A$ or*

$$\sqrt{m(r)} = K\frac{\sinh(2r+1)B}{\sinh B} \quad \Big(A \ge 1, \quad B > 0, \quad K = [\sup_{r \in Q} |P_r(t)|]^3 \Big).$$

Indeed, it is necessary to check inequality (3.19). We have

$$\sum_{r=|p-q|}^{p+q} \left| \int P_p(t) P_q(t) P_r(t) d\sigma(t) \right| K(2r+1)^A$$

$$\le K^2 \sum_{r=|p-q|}^{p+q} (2r+1)^A \le \sqrt{m(p)m(q)} \quad (p, q \in \mathbb{N}_0).$$

For the multiplicative measure

$$\sqrt{m(r)} = K\frac{\sinh(2r+1)B}{\sinh B},$$

inequality (3.19) can be verified similarly.

■

For hypercomplex systems with real structure constants, Theorem 3.3 remains true. To prove this it suffices to repeat the proof of Theorem 3.3, changing the expression for $P_p(t)$ as follows:

$$P_p(t) = \chi_t(t)\sqrt{m(p)}\,\mathrm{sign}\prod_{k=1}^{p} c(k-1, 1, k).$$

In this case, in the recurrence relation (3.11), we have

$$\alpha_p = \frac{|c(p, 1, p+1)|}{m(1)}\sqrt{\frac{m(p+1)}{m(p)}} \ge 0, \qquad \beta_p = \frac{c(p, 1, p)}{m(1)} \in \mathbb{R}^1,$$

and the other arguments remain the same. We only note that the structure constants $c(p, q, r)$ are calculated by formulas (3.20) with $R_p(t) = \chi_t(p)/\sqrt{m(p)}$. Thus, a hypercomplex system with real structure constants which satisfies the conditions of Theorem 3.3 can be uniquely recovered from a sequence of orthogonal polynomials $P_p(t)$, multiplicative weight $m(p)$, and a given sequence of numbers ε_p, where $\varepsilon_p = \pm 1$.

■

Theorem 3.4 also remains true for hypercomplex systems with real structure constants because, in the proof of an analog of Theorem 3.3 for such hypercomplex systems, it was shown that $c(p, 1, p+1) \ne 0$ $(p \in \mathbb{N}_0)$.

3.7. Examples of Hypercomplex Systems with Real Structure Constants. Let us give examples of hypercomplex systems with real structure measure associated with orthogonal polynomials. According to Subsection 3.6, it is sufficient to prove that the orthogonal polynomials $P_r(t)$ are uniformly bounded with respect to $r \in Q$ and $t \in E$.

First, let us show *that the Jacobi polynomials* $P_n^{(\alpha,\beta)}(t)$ $(-1 < \alpha, \beta \le -1/2)$ *are uniformly bounded on* $[-1, 1]$. Indeed, for the Jacobi polynomials $\tilde{P}_n^{(\alpha,\beta)}(t)$ $(\alpha, \beta > -1)$ normalized by the condition $\tilde{P}_n^{(\alpha,\beta)}(1) = \binom{n+\alpha}{n}$, the following asymptotic estimate holds (see [Sze, p. 75]):

$$\max_{-1\le t\le 1} \left|\tilde{P}_n^{(\alpha,\beta)}(t)\right| \sim n^{-1/2} \quad (-1 < \alpha, \beta \le -1/2). \tag{3.21}$$

Since

$$P_n^{(\alpha,\beta)}(t) = \left[\frac{2n+\alpha+\beta+1}{2^{\alpha+\beta+1}}\,\frac{\Gamma(n+1)\Gamma(n+\alpha+\beta+1)}{\Gamma(n+\alpha+1)\Gamma(n+\beta+1)}\right]^{1/2} \tilde{P}_n^{(\alpha,\beta)}(t)$$

(see [Sze]), by using the Stirling formula for the γ-function, we get

$$P_n^{(\alpha,\beta)}(t) \sim n^{1/2}\tilde{P}_n^{(\alpha,\beta)}(t). \tag{3.22}$$

It follows from (3.21) and (3.22) that

$$\max_{-1\le t\le 1} \left|\tilde{P}_n^{(\alpha,\beta)}(t)\right| \sim 1 \quad (-1 < \alpha, \beta \le -1/2). \tag{3.23}$$

■

Consider the hypercomplex systems $l_1\left((1-t)^\alpha(1+t)^\beta dt,\ K^2(2p+1)^{2A}\right)$ and

$$l_1\left((1-t)^\alpha(1+t)^\beta dt,\ K^2\frac{\sinh^2(2p+1)B}{\sinh^2 B}\right).$$

Let us show that *the spaces of characters of these hypercomplex systems are the segment* $[-1, 1]$ *and the region of the complex plane given by* $|z + (z^2-1)^{1/2}| \le e^{2B}$, *respectively.* Indeed, for the Jacobi polynomials $\tilde{P}_n^{(\alpha,\beta)}(t)$, the following asymptotic relation is true (see [Sze]) for z lying outside the closed interval $[-1, 1]$:

$$\tilde{P}_n^{(\alpha,\beta)}(z) \approx (z-1)^{-\alpha/2}(z+1)^{-\beta/2}[(z+1)^{1/2}+(z-1)^{1/2}]^{\alpha+\beta}$$

$$\times (z^2-1)^{-1/4}(2\pi n)^{-1/2}[z+(z^2-1)^{1/2}]^{n+1/2}. \tag{3.24}$$

It follows from (3.22) and (3.24) that, for fixed $z \in [-1, 1]$,

$$\frac{P_n^{(\alpha,\beta)}(z)}{K(2n+1)^A} \sim \frac{[z+(z^2-1)^{1/2}]^{n+1/2}}{K(2n+1)^A}. \tag{3.25}$$

Since $|z+(z^2-1)^{1/2}| > 1$ for $z \notin [-1, 1]$, it follows from (3.25) that, in the first case, the space of maximal ideals is the segment $[-1, 1]$. The second case is considered by analogy.

■

We can now show that *it is always possible to construct a hypercomplex system with discrete basis associated with the Jacobi polynomials* $\tilde{P}_n^{(\alpha,\beta)}(t)$ $(\alpha, \beta > -1)$ (in fact, this result was first obtained by Askey and Gasper [AsG1]). Indeed, let us first consider the case $\alpha \geq \beta$. Denote $a = \alpha + \beta + 1$ and $b = \alpha - \beta$. As follows from the results of Gasper [Gas1], for the structure constants $c(p, q, r)$ to be nonnegative, it is necessary and sufficient that the inequality $a(a+5)(a+3)^2 \geq (a^2 - 7a - 24)b^2$ hold. If this inequality does not hold, then by virtue of Theorem 1 in Gasper [Gas3], we get

$$\sum_r \left|\int P_p^{(\alpha,\beta)}(t)P_q^{(\alpha,\beta)}(t)P_r^{(\alpha,\beta)}(t)(1-t)^\alpha(1+t)^b dt\right| P_r^{(\alpha,\beta)}(1) \;\leq\; CP_p^{(\alpha,\beta)}(1)P_q^{(\alpha,\beta)}(1)$$

for $\alpha \geq \beta > -1$ and $\alpha \geq -1/2$. Consequently, inequality (3.19) holds in this case, and we can choose the sequence $m(r) = [\,P_r^{(\alpha,\beta)}(1)\,]^2$ to be a multiplicative measure. The case $\alpha, \beta < -1/2$ has just been considered. It should be noted that the results of Askey and Gasper [Gas1] imply that, in this case, one can also consider the hypercomplex system with the structure constant

$$c(p, q, r) = \frac{P_p^{(\alpha,\beta)}(c)P_p^{(\alpha,\beta)}(c)}{P_r^{(\alpha,\beta)}(c)} \int P_p^{(\alpha,\beta)}(t)P_q^{(\alpha,\beta)}(t)P_r^{(\alpha,\beta)}(t)d\sigma(t)$$

and the multiplicative weight $m(r) = [\,P_r^{(\alpha,\beta)}(c)\,]^2$, where $c > 1$.

Now let $\beta \geq \alpha$. By virtue of the well-known identity for Jacobi polynomials $P_r^{(\alpha,\beta)}(-t) = (-1)^r P_r^{(\beta,\alpha)}(t)$, it is easy to see that we have the situation described in Remark 1 to Theorem 3.1. Consequently, in this case, it is also always possible to construct a hypercomplex system by replacing $P_r^{(\alpha,\beta)}(1)$ with $P_r^{(\alpha,\beta)}(-1)$ in all reasonings.

■

In order to consider polynomials more general than Jacobi polynomials we use an extension of the Koraus theorem on bounded orthonormal polynomials. Before formulating the required result, we establish an identity necessary for what follows.

Let $P_r(t)$ and $\tilde{P}_r(t)$ $(r \in \mathbb{N}_0)$ be systems of polynomials orthonormal with respect to measures $d\sigma(t)$ and $h(t)d\sigma(t)$, respectively, where $h(t)$ is a nonnegative function. We set

$$K_n(z, s) = \sum_{k=0}^{n} P_k(z)P_k(s)$$

Since every polynomial $\tilde{P}_n(t)$ is expressed in terms of the first $n+1$ polynomials of the system $\{P_r(t)\}$, for any complex z, we have

$$\tilde{P}_n(z) = \sum_{k=0}^{n} \int \tilde{P}_n(s)P_k(s)d\sigma(s)P_k(z) = \int K_n(z, s)\tilde{P}_n(s)\, d\sigma(s). \tag{3.26}$$

It is clear that $K_n(z, s) = P_n(z)P_n(s) + K_{n-1}(z, s)$. Hence, it follows from (3.26) that

$$h(t)\,\tilde{P}_n(z) = h(t)P_n(z)\int P_n(s)\tilde{P}_n(s)\, d\sigma(s) + h(t)\int K_{n-1}(z, s)\tilde{P}_n(s)\, d\sigma(s)$$

$$= h(t)P_n(z)\int P_n(s)\tilde{P}_n(s)\, d\sigma(s) + \int K_{n-1}(z, s)\tilde{P}_n(s)\,[h(t)-h(s)]\,d\sigma(s) \tag{3.27}$$

for any $t \in E$. This equality is true because, for fixed z, $K_{n-1}(z, s)$ regarded as a polynomial of degree $n-1$ in s is orthogonal to the polynomial $\tilde{P}_n(s)$ with respect to the measure $h(s)d\sigma(s)$. The Cristoffel–Darboux identity (see [Sze, p. 56]) yields

$$K_{n-1}(z, s) = \frac{a_{n-1}}{a_n}\,\frac{P_n(z)P_{n-1}(s) - P_{n-1}(z)P_n(s)}{z-s}$$

(a_j is the leading coefficient of P_j). For any $z \in \mathbb{C}$ and $t \in E$ such that $h(t) \neq 0$, we conclude from (3.27) that

$$\tilde{P}_n(z) = A_n(z, t)P_n(z) + B_n(z, t)P_{n-1}(z) \quad (n \in \mathbb{N}_0), \tag{3.28}$$

where

$$A_n(z, t) = \int P_n(s)\tilde{P}_n(s)d\sigma(s) + \frac{a_{n-1}}{a_n}\,\frac{1}{h(t)}\int \frac{h(t)-h(s)}{z-s}\,\tilde{P}_n(s)\tilde{P}_{n-1}(s)d\sigma(s),$$

$$B_n(z, t) = -\frac{a_{n-1}}{a_n}\,\frac{1}{h(t)}\int \frac{h(t)-h(s)}{z-s}\,\tilde{P}_n(s)P_n(s)d\sigma(s) \quad (n \in \mathbb{N}_0;\ P_{-1}(s) \equiv 0).$$

This is the just the required identity.

Lemma 3.1. *If $h(t) > 0$ $(t \in E)$ satisfies the Lipschitz condition, then $A_n(t, t)$, $B_n(t, t)$ and $A_n(z_0, t)$, $B_n(z_0, t)$ are uniformly bounded with respect to n and $t \in E$ for fixed z_0 lying outside E.*

Proof. Let us estimate a_{n-1}/a_n and $J_{nk} = \int |P_n(s)\tilde{P}_k(s)|\,d\sigma(s)$. It is known [Sze, p. 55] that $a_{n-1}/a_n = \int sP_{n-1}(s)P_n(s)d\sigma(s)$, whence, by using the Cauchy–Buniakowski inequality, we establish that a_{n-1}/a_n are bounded by a certain constant α. To estimate J_{nk} we can also use the Cauchy–Buniakowski inequality and get

$$J_{nk} = \int \frac{1}{\sqrt{h(s)}}\left|P_n(s)\tilde{P}_k(s)\right|\sqrt{h(s)}\,d\sigma(s)$$

$$\leq \frac{1}{\sqrt{h}}\left[\int |P_n(s)|^2 d\sigma(s)\right]^{1/2}\left[\int \left|\tilde{P}_k(s)\right|^2 h(s)d\sigma(s)\right]^{1/2} \leq \beta$$

$$\left(h = \min_{s \in E} h(s);\ n, k \in \mathbb{N}_0;\ \beta \geq \frac{1}{\sqrt{h}}\right).$$

By using these estimates and the inequalities

$$\left|\frac{h(t) - h(s)}{t - s}\right| \leq \gamma, \qquad h \leq h(s) \leq H,$$

we obtain

$$|A_n(t, t)| \leq J_{nn} + \left|\frac{a_{n-1}}{a_n}\right|\frac{1}{h}\int\left|\frac{h(t) - h(s)}{t - s}\right|\left|\tilde{P}_n(s)P_{n-1}(s)\right|d\sigma(s) \leq \delta,$$

$$|B_n(t, t)| \leq \left|\frac{a_{n-1}}{a_n}\right|\frac{1}{h}\int\left|\frac{h(t) - h(s)}{t - s}\right|\left|\tilde{P}_n(s)P_n(s)\right|d\sigma(s) \leq \varepsilon.$$

The boundedness of $A_n(z_0, t)$ and $B_n(z_0, t)$ can be established similarly if we note that $\left|\dfrac{1}{z_0 - s}\right| \leq \kappa$ $(s \in E)$.

∎

This and (3.28) imply the following statement:

Corollary (Koraus theorem). *If the polynomials $\{P_n(t)\}$ are uniformly bounded, then the polynomials $\{\tilde{P}_n(t)\}$ are also uniformly bounded.*

Consider the system of polynomials $P_n^{(\alpha,\beta)}(t)$ $(-1<\alpha, \beta \leq -1/2)$ orthonormal on the segment $[-1, 1]$ with respect to the measure $(1-t)^{\alpha}(1+t)^{\beta}h(t)dt$, where $h(t)>0$ satisfies the Lipschitz condition. By using the Koraus theorem, we conclude that these polynomials are uniformly bounded and, hence, it is reasonable to consider the hypercomplex systems $l_1((1-t)^{\alpha}(1+t)^{\beta}dt,\ K^2(2p+1)^{2A})$ and

$$l_1\left((1-t)^{\alpha}(1+t)^{\beta}h(t)dt,\ K^2\frac{\sinh^2(2p+1)B}{\sinh^2 B}\right).$$

The spaces of characters of these hypercomplex systems are the segment $[-1, 1]$ *and the region in the complex plane defined by the inequality* $|z+(z^2-1)^{1/2}| \leq e^{2B}$.

Indeed, we use Lemma 3.1, taking $P_n^{(\alpha,\beta,h)}(t)$ and $P_n^{(\alpha,\beta)}(t)$ as the systems $P_n(t)$ and $\tilde{P}_n(t)$, respectively (this is possible because $1/h(t)>0$ satisfies the Lipschitz condition). Let z_0 be a fixed point of the complex plane. By virtue of (3.28), we have

$$\frac{P_n^{(\alpha,\beta)}(z_0)}{(2n+1)^A} \leq |A_n(z_0,t)|\frac{|P_n^{(\alpha,\beta,h)}(z_0)|}{(2n+1)^A} + |B_n(z_0,t)|\frac{|P_{n-1}^{(\alpha,\beta,h)}(z_0)|}{(2n+1)^A}$$

$$\leq C\left[\frac{|P_n^{(\alpha,\beta,h)}(z_0)|}{(2n+1)^A} + \frac{|P_{n-1}^{(\alpha,\beta,h)}(z_0)|}{(2n+1)^A}\right].$$

Thus, if

$$\limsup_{n\to\infty}\frac{|P_n^{(\alpha,\beta,h)}(z_0)|}{(2n+1)^A} < \infty,$$

i.e., if z_0 generates a character of the hypercomplex system $l_1((1-t)^{\alpha}(1+t)^{\beta}h(t)dt,$ $K^2(2p+1)^{2A})$, then

$$\limsup_{n\to\infty}\frac{|P_n^{(\alpha,\beta)}(z_0)|}{(2n+1)^A} < \infty,$$

i.e., z_0 generates a character of the hypercomplex system $l_1((1-t)^{\alpha}(1+t)^{\beta}dt,$ $K^2(2p+1)^{2A})$ and, hence, $z_0 \in [-1, 1]$. The first part of the statement is proved. The second part can be proved similarly.

■

Since the series

$$\sum_{p=0}^{\infty} |x_p|(2p+1)^A, \qquad \sum_{p=0}^{\infty} |x_p| \frac{\sinh(2p+1)B}{\sinh B}$$

converge or diverge together with the series

$$\sum_{p=0}^{\infty} |x_p| p^A, \quad \sum_{p=0}^{\infty} |x_p| e^{2pB},$$

we obtain the following analog of the Wiener–Levi theorem:

Theorem 3.8. *If the function*

$$x(t) = \sum_{p=0}^{\infty} x_p P_p^{(\alpha,\beta,h)}(t) \quad (t \in [-1,1])$$

is such that the series

$$\sum_{p=0}^{\infty} |x_p| p^A, \quad A \geq 1 \quad \left(\sum_{p=0}^{\infty} |x_p| e^{pB}, \ B > 0 \right),$$

is convergent, then the function $f(x(t))$, *where* $f(z)$ *is regular on the range of values of* $x(t)$, $t \in [-1,1]$,

$$\left(x(t) = \sum_{p=0}^{\infty} x_p P_p^{(\alpha,\beta,h)}(t), \ \left| t + (t^2-1)^{1/2} \right| \leq e^B \right),$$

also possesses this property.

3.8. Transmutation Operators. Here, we consider transmutation operators for hypercomplex systems associated with orthogonal polynomials. Consider two hypercomplex systems $l_1(d\sigma, m)$ and $l_1(d\tilde{\sigma}, \tilde{m})$ with the same set E. *An operator* U *that acts from* $l_1(d\sigma, m)$ *into* $l_1(d\tilde{\sigma}, \tilde{m})$ *and is defined by the matrix* $(u_{pq})_{p,q=0}^{\infty} = (\int \tilde{P}_q(t) P_q(t) d\tilde{\sigma}(t)$ *according to the formula*

$$(Ux)(p) = \frac{1}{\sqrt{\tilde{m}(p)}} \sum_q u_{pq} x(q) \sqrt{m(q)} \quad \Big(p \in Q\,,\; x(p) \in l_1(d\sigma, m)\Big)$$

is called a transmutation operator. It is not difficult to show that the operator U is continuous if and only if the inequalities

$$\sum_p |u_{pq}| \sqrt{\tilde{m}(p)} \le A \sqrt{m(q)} \quad (q \in Q)$$

hold. Denote by V a transmutation operator acting from $l_1(d\tilde{\sigma}, \tilde{m})$ into $l_1(d\sigma, m)$. Clearly, if both operators U and V are continuous, then $l_1(d\sigma, m)$ and $l_1(d\tilde{\sigma}, \tilde{m})$ are topologically equivalent.

Before formulating the main theorem about transmutation operators, we recall that it was shown in Section 3.3 in Chapter 1 that the discrete hypercomplex system $l_1(m)$ is isometrically isomorphic to the algebra of continuous functions on X_h which can be expanded into absolutely convergent series in the complete orthonormal system of functions $p(\chi)$ $(\chi \in X_h,\ p \in Q)$ with ordinary algebraic operations. Therefore, the notation $x(t) \in l_1(d\sigma, m)$ makes sense and means that

$$x(t) = \sum_{p \in Q} x(p) P_p(t) = \sum_{p \in Q} x(p) \sqrt{m(p)}\, Q_p(t)$$

with

$$\|x(p)\|_{l_1(d\sigma, m)} = \sum_{p \in Q} x(p) \sqrt{m(p)} < \infty.$$

Consider the hypercomplex system $l_1(g(t)dt, m)$ associated with polynomials orthogonal with respect to the measure $g(t)dt$ (here, $g(t) \ge 0$ is an integrable function).

Theorem 3.9. *Suppose that the hypercomplex system* $l_1(g(t)dt, m)$, $h(t) > 0$ $(t \in E)$ *has a bounded first-order derivative,* $1/h(t) \in l_1(g(t)dt, m)$, *and*

$$\left| \int \frac{h(t) - h(s)}{t - s} P_p(t) g(t) dt \right| \sqrt{m(p)} \le k_p \quad (p \in Q) \tag{3.29}$$

uniformly in $s \in E$, *where* $\sum_p k_p < \infty$. *Then the transmutation operators* U *and* V *for the spaces* $l_1(g(t)dt, m)$ *and* $l_1(h(t)g(t)dt, m)$ *are continuous* (it readily follows from (3.28) and Lemma 3.1 that $\tilde{P}_p(t)/\sqrt{m(p)}$ are uniformly bounded and, hence, the space $l_1(h(t)g(t)dt, m)$ exists). *If the polynomials $P_p(t)$ are uniformly bounded, then condition (3.29) can be replaced by the following weaker condition:*

$$\sum_p \left| \int \frac{h(t)-h(s)}{t-s} P_p(t)g(t)dt \right| \sqrt{m(p)} \le K < \infty \quad (s \in E) .$$

A proof of this theorem can be found in [Ber4].

Let us now consider the ultraspherical polynomials $U_n^{(\lambda)}(t)$ $(\lambda=1, 2,\ldots, -1 \le t \le 1)$ orthonormal with respect to the measure $(1-t^2)^{\lambda/2-1}$. It is known (see, e.g., [Vil]) that they coincide to within a factor with zonal spherical functions for the Gelfand pair $(SO(n), SO(n-1))$. As indicated in Subsection 3.4, we can consider the hypercomplex system $l_1((1-t^2)^{\lambda/2-1}dt, m)$, where $m(p) = [U_p^{\lambda/2-1}(1)]^2$. By using the known relations for Jacobi polynomials (see [Sze, pp.70, 80]), one can easily find that

$$m(p) = [U_p^{(\lambda)}(1)]^2 = \frac{1}{2^{(\lambda-1)/2}\Gamma(\lambda/2)}\left[\frac{1}{p!}(2p+\lambda-1)\Gamma(p+\lambda-1)\right] \sim p^{\lambda-2}$$

and, hence, the space $l_1((1-t^2)^{\lambda/2-1}dt, m)$ is topologically equivalent to the space $l_1((1-t^2)^{\lambda/2-1}dt, p^{\lambda-1})$. By using Theorem 3.9, one can prove the following statement:

Theorem 3.10. *Let $h(t) > 0$ and let U and V be the transmutation operators constructed for the spaces $l_1((1-t^2)^{\lambda/2-1}dt, p^{\lambda-1})$ and $l_1(h(t)(1-t^2)^{\lambda/2-1}dt, p^{\lambda-1})$, where $\lambda=1, 2,\ldots$. For $\lambda=2, 3,\ldots$, the operators U and V are continuous if $h(t)$ has a bounded derivative of order $[(\lambda+5)/2]$. For $\lambda=1$, these operators are continuous if $h(t)$ has a bounded derivative of the second order.*

A proof of this theorem can also be found in [Ber4].

Let $U_p^{(\lambda,h)}(t)$ $(-1 \le t \le 1)$ be polynomials orthonormal with respect to the measure $h(t)(1-t^2)^{\lambda/2-1}dt$. The results obtained above imply the following statement:

Theorem 3.11. *Let $\lambda=1,2, \ldots$ and let $h(t)>0$ $(-1 \le t \le 1)$ have a bounded derivative of the second order (for $\lambda=1$) or of order $[(\lambda+5)/2]$ (if $\lambda = 2, 3,\ldots$). If*

$$\hat{x}(t) = \sum_{r \in Q} x(r) U_r^{(\lambda)}(t) = \sum_{r \in Q} x^{(h)}(r) U_r^{(\lambda,h)}(t),$$

then the series

$$\sum_r |x(r)| r^{(\lambda-1)/2} \quad \textit{and} \quad \sum_r \left|x^{(h)}(r)\right| r^{(\lambda-1)/2}$$

are either simultaneously convergent or simultaneously divergent.

This theorem, in particular, yields the Wiener–Levi theorem for expansions in the orthonormal polynomials $U_r^{(\lambda,h)}(t)$.

4. Hypercomplex Systems Constructed for the Sturm–Liouville Equation

Many hypercomplex systems considered above (e.g., hypercomplex systems associated with symmetric Riemann spaces of rank one, compact hypercomplex systems constructed from Jacobi polynomials, and others) have the following property: characters of these hypercomplex systems are eigenfunctions of a certain differential equation of the Sturm–Liouville type. In the present section, we describe hypercomplex systems constructed for the Sturm–Liouville equation on a semiaxis by the method of Levitan and Povzner and present a survey of the related results.

4.1. Riemann Function. First, we present preliminary results necessary for our subsequent discussion. Consider the Sturm–Liouville equation on a semiaxis

$$L[y] = \frac{d^2y}{dt^2} - q(t)y = -\lambda y \quad (0 \le t < \infty), \tag{4.1}$$

where $q(t)$ is a continuous function. Denote by $\varphi(t, \lambda)$ a solution of equation (4.1) satisfying the initial conditions $\varphi(0, \lambda) = 1$ and $\varphi_t'(0, \lambda) = 0$. We extend the function $q(t)$ to the entire axis by evenness and consider the following partial differential equation:

$$\frac{\partial^2 u}{\partial t^2} - q(t)u = \frac{\partial^2 u}{\partial s^2} - q(s)u \quad (-\infty \le t, s < \infty). \tag{4.2}$$

The solution $u(t, s)$ of equation (4.2) is completely determined by the initial conditions of the form

$$u(t, 0) = f(t) \quad \text{and} \quad \frac{\partial u(t, 0)}{\partial s} = 0, \tag{4.3}$$

where $f(t)$ is a twice continuously differentiable function which can be found by using the Riemann formula. For convenience, we now deduce this formula as applied to equation (4.2) with initial conditions (4.3). We set

$$\mathfrak{L}[u] = L_t[u] - L_s[u] = \frac{\partial^2 u}{\partial t^2} - \frac{\partial^2 u}{\partial s^2} - [q(t) - q(s)]\,u.$$

For any two differentiable functions u and v, it is easy to see that

$$v\mathfrak{L}[u]-u\mathfrak{L}[v]=\frac{\partial}{\partial t}\left(v\frac{\partial u}{\partial t}-u\frac{\partial v}{\partial t}\right)-\frac{\partial}{\partial s}\left(v\frac{\partial u}{\partial s}-u\frac{\partial v}{\partial s}\right).$$

By integrating this equality over the domain G bounded by a piecewise smooth curve Γ and using the Green formula, we obtain

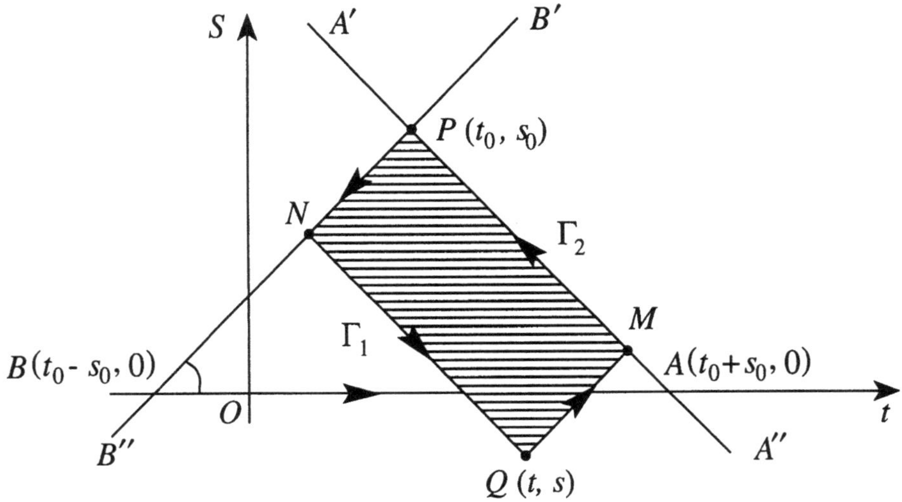

Fig. 1.

$$\iint_G\{v\mathfrak{L}[u]-u\mathfrak{L}[v]\}\,dt\,ds=\int_\Gamma\left(v\frac{\partial u}{\partial s}-u\frac{\partial v}{\partial s}\right)dt+\left(v\frac{\partial u}{\partial t}-u\frac{\partial v}{\partial t}\right)ds. \tag{4.4}$$

Let $P(t_0,s_0)$ $(s_0\neq 0)$ be a fixed point in the plane. We construct a triangle PAB as shown in Fig. 1 and apply relation (4.4) to this triangle by assuming that u is a solution of the equation $\mathfrak{L}[u]=0$ with the indicated initial conditions and v is, for the present, an arbitrary function. After simple calculations, we arrive at the identity (for the sake of definiteness, we assume that $s_0>0$):

$$\iint_G u\mathfrak{L}[v]\,dt\,ds=2u(P)v(P)-u(A)v(A)-u(B)v(B)$$

$$+2\int_{\Gamma_1}u\left(\frac{\partial v}{\partial s}-\frac{\partial v}{\partial t}\right)dt+2\int_{\Gamma_3}u\left(\frac{\partial v}{\partial s}+\frac{\partial v}{\partial t}\right)dt+\int_{t_0-s_0}^{t_0+s_0}f(t)\left.\frac{\partial v}{\partial s}\right|_{t=0}dt. \tag{4.5}$$

Suppose that the function $v(t,s)=v(t_0,s_0;t,s)$ $(-\infty\leq t,s<\infty)$ is chosen so that

(a) $v(t_0, s_0; t_0, s_0) = 1;$

(b) $\dfrac{\partial v}{\partial t} = \dfrac{\partial v}{\partial s}$ on PA and $\dfrac{\partial v}{\partial t} = -\dfrac{\partial v}{\partial s}$ on PB;

(c) $\mathfrak{L}[v] = 0.$

It follows from (b) that v is constant on PA and PB and, hence, it is equal on these lines to 1. For this choice of the function v, relation (4.5) implies the formula

$$u(t_0, s_0) = \frac{1}{2}[f(t_0 - s_0) + f(t_0 + s_0)] - \frac{1}{2}\int_{t_0 - s_0}^{t_0 + s_0} \frac{\partial v(t_0, s_0; t, 0)}{\partial s} f(t)dt, \quad (4.6)$$

for the solution of equation (4.2) satisfying the initial conditions. The function $v(t_0, s_0; t, s) = v(P; Q)$ is called a *Riemann function.* Its existence and uniqueness can be proved as follows: Since $\mathfrak{L}[v] = 0$, we can write

$$\frac{\partial^2 v}{\partial t^2} - \frac{\partial^2 v}{\partial s^2} = [q(t) - q(s)]\, v \qquad (-\infty \le t, s < \infty).$$

By applying the Green formula to this relation with $PMQN = G$ (see Fig. 1), we easily obtain

$$\iint_G [q(t) - q(s)]\, v\, dt\, ds = \int_\Gamma \frac{\partial v}{\partial s} dt + \frac{\partial v}{\partial t} ds$$

$$= v(P; M) + v(P; N) - 2v(P; Q) = 2 - 2v(P; Q),$$

whence we conclude that, for fixed P, the Riemann function satisfies the integral equation

$$v(P; Q) = 1 + \frac{1}{2}\iint_G [q(s) - q(t)]\, v\, dt\, ds. \quad (4.7)$$

Conversely, it is clear that each solution of (4.7) is a Riemann function. At the same time, in the coordinate system $A''PB''$, (4.7) is a two-dimensional Volterra equation. Therefore, its solution $v(P; Q)$ always exists is unique and defined for all Q. This solution can be obtained by using the Neumann series

$$v(P; Q) = 1 + \frac{1}{2}\iint_G [q(s) - q(t)]\, dt\, ds + \ldots, \quad (4.8)$$

where dots stand for the sum of iterations of the kernel of (4.7) applied to 1.

We need a formula for the derivative $dv(t_0, s_0; t, 0)/ds$ with $t_0 - s_0 \le t \le t_0 + s_0$ ($s_0 > 0$). By using (4.7), we obtain (in the notation of Fig. 2)

$$-\frac{1}{h}[v(t_0, s_0; t, -h) - v(t_0, s_0; t, 0)]$$

$$= -\frac{1}{2h}\iint_{G_1}[q(s) - q(t)]\,v\,dt\,ds - \frac{1}{2h}\iint_{G_2}[q(s)-q(t)]\,v\,dt\,ds$$

$$-\frac{1}{2h}\iint_{G_3}[q(s)-q(t)]\,v\,dt\,ds$$

$$= -\frac{1}{2h}\iint_{G_2}[q(s)-q(t)]\,v\,dt\,ds$$

$$-\frac{1}{4}(t_0+s_0-t)\frac{2}{(t_0+s_0-t)h}\iint_{G_1}[q(s)-q(t)]\,v\,dt\,ds$$

$$-\frac{1}{4}(t-t_0+s_0)\frac{2}{(t-t_0+s_0)h}\iint_{G_3}[q(s)-q(t)]\,v\,dt\,ds.$$

By passing to the limit as $h \to 0$, we get

$$\frac{dv(t_0, s_0; t, 0)}{ds} = -\frac{1}{4}\left\{(t_0+s_0-t)\int_Q^M[q(s)-q(t)]v\,dl + (t-t_0+s_0)\int_Q^N[q(s)-q(t)]v\,dl\right\}, \tag{4.9}$$

where dl is a differential of the length of the arc.

Relations (4.8) and (4.9) yield some important properties of the Riemann function. First, *if $q(t)$ does not increase as $t \to +\infty$, then the Riemann function $v(t_0, s_0; t, s)$ is nonnegative for* $0 < s_0 \le t_0$ and $0 \le s \le t$. Indeed, if equation (4.7) is considered in the region $0 \le s \le t$, then its kernel is nonnegative. Hence, all iterations of this kernel are also nonnegative but, in this case, relation (4.8) implies that $v(P; Q)$ is nonnegative.

■

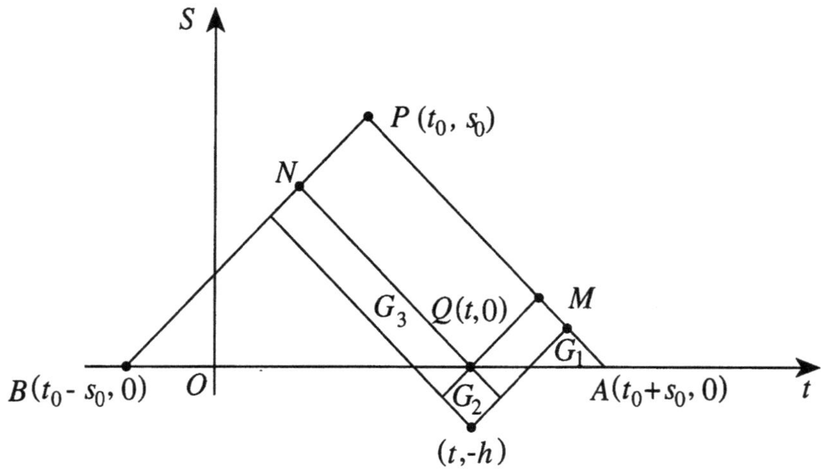

Fig. 2

Let $0 < s_0 \le t_0$. *If* $q(t)$ *is a function nonincreasing as* $t \to +\infty$, *then the derivative* $dv(t_0, s_0; t, 0)/ds$ *is nonpositive for* $t_0 - s_0 \le t \le t_0 + s_0$.

This assertion directly follows from (4.9) because the Riemann function is nonnegative in the triangle *PAB*.

■

Up to now, we did not impose any restrictions on the function $f(t)$. Let us now prove that *if the function* $f(t)$ *is even, then the solution* $u(t, s)$ *of equation (4.2) satisfying the initial conditions (4.3) is symmetric about the coordinate axes and the bisectrices of the coordinate angles.*

It is easy to see that the fact that the solution is symmetric about the coordinate axes follows from the uniqueness of the solution of problem (4.2) –(4.3) and the fact that the functions $q(t)$ and $f(t)$ are even. Let us prove that $u(t, s) = u(s, t)$. For this purpose, we consider equation (4.1) in the entire axis ($q(t)$ is extended by evenness) and denote by $\varphi(t, \lambda)$ its solution satisfying the initial conditions $\varphi(t, 0) = 1$ and $d\varphi(0, \lambda)/dt = 0$. It is clear that $\varphi(t, \lambda) = \varphi(-t, \lambda)$. The function $\varphi(t, \lambda)\varphi(s, \lambda)$ is a solution of problem (4.2)–(4.3) under the condition $f(t) = \varphi(t, \lambda)$ and hence, if

$$f(t) = \sum_{j=1}^{n} c_j \varphi(t, \lambda_j),$$

then the corresponding solution has the form

$$u(t, s) = \sum_{j=1}^{n} c_j \varphi(t, \lambda_j)\varphi(s, \lambda_j).$$

This implies that, under the indicated initial conditions, the function $u(t, s)$ is symmetric.

We fix t and s. It follows from formula (4.6) that, first, the value of $u(t, s)$ depends only on the values of the initial function $f(r)$ on the interval $[t-s, t+s]$ and, second, that if $f_n(\tau) \to f(\tau)$ uniformly on $[t-s, t+s]$, then $u_n(t, s) \to u(t, s)$. Thus, to prove the required assertion in the general case, it suffices to show that any twice continuously differentiable even function can be approximated by functions of the form $\sum_{j=1}^{n} c_j \varphi(\tau, \lambda_j)$ uniformly on $[t-s, t+s]$. But this is an immediate consequence of the theory of the Sturm–Liouville equation. Indeed, we choose $a > 0$ so large that $[t-s, t+s] \subset (-a, a)$ and consider the Sturm–Liouville problem on the interval $(0, a)$ with boundary conditions $y(0) = 1$, $y'(0) = 0$, and $y(a) = 0$. For some values of the parameter λ, this problem has eigenfunctions $\varphi(t, \lambda)$. By using linear combinations of these functions, one can construct uniform (on $(0, a)$) approximations of any twice continuously differentiable function satisfying the boundary conditions. Thus, to complete the proof, it remains to perform an appropriate change of values of the function $f(\tau)$ outside the segment $[t-s, t+s]$ and use the fact that the functions $f(\tau)$ and $\varphi(t, \lambda)$ are even. This proves that the solution $u(t, s)$ is symmetric about the bisectrix of the first quadrant. The fact that it is symmetric about the second bisectrix follows this fact and from the symmetry about the coordinate axes.

■

4.2. Hypercomplex Systems Constructed for the Sturm–Liouville Equation. In the present subsection, we construct a hypercomplex system associated with the Sturm–Liouville equation. The potential $q(t)$ satisfies the conditions guaranteeing that the structure measure of the relevant hypercomplex system is nonnegative.

Suppose that $q(t)$ is a nonnegative nonincreasing function. Denote by $\mu(t)$ $(0 \le t < \infty)$ the solution of equation (4.1) with $\lambda = 0$ satisfying the initial conditions $\mu(0) = 1$ and $\mu'(0) = 0$. We always have $\mu'(t) \ge 0$ and, therefore, the *function* $\mu(t)$ *is nonincreasing and bounded from below by one.*

Indeed, equation (4.1) implies the following integral Volterra equation for the function $\mu'(t)$:

$$y'(t) = \int_0^t q(t)y(s)\,ds = \int_0^t q(s)\left[1 + \int_0^s y'(r)dr\right]ds$$

$$= \int_0^t q(t)ds + \int_0^t y'(r)\int_r^t q(s)\,ds\,dr = p(t) + \int_0^t [p(t)-p(r)]\,y'(r)\,dr, \quad (4.10)$$

where

$$p(t) = \int_0^t q(s)\,ds.$$

The solution $\mu'(t)$ of equation (4.10) can by obtained by using the Neumann series

$$\mu'(t) = p(t) + \int_0^t [p(t) - p(r)]p(r)dr + \dots, \tag{4.11}$$

and, hence, the nonnegativity of $\mu'(t)$ follows from the nonnegativity of the kernel of the integral equation (4.10) and its iterations.

■

For a twice continuously differentiable even function $x(t)$, we introduce generalized translation operators by setting $(R_s x)(t) = u(t, s)/\mu(s)\,\mu(t)$, where $u(t, s)$ is a solution of equation (4.2) satisfying the initial conditions

$$u(t, 0) = x(t)\mu(t) \quad \text{and} \quad \frac{\partial u(t, 0)}{\partial s} = 0 \quad (-\infty < t, s < \infty,\ \mu(-t) = \mu(t)).$$

Let us now rewrite this solution in terms of the Riemann formula. We get

$$(R_s x)(t) = \frac{1}{2\mu(t)\mu(s)}[x(t-s)\mu(t-s) + x(t+s)\mu(t+s)]$$

$$+ \frac{1}{2\mu(t)\mu(s)} \int_{t-s}^{t+s} w(t, s, \tau)x(\tau)\mu(\tau)d\tau, \tag{4.12}$$

where

$$w(t, s, \tau) = -\left.\frac{\partial v(t, s; \tau, \sigma)}{\partial \sigma}\right|_{\sigma=0}.$$

By using relation (4.12), we can now define generalized translation operators R_s for any locally integrable functions $x(t)$ defined for $t \geq 0$ and extended as even functions to the entire axis.

Theorem 4.1. *Let $q(t)$ be a nonnegative nonincreasing function and let $\mu(t)$ be a solution of equation (4.1) with $\lambda = 0$ satisfying the initial conditions $y(0) = 1$ and*

$y'(0)=0$. The space $L_1([0,\infty), \mu^2(t), dt)$ of functions summable with respect to the measure $\mu^2(t)$ on the semiaxis $[0,\infty)$ with the operation of convolution defined by the formula

$$(x * y)(t) = \int_0^\infty (R_s x)(t) y(s) \mu^2(s)\,ds$$

$$= \frac{1}{2\mu(t)} \int_0^\infty [x(t-s)\mu(t-s) + x(t+s)\mu(t+s)]\, y(s)\mu(s)\,ds$$

$$+ \frac{1}{2\mu(t)} \int_0^\infty \int_{t-s}^{t+s} w(t,s,\tau) x(\tau)\mu(\tau)\,d\tau\, y(s)\mu(s)\,ds \tag{4.13}$$

is a commutative Hermitian hypercomplex system with basis unit $e=0$ and nonnegative structure measure vanishing for large t. Moreover, this hypercomplex system satisfies the condition of separate continuity.

We first prove several useful lemmas. The only assumption imposed on the function $q(t)$ $(0 \le t < \infty)$ in these lemmas is its continuity.

Lemma 4.1. *Let $u(t,s)$ and $v(t,s)$ be, respectively, the solutions of problem (4.2) with even twice continuously differentiable and compactly supported initial functions $f(t)$ and $g(t)$. Then*

$$\int_{-\infty}^\infty u(t,s) g(t)\,dt = \int_{-\infty}^\infty f(t) v(t,s)\,dt \qquad (-\infty < s < \infty). \tag{4.14}$$

Proof. First, we note that the conditions imposed on the function $f(t)$ and relation (4.6) imply that $u(t,s)$ vanishes for large $|t|$. Hence, the function

$$y(s) = \int_{-\infty}^\infty u(t,s)\varphi(t,\lambda)\,dt$$

is well defined and satisfies the initial conditions

$$y(0) = \int_{-\infty}^\infty f(t)\varphi(t,\lambda)\,dt \quad \text{and} \quad \frac{dy(0)}{ds} = 0. \tag{4.15}$$

The function $y(s)$ satisfies the differential equation (4.1). Indeed, by using the Green formula, we get

$$L_s[y] = \int_{-\infty}^{\infty} L_s[u(t,s)]\,\varphi(t,\lambda)\,dt = \int_{-\infty}^{\infty} L_t[u(t,s)]\,\varphi(t,\lambda)\,dt$$

$$= \int_{-\infty}^{\infty} u(t,s)\,L_t[\varphi(t,\lambda)]\,dt = -\lambda y.$$

On the other hand, the function $\int_{-\infty}^{\infty} f(t)\varphi(t,\lambda)\,\varphi(s,\lambda)\,dt$ also satisfies equation (4.1) with the initial conditions (4.15) and, hence, by the uniqueness theorem,

$$\int_{-\infty}^{\infty} u(t,s)\,\varphi(t,\lambda)\,dt = \int_{-\infty}^{\infty} f(t)\varphi(t,\lambda)\,\varphi(s,\lambda)\,dt \quad (-\infty < s < \infty). \tag{4.16}$$

It is clear that equality (4.16) also holds for linear combinations of the functions $\varphi(t,\lambda)$. By constructing uniform approximations of the twice continuously differentiable even function $g(t)$ on its compact support $\operatorname{supp} g$ by sums of the form

$$g_n(t) = \sum_{k=1}^{n} c_k \varphi(t,\lambda_k)$$

and passing to the limit, we arrive at relation (4.14).

■

Lemma 4.2. *Any three functions x, y, and z from the collection $\mathfrak{N}$ of functions finite and bounded on $[0,\infty)$ satisfy the relations of commutativity and associativity*

$$x * y = y * x, \quad (x * y) * z = x * (y * z). \tag{4.17}$$

and, moreover, $x * y \in \mathfrak{N}$.

Proof. If $f(t), g(t) \in \mathfrak{N}$ are twice continuously differentiable functions, then it follows from (4.14) that

$$(f * g)(t) = \int_{0}^{\infty} (R_s f)(t)\,g(s)\,\mu^2(s)\,ds = \frac{1}{\mu(t)}\int_{0}^{\infty} u(t,s)\,g(s)\,\mu(s)\,ds$$

$$= \frac{1}{\mu(t)} \int_0^\infty f(s)\mu(s)v(t,s)\,ds = g*f(t),$$

where $u(t,s)$ and $v(t,s)$ denote the solutions of differential equation (4.2) satisfying the initial conditions $u(t,0) = f(t)\mu(t)$ and $du(t,0)/ds = 0$ and $v(t,0) = g(t)\mu(t)$ and $dv(t,0)/ds = 0$, respectively. By approximating (almost everywhere) arbitrary functions $x, y \in \mathfrak{N}$ by twice continuously differentiable functions f_n and g_n bounded uniformly in n and passing to the limit in (4.13), we conclude that $x * y = y * x$. It also follows from relation (4.13) that $x * y \in \mathfrak{N}$ for any $x, y \in \mathfrak{N}$.

Relation (4.13) implies the existence of a convolution of the form $x * \psi$, where $x \in \mathfrak{N}$ and $\psi(t)$ is a function summable on every bounded interval. For a twice continuously differentiable function $x \in \mathfrak{N}$, we can write

$$\left(x * \frac{\varphi(\cdot,\lambda)}{\varphi(\cdot)}\right)(t) = \frac{1}{\mu(t)} \int u(t,s)\varphi(s,\lambda)\,ds = \frac{\varphi(t,\lambda)}{\mu(t)} \int x(s)\varphi(s,\lambda)\mu(s)\,ds. \quad (4.18)$$

By approximating (almost everywhere) an arbitrary function $x \in \mathfrak{N}$ by twice continuously differentiable finite functions bounded uniformly in n, we prove that (4.18) holds for any $x \in \mathfrak{N}$. By using (4.18), we obtain

$$\left((x*y) * \frac{\varphi(\cdot,\lambda)}{\varphi(\cdot)}\right)(t) = \int_0^\infty (x*y)(s)\,\frac{\varphi(t,\lambda)}{\mu(t)}\,\varphi(s,\lambda)\mu(s)\,ds$$

$$= \frac{\varphi(t,\lambda)}{\mu(t)} \int_0^\infty\int_0^\infty (R_r x)(s)\,y(r)\mu^2(r)\,dr\,\varphi(s,\lambda)\mu(s)\,ds$$

$$= \frac{\varphi(t,\lambda)}{\mu(t)} \int_0^\infty y(r) \int_0^\infty (R_r x)(s)\,\frac{\varphi(s,\lambda)}{\mu(s)}\,\mu^2(s)\,ds\,\mu^2(r)\,dr$$

$$= \frac{\varphi(t,\lambda)}{\mu(t)} \int_0^\infty y(r) \int_0^\infty x(s)\varphi(s,\lambda)\varphi(r,\lambda)\,ds\,\mu(r)\,dr$$

$$= \frac{\varphi(t,\lambda)}{\mu(t)} \int_0^\infty x(s)\varphi(s,\lambda)\mu(s)\,ds \int_0^\infty y(r)\varphi(r,\lambda)\mu(r)\,dr.$$

Consequently,

$$\left((x*y) * \frac{\varphi(\cdot,\lambda)}{\mu(\cdot)}\right) = x * \left(y * \frac{\varphi(\cdot,\lambda)}{\mu(\cdot)}\right),$$

whence we arrive at the relation of associativity for

$$z(t) = \frac{1}{\mu(t)} \sum_{k=1}^{n} c_k \varphi(t, \lambda_k).$$

By passing to the limit, we establish this relation first for functions $z \in \mathfrak{N}$ whose second derivatives are continuous and then for arbitrary $z \in \mathfrak{N}$.

■

Corollary 4.1. *The function*

$$c(A, B, r) = (\kappa_A * \kappa_B)(r) \quad (A, B \in \mathcal{B}([0, \infty)),\ t \geq 0)$$

is a structure measure. The operation of convolution with respect to this measure has the form (4.13).

Indeed, this assertion is a direct consequence of Lemma 4.2 and relation (4.13) from which it follows that the function $x * y(t)$ is finite and continuous for $x, y \in \mathfrak{N}$.

■

Proof of Theorem 4.1. Since $q(t)$ is nonincreasing as $t \to +\infty$, the function $w(t, s, \tau)$ is nonnegative for $|t-s| \leq \tau \leq t+s$. Let us check relation (1.2) in Chapter 1 for the measure $dm(t) = \mu^2(t)dt$. It follows from (4.13) that there exists a convolution $x * \psi$, where $x \in \mathfrak{N}$ and ψ is a locally integrable function. For this convolution, relation (4.14) remains valid. Thus,

$$\int_0^\infty c(A, B, t)\mu^2(t)dt = \int_0^\infty \kappa_B(s)\mu^2(s) \int_0^\infty (R_s \kappa_A)(t)\mu^2(t)\,dt\,ds$$

$$= \int_0^\infty \kappa_B(s) \int_0^\infty [\kappa_A(s)\mu(t)]\,\mu(t)\mu(s)\,dt\,ds$$

$$= \int_0^\infty \kappa_A(t)\mu^2(t)dt \int_0^\infty \kappa_B(s)\mu^2(s)\,ds = m(A)m(B).$$

To show that the hypercomplex system $L_1([0, \infty), \mu^2(t), dt)$ is Hermitian, we write

$$c(A, B, C) = \int_0^\infty \kappa_B(s) \int_0^\infty (R_s \kappa_A)(t)\kappa_C(t)\mu^2(t)\,dt\,\mu^2(s)\,ds$$

$$= \int_0^\infty \kappa_B(s) \int_0^\infty (R_t \kappa_A)(s) \kappa_C(t) \mu^2(t) dt \, \mu^2(s) ds$$

$$= \int_0^\infty \kappa_B(s) \int_0^\infty \kappa_A(s)(R_t \kappa_C)(s) \mu^2(t) dt \, \mu^2(s) ds$$

$$= \int_0^\infty \kappa_A(t) \int_0^\infty \kappa_B(s)(R_s \kappa_C)(t) \mu^2(s) ds \, \mu^2(t) dt = c(C, B, A)$$

$$(A, B, C \in \mathcal{B}_0([0, \infty)).$$

Since

$$c(A, B, 0) = \int_0^\infty (R_s \kappa_A)(0) \kappa_B(s) \mu^2(s) ds$$

$$= \int_0^\infty (R_0 \kappa_A)(s) \kappa_B(s) \mu^2(s) ds$$

$$= \int_0^\infty \kappa_A(s) \kappa_B(s) \mu^2(s) ds = m(A \cap B),$$

we conclude that 0 is the basis unit of this hypercomplex system.

4.3. Structure Measure which Is Not Necessarily Nonnegative. In the present subsection, we construct a hypercomplex system whose structure measure is not necessarily nonnegative (see Subsection 2.4 in Chapter 1). Assume that $q(t)$ is a continuous function of bounded variation on the semiaxis $[0, \infty)$. Let $V(t)$ be the variation of the function $q(t)$ on the interval (t, ∞), i.e., $V(t) = \operatorname{Var}_t^\infty q$. The function $V(t)$ is continuous and nonnegative, moreover, it approaches 0 as $t \to \infty$. Since

$$\operatorname{Var}_a^b q = \operatorname{Var}_a^\infty q - \operatorname{Var}_b^\infty q = V(a) - V(b) \quad (a \le b),$$

we conclude that

$$|q(t') - q(t'')| \le V(t') - V(t'') \quad (0 \le t' \le t'' < \infty).$$

Let $\tilde{q}(t)$ $(0 \le t < \infty)$ be a continuous nonnegative nonincreasing function such that

$$|q(t') - q(t'')| \le \tilde{q}(t') - \tilde{q}(t'') \quad (0 \le t' \le t'' < \infty). \tag{4.19}$$

(in particular, we can assume that $\tilde{q}(t)$ is equal to $V(t)$; we always have $\tilde{q}(t) \ge V(t)$). Parallel with (4.1), we consider the equation

$$y'' - \tilde{q}(t)y = -\lambda y. \tag{4.20}$$

According to Theorem 4.1, equation (4.20) can be associated with a hypercomplex system $L_1([0, \infty), \tilde{\mu}^2(t), dt)$ with the nonnegative measure $\tilde{c}(A, B, t)$, where $\tilde{\mu}(t)$ is a solution of (4.20) with $\lambda = 0$ satisfying the initial conditions $y(0) = 1$ and $y'(0) = 0$.

Theorem 4.2. *Let $q(t)$ be a function continuous of bounded variation on the semiaxis $[0, \infty)$, let $\tilde{q}(t)$ $(0 \le t < \infty)$ be a fixed continuous nonnegative nonincreasing function satisfying inequality (4.19), and let $\tilde{\mu}(t)$ be a solution of equation (4.20) with $\lambda = 0$ satisfying the initial conditions $y(0) = 1$ and $y'(0) = 0$. Then the space $L_1([0, \infty), \tilde{\mu}^2(t), dt)$ with the operation of multiplication given by (4.13) is a commutative hypercomplex system whose structure measure is not, generally speaking, positive (it is assumed that the function μ in relation (4.13) is replaced by $\tilde{\mu}$).*

Proof. Let $c(A, B, t) = (\kappa_A * \kappa_B)(t)$. By virtue of Lemmas 4.1 and 4.2, $c(A, B, t)$ is a structure measure which induces convolution (4.13). To prove the theorem, it suffices to check inequality (2.14) in Chapter 1. This inequality follows from the inequality

$$|c(A, B, t)| \le \tilde{c}(A, B, t) \quad (A, B \in \mathcal{B}([0, \infty)),\ 0 \le t < \infty). \tag{4.21}$$

Indeed, since $\tilde{c}(A, B, t) \ge 0$, it follows from (4.21) that $V(A, B, t) \le \tilde{c}(A, B, t)$. Thus,

$$\int_0^\infty V(A, B, t)\tilde{\mu}^2(t)dt \le \int_0^\infty \tilde{c}(A, B, t)\tilde{\mu}^2(t)dt = \int_A \tilde{\mu}^2(t)dt \int_B \tilde{\mu}^2(s)ds,$$

as required. Hence, it remains to prove inequality (4.21). Since $(R_s\kappa_A)(t) = (R_t\kappa_A)(s)$, we get

$$|c(A, B, t)| = \left| \int_0^\infty (R_s\kappa_A)(t)\kappa_B(s)\mu^2(s)ds \right|$$

$$\le \left| \int_0^t (R_s\kappa_A)(t)\kappa_B(s)\mu^2(s)ds \right| + \left| \int_t^\infty (R_t\kappa_A)(s)\kappa_B(s)\tilde{\mu}^2(s)ds \right|$$

$$\leq \frac{1}{2\tilde{\mu}(t)} \int_0^\infty [\kappa_A(t-s)\tilde{\mu}(t-s) + \kappa_A(t+s)\tilde{\mu}(t+s)]\kappa_B(s)\tilde{\mu}(s)ds$$

$$+ \frac{1}{2\tilde{\mu}(t)} \int_0^t \int_{t-s}^{t+s} |w(t, s, \tau)|\kappa_A(\tau)\tilde{\mu}(\tau)d\tau\,\kappa_B(s)\tilde{\mu}(s)ds$$

$$+ \frac{1}{\tilde{\mu}(t)} \int_t^\infty \int_{s-t}^{t+s} |w(t, s, \tau)|\kappa_A(\tau)\tilde{\mu}(\tau)d\tau\,\kappa_B(s)\tilde{\mu}(s)ds.$$

Therefore, it suffices to check that $|w(t, s, r)| \leq \tilde{w}(t, s, r)$ $(0 \leq s \leq t < \infty,\ t-s \leq r \leq t+s)$, where $\tilde{w}(t, s, r)$ is a function constructed by using equation (4.20) in exactly the same way as the function $w(t, s, r)$ was constructed by using equation (4.1).

Let $t_0 \geq s_0 > 0$. Riemann functions satisfy the inequality

$$|v(t_0, s_0; t, s)| \leq \tilde{v}(t_0, s_0; t, s) \quad (0 < s_0 \leq t,\ 0 \leq s \leq t). \tag{4.22}$$

Indeed, we fix a point (t_0, s_0) $(0 < s_0 \leq t_0)$ and denote by $K(t, s; \tau, \sigma)$ and $\tilde{K}(t, s; \tau, \sigma)$, respectively, the kernels of the integral equation (4.7) considered in the region $0 \leq s \leq t$ and the same integral equation with q replaced by $\tilde{q}$. Clearly,

$$K(t, s; \tau, \sigma) = \begin{cases} \frac{1}{2}[q(\sigma) - q(\tau)], & (\tau, \sigma) \in G, \\ 0, & (\tau, \sigma) \notin G, \end{cases}$$

and a similar relation holds for $\tilde{K}(t, s; \tau, \sigma)$. This and relation (4.19) imply the inequality $|K(t, s; \tau, \sigma)| \leq \tilde{K}(t, s; \tau, \sigma)$. But then the same relations hold for the iterated kernels and, thus, inequality (4.22) follows from the expansion of (4.8) of the Riemann function in the Neumann series.

By using inequalities (4.19) and (4.22) and relation (4.9), for $0 \leq s_0 \leq t_0$ and $t_0 - s_0 \leq t \leq t_0 + s_0$ (see Fig. 2), we obtain

$$\left|\frac{\partial v(t_0, s_0; t, 0)}{\partial s}\right| = \frac{1}{4}\left|(t_0 + s_0 - t) \int_Q^M [q(s) - q(t)]v\,dl \right.$$

$$\left. + (t - t_0 + s_0) \int_Q^N [q(s) - q(t)]v\,dl \right|$$

$$\leq \frac{1}{4}\left\{(t_0 + s_0 - t)\int\limits_Q^M |q(s) - q(t)||v|dl + (t - t_0 + s_0)\int\limits_Q^N |q(s) - q(t)||v|dl\right\}$$

$$\leq \frac{1}{4}\left\{(t_0 + s_0 - t)\int\limits_Q^M [\tilde{q}(s) - \tilde{q}(t)]\tilde{v}dl + (t - t_0 + s_0)\int\limits_Q^N [\tilde{q}(s) - \tilde{q}(t)]\tilde{v}dl\right\}$$

$$= -\frac{\partial\tilde{v}(t_0, s_0; t, 0)}{\partial s}.$$

This inequality actually means that $|w(t, s, r)| \leq \tilde{w}(t, s, r)$ $(0 \leq s \leq t < \infty,\ t - s \leq r \leq t + s)$.

■

It should be noted that the function $q(t)$ can be nonincreasing and nonnegative itself; in particular, if $\tilde{q}(t) = q(t)$, we arrive at the case considered above.

Theorem 4.3. *Characters of the hypercomplex system* $L_1([0, \infty), \tilde{\mu}^2(t), dt)$ *have the form* $\chi(t, \lambda) = \varphi(t, \lambda)/\tilde{\mu}(t)$, *where* $\varphi(t, \lambda)$ *is a solution of equation (4.1) satisfying the initial conditions* $\varphi(0, \lambda) = 1$ *and* $\varphi'(0, \lambda) = 0$ *and the complex number* λ *is such that* $\varphi(t, \lambda) = O(\tilde{\mu}(t))$.

Proof. In one direction the proof is evident. Indeed, if λ is such that $\varphi(t, \lambda) = O(\tilde{\mu}(t))$, then $\chi(t, \lambda)$ is bounded. Moreover,

$$\int\limits_0^\infty c(A, B, t)\chi(t, \lambda)\tilde{\mu}^2(t)dt = \int\limits_0^\infty \kappa_B(s)\int\limits_0^\infty (R_t\kappa_A)(s)\varphi(t, \lambda)\tilde{\mu}(t)dt\,\tilde{\mu}^2(s)ds$$

$$= \int\limits_0^\infty \kappa_B(s)\frac{\varphi(s, \lambda)}{\tilde{\mu}(s)}\int\limits_0^\infty \kappa_A(t)\varphi(t, \lambda)\tilde{\mu}(t)dt\,\tilde{\mu}^2(s)ds$$

$$= \int\limits_A \chi(t, \lambda)\tilde{\mu}^2(t)dt\int\limits_B \chi(s, \lambda)\tilde{\mu}^2(s)ds$$

$$= \chi(A, \lambda)\chi(B, \lambda) \quad (A, B \in \mathcal{B}_0([0, \infty))).$$

Let us prove the converse assertion. First, we show that $\chi(t)$ is twice continuously differentiable. Let $f(t)$ be a twice continuously differentiable even function with compact support. Then $(R_sf)(t) = u(t, s)/\dot{\tilde{\mu}}(s)\tilde{\mu}(t)$, where $u(t, s)$ is a solution of equation (4.2) with the initial conditions $u(t, 0) = f(t)\tilde{\mu}(t)$ and $\partial u(t, 0)/\partial s = 0$, whence it follows that $(R_s f)(t)$ is twice continuously differentiable with respect to s and, for s varying within a finite interval, $(R_s f)(t)$ regarded as a function of t is finite. By passing to the limit in the equality

$$\int_0^\infty (f * g_n)\chi(t)\tilde{\mu}^2(t)dt = \int_0^\infty f(t)\chi(t)\tilde{\mu}^2(t)dt \int_0^\infty g_n(t)\chi(s)\tilde{\mu}^2(s)ds,$$

where g_n is a sequence of functions which converge to δ_s, we get

$$\int_0^\infty (R_s f)(t)\chi(t)\tilde{\mu}^2(t)dt = \chi(s) \int_0^\infty f(t)\chi(t)\tilde{\mu}^2(t)dt \quad (s \in [0, \infty)).$$

We choose $f(t)$ such that the integral on the right-hand side is nonzero. Since integration on the left-hand side is actually carried out over a finite interval, this integral is twice continuously differentiable with respect to s, which implies the differentiability of the function $\chi(s)$. This means that the function

$$\tilde{\mu}(t)\tilde{\mu}(s)(R_s\chi)(t) = u(t, s) = \tilde{\mu}(t)\tilde{\mu}(s)\chi(t)\chi(s) = \varphi(t)\varphi(s) \quad (\varphi(t) = \tilde{\mu}(t)\chi(t))$$

is a solution of equation (2) with the initial conditions $u(t, 0) = \varphi(t)$ and $\partial u(t, 0)/\partial s = 0$. Since $\varphi(t) \neq 0$, these initial conditions imply that $\varphi(0) = 1$ and $\varphi'(0) = 0$. Further, by substituting $u(t, s) = \varphi(t)\varphi(s)$ in (4.2), we find that

$$[\varphi''(t) - q(t)\varphi(t)]\,\varphi(s) = [\varphi''(s) - q(s)\varphi(s)]\,\varphi(t),$$

whence it follows that $\varphi''(t) - q(t)\varphi(t) = -\lambda\varphi(t)$, where λ is a constant, i.e. $\varphi(t) -$ $\varphi(t, \lambda)$. Finally, $\varphi(t, \lambda) = O(\tilde{\mu}(t))$ because $|\chi(t)| \leq 1$.

■

Thus, the space of characters coincides with a certain set X of values of the parameter λ lying in the complex plane.

Theorem 4.4. *The topology of the space of characters coincide with the ordinary topology in the complex plane. The point* ∞ *corresponds to the maximal ideal* M_∞.

Lemma 4.3. *If* $|\lambda_n| \to \infty$ $(\lambda_n \in X)$, *then, for any function* $x(t) \in L_1([0, \infty), \tilde{\mu}^2(t), dt)$, *we have*

$$\int_0^\infty x(t)\chi(t, \lambda_n)\mu^2(t)dt \to 0.$$

Proof. Let $a<b$ and $0<h<b-a$ be arbitrary numbers. By integrating the equality $\varphi(t, \lambda_n) = -[\varphi''(t, \lambda_n) - q(t)\varphi(t, \lambda_n)]/\lambda_n$, we obtain

$$\left|\frac{1}{h}\int_a^b d\tau \int_\tau^{\tau+h} \varphi(t, \lambda_n)dt\right| = \frac{1}{h|\lambda_n|}\left|\left[\int_a^b d\tau \int_\tau^{\tau+h} \varphi''(t, \lambda_n)dt \int_a^b d\tau \int_\tau^{\tau+h} q(t)\varphi(t, \lambda_n)dt\right]\right|$$

$$\leq \frac{1}{h|\lambda_n|}\left|\int_a^b [\varphi'(\tau+h, \lambda_n) - \varphi'(\tau, \lambda_n)]d\tau\right|$$

$$+ \frac{1}{|\lambda_n|}\sup_{0\leq t<\infty}|q(t)|\int_a^b \tilde{\mu}(\tau+h)d\tau$$

$$= \frac{1}{h|\lambda_n|}[\varphi(b+h, \lambda_n) - \varphi(a+h, \lambda_n) - \varphi(b, \lambda_n) + \varphi(a, \lambda_n)]$$

$$+ \frac{1}{|\lambda_n|}\sup_{0\leq t<\infty}|q(t)|\int_a^b \tilde{\mu}(\tau+h)d\tau$$

$$= \frac{2}{h|\lambda_n|}[\tilde{\mu}(a+h) + \tilde{\mu}(b+h)]$$

$$+ \frac{1}{|\lambda_n|}\sup_{0\leq t<\infty}|q(t)|\int_a^b \tilde{\mu}(\tau+h)d\tau.$$

This implies that the integral on the left-hand side of this inequality tends to zero as $|\lambda_n| \to \infty$. By changing the order of integration, we easily see that this integral is equal to

$$\int_0^\infty \frac{x(t; a, b, h)}{\tilde{\mu}(t)} \chi(t, \lambda_n)\tilde{\mu}^2(t)dt,$$

where $x(t; a, b\, h)$ is a function depicted in Fig. 3. Thus, Lemma 4.3 is proved for the functions $x(t; a, b, h)/\mu(t) \in L_1([0, \infty), \tilde{\mu}^2(t), dt)$ and their linear combinations $y(t)$. Now let $x(t)$ be an arbitrary function from $L_1([0, \infty), \tilde{\mu}^2(t), dt)$. We choose $y(t)$ such that

$$\int_0^\infty |x(t) - y(t)|\tilde{\mu}^2(t)dt < \varepsilon.$$

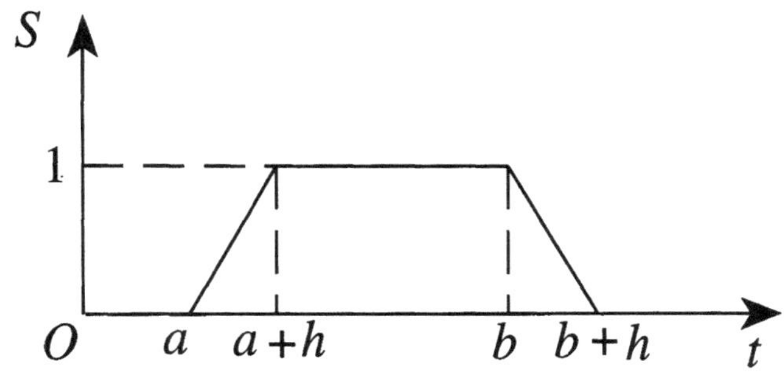

Fig. 3

Then

$$\left|\int_0^\infty x(t)\chi(t,\lambda_n)\tilde{\mu}^2(t)\,dt\right|$$

$$\leq \left|\int_0^\infty [x(t)-y(t)]\chi(t,\lambda_n)\tilde{\mu}^2(t)\,dt\right| + \left|\int_0^\infty y(t)\chi(t,\lambda_n)\tilde{\mu}^2(t)\,dt\right|$$

$$< \varepsilon + \left|\int_0^\infty y(t)\chi(t,\lambda_n)\tilde{\mu}^2(t)\,dt\right| \to \varepsilon.$$

■

To prove the theorem, it is necessary to show that the space $\mathfrak{M}$ of maximal ideals of the algebra $\mathfrak{L}_1([0,\infty),\mu^2(t),dt)$ is homeomorphic to $X\cup\{\infty\}$. Since the relevant mapping is, clearly, bijective and $\mathfrak{M}$ and $X\cup\{\infty\}$ are compact, it suffices to prove that convergence in $X\cup\{\infty\}$ implies convergence in $\mathfrak{M}$. Let $\lambda_n\to\lambda\neq\infty$. Since $\chi(t,\lambda_n)\to\chi(t,\lambda)$ pointwise and $x(t)\in L_1([0,\infty),\tilde{\mu}^2(t),dt)$, we have

$$\int_0^\infty x(t)\chi(t,\lambda_n)\tilde{\mu}^2(t)dt \to \int_0^\infty x(t)\chi(t,\lambda_n)\tilde{\mu}^2(t)dt.$$

If $\lambda_n\to\infty$, then, by virtue of Lemma 4.3,

$$\int_0^\infty x(t)\chi(t,\lambda_n)\mu^2(t)dt \to 0.$$

■

Since $\chi(t, \lambda)$ is real if and only if λ is real, $\chi(t, \lambda)$ is Hermitian if and only if $\lambda \in X \cap \mathbb{R}^1 = X_h$.

Since $\tilde{q}(t)$ is monotone, there exists a finite limit $c = \lim\limits_{t \to \infty} \tilde{q}(t) \geq 0$. Inequality (4.19) remains valid with $\tilde{q}(t)$ replaced by $\tilde{q}(t) - c$ and, in addition, $\tilde{q}(t) \xrightarrow[t \to \infty]{} 0$. Hence, without loss of generality, we can always assume that this condition is satisfied.

4.4. Set of Characters of the Hypercomplex System Associated with the Sturm–Liouville Equation. Let us now describe the structure of the space of characters. In fact, we want to show that this space coincides with the spectrum of the operator acting in the space $L_2([0, \infty), dt)$ and generated by the differential expression L with the boundary condition $y'(0) = 0$ (for details, see [Ber10], [Gla], and [LeS]). First, we formulate several auxiliary lemmas concerning the Sturm–Liouville equation (14.1) with continuous real potential $q(t)$ approaching zero as $t \to \infty$. Denote $\lambda = s^2$, where $s = \sigma + i\tau$ varies in the upper half plane.

Lemma 4.4. *The solution* $\varphi(t, \lambda)$ *of equation (4.1) admits the following estimate:*

$$|\varphi(t, \lambda)| \leq \exp\left(t\tau + \frac{1}{|s|}\int_0^t |q(\alpha)| d\alpha\right) \quad (0 \leq t < \infty). \tag{4.23}$$

Proof. For $\varphi(t, \lambda)$, we obtain

$$\varphi(t, \lambda) = \cos st + \frac{1}{s}\int_0^t \sin s(t - \alpha) q(\alpha) \varphi(\alpha, \lambda) d\alpha. \tag{4.24}$$

By setting $\varphi_1(t, \lambda) = \varphi(t, \lambda) e^{-\tau t}$, in view of relation (4.24), we get

$$\varphi_1(t, \lambda) = \cos st e^{-\tau t} + \frac{1}{s}\int_0^t \sin s(t - \alpha) e^{-\tau(t-\alpha)} q(\alpha) \varphi_1(\alpha, \lambda) d\alpha,$$

whence

$$|\varphi_1(t, \lambda)| \leq 1 + \frac{1}{|s|}\int_0^t q(\alpha) |\varphi_1(\alpha, \lambda)| d\alpha.$$

By using the well-known lemma (see [LeS, Chapter 4, Section3]), we conclude that

$$|\varphi_1(t,\lambda)| \le \exp\left\{\frac{1}{|s|}\int_0^t |q(\alpha)|d\alpha\right\},$$

which implies (4.23).

■

Lemma 4.5. *Assume that* $q(t) \xrightarrow[t\to\infty]{} 0$. *Let* $\rho > 0$ *and* $\varepsilon > 0$ *be given numbers such that* $\varepsilon < \sqrt{\rho}$. *Then, for any* λ *from the interval* $(-\infty, -\rho)$ *there are two linearly independent solutions of equation (4.1)* $y_\lambda(t)$ *and* $z_\lambda(t)$ *such that* $|y_\lambda(t)| \ge C_1(\lambda)\, e^{(\sqrt{|\lambda|}-\varepsilon)t}$ *and* $|z_\lambda(t)| \le C_2(\lambda)\, e^{(\sqrt{|\lambda|}+\varepsilon)t}$ $(0 \le t \le \infty,\ -\infty < \lambda < -\rho)$. *Moreover, for any fixed* t, *the functions* $y_\lambda(t)$, $dy_\lambda(t)/dt$, $z_\lambda(t)$ *and* $dz_\lambda(t)/dt$ *are analytic functions of* λ *in the interval* $(-\infty, -\rho)$.

Proof. We now choose sufficiently large t_0 such that, for $t \ge t_0$, we have $|q(t)| < 2\varepsilon\sqrt{\rho} - \varepsilon^2$ (this is possible since, by the condition, $2\varepsilon\sqrt{\rho} - \varepsilon^2 = \varepsilon(2\sqrt{\rho} - \varepsilon) > 0$). Denote by $y_\lambda(t)$ the solution of (4.1) $(0 \le t < \infty,\ -\infty < \lambda < -\rho)$ satisfying the initial conditions $y_\lambda(t_0) = 1$ and $y'_\lambda(t_0) = 0$. Let us show that this is the first required solution. For $t \ge t_0$ and $\lambda < -\rho$, we have $|q(t)| < 2\varepsilon\sqrt{\rho} - \varepsilon^2 < 2\varepsilon\sqrt{|\lambda|} - \varepsilon^2$ and, hence, $-\lambda + q(t) \ge |\lambda| - 2\varepsilon\sqrt{|\lambda|} + \varepsilon^2 = (\sqrt{|\lambda|} - \varepsilon)^2$. By using this inequality, we obtain

$$\begin{aligned}
y_\lambda(t) &= 1 + \int_{t_0}^{t} (t-\alpha)[-\lambda + q(\alpha)]\, y_\lambda(\alpha)\, d\alpha \\
&= 1 + \int_{t_0}^{t} (t-\alpha)[-\lambda + q(\alpha)]\, d\alpha \\
&\quad + \int_{t_0}^{t} (t-\alpha)[-\lambda + q(\alpha)]\, d\alpha \int_{t_0}^{\alpha} (\alpha-\beta)[-\lambda + q(\beta)]\, d\beta \ldots \\
&\ge 1 + \int_{t_0}^{t} (t-\alpha)(\sqrt{|\lambda|} - \varepsilon)^2 d\alpha \\
&\quad + \int_{t_0}^{t} (t-\alpha)(\sqrt{|\lambda|} - \varepsilon)^2 d\alpha \int_{t_0}^{\alpha} (\alpha-\beta)(\sqrt{|\lambda|} - \varepsilon)^2 d\beta + \ldots = \eta(t)
\end{aligned}$$

$$(t \ge t_0),$$

where $\eta(t)$ is a solution of the equation $y'' = (\sqrt{|\lambda|} - \varepsilon)^2 y$ satisfying the initial conditions $y(t_0) = 1$ and $y'(t_0) = 0$, i.e., $\eta(t) = \cosh((t - t_0)(\sqrt{|\lambda|} - \varepsilon))$. This implies that there exists a constant C_1 which, generally speaking, depends on λ such that $|y_\lambda(t)| \geq C_1 e^{(\sqrt{|\lambda|}-\varepsilon)t}$ $(0 \leq t < \infty)$. Since $y_\lambda(t)$ satisfies the initial conditions $y_\lambda(t_0) = 1$ and $y'_\lambda(t_0) = 0$, for any fixed t, it is an analytic function of λ in the interval $(-\infty; -\rho)$.

To find the second partial solution, it is necessary to establish an estimate for $dy_\lambda(t)/dt = y'_\lambda(t)$. By using the same notation as above, we obtain

$$y'_\lambda(t) = \int_{t_0}^{t} [-\lambda + q(\alpha)]\, y_\lambda(\alpha)\, d\alpha = \int_{t_0}^{t} [-\lambda + q(\alpha)] \left[1 + \int_{t_0}^{\alpha} y'_\lambda(\beta) d\beta \right] d\alpha$$

$$= \int_{t_0}^{t} [-\lambda + q(\alpha)]\, d\alpha + \int_{t_0}^{t} \int_{\beta}^{t} [-\lambda + q(\alpha)]\, d\alpha \int_{t_0}^{\beta} [-\lambda + q(\gamma)]\, d\gamma\, d\beta + \ldots$$

$$\geq \int_{t_0}^{t} (\sqrt{|\lambda|} - \varepsilon)^2 d\alpha + \int_{t_0}^{t} \int_{\beta}^{t} (\sqrt{|\lambda|} - \varepsilon)^2 d\alpha \int_{t_0}^{\beta} (\sqrt{|\lambda|} - \varepsilon)^2 d\gamma\, d\beta + \ldots$$

$$= \eta'(t) = (\sqrt{|\lambda|} - \varepsilon) \sinh (t - t_0)(\sqrt{|\lambda|} - \varepsilon),$$

whence it follows that there is a constant C_3 which, generally speaking, depends on λ such that $|y'_\lambda(t)| \geq C_3 e^{(\sqrt{|\lambda|}-\varepsilon)t}$ $(0 \leq t)$. The second partial solution $z_\lambda(t)$ defined on the interval (t_0, ∞) is now sought in the form $z_\lambda(t) = u_\lambda(t) y_\lambda(t)$. By substituting this relation in (4.1), we find that $u_\lambda(t)$ is a solution the equation $u''_\lambda y_\lambda + 2u'_\lambda y'_\lambda = 0$, whence

$$u_\lambda(t) = A \int_{t_0}^{\beta} \frac{d\alpha}{y_\lambda^2(\alpha)} + B$$

(for $t \geq t_0$, $y_\lambda(t) > 0$). We set $A = 1$ and $B = -\int_{t_0}^{\infty} \frac{d\alpha}{y_\lambda^2(\alpha)}$. Then

$$z_\lambda(t) = y_\lambda(t) \int_{t}^{\infty} \frac{d\alpha}{y'_\lambda(\alpha)} \qquad (t \geq t_0).$$

By using the Cauchy formula, we find a point $r_t > t$ such that

$$z_\lambda(t) = \int_t^\infty \frac{d\alpha}{y_\lambda'(\alpha)}\left(\frac{1}{y_\lambda(t)}\right)^{-1} = \frac{-1/y_\lambda^2(r_t)}{-y'(r_t)/y_\lambda^2(r_t)} = \frac{1}{y'(r_t)}.$$

This gives the estimate

$$|z_\lambda(t)| = \left|(y_\lambda'(r_t))^{-1}\right| \le \frac{1}{C_3(\lambda)}e^{-(\sqrt{|\lambda|}-\varepsilon)r_t} < \frac{1}{C_3(\lambda)}e^{(-\sqrt{|\lambda|}+\varepsilon)t} \quad (t_0 \le t < \infty).$$

It follows from (4.24) that the solution $z_\lambda(t)$ satisfies the initial conditions

$$z_\lambda(t_0) = \int_{t_0}^\infty \frac{d\alpha}{y_\lambda^2(\alpha)} \quad \text{and} \quad z_\lambda'(t_0) = -1. \tag{4.25}$$

We use these initial conditions to extend $z_\lambda(t)$ to the interval $(0, t_0)$ and to construct a solution $z_\lambda(t)$ such that $|z_\lambda(t)| \ge C_2(t)\,e^{(-\sqrt{|\lambda|}+\varepsilon)t}$ $(0 \le t < \infty,\ -\infty < \lambda < -\rho)$. The function $z_\lambda(t_0)$ analytically depends on λ $(-\infty < \lambda < -\rho)$ because the integrand $1/y_\lambda^2(t)$ possesses the indicated property; moreover, we have

$$y_\lambda(t) \ge \cosh(t-t_0)(\sqrt{|\lambda|}-\varepsilon) > \frac{1}{2}e^{(t-t_0)(\sqrt{|\lambda|}-\varepsilon)} > \frac{1}{2}e^{(t-t_0)(\sqrt{\rho}-\varepsilon)} \quad (t_0 \le t < \infty).$$

Thus, $1/y_\lambda^2(t) < 2\,e^{-(\sqrt{\rho}-\varepsilon)(t-t_0)}$ and the integral $\int_{t_0}^\infty e^{-(\sqrt{\rho}-\varepsilon)(t-t_0)}dt$ is convergent. Hence, the initial conditions (4.25) are analytic functions of λ $(-\infty < \lambda < -\rho)$ and, consequently, the solution $z_\lambda(t)$ and its derivative are analytic in λ for all fixed t.

■

We say that a function $f(t)$ $(0 \le t < \infty)$ grows slower than any exponential function if $f(t) = O(e^{\varepsilon t})$ as $t \to \infty$ for any $\varepsilon > 0$.

Lemma 4.6. *Let* $q(t) \xrightarrow[t\to\infty]{} 0$. *There exists an (at most countable) family of negative values of* λ *bounded from below and having at most one limiting point (whose role may be played only by zero) for which* $\varphi(t,\lambda)$ *grows slower than any exponential function. Then* $\varphi(t,\lambda) = O(e^{(-\sqrt{|\lambda|}+\varepsilon)t})$ *as* $t \to \infty$ *for all* λ *from this family and any sufficiently small* $\varepsilon > 0$.

Proof. Let $\delta > 0$ be given and $\varepsilon > 0$ be such that $\varepsilon < \sqrt{\delta/2}$. It suffices to show that the number of negative $\lambda < -\delta$ such that $\varphi(t,\lambda)$ grows slower than any exponential function is finite and $\varphi(t,\lambda) = O(e^{(-\sqrt{|\lambda|}+\varepsilon)t})$ for each λ of this sort. In Lemma

4.5, we now set $\rho = \delta/2$. The solutions $y_\lambda(t)$ and $z_\lambda(t)$ and their derivatives with respect to t are analytic in λ in the interval $(-\infty, -\delta/2)$. Because $y_\lambda(t)$ and $z_\lambda(t)$ are linearly independent for all λ, we have

$$\varphi(t, \lambda) = A(\lambda) y_\lambda(t) + B(\lambda) z_\lambda(t) \quad \left(0 \le t < \infty;\ -\infty < \lambda < \frac{\delta}{2}\right). \tag{4.26}$$

By differentiating (4.26) with respect to t, we get

$$\varphi'(t, \lambda) = A(\lambda)\, y'_\lambda(t) + B(\lambda)\, z'_\lambda(t) \quad \left(0 \le t < \infty;\ -\infty < \lambda < \frac{\delta}{2}\right). \tag{4.27}$$

By setting $t = 0$ in (4.26) and (4.27), we obtain the following system of equations:

$$\begin{aligned} A(\lambda) y_\lambda(0) \ +\ B(\lambda) z_\lambda(0) \ &=\ \varphi(0, \lambda) \\ A(\lambda)\, y'_\lambda(0) \ +\ B(\lambda)\, z'_\lambda(0) \ &=\ \varphi'(0, \lambda) \quad \left(-\infty < \lambda < \frac{\delta}{2}\right) \end{aligned} \tag{4.28}$$

For all λ, the determinant of this system is nonzero because the solutions $y_\lambda(t)$ and $z_\lambda(t)$ are linearly independent. Since all coefficients of system (4.28) are analytic functions in the interval $(-\infty, -\delta/2)$, its solutions $A(\lambda)$ and $B(\lambda)$ are also analytic.

It is easy to see that $A(\lambda)$ cannot have zeros with arbitrarily large absolute values. Indeed, if $A(\lambda_0) = 0$ $(-\infty < \lambda_0 < \delta/2)$, then

$$|\varphi(t, \lambda_0)| = |B(\lambda_0) z_{\lambda_0}(t)| = O(e^{(-\sqrt{|\lambda_0|}+\varepsilon)t}),$$

and, hence, $\varphi(t, \lambda_0) \in L_1(0, \infty) \cap L_2(0, \infty)$. Therefore, by integrating the equality

$$\varphi''(t, \lambda_0) = [q(t) - \lambda_0]\, \varphi(t, \lambda_0)$$

we conclude that $\varphi'(t, \lambda_0) = O(1)$, whence, by integrating by parts, we obtain

$$\int_0^\infty \varphi''(t, \lambda_0) \varphi(t, \lambda_0) dt = -\int_0^\infty (\varphi')^2(t, \lambda_0) dt \le 0.$$

By multiplying (4.29) by $\varphi(t, \lambda_0)$, integrating, and using the relation established above, we get

$$\int_0^\infty [q(t) - \lambda_0]\, \varphi^2(t, \lambda_0)\, dt \le 0,$$

for arbitrarily large $-\lambda_0$, which is absurd.

We can now complete the proof of the lemma as follows: It is clear that $\varphi(t, \lambda)$ grows slower than any exponential function if and only if λ is a zero of the function $A(\lambda)$ analytic in the interval $(-\infty, -\delta/2)$. Since zeros of the function $A(\lambda)$ are bounded from below, they cannot have limiting points other than $-\delta/2$.

■

Corollary 4.2. *The following statements are equivalent:*

(i) a point $\lambda < 0$ belongs to the spectrum of the differential operator (4.1) with the boundary condition $y'(0) = 0$;

(ii) the point $\lambda < 0$ is such that $\varphi(t, \lambda)$ grows slower than any exponential function as $t \to \infty$;

(iii) the point $\lambda < 0$ is such that, for any $0 < \varepsilon < \sqrt{|\lambda|}$,

$$\varphi(t, \lambda) = O(e^{(-\sqrt{|\lambda|}+\varepsilon)t}), \quad t \to \infty. \tag{4.29}$$

Proof. The equivalence of (ii) and (iii) has already been proved. Now let $\lambda < 0$ be a point of the spectrum. Since the spectrum is discrete for negative values of λ (see, e.g., Glazman [Gla, Chap. 2, Sec. 27]), we conclude that λ is necessarily an eigenvalue. By virtue of (4.29) and Lemma 4.5, $A(\lambda) = 0$, which yields (ii) and (iii). Conversely, if (ii) and (iii) hold, then $A(\lambda) = 0$, i.e., λ is an eigenvalue because $\varphi(t, \lambda) \in L_2(0, \infty)$.

■

Lemma 4.7. *Let $q(t)$ $(0 \le t < \infty)$ be bounded. If, for some λ, $\varphi(t, \lambda) = O(e^{\varepsilon t})$ for any $\varepsilon > 0$, then λ is real.*

Proof. We set

$$\frac{1}{2}[f(r-s) + f(r+s)] + \frac{1}{2}\int_{r-s}^{r+s} w(r, s, t) f(t)\, dt = \int_0^\infty f(t)\, d_t W(r, s, t).$$

For $N > r$, we have

$$\int_0^N \overline{\varphi(s, \lambda)}\, d_s \int_0^N \varphi(t, \lambda)\, d_t W(r, s, t)$$

$$= \int_0^N \overline{\varphi(s,\lambda)}\,d_s \int_0^\infty \varphi(t,\lambda)\,d_t W(r,s,t) - \int_0^N \overline{\varphi(s,\lambda)}\,d_s \int_N^\infty \varphi(t,\lambda)\,d_t W(r,s,t)$$

$$= \varphi(r,\lambda) \int_0^N |\varphi(s,\lambda)|^2\,ds - \int_{N-r}^N \overline{\varphi(s,\lambda)}\,d_s \int_N^{N+r} \varphi(t,\lambda)\,d_t W(r,s,t). \qquad (4.30)$$

The left-hand side of this equality is real because it is a composition of a function equal to $\varphi(t,\lambda)$ for $0 \le t \le N$ and to 0 for $t > N$ with the conjugate function (for $\mu(t) = 1$). Thus, by dividing (4.30) by $\int_0^N |\varphi(s,\lambda)|^2\,ds$ and passing to the limit as $N \to \infty$, we conclude that $\varphi(r,\lambda)$ is a real function provided that a sequence $N_k \to \infty$ is chosen so that

$$\frac{1}{\int_0^{N_k} |\varphi(s,\lambda)|^2\,ds} \int_{N_k-r}^{N_k} \overline{\varphi(s,\lambda)}\,d_s \int_{N_k}^{N_k+r} \varphi(t,\lambda)\,d_t W(r,s,t) \xrightarrow[k\to\infty]{} 0. \qquad (4.31)$$

We consider N_k so large that $N_k > 2r$. Then $0 < s - r < N_k$ and $r + s \ge N_k$ for $N_k - r \le s \le N_k$. Thus, in view of the fact that $|W(r,s,t)| \le C$ $(0 \le t, s < \infty)$, we obtain

$$\left| \int_{N_k-r}^{N_k} \overline{\varphi(s,\lambda)}\,d_s \int_{N_k}^{N_k+r} \varphi(t,\lambda)\,d_t W(r,s,t) \right|$$

$$= \left| \int_{N_k-r}^{N_k} \overline{\varphi(s,\lambda)}\,d_s \left[\frac{1}{2}\varphi(r+s,\lambda) + \frac{1}{2} \int_{N_k}^{N_k+r} w(r,s,t)\varphi(t,\lambda)\,dt \right] ds \right|$$

$$\le \frac{1}{2} \left| \int_{N_k-r}^{N_k} \overline{\varphi(s,\lambda)}\varphi(r+s,\lambda)\,ds \right| + \frac{1}{2} \left| \int_{N_k-r}^{N_k} \overline{\varphi(s,\lambda)} \int_{N_k}^{N_k+r} w(r,s,t)\varphi(t,\lambda)\,dt\,ds \right|$$

$$\le \frac{1}{2} \left[\int_{N_k-r}^{N_k} |\varphi(s,\lambda)|^2\,ds \right]^{1/2} \left[\int_{N_k-r}^{N_k} |\varphi(r+s,\lambda)|^2\,ds \right]^{1/2}$$

$$+ \frac{Cr}{2} \left[\int_{N_k-r}^{N_k} |\varphi(s,\lambda)|^2\,ds \right]^{1/2} \left[\int_{N_k}^{N_k+r} |\varphi(t,\lambda)|^2\,dt \right]^{1/2}$$

$$= \frac{1}{2}(1+Cr) \left[\int_{N_k-r}^{N_k} |\varphi(s,\lambda)|^2\,ds \right]^{1/2} \left[\int_{N_k}^{N_k+r} |\varphi(t,\lambda)|^2\,dt \right]^{1/2}. \qquad (4.32)$$

Since

$$\int_{N_k-r}^{N_k} |\varphi(s,\lambda)|^2 ds \le \int_0^{N_k} |\varphi(s,\lambda)|^2 ds,$$

it follows from (4.32) that, in order to prove (4.31), it suffices to show that

$$\left(\int_0^{N_k} |\varphi(s,\lambda)|^2 ds\right)^{-1} \int_{N_k}^{N_k+r} |\varphi(s,\lambda)|^2 ds \xrightarrow[k\to\infty]{} 0.$$

We set

$$\alpha_k = \int_{(k-1)r}^{kr} |\varphi(s,\lambda)|^2 ds.$$

Let us show that there exists a sequence k_n such that $\dfrac{\alpha_{k_{n+1}}}{\alpha_1+\ldots+\alpha_{k_n}} \to 0$ as $n\to\infty$. Assume the contrary. Then there exists $\beta>0$ such that $\dfrac{\alpha_{k+1}}{\alpha_1+\ldots+\alpha_k} \ge \beta$ $(k=1, 2,\ldots)$. These inequalities easily imply that $\alpha_k > \alpha_1\beta(1+\beta)^{k-1} = C_1(1+\beta)^k = C_1 e^{k\ln(1+\beta)}$ $(k=1,2,\ldots)$. At the same time,

$$\alpha_k \le C_2 \int_{(k-1)r}^{kr} e^{2\varepsilon s} ds = \frac{C_2}{2\varepsilon}\left[e^{2\varepsilon rk} - e^{2\varepsilon rk}e^{-2\varepsilon r}\right] < C_3 e^{2\varepsilon rk} \quad (k=1,2,\ldots),$$

and, thus, by choosing $\varepsilon>0$ so small that $2\varepsilon r < \ln(1+\beta)$, we arrive at a contradiction. Hence, (4.32) is true and, therefore, $\varphi(r,\lambda)$ is real, i.e., λ is real.

■

Theorem 4.5. If $\tilde{q}(t) \xrightarrow[t\to\infty]{} 0$, *then the space of characters coincides with the spectrum of equation (4.1) with the boundary condition* $y'(0)=0$, *i.e., this space consists of all nonnegative* λ *and of at most countably many negative* λ *bounded from below and having at most one limiting point whose role may be played only by zero.*

Proof. First, we note that if $\tilde{q}(t) \xrightarrow[t\to\infty]{} 0$, then $\tilde{\mu}(t)$ grows slower than any exponential function. Indeed, $\tilde{\mu}(t)$ is a solution of equation $y'' - \tilde{q}(t)\,y = 0$ or $y'' +$

$\sigma^2 y = [\tilde{q}(t) + \sigma^2] y$, where $\sigma > 0$ is an arbitrary number satisfying the boundary conditions $y(0) = 1$ and $y'(0) = 0$. It follows from the last equation and Lemma 4.4 that

$$\tilde{\mu}(t) < \exp\left\{\frac{1}{\sigma}\int_{t_0}^{t}[\tilde{q}(\alpha)+\sigma^2]d\alpha\right\}\exp\left\{-\frac{1}{\sigma}\int_{0}^{t_0}[\tilde{q}(\alpha)+\sigma^2]d\alpha\right\}$$

$$= C_1 \exp\left\{\int_{t_0}^{t}\left[\frac{\tilde{q}(\alpha)}{\sigma}+\sigma\right]d\alpha\right\}.$$

We choose t_0 so large that $\tilde{q}(\alpha) < \sigma^2$ for $\alpha \geq t_0$. Then $\tilde{\mu}(t) < C_1 e^{2\sigma(t-t_0)} = C_2 e^{2\sigma t}$ $(t \geq 0)$. Since σ is arbitrary, the assertion is proved.

The collection of characters consists of those and only those points λ for which $\varphi(t, \lambda) = O(\tilde{\mu}(t))$. By virtue of the already proved result, this means that $\varphi(t, \lambda)$ also grows slower than any exponential function but, in this case, in view of Lemma 4.7, this λ is real. Lemma 4.6 and its corollary describe the structure of the negative part of the set X. It remains to prove that every nonnegative λ lies in X.

It is known ([Gla, Chap. 5, Sec. 55] and [Ber10, Chap. 6, Subsection 2.10]) that, for any $\varepsilon > 0$ and almost all $\lambda \geq 0$ (with respect to the spectral measure), the following inequality is true $|\varphi(t, \lambda)| \leq C(\lambda, \varepsilon)|t|^{1/2+\varepsilon}$ $(t \in [0, \infty))$. On the other hand, if $0 \leq \tilde{q}(t) \not\equiv 0$, then it follows from (4.11) that $\tilde{\mu}'(t) \geq a > 0$ and, therefore, $\tilde{\mu}(t) \geq at$ $(t \in [0, \infty))$. Thus, for the indicated values of λ and sufficiently small $\varepsilon > 0$, we have $\varphi(t, \lambda) = O(\tilde{\mu}(t))$, i.e., these λ lie in X. At the same time, the collection of these λ is dense in $[0, \infty)$ and X, as the space of maximal ideals, is closed in the ordinary topology of the complex plane. Hence, $[0, \infty) \subset X$. In the case where $\tilde{q}(t) \equiv 0$, the last inclusion is obvious.

■

Corollary 4.3. *For all functions $\varphi(t, \lambda)$, where λ belongs to the spectrum of equation (4.1) with the boundary condition $y'(0) = 0$, the following inequality holds:*

$$|\varphi(t, \lambda)| \leq \tilde{\mu}(t),$$

where $\tilde{\mu}(t)$ is a solution of the equation $y'' - [\operatorname{Var}_t^{\infty} q]\, y = 0$ satisfying the initial conditions $\tilde{\mu}(0) = 1$ and $\tilde{\mu}'(0) = 0$.

Proof. This statement follows from the fact that each character of a hypercomplex system is bounded by one.

■

By applying Theorems 3.2 and 3.3 from Chapter 1 to the hypercomplex system $L_1([0,\infty), \mu^2(t), dt)$, we obtain the well-known Plancherel theorem and the inversion formula for the Sturm–Liouville equation (although for a fairly poor set of potentials; this fact serves as an additional explanation of the necessity of generalization of the concept of hypercomplex systems; in this connection, see [Vai6], [Vai7]). The duality Theorem 3.5, generally speaking, is not true for the hypercomplex system $L_1([0,\infty),$ $\mu^2(t), dt)$.

4.5. Survey of Related Results. Chebli [Che1]–[Che3] and Achour and Trimeche [AcT2] studied hypercomplex systems associated with the operator

$$Ly = -\frac{1}{A(t)}\frac{d}{dt}\left(A(t)\frac{dy}{dt}\right) \tag{4.33}$$

in the space $H = L_2([0,\infty), A(t)dt)$, where the function $A(t)$ satisfies the following conditions:

1) $A(0) = 0$ and $A(t) > 0$ for $t > 0$;

2) in a neighborhood of zero, $A'(t)/A(t) = (\alpha/t) + B(t)$, where $\alpha > 0$, $B \in C^\infty(\mathbb{R})$;

3) the function $A(t)$ is increasing and $A'(t)/A(t)$ is decreasing as t approaches ∞.

Chebli [Che1] showed that the operator L is self-adjoint in the region $\mathfrak{D}_L = \{y \in H \mid Ly \in H \text{ and } \lim\limits_{t\to\infty} A(t)y'(t) = 0\}$ and has a simple spectrum whose continuous part is $[\rho^2, \infty)$, where

$$\rho = \lim_{t\to\infty}\frac{1}{2}\frac{A'(t)}{A(t)}.$$

The eigenfunctions $\varphi(t,\lambda)$ of the problem $Ly = \lambda y$, $y(0) = 1$, and $y'(0) = 0$ are equibounded, i.e., $|\varphi(t,\lambda)| \le 1$, and the generalized translation operators R_s associated with the operator L by the method of Levitan and Povzner preserve positivity (this result follows from the maximum principle established in [Wei]). It is not difficult to show that the generalized translation operators R_s satisfy all conditions of Theorem 2.1 in Chapter 1. Thus, $L_1([0,\infty), A(t)dt)$ is a commutative Hermitian hypercomplex system with basis unity $e = 0$ satisfying the condition of separate continuity. In [Che2], it was proved that the operator L is unitarily equivalent to the operator of multiplication by an independent variable in the space $L_2(\hat{Q}, \hat{m})$, where $\hat{m}$ is a Plancherel measure, and an analog of the Paley–Wiener theorem for the Fourier transformation over the characters of the hypercomplex system $L_1([0,\infty), A(t)dt)$ was established (see also [Fle]). For the further development of harmonic analysis for operator (4.33), see [AcT2].

Note that the radial part of the Laplace–Beltrami operator acting in noncompact Riemannian symmetric spaces of rank 1 (i.e., in the Euclidean spaces and Riemannian noncompact-type symmetric spaces of rank I) has the form (4.33) with the function $A(t)$ satisfying conditions 1)–3). In particular, the case $A(t) = t^{2\alpha+1}$ leads to the Hankel transformation (this case was already analyzed in [Del2] and [Del3]).

The operator L given by (4.33) on the compact set $[0, \pi/2]$, where $A(t) = (\sin 2t)^{2\alpha+1} B(t)$ and $B(t) \in C^\infty([0, \pi/2])$ is a strictly positive function satisfying the conditions: 1) $B'(0) = 0$, $B(t) = B((\pi/2) - t)$; 2) $A(t)$ is an increasing function in the interval $(0, \pi/2]$; 3) $A'(t)/A(t)$ decreases in the interval $(0, \pi/2)$, was studied in [AcT2]. The relevant generalized translation operators satisfy all conditions of Theorem 2.1 in Chapter 1 and $L_1([0, \pi/2], A(t)dt)$ is an Hermitian hypercomplex system with basis unity $e = 0$ satisfying the condition of separate continuity. If $B(t) = 2^{\alpha-\beta+1} \sin^{2\beta-2\alpha} t$, then the Jacobi polynomials $R_k^{(\alpha,\beta)}(x)$ normalized by the condition $R_k^{(\alpha,\beta)}(1) = 1$ are the eigenfunctions of the problem under consideration (the change of variables $x = \cos 2t$ transforms the corresponding equation into the ordinary differential equation for the Jacobi polynomials).

If, in (4.33), we set $A(t) = (2\alpha+1)/t$ $(\alpha > -1/2)$ then the eigenfunctions of the boundary-value problem

$$Ly(t) = \lambda^2 y(t), \quad y(0) = 1, \quad y(1) = 0, \quad y'(0) = 0 \quad (t \in [0,1]) \tag{4.34}$$

are the *Fourier–Bessel functions* $J_k^{(\alpha)}(t) = j_\alpha(\lambda_k t)$, where $0 < \lambda_1 < \ldots < \lambda_k < \ldots$ is the set of positive zeros of the normalized Bessel function $j_\alpha(t) = 2^\alpha \Gamma(\alpha+1) t^{-\alpha} J_\alpha(t)$, written in the order of increasing. It is known that the Fourier–Bessel functions form a complete orthonormalized system in $L_2([0,1], t^{2\alpha+1}dt)$. Consider the functions $\tilde{J}_k^{(\alpha)}(t) = J_k^{(\alpha)}(t)/J_0^{(\alpha)}(t)$. The functions $\tilde{J}_k^{(\alpha)}(t)$ are solutions of the boundary-value problem (4.34) with $A(t) = [J_0^{(\alpha)}(t)]^2 t^{2\alpha+1}$ and the relevant eigenvalues are $\lambda^2 = \lambda_k^2 - \lambda_0^2$. It is not difficult to calculate that $\tilde{J}_k^{-1/2}(t) = R_k^{(-1/2,-1/2)}(\cos \pi t)$ and $\tilde{J}_k^{(1/2)}(t) = R_k^{(1/2,1/2)}(\cos \pi t)$ (here $R_k^{(\alpha,\beta)}(t)$ are normalized Jacobi polynomials with $R_k^{(\alpha,\beta)}(1) = 1$). In other cases, $\tilde{J}_k^{(\alpha)}(t)$ are not reduced to orthogonal polynomials. Generalized translation operators associated with problem (4.34) for $A(t) = [J_0^{(\alpha)}(t)]^2 t^{2\alpha+1}$ were studied in Markett [Mar4]. These generalized translation operators have the form

$$R_t f(s) = \int_0^1 f(r) K^{(\alpha)}(t,s,r) [J_0^{(\alpha)}(r)]^2 r^{2\alpha+1} dr.$$

For $\alpha = n + 1/2$ $(n = -1, 0, 1, \dots)$, the explicit form of the kernel $K^{(\alpha)}(t, s, r)$ was found. If $\alpha \in \{1/2, 3/2, 5/2\}$, the kernel $K^{(\alpha)}(t, s, r)$ is expressed in terms of elementary functions, and the corresponding generalized translation operators preserve positiveness. This implies that, for $\alpha \in \{1/2, 3/2, 5/2\}$, $L_1([0, 1], [J_0^{(\alpha)}(t)]^2 t^{2\alpha+1} dt)$ is an Hermitian hypercomplex system with basis unity $e = 0$ satisfying the condition of separate continuity, and the functions $\tilde{J}_k^{(\alpha)}(t)$ are characters of this hypercomplex system.

In [CoS3] (see also [CoS4]), Connett and A. Schwartz studied the properties of the characters and the Plancherel measure of hypergroups generated by equation (4.33) on the segment $[0, \pi]$ with the boundary conditions $y'(0) = 0$, $\lim\limits_{y \to \pi-} [C_1 y(t) + C_2 A(t) y'(t)] = 0$ $(C_1^2 + C_2^2 \neq 0)$. Here, the function $A(t)$ satisfies the following conditions: (i) $A(t) > 0$; (ii) $A(t) \in C^1(0, \pi)$; (iii) for some $\alpha, \beta \in \mathbb{R}$, there exists positive limits $\lim\limits_{t \to 0+} t^{-2\alpha-1} A(t)$ and $\lim\limits_{t \to 0+} (\pi - t)^{-2\beta-1} A(t)$. This class of hypergroups, called by the authors "Jacobi-type hypergroups," includes hypergroups associated with Jacobi polynomials and Fourier–Bessel functions; it also includes hypergroups obtained for ultraspherical polynomials by perturbation of a differential equation ([CoS5]).

Generalized translation operators associated with the regular Sturm–Liouville problem

$$Ly = -y'' + q(t)y = \lambda y \quad (0 \leq t \leq \pi), \tag{4.35}$$

$$y'(0) = hy(0), \quad y'(\pi) = h_1 y(\pi), \tag{4.36}$$

were studied by Levitan [Lev9] and Vainerman [Vai2], [Vai3] (see also Section 3 in Chapter 3). Vainerman [Vai2], [Vai3] showed that these generalized translation operators can be associated with a real hypercomplex system with compact basis $[0, \pi]$ defined in the sense of Subsection 5.5 in Chapter 1. The hypercomplex system with discrete basis dual to this system is determined by the structure constants

$$c_{ij}^k = \int_0^\pi \frac{\varphi(r, \lambda_i)\varphi(r, \lambda_j)\varphi(r, \lambda_k)}{\|\varphi(\cdot, \lambda_k)\|_2^2} dr,$$

where $i, j, k \in \mathbb{N}_0$ and $\varphi(r, \lambda_k)$ is an eigenfunction of problem (4.35), (4.36), and by the Plancherel measure $\|\varphi(\cdot, \lambda_k)\|_2^{-2}$.

By virtue of the theorem on approximation, linear combinations of eigenfunctions are uniformly dense in $C([0, \pi])$. The application of the duality principle leads to the following version of the inverse problem for the Sturm–Liouville operator on a segment: The parameters $q(t)$, h, and h_1 of problem (4.35), (4.36) are uniquely recovered from the "spectral data" c_{ij}^k and μ_k [Vai3]. It is interesting to clarify under what conditions on the potential these generalized translation operators preserve positiveness.

Trimeche [Tri1], [Tri2] and Markett [Mar2] considered the operator

$$L = \frac{1}{A(t)}\frac{d}{dt}\left(A(t)\frac{d}{dt}\right) + q(t)$$

on the semiaxis and on a finite interval and studied generalized translation operators associated with it; here, $A(t) = t^{2\alpha+1}B(t)$ $(\alpha > -1/2)$, $B(t) \in C^\infty(\mathbb{R})$ is an even strictly positive function, and $q(t) \in C^\infty(\mathbb{R})$. A particular case where $B(t) \equiv 1$ was investigated by Markett [Mar1], [Mar2], Braaksma [Bra], and Braaksma and de Snoo [BrS]. In particular, Braaksma and de Snoo [BrS] established sufficient conditions for the corresponding generalized translation operators to preserve positiveness, and Markett [Mar2] obtained estimates for the norms of these generalized translation operators.

If one takes $B(t) \equiv 1$ and $q_1(t) = t^2$ or

$$q_2(t) = \left(\frac{1}{4} - \alpha^2\right)(t^{-2} - \sinh^2 t) + \left(\frac{1}{4} - \beta^2\right)\cosh^{-2} t \quad \left(\alpha \geq \beta \geq -\frac{1}{2}\right),$$

then the eigenfunctions of the problem $Ly = \lambda y$, $y(0) = 1$, $y'(0) = 0$, are, respectively, the normalized Wittaker functions $\varphi_1(t, \lambda) = t^{-\alpha-1}M_{\lambda^2/4, \alpha/4}(t^2)$ and normalized Jacobi functions

$$\varphi_2(t, \lambda) = \left(\frac{\sinh t}{t}\right)^{(\alpha+1)/2}(\cosh t)^{(\beta+1)/2}\varphi_\lambda^{(\alpha,\beta)}(t).$$

Generally speaking, unbounded generalized translation operators associated with the equation

$$\frac{d}{dt}\left(p(t)\frac{dy}{dt}\right) + q(t)y = \lambda r(t)y,$$

where $p(t)$ and $r(t)$ are continuous positive functions on the semiaxis, were studied by Vainerman [Vai6], [Vai7]. For these generalized translation operators, an unbounded Hilbert algebra was constructed (see the cited works), and the Plancherel theorem and inversion formula were established for the generalized Fourier transformation related to these generalized translation operators.

3. ELEMENTS OF LIE THEORY FOR GENERALIZED TRANSLATION OPERATORS

As in the case of topological groups, there are two approaches to the investigation of generalized translation operators—global and infinitesimal. The first one includes the theory of representations, harmonic analysis, the theory of almost periodic functions, etc. For groups, the second approach reduces to the construction of Lie theory, i.e., to the analysis of, generally speaking, nonlinear law of multiplication in a certain neighborhood of the identity element of the Lie group in terms of a linear object, i.e., in terms of the Lie algebra. In the present chapter, we construct elements of Lie theory for generalized translation operators. It is worth noting that the Lie theory of generalized translation operators is far from being complete.

In the first section we show that, as in the case of Lie groups, generalized translation operators R_s ($s \in Q$, Q is a smooth manifold) are associated with infinitesimal operators (generators)

$$(R_\alpha f)(t) = D_s^{(\alpha)}(R_s f)(t)\big|_{s=e}$$

$$\left(\alpha = (\alpha_1, \ldots, \alpha_n);\ f \in C^\infty(Q),\ D_s^{(\alpha)} = \frac{\partial^{|\alpha|}}{\partial s_1^{\alpha_1} \ldots \partial s_n^{\alpha_n}}\right).$$

The algebra generated by all generators of generalized translation operators is an analog of the universal enveloping algebra of the Lie algebra. However, unlike the case of Lie groups, for generalized translation operators, there is no general concept similar to the concept of Lie algebras. This is explained by the fact that generators corresponding to generalized translation operators may be degenerate and nondegenerate generators satisfy much more general relations than in the case of Lie algebras. Therefore, it is interesting to analyze the problem of selection of various special classes of generalized translation operators for which it is possible to construct meaningful Lie theory.

In the second section, we describe three classes of generalized translation operators of the indicated type associated with Lie groups: Delsarte generalized translation operators, Delsarte-type generalized translation operators, and generalized translation operators associated with the hypercomplex system $L_1(G, H)$ with basis $Q = G//H$, where G is a Lie group and H is its compact subgroup.

In the third section, we present another point of view on the generators of one-dimensional commutative hypercomplex systems. Namely, the first Nth-order nonzero generator X_N of the one-dimensional hypercomplex system $L_1(Q, m)$ ($Q = [a, b]$, $-\infty \leq a < b \leq \infty$) is associated with a self-adjoint operator X acting in $L_2(Q, m)$ and such that $(Xf)(t) = (X_N f)(t)$ for all $f \in C^\infty(Q) \cap \mathcal{D}(X) \neq \varnothing$. By using the decomposition of the operator X in generalized eigenfunctions $\chi(t, \tau)$ ($t \in Q$; $\tau \in \hat{Q}$), we determine the relevant collection of generalized translation operators by the formula $(R_s f)(t) = (\chi(s, X)f)(t)$ ($t, s \in X$), where $\chi(t, \tau)$ regarded as functions of t are characters of the hypercomplex system $L_1(Q, m)$ and, as functions of τ, they are characters of the dual hypercomplex system. In the third section, we also establish an "infinitesimal analog of Pontryagin's duality," i.e., for a given generator X of the hypercomplex system $L_1(Q, m)$, we construct the "dual" generator of the hypercomplex system $L_1(\hat{Q}, \hat{m})$, and vice versa.

1. Basic Concepts

It is well known that the Lie agebra $\mathfrak{g}$ of a Lie group G, i.e., the tangent space at the unit element of the group endowed with the operation of commutation generated by the operation of multiplication in the group G, is its infinitesimal object. The Lie algebra $\mathfrak{g}$ is also isomorphic to the algebra of the first-order group generators which are differential operators of the first order (with respect to the ordinary operation of commutation of operators). The notion of generator is one of the most important notions in the infinitesimal theory of generalized translation operators (the precise definition of generators of generalized translation operators is given in what follows).

For arbitrary generalized translation operators R_s ($s \in Q$), the situation is more complicated than in the case of Lie groups: First, in many interesting cases (e.g., $Q = G//H$, where G is a Lie group and H is its compact subgroup), the set Q is, generally speaking, not a manifold. Second, all generators of generalized translation operators under consideration may be degenerate up to a certain fixed order and nondegenerate generators are, generally speaking, not differential operators. Finally, the relationships between nondegenerate generators are, in general, more complicated than in the case of Lie algebras. This explains the necessity of the investigation of nonreduced generalized translation operators on smooth manifolds.

Let us show that the algebra generated by all generators is the infinitesimal object for generalized translation operators of this type and the subalgebra generated by all right generators is the infinitesimal object for reduced generalized translation operators (in the case of Lie groups, this means that, instead of Lie algebras, the role of the infinitesimal object is played by universal enveloping algebras of the corresponding Lie algebras).

We also study general properties of the generators of generalized translation operators and develop a pure algebraic approach to the infinitesimal theory of formal generalized translation operators. Subsection 3.2 deals with the theory of topological bialgebras. It is included in the book because constructions presented there are helpful in understanding the algebraic nature of the objects studied in Section 3 (thus, it is shown that generalized translation operators can be regarded as coassociative comultiplication with right counit). At the end of the section we present necessary facts from the theory of topological vector spaces. In our presentation, we mainly follow the survey [Lit3].

1.1. Hypergroup Algebra of Infinitely Differentiable Generalized Translation Operators. In this subsection, we show that every smooth family of (nonreduced; see Section 1 in Chapter 2) generalized translation operators in $C^\infty(Q)$ (Q is a smooth manifold) admits a unique description by endowing $\mathcal{D}(Q) = (C^\infty(Q))'$ with the structure of associative topological algebra with right identity δ_e ($e \in Q$, $R_e = I$). This result is similar to Theorem 2.1 in Chapter 1 (also see remarks to this theorem). In this case, reduced generalized translation operators are characterized with the help of a subalgebra of the algebra $\mathcal{D}(Q)$. We recall that the space $\mathcal{D}(Q)$ is equipped with the Mackey topology.

Consider a family of generalized translation operators $C^\infty(Q) \ni f \mapsto R_s f(t) \in C^\infty(Q)$ given in the space $C^\infty(Q)$, ($s \in Q$; we do not assume that the generalized translation operators R_s are reduced). The family of generalized translation operators R_s is called *infinitely differentiable* if the correspondence $f(t) \mapsto (R_s f)(t)$ is a continuous mapping from $C^\infty(Q)$ into $C^\infty(Q \times Q)$. This implies that, for any $F \in \mathcal{D}(Q)$ and $f \in C^\infty(Q)$, the functions $Q \ni s \mapsto \langle F, R_s f\rangle$ and $Q \ni s \mapsto \langle F, L_s f\rangle$ belong to the space $C^\infty(Q)$.

For any $F \in \mathcal{D}(Q)$, we introduce operators $R(F)$ and $L(F)$ acting in the space $C^\infty(Q)$ by setting

$$L(F)f(t) = \langle F, R_t f\rangle \quad \text{and} \quad R(F)f(t) = \langle F, L_t f\rangle \quad (t \in Q). \tag{1.1}$$

for all functions $f \in C^\infty(Q)$. The operators $L(F)$ and $R(F)$ are continuous in view of the continuity of R_t and L_t as operators in $C^\infty(Q)$.

We introduce the operation of convolution in $\mathcal{D}(Q)$ as follows: For any $F, G \in \mathcal{D}(Q)$ and $f \in C^\infty(Q)$, we set $\langle F * G, f\rangle = \langle F, R(G)f\rangle$. Since the operator $R(G)$ is continuous, we conclude that $F * G \in \mathcal{D}(Q)$.

Since elements of the space $\mathcal{D}(Q)$ satisfy an analog of the Fubini theorem (see [Rha]), we can write

$$\langle F * G, f\rangle = \langle F^t, \langle G^s, (R_s f)(t)\rangle\rangle = \langle G, L(F)f\rangle$$

(as in Section 2 in Chapter 1, the superscript of the functional denotes the variable upon which it acts). As a result, we obtain

$$\langle F * G, f\rangle = \langle F, R(F)f\rangle = \langle G, L(G)f\rangle \quad (F, G \in \mathcal{D}(Q);\ f \in C^\infty(Q)). \tag{1.2}$$

Since the operator $F \mapsto F * G$ in $\mathcal{D}(Q)$ is adjoint to the continuous operator $R(G)$ in $C^\infty(Q)$, the operation of convolution is separately continuous in $\mathcal{D}(Q)$. In view of the fact that the Mackey topology is stronger than the weak topology, this operation is also separately continuous in the topology $\sigma(\mathcal{D}(Q), C^\infty(Q))$. By virtue of the result presented at the end of Subsection 1.6, the convolution $F * G$ is continuous in $\mathcal{D}(Q)$.

The mapping $\mathcal{D}(Q) \ni F \mapsto R(F)$ *is a representation of the algebra* $\mathcal{D}(Q)$ *and the mapping* $\mathcal{D}(Q) \ni F \mapsto L(F)$ *is its antirepresentation.* Indeed, for any $F, G \in \mathcal{D}(Q)$, $f \in C^\infty(Q)$ we have

$$\begin{aligned}
(R(F)R(G)f)(t) &= \langle F, L_t R(G)f\rangle = \langle L_t^* F, R(G)f\rangle \\
&= \langle (L_t^* F)^s, \langle G^r, (L_s f)(r)\rangle\rangle = \langle G^r, \langle F^s, (L_t^s(L_s f))(r)\rangle\rangle \\
&= \langle G^r, \langle F^s, L_s^r(L_t f)(r)\rangle\rangle = \langle F^s, \langle G^r, (L_s^r(L_t f))(r)\rangle\rangle \\
&= \langle F, R(G)L_t f\rangle = \langle F * G, L_t f\rangle = (R(F * G)f)(t),
\end{aligned}$$

where $t \in Q$ and L_t^* is the operator adjoint to the operator L_t.

Similarly,

$$(L(F)L(G)f)(t) = (L(G * F)f)(t).$$

■

Theorem 1.1 (Litvinov [Lit1], [Lit3]). *The space* $\mathcal{D}(Q)$ *with operation* $*$ *is an associative topological algebra with continuous multiplication and right identity* δ_e *which is a multiplicative functional on* $C^\infty(Q)$.

Conversely, suppose that $\mathcal{D}(Q)$ *is endowed with the structure of an associative algebra and there is a multiplicative functional which is the right identity in* $\mathcal{D}(Q)$. *If the multiplication in* $\mathcal{D}(Q)$ *is separately continuous in some locally convex topology compatible with the duality* $\langle \mathcal{D}(Q), C^\infty(Q)\rangle$, *then this multiplication is continuous in the Mackey topology on* $\mathcal{D}(Q)$. *In this case, there exists a unique infinitely differentiable family of generalized translation operators* R_s $(s \in Q)$ *such that the multiplication in* $\mathcal{D}(Q)$ *coincides with the relevant operation of convolution.*

Proof. Let us prove the first assertion of the theorem. The fact that the algebra $\mathcal{D}(Q)$ is associative is a consequence of the following chain of equalities which is true for all $F, G, H \in \mathcal{D}(Q)$ and $f \in C^\infty(Q)$:

$$\langle (F * G) * H, f\rangle = \langle F * G, R(H)f\rangle = \langle F, R(G)R(H)f\rangle$$

$$= \langle F, R(G * H)f\rangle = \langle F * (G * H), f\rangle.$$

The continuity of multiplication in $\mathcal{D}(Q)$ has already been proved. Since

$$\langle F * \delta_e, f\rangle = \langle F, R(\delta_e)f\rangle = \langle F^t, \langle \delta_e, L_t f\rangle\rangle$$

$$= \langle F, (L_t f)(e)\rangle = \langle F, R_e f\rangle = \langle F, f\rangle,$$

the multiplicative functional δ_e is the right identity of the algebra $\mathcal{D}(Q)$.

Let us prove the converse assertion. Suppose that $\mathcal{D}(Q)$ is an associative algebra with multiplication separately continuous in some topology compatible with the duality $\langle \mathcal{D}(Q), C^\infty(Q)\rangle$ and right identity which is a multiplicative functional on $C^\infty(Q)$. Since multiplication in $\mathcal{D}(Q)$ is separately continuous in a topology compatible with the duality $\langle \mathcal{D}(Q), C^\infty(Q)\rangle$, it is separately continuous in the weak topology on $\mathcal{D}(Q)$. Since $C^\infty(Q)$ is a Fréchet space, this multiplication is also continuous in the strong topology $\beta(\mathcal{D}(Q), C^\infty(Q))$ on $\mathcal{D}(Q)$, which coincides with the Mackey topology in the case under consideration. Thus there exists a continuous mapping $\mathcal{D}(Q) \hat{\otimes} \mathcal{D}(Q) \to \mathcal{D}(Q)$. Let

$$\Delta : C^\infty(Q) \to C^\infty(Q) \hat{\otimes} C^\infty(Q) = C^\infty(Q \times Q)$$

be the dual mapping. Since $C^\infty(Q)$ and $C^\infty(Q \times Q)$ are reflexive, the mapping Δ is continuous. For any $s \in Q$, we set $\langle F, (R_s f)(t)\rangle = \langle F * \delta_s, f\rangle$, i.e., we define the operator R_s in $C^\infty(Q)$ as an operator adjoint to the operator of right multiplication by δ_s in $\mathcal{D}(Q)$. Since the mapping $f(t) \mapsto u(t, s) = (R_s f)(t)$ coincides with Δ, the family of operators R_s is infinitely differentiable.

Let us show that the operators R_s satisfy the axioms of generalized translation operators. First, we demonstrate that there exists a point $e \in Q$ such that the right identity in $\mathcal{D}(Q)$ is equal to δ_e. Indeed, let $\Phi \in \mathcal{D}(Q)$ be a multiplicative functional and let K be its support. We show that there exists a point $t_0 \in Q$ such that $\Phi = \delta_{t_0}$. Suppose that K contains at least two different points $t, s \in K$. Since Q is the Hausdorff space, there are disjoint neighborhoods U_t and U_s of the points t and s, respectively, and functions $f, g \in C^\infty(Q)$ such that $\operatorname{supp} f \subset U_t$, $\operatorname{supp} g \subset U_s$ and $\Phi(f)\Phi(g) \neq 0$. However, in this case, by virtue of the multiplicativity of Φ, we get

$$0 = \langle \Phi, fg \rangle = \Phi(f)\Phi(g) = 0$$

because $fg \equiv 0$. Consequently $K = \{t_0\}$. In a neighborhood U of the point t_0, we introduce coordinates φ_U such that $\varphi_U(t_0) = 0$. By virtue of the well-known theorem, in these coordinates, the functional Φ can be rewritten as

$$\langle \Phi, f \rangle = \sum_{|\alpha| \le k} c_\alpha \left\langle \delta_0^{(\alpha)}, f \circ \varphi_U^{-1} \right\rangle$$

where $k < \infty$, $\alpha = (\alpha_1, \ldots, \alpha_n)$ is the multiindex, n is the dimensionality of the manifold Q, and $c_\alpha \in \mathbb{C}$.

A functional of the indicated form is multiplicative only in the case where

$$c_\alpha = \begin{cases} 0, & \alpha \neq (0, \ldots, 0), \\ 1, & \alpha = (0, \ldots, 0). \end{cases}$$

Since the right identity Φ_e of the algebra $\mathcal{D}(Q)$ is a multiplicative functional, we conclude that $\Phi_e = \delta_e$, where e is a point from Q. This implies that R_e is the identity operator.

Let L_t be an operator in $C^\infty(Q)$ adjoint to the operator of left multiplication by δ_t, i.e., $\langle F, L_t f \rangle = \langle \delta_t * F, f \rangle$. Let us show that $(L_t f)(s) = (R_s f)(t)$. Indeed,

$$(R_s f)(t) = \langle \delta_t, R_s f \rangle = \langle \delta_t * \delta_s, f \rangle$$

$$= \langle \delta_s, L_t f \rangle = L_t f(s) \quad (t, s \in Q;\ f \in C^\infty(Q)),$$

whence it follows that, for any $t, s, r \in Q$, we have

$$\left(R_r^t(R_s f)\right)(t) = \langle (\delta_t * \delta_r) * \delta_s, f \rangle = \langle \delta_t * (\delta_r * \delta_r), f \rangle$$

$$= \langle \delta_r * \delta_s, L_t f \rangle = \langle (\delta_r * \delta_s)^\lambda, (R_\lambda f)(t) \rangle$$

$$= \left\langle \delta_r^\lambda, \left(R_s^\lambda (R_\lambda f)\right)(t) \right\rangle = \left(R_s^r (R_r f)\right)(t) \quad (f \in C^\infty(Q)),$$

which yields the associativity of the operators R_s $(s \in Q)$. Hence, R_s is a family of infinitely differentiable generalized translation operators. Since $\Delta f(t, s) = (R_s f)(t)$, the operation of multiplication in $\mathcal{D}(Q)$ coincides with the operation of generalized convolution defined for the family of generalized translation operators R_s according to (1.2).

■

Let $\tilde{C}^\infty(Q) = L_e(C^\infty(Q))$ be the principal space of functions where the reduced generalized translation operators act. The projector $\mathcal{D}(Q) \ni F * \delta_e \mapsto F \in \mathcal{D}(Q)$ is adjoint to the projector L_e. Denote by $\tilde{\mathcal{D}}(Q)$ the image of this operator.

The space $\tilde{\mathcal{D}}(Q)$ is a close subalgebra of $\mathcal{D}(Q)$ with two-sided identity δ_e and, moreover, $\tilde{\mathcal{D}}(Q)$ can be identified with the quotient algebra (endowed with the quotient topology) of the algebra $\mathcal{D}(Q)$ with respect to the ideal orthogonal to the principal space $\tilde{C}^\infty(Q)$.

Indeed, the first part of this statement is evident. Let

$$I = \{F \in \mathcal{D}(Q) \mid \langle F, \tilde{f} \rangle = 0 \quad \text{for all} \quad \tilde{f} \in \tilde{C}^\infty(Q)\}.$$

Since, for any $F \in \mathcal{D}(Q)$, we have $\langle \delta_e * F, f \rangle = \langle F, L_e f \rangle$ $(t, s \in Q;\ f \in C^\infty(Q))$, we conclude that $F \in I$ if and only if $\delta_e * F = 0$. Let us show that I is a two-sided ideal. Indeed, for any $F \in I$ and $G \in \mathcal{D}(Q)$, we can write

$$\delta_e * (F * G) = (\delta_e * F) * G = 0$$

and

$$\delta_e * (G * F) = \delta_e * G * \delta_e * F = 0$$

because δ_e is the right identity in $\mathcal{D}(Q)$. The equality $\tilde{\mathcal{D}}(Q) = \mathcal{D}(Q)/I$ is obvious.

■

Hence, $\tilde{\mathcal{D}}(Q)$ coincides with the space dual to $\tilde{C}^\infty(Q)$. In what follows, the algebra $\mathcal{D}(Q)$ is called the *hypergroup algebra* and $\tilde{\mathcal{D}}(Q)$ is called the *hypergroup algebra of reduced generalized translation operators.* It should be noted that if $\tilde{C}^\infty(Q)$ does not coincide with $C^\infty(Q)$, then the hypergroup algebra $\mathcal{D}(Q)$ is noncommutative even in the case where the generalized translation operators R_s $(s \in Q)$ are commuting (see Example 2 in Subsection 3.3). In this case, the hypergroup algebra of reduced generalized translation operators is commutative.

Note that we are interested in the case of reduced generalized translation operators and, therefore, focus our attention on the algebra $\tilde{\mathcal{D}}(Q)$ (although it is clear that $\mathcal{D}(Q)$

contains information both about $\tilde{\mathcal{D}}(Q)$ and about the process of reduction, the algebra $\tilde{\mathcal{D}}(Q)$ is much simpler).

1.2. Topological Bialgebras. In the case under consideration, topological bialgebras play the same role as Hilbert bialgebras in the theory of quantized hypercomplex systems. In what follows, we are mainly interested in the infinitesimal object for generalized translation operators and, therefore, only outline problems related to the theorem on realization and duality. Our attention is mainly focused on the pure algebraic scheme.

Let $\mathcal{F}$ be a locally convex space and let $\mathcal{F}'$ be its dual space. The spaces $\mathcal{F}$ and $\mathcal{F}'$ are equipped with the Mackey topologies $\tau(\mathcal{F}, \mathcal{F}')$ and $\tau(\mathcal{F}', \mathcal{F})$, respectively. If $\mathcal{F}$ and $\mathcal{F}'$ are associative topological algebras with separately continuous multiplication, then the pair $\mathfrak{U} = (\mathcal{F}, \mathcal{F}')$ is called a *topological bialgebra.* Multiplication in the algebra $\mathcal{F}$ is denoted by $\varphi \dot{\cap} \psi$ $(\varphi, \psi \in \mathcal{F})$ and multiplication in the algebra $\mathcal{F}'$ is denoted by $F \dot{\cup} G$ $(\varphi, \psi \in \mathcal{F}')$.

Let $\mathfrak{U} = (\mathcal{F}, \mathcal{F}')$ be a topological bialgebra and let $\mathcal{F}$ be a commutative Arens–Michael algebra (i.e., $\mathcal{F}$ is a complete Hausdorff topological space such that the inequality $p(\varphi \dot{\cap} \psi) \leq p(\varphi)p(\psi)$ holds for all seminorms p from some defining system of seminorms and any $\varphi, \psi \in \mathcal{F}$).

We denote the subset of $\mathcal{F}'$ of all multiplicative functionals on $\mathcal{F}$ by $Q = Q(\mathcal{F})$ and equip it with the induced topology. Unlike the theory of Banach algebras, Q is not necessarily locally compact. As is known (see, e.g., [Khe]), the *Gelfand transformation* $\varphi \mapsto \langle t, \varphi \rangle = \varphi(t)$ $(\varphi \in \mathcal{F},\ t \in Q)$ maps the space $\mathcal{F}$ into the space $C(Q)$ of continuous functions on Q. Since $\mathcal{F}$ is an Arens–Michael algebra, its radical coincides with the kernel of the Gelfand transformation.

Let $\mathcal{F}$ be semisimple. Then $\mathcal{F}$ is (algebraically) isomorphic to a certain subalgebra $S(Q)$ of the algebra $C(Q)$. For any $t \in Q$, we denote by R_t (L_t) the operator in $S(Q)$ adjoint to the operator of right (left) multiplication by $t \in Q$ in $\mathcal{F}'$.

Assume that, in a topological bialgebra $\mathfrak{U} = (\mathcal{F}, \mathcal{F}')$, $\mathcal{F}$ *is a commutative semisimple Arens–Michael algebra and the algebra* $\mathcal{F}'$ *possesses the right identity which is a multiplicative functional. Then* R_t $(t \in Q)$ *are generalized translation operators.*

Indeed, let $e \in Q$ be a multiplicative functional which is the right identity of $\mathcal{F}'$. It is clear that $R_e = \mathrm{id}$. In view of the fact that

$$(R_s\varphi)(t) = \langle \varphi, t \dot{\cup} s \rangle = (L_t\varphi)(s) \quad (\varphi \in \mathcal{F};\ t, s \in Q),$$

we have $(R_s\varphi)(t) \in S(Q)$ for any fixed t. The associativity of the operators R_s is checked in exactly the same way as in Theorem 1.1. Hence, R_s $(s \in Q)$ are generalized translation operators.

■

If all conditions of the previous assertion are satisfied, then all operators R_s ($s \in Q$) are endomorphisms of the algebra $\mathcal{F}$ if and only if the operation of multiplication in $\mathcal{F}'$ induces the structure of a semigroup in Q. In this case, $R_s f(t) = f(t \cup s)$.

Indeed, assume that all R_s ($s \in Q$) are endomorphisms of the algebra $\mathcal{F}$. Then

$$\langle t \cup s, \varphi \cap \psi \rangle = (R_s(\varphi \cap \psi))(t) = (R_s \varphi)(t)(R_s \psi)(t) = \langle t \cup s, \varphi \rangle \langle t \cup s, \psi \rangle$$

$$(t, s \in Q),$$

whence it follows that $t \cup s = r$ is a multiplicative functional and Q is a semigroup.

By introducing evident modifications in this chain of equalities, we arrive at the converse assertion.

■

To transform the semigroup Q into a group, it suffices to require that the right identity $e \in Q$ of the algebra $\mathcal{F}'$ be two-sided and find an involutive automorphism β : $\mathcal{F} \to \mathcal{F}$ such that $\langle t, \beta(R_t \varphi) \rangle = \langle e, \varphi \rangle$ for any $e \in Q$ and $\varphi \in \mathfrak{F}$.

The topological bialgebra $\mathfrak{U}' = (\mathcal{F}, \mathcal{F}')$ is called *dual* to the topological bialgebra $\mathfrak{U} = (\mathcal{F}, \mathcal{F}')$. If both $\mathcal{F}$ and $\mathcal{F}'$ are commutative semisimple Arens–Michael algebras with identities which are multiplicative functionals on $\mathcal{F}'$ and $\mathcal{F}$, respectively, then, in the space $\hat{Q} \subset \mathcal{F}$ of multiplicative functionals on $\mathcal{F}'$, one can also construct a family of generalized translation operators $R_{\hat{s}}$ ($\hat{s} \in \hat{Q}$) which is naturally regarded as *dual* to R_t ($t \in Q$). The generalized translation operators R_t ($t \in Q$) and $R_{\hat{t}}$ ($\hat{t} \in \hat{Q}$) are commutative. Since $(\mathfrak{U}')' = \mathfrak{U}$, we have $\hat{\hat{Q}} = Q$. It should be noted that, generally speaking, this duality does not coincide with the Pontryagin duality.

Thus, let $\mathcal{F}$ coincide with $C^{\infty}(\mathbb{R})$ with the ordinary multiplication of functions and let $\mathcal{F}' = \mathcal{D}(\mathbb{R})$ be the group algebra of the group $\mathbb{R}$. Then $Q = \mathbb{R}$ and the generalized translation operators R_s ($s \in Q$) coincide with ordinary right translations and $\hat{Q}$ consists of all functions $e^{\lambda s}$ ($\lambda \in \mathbb{C}$). Hence, $\hat{Q}$ can be identified with the group $\mathbb{C}$. The Gelfand transformation on $\mathcal{D}(\mathbb{R})$ coincides with the Fourier–Laplace transformation and maps $\mathcal{D}(\mathbb{R})$ into an algebra of analytic functions on $\mathbb{C}$.

In the case where $\mathfrak{U} = (\mathcal{D}(Q), C^{\infty}(Q))$, the operation of multiplication (generalized convolution) in $\mathcal{D}(Q)$ is automatically continuous and can be extended to a continuous mapping $\mathcal{D}(Q) \hat{\otimes} \mathcal{D}(Q) \to \mathcal{D}(Q)$. The adjoint mapping

$$\Delta \colon C^{\infty}(Q) \to C^{\infty}(Q) \hat{\otimes} C^{\infty}(Q) = C^{\infty}(Q \times Q)$$

is called *comultiplication*.

If $\mathcal{D}(Q)$ possesses the right identity δ_e $(e \in Q)$, then, by Theorem 1.1, the comultiplication Δ coincides with the generalized translation, i.e.,

$$\Delta f(t, s) = R_s f(t).$$

The associativity of multiplication in $\mathcal{D}(Q)$ yields the *coassociativity of comultiplication* Δ, namely,

$$(\mathrm{id} \otimes \Delta)\Delta f = (\Delta \otimes \mathrm{id})\Delta f \quad (f \in C^\infty(Q))$$

or, equivalently, the commutativity of the diagram

$$\begin{array}{ccc} C^\infty(Q) & \xrightarrow{\Delta} & C^\infty(Q \times Q) \\ \Delta \downarrow & & \downarrow \Delta \otimes \mathrm{id} \\ C^\infty(Q \times Q) & \xrightarrow{\mathrm{id} \otimes \Delta} & C^\infty(Q \times Q \times Q) \end{array}.$$

A linear functional $\varepsilon: C^\infty(Q) \to \mathbb{C}$ is called the *right counit* with respect to comultiplication Δ if $(\mathrm{id} \otimes \varepsilon)\Delta = \mathrm{id}$. It is clear that δ_e is the right counit for comultiplication under consideration. It is also clear that the algebra $C^\infty(Q)$ with the ordinary multiplication of functions and coassociative comultiplication with right counit contains the entire body of information about the topological bialgebra $\mathfrak{U} = (\mathcal{D}(Q), C^\infty(Q))$. Similarly, the operation of multiplication of functions in $C^\infty(Q)$ generates the operation of comultiplication in $\mathcal{D}(Q)$ and the algebra $\mathcal{D}(Q)$ with generalized convolution as multiplication and comultiplication generated by the ordinary multiplication of functions in $C^\infty(Q) = (\mathcal{D}(Q))'$ contains all necessary information about the dual topological bialgebra $\mathfrak{U}' = (\mathcal{D}(Q), C^\infty(Q))$. This construction appears to be fairly general and can be successfully applied in the theory of duality of locally compact groups and hypercomplex systems, theory of quantum groups, and other problems of analysis.

1.3. Infinitesimal Object for Generalized Translation Operators. Let us introduce the notion of infinitesimal object for infinitely differentiable generalized translation operators. We give two definitions. In the first definition, the infinitesimal object is understood as a subalgebra (infinitesimal algebra) of the hypergroup algebra. The second definition is based on the notion of generators of generalized translation operators. We establish that these definitions are equivalent and show that the linear space of an infinitesimal algebra can be regarded as the tangent space to the infinite-dimensional manifold $\mathcal{D}(Q)$ at the point δ_e. In the sequel, we always assume that coordinates φ in the neighborhood of the point $e \in Q$ are chosen so that $\varphi(e) = (0, \ldots, 0)$.

Let R_s $(s \in Q)$ be infinitely differentiable generalized translation operators on a smooth manifold Q, let $\mathcal{D}(Q)$ be the corresponding hypergroup algebra, and let $\tilde{\mathcal{D}}(Q)$

be the hypergroup algebra of reduced generalized translation operators. The *infinitesimal algebra* of the generalized translation operators R_s $(s \in Q)$ is defined as a subalgebra $\mathcal{D}_e(Q) \subset \mathcal{D}(Q)$ generated by generalized functions whose supports are concentrated at the unit element $e \in Q$. The subalgebra $\tilde{\mathcal{D}}(Q) = \mathcal{D}_e(Q) \cap \tilde{\mathcal{D}}(Q)$ of the algebra $\mathcal{D}_e(Q)$ is called the *infinitesimal algebra for reduced generalized translation operators.* It is clear that $\tilde{\mathcal{D}}(Q) = \delta_e * \mathcal{D}_e(Q)$. The algebras $\mathcal{D}_e(Q)$ and $\tilde{\mathcal{D}}(Q)$ are regarded as the infinitesimal objects for infinitely differentiable and reduced generalized translation operators, respectively.

We now give another definition of the infinitesimal object.

The *right generator* of order $\alpha = (\alpha_1, \dots, \alpha_n)$ of generalized translation operators R_s $(s \in Q)$ in $C^\infty(Q)$ is defined by the formula

$$(R_\alpha f)(t) = D_s^{(\alpha)}(R_s f)(t)\big|_{s=e} \quad (f \in C^\infty(Q)). \tag{1.3}$$

It is clear that, on the right-hand side of (1.3), $t, s, e \in Q$ mean, respectively, $\varphi(t)$, $\varphi(s)$, and $\varphi(e) = (0, \dots, 0) \in \mathbb{R}^n$, while $f \in C^\infty(Q)$ means $f \circ \varphi^{-1}$, where φ are some fixed coordinates in a certain neighborhood of the point e. However, we prefer the simpler notation and hope that this would not lead to misunderstanding.

The *left generator* is defined similarly:

$$(L_\alpha f)(t) = D_t^{(\alpha)}(R_s f)(t)\big|_{t=e} \quad (f \in C^\infty(Q)). \tag{1.4}$$

The algebra $\mathfrak{L}$ generated by all left generators is regarded as the infinitesimal object for infinitely differentiable generalized translation operators R_s $(s \in Q)$ and the algebra $\mathcal{R}$ generated by all right generators is assumed to be the infinitesimal object for reduced generalized translation operators.

Theorem 1.2 (Litvinov [Lit3]).

(i) The algebras $\mathcal{D}_e(Q)$ and $\mathfrak{L}$ are antiisomorphic.

(ii) The algebras $\tilde{\mathcal{D}}_e(Q)$ and $\mathcal{R}$ are isomorphic.

Proof. Consider the collection of operators $R(F)$ $(F \in \mathcal{D}_e(Q))$. Since $\mathcal{D}_e(Q) \ni F \mapsto R(F)$ is a representation of the algebra $\mathcal{D}_e(Q)$, we conclude that

$$\mathcal{D}_e^R(Q) = \{R(F) \,|\, F \in \mathcal{D}_e(Q)\}$$

is an algebra of operators called the *algebra of infinitesimal right translations.*

In a similar way, we define the algebra of *infinitesimal left translations* as

$$\mathcal{D}_e^L(Q) = \{L(F) \mid F \in \mathcal{D}_e(Q)\}$$

The algebras $\mathcal{D}_e(Q)$ and $\mathcal{D}_e^L(Q)$ are antiisomorphic.

Indeed, assume that an element $F \in \mathcal{D}_e(Q)$ is such that $L(F) = \mathrm{id}$. Then, for any $f \in C^\infty(Q)$, we can write

$$f(e) = \langle \delta_e, L(F)f \rangle = \langle F * \delta_e, f \rangle = \langle F, f \rangle$$

whence it follows that $F = \delta_e$, and the required assertion follows from the fact the mapping $\mathcal{D}_e(Q) \ni F \mapsto L(F) \in \mathcal{D}_e^L(Q)$ is an antirepresentation.

The algebra $\mathcal{D}_e^R(Q)$ is isomorphic to the algebra $\tilde{\mathcal{D}}_e(Q) = \delta_e * \mathcal{D}_e(Q)$. Indeed, in view of the fact that $\delta_e * F = F$ for any $F \in \tilde{\mathcal{D}}_e(Q)$, by assuming that $R(F) = \mathrm{id}$, we obtain

$$f(e) = \langle \delta_e, R(F)f \rangle = \langle \delta_e * F, f \rangle = \langle F, f \rangle,$$

i.e., $F = \delta_e$.

Clearly,

$$R_\alpha = R\left(\delta_e^{(\alpha)}\right) \quad \text{and} \quad L_\alpha = L\left(\delta_e^{(\alpha)}\right). \tag{1.5}$$

Since $\delta_e^{(\alpha)}$ $(\alpha = (\alpha_1, \dots, \alpha_n)$, $\alpha_i \in \mathbb{N} \cup \{0\})$ form a system of generating elements in $\mathcal{D}_e(Q)$, the generators L_α and L_e (R_α and $R_e = \mathrm{id}$, respectively) generate the algebra of infinitesimal left (right) translations $\mathcal{D}_e^L(Q)$ $\left(\mathcal{D}_e^R(Q)\right)$. Consequently, $\mathfrak{L} = \mathcal{D}_e^L(Q)$ and $\mathcal{R} = \mathcal{D}^R(Q)$.

■

We consider several simple examples.

Example 1. If $Q = G$ is a Lie group, the Schwartz theorem implies that the algebra $\mathcal{D}_e(Q) = \tilde{\mathcal{D}}_e(Q)$ is isomorphic to the universal enveloping algebra $U(\mathfrak{g})$ of the Lie algebra $\mathfrak{g}$ of the group G.

By the Poincarè–Birkhoff–Witt theorem (PBW), elements of the form

$$X_{i_1} \dots X_{i_p}$$

$(1 \le i_1 \le \dots \le i_p \le n;\ X_1 \dots X_n$ is a basis in $\mathfrak{g})$ form a basis in $U(\mathfrak{g})$.

Example 2. Let $Q = \mathbb{R}^1$ and let

$$(R_s f)(t) = \frac{1}{2}[f(t+s)+f(t-s)]$$

Then

$$R_1 = R_3 = \ldots R_{2k+1} = 0, \quad R_2 = \frac{d^2}{dt^2}, \quad R_{2k} = (R_2)^k,$$

and the algebra $\mathcal{D}_e^R(Q)$ is commutative and generated by a single generator R_2. The algebra of infinitesimal left translations $\mathcal{D}_e^L(Q)$ is characterized by a more complicated structure. The generators

$$L_n = L_0 \frac{d^n}{dt^n},$$

where

$$(L_0 f)(t) = \frac{1}{2}(f(t) + f(-t))$$

form a basis in $\mathcal{D}_e^L(Q)$. Since $L_{2k+1}L_n = 0$ and $L_{2k}L_n = L_{2k+n}$, we have $L_{2k} = (L_2)^k$ and $L_{2k+1} = (L_2)^k L_1$, whence it follows that L_1, L_2, and L_0 are generating elements in $\mathcal{D}_e^L(Q)$. They satisfy the relations

$$L_1 L_2 = L_1^2 = L_1 L_0 = 0, \quad L_0 L_2 = L_2 L_0 = L_2, \quad L_0^2 = L_0, \quad L_0 L_1 = L_1$$

and, therefore, $\mathcal{D}_e^L(Q)$ is noncommutative although the generalized translation operators R_s are commuting.

Thus, to construct the direct Lie theory of generalized translation operators, it is necessary to describe the infinitesimal algebra $\mathcal{D}_e(Q)$ (or $\tilde{\mathcal{D}}_e(Q)$ if we are interested only in the reduced case), i.e., to find the minimal collection of generating elements and establish relationships between them. The subject of the converse Lie theory can be described as the construction of the "global" object—a family of generalized translation operators—in terms of the relevant infinitesimal object, i.e., in terms of the algebra $\mathcal{D}_e(Q)$ (or $\tilde{\mathcal{D}}_e(Q)$).

As follows from Example 2, to reconstruct the original generalized translation operators in terms of a given infinitesimal object, it is not sufficient to know the algebra $\tilde{\mathcal{D}}_e(Q)$. Indeed, the algebra $\tilde{\mathcal{D}}_e(Q)$ in Example 2 is the infinitesimal algebra of right translations both for ordinary group translations on $\mathbb{R}^1$ and for many other generalized translation operators. However, if we *a priori* know something about the required generalized translation operators (e.g., the method of reduction or the explicit form of generators), then we can reconstruct the family of generalized translation operators for a given algebra $\tilde{\mathcal{D}}_e(Q)$ in a unique way. Thus, local Lie groups can be reconstructed in a unique way for given Lie algebras (also see Sections 2 and 3).

In numerous examples, $\mathcal{D}_e(Q)$ coincides with the set of all functionals whose supports are concentrated at the point e and, therefore, up to the end of this section, we assume that $\mathcal{D}_e(Q)$ possesses the indicated property, i.e.,

$$\mathcal{D}_e(Q) = \left\{ \sum_{|\alpha| \le k} c_\alpha \delta_e^{(\alpha)} \mid \alpha = (\alpha_1, \dots, \alpha_n),\ \alpha_1, \dots, \alpha_n,\ k \in \mathbb{N} \cup \{0\},\ c_\alpha \in \mathbb{C} \right\}.$$

We also present another definition of the infinitesimal object of a family of infinitely differentiable generalized translation operators. It is known that the tangent space $T_{t_0}M$ tangent to a manifold M at a point t_0 can be obtained as a set of all tangent vectors to smooth curves from M passing through the point t_0 at this point. The linear space of the Lie algebra of a Lie group G can be regarded as the space tangent to the Lie group G at the identity element of the group. Let us show that the linear space of the algebra $\mathcal{D}_e(Q)$ admits the same interpretation. Since some generators are degenerate, it is *a priori* impossible to consider only the tangent space T_eQ or its finite tensor power. Thus, one must take the infinite-dimensional space $\mathcal{D}(Q)$ as the relevant manifold.

A *path* with the initial point at $\delta_e \in \mathcal{D}(Q)$ is defined as a mapping $\mathbb{R}^1 \ni t \mapsto F_t \in \mathcal{D}(Q)$ satisfying the conditions:

1) $F_0 = \delta_e$.

2) The vector-valued function $\mathbb{R}^1 \ni t \mapsto F_t$ is infinitely differentiable.

3) The mapping $\mathbb{R}^1 \ni t \mapsto \operatorname{supp} F_t$ is continuous in the Michael topology [Mic] (see Subsection 2.3 in Chapter 1).

Two paths F_t and G_t are called *equivalent* whenever

$$|\langle F_t, f\rangle - \langle G_t, f\rangle| = o(t) \quad \text{as} \quad t \to 0$$

for all $f \in C^\infty(Q)$.

The collection of classes of equivalent paths with the initial point at δ_e is called the *tangent space* $T_e\mathcal{D}(Q)$ to the manifold $\mathcal{D}(Q)$ at the point δ_e. It is easy to see that the *tangent space* $T_e\mathcal{D}(Q)$ *coincides with the set of all functionals (tangent vectors)* $\hat{F} \in \mathcal{D}(Q)$ *of the form*

$$\langle \hat{F}, f \rangle = \frac{d}{dt} \langle F_t, f \rangle |_{t=0} \quad (f \in C^\infty(Q)). \tag{1.6}$$

Indeed, let us show that the mapping which associates a tangent vector $\hat{F}$ with a path F_t according to (1.6) is defined on the set of classes of equivalent paths and injective. If $F_t \sim G_t$, then, for all $f \in C^\infty(Q)$, we have

$$\frac{d}{dt} [\langle F_t, f \rangle - \langle G_t, f \rangle] |_{t=0} \le \lim_{t \to 0} \frac{1}{t} |\langle F_t - G_t, f \rangle| = 0,$$

whence it follows that $\hat{F} = \hat{G}$. Conversely, if $\hat{F} = \hat{G}$, then

$$0 = \langle \hat{F} - \hat{G}, f \rangle = \lim_{t \to 0} \frac{1}{t} [\langle F_t, f \rangle - \langle G_t, f \rangle]$$

for all $f \in C^\infty(Q)$ and, hence, $F_t \sim G_t$.

■

The tangent space $T_e \mathcal{D}(Q)$ is linear. If F_t and G_t are two paths with the initial point δ_e and $\hat{F}$ and $\hat{G}$ are the relevant tangent vectors, then the vector $\lambda \hat{F} + \mu \hat{G}$ $(\lambda, \mu \in \mathbb{C})$ is a tangent vector to the path

$$H_t = \frac{1}{2} (F_{2\lambda t} + G_{2\mu t}).$$

The tangent space $T_e \mathcal{D}(Q)$ coincides with the set $\mathcal{D}_e(Q)$ of functionals whose supports are concentrated at the point e.

Indeed, if $\hat{F} \in \mathcal{D}_e(Q)$, then $\hat{F}$ is a vector tangent to the path $F_t = \delta_e + t\hat{F}$, i.e., $\mathcal{D}_e(Q) \subset T_e \mathcal{D}(Q)$.

Let us prove the inverse inclusion. If $p \in Q$, $p \neq e$, then, in view of condition 3) of the definition of the path F_t and the fact that the Michael topology is a Hausdorff topology, one can indicate a precompact neighborhood U_p of the point p whose intersection with e is empty and such that $\operatorname{supp} F_t \cap U_p = \varnothing$ for all t sufficiently close to zero.

Let $\hat{F}$ be a vector tangent to the path F_t. Then

$$\langle \hat{F}, f \rangle = \lim_{t \to \infty} \frac{1}{t} [\langle F_t, f \rangle - \langle \delta_e, f \rangle] = 0$$

whenever $\operatorname{supp} f \subset U_p$ which means that $p \notin \operatorname{supp} \hat{F}$. Since the point $p \in Q$, $p \neq e$, is arbitrary, we arrive at the required assertion.

■

1.4. General Properties of Generators of Generalized Translation Operators. As shown in Example 2 in Subsection 1.3, even the structure of infinitesimal algebras of very simple generalized translation operators may be quite complicated. This explains the fact that the number of established general properties of generators of arbitrary generalized translation operators is relatively small. In what follows, we establish analogs of the first direct and converse Lie theorems on the construction of a system of differential equations for a given group translation and on the reconstruction of local Lie groups according to the indicated systems of differential equations.

By differentiating the relation of associativity

$$\big(R_r^t(R_s f)\big)(t) = \big(R_s^r(R_r f)\big)(t) \tag{1.7}$$

α_1 times with respect to r_1, α_2 times with respect to $r_2, \ldots,$ and α_n times with respect to r_n and setting $r = e$, we get

$$\begin{aligned} R_\alpha^t R_s\, f(t) &= D_r^{(\alpha)}\big(R_r^t(R_s\, f)\big)(t)\,|_{r=e} = D_r^{(\alpha)}\big(R_s^r(R_r\, f)\big)(t)\,|_{r=e} \\ &= D_r^{(\alpha)}\big(R_s^r(L_t\, f)\big)(r)\,|_{r=e} = \big(L_\alpha^s(L_t\, f)\big)(s) = \big(L_\alpha^s(R_s\, f)\big)(t), \end{aligned}$$

where the superscript of the generator denotes the variable upon which it acts. Hence, the *function* $u(t, s) = (R_s f)(t)$ *satisfies the system of equations*

$$\big(R_\alpha^t u\big)(t, s) = \big(L_\alpha^s u\big)(t, s) \quad (\alpha = (\alpha_1, \ldots, \alpha_n);\ \alpha_i \in \mathbb{N} \cup \{0\}). \tag{1.8}$$

This statement is an analog of the first direct Lie theorem for generalized translation operators. Since, by assumption, $\mathcal{D}_e(Q)$ coincides with the set of generalized functions whose supports are concentrated at the point e, it follows from (1.5) that

$$(R(F)_t u)(t, s) = (L(F)_s u)(t, s) \quad (F \in \mathcal{D}_e(Q),\ u(t, 0) = f(t)).$$

Let us now establish some properties of right and left generators.

The right and left generators satisfy the following commutation relations:

$$\big(R_\alpha^s(R_s f)\big)(t) = (R_s(R_\alpha f))(t), \tag{1.9}$$

$$\big(L_\alpha^t(R_s f)\big)(t) = (R_s(L_\alpha f))(t). \tag{1.10}$$

Thus, by applying the operator $D_s^{(\alpha)}$ to both sides of relation (1.7) and setting $s = e$, we obtain (1.9). Indeed,

$$\begin{aligned}
\big(R_\alpha^r(R_r f)\big)(t) &= D_s^{(\alpha)}\big(R_s^r(R_r f)\big)(t)|_{s=e} = D_s^{(\alpha)}\big(R_r^t(R_s f)\big)(t)|_{s=e} \\
&= \Big\langle\big(\delta_e^{(\alpha)}\big)^s, \big(R_r^t(R_s f)\big)(t)\Big\rangle = \Big\langle\big(\delta_e^{(\alpha)}\big)^s, \langle\delta_t, R_r R_s f\rangle\Big\rangle \\
&= \Big\langle\big(\delta_e^{(\alpha)}\big)^s, \langle\delta_t * \delta_r, R_s f\rangle\Big\rangle = \Big\langle\delta_t * \delta_r, \Big\langle\big(\delta_e^{(\alpha)}\big)^s, R_s f\Big\rangle\Big\rangle \\
&= \langle\delta_t * \delta_r, R_\alpha f\rangle = (R_r(R_\alpha f))(t).
\end{aligned}$$

By applying the operator $D_t^{(\alpha)}$ to both sides of relation (1.7) and setting $t = e$, we arrive at relation (1.10), namely,

$$\begin{aligned}
L_\alpha^r R_s f(r) &= D_t^{(\alpha)} R_r^t R_s f(t)|_{t=e} = D_t^{(\alpha)}\big(R_s^r(R_r f)\big)(t)|_{t=e} \\
&= \Big\langle\big(\delta_e^{(\alpha)}\big)^t, \big(R_s^r(R_r f)\big)(t)\Big\rangle = \Big\langle\big(\delta_e^{(\alpha)}\big)^t, \Big\langle(\delta_r * \delta_s)^\lambda, (R_\lambda f)(t)\Big\rangle\Big\rangle \\
&= \Big\langle(\delta_r * \delta_s)^\lambda, \Big\langle\big(\delta_e^{(\alpha)}\big)^t, (R_\lambda f)(t)\Big\rangle\Big\rangle \\
&= \langle\delta_r * \delta_s, L_\alpha f\rangle = (R_s(L_\alpha f))(t).
\end{aligned}$$

■

Right and left generators (not necessarily of the same order) commute, i.e.,

$$[R_\alpha, L_\beta]f = 0 \quad (f \in C^\infty(Q)).$$

Indeed, to prove this assertion, it suffices to apply the operator $D_t^{(\beta)}$ to both sides of relation (1.9) and set $t = e$.

■

It is often necessary to know whether the generators of generalized translation operators are linearly independent. Since e is an interior point of the manifold Q, we have

$$(L_\alpha f)(e) = D_t^{(\alpha)}(R_e f)(t)|_{t=e} = \big(D^{(\alpha)} f\big)(e).$$

This means that left generators are linearly independent. At the same time, we are much more interested in the algebra of infinitesimal right translations. In this case, the problem of linear independence of right generators is nontrivial.

Assume that infinitely differentiable generalized translation operators R_s $(s \in Q)$ *are reduced. Then all right generators are nondegenerate and linear independent.*

Indeed,

$$(R_\alpha f)(e) = D_s^{(\alpha)}(R_s f)(e)|_{s=e} = D_s^{(\alpha)}(L_e f)(s)|_{s=e} = D_s^{(\alpha)} f(s)|_{s=e} = \big(D^{(\alpha)} f)\big)(e).$$

■

It worth noting that the case considered in this assertion is rather rare.

Let us now outline the procedure of reconstruction of generalized translation operators for a given infinitesimal object. Suppose that right and left generators acting in the space of formal power series in variables $t_1, \dots, t_n$ are known. We solve system (1.8) and find the function $(R_s f)(t) = u(t, s)$. The operators R_s are regarded as natural candidates for the role of the required generalized translation operators. As a rule, it is not necessary to use all equations of system (1.8) for finding $u(t, s)$. Thus, in the case of Lie groups, generators of the first order uniquely determine group translations.

This heuristic construction is made rigorous by the following theorem:

Theorem 1.3. *Assume that there exists a finite family of operators* R_α, L_α $(\alpha = (\alpha_1, \dots, \alpha_n) \in \Gamma,\ \alpha_i \in \mathbb{N} \cup \{0\},\ i = \overline{1, n})$ *acting in the space of formal power series in variables* $t_1, \dots, t_n$ *and satisfying the following conditions:*

(i) $[R_\alpha, L_\beta] = 0 \quad (\alpha, \beta \in \Gamma)$;

(ii) $(L_\alpha f)(0) = (R_\alpha f)(0) = \big(D^{(\alpha)} f\big)(0) \quad (\alpha \in \Gamma)$ *for any formal power series* $f(t)$;

(iii) the system of equations

$$\big(R_\alpha^t u\big)(t, s) = \big(L_\alpha^s u\big)(t, s) \quad (\alpha \in \Gamma;\ t = (t_1, \dots, t_n),\ s = (s_1, \dots, s_n)) \tag{1.11}$$

with the initial condition

$$u(t, 0) = f(t) \tag{1.12}$$

is uniquely solvable in the class of formal power series.

Then the operators R_s $(s = (s_1, \dots, s_n))$ *defined in the space of formal power series by the formula* $(R_s f)(t) = u(t, s)$ *are generalized translation operators.*

Proof. By definition, $R_0 = \mathrm{id}$. Moreover, if $u(t, s)$ is a solution of system (1.11)–(1.12), then by virtue of (ii),

$$(R_\alpha f)(t) = \left(R_\alpha^t u\right)(t, 0) = \left(L_\alpha^s u\right)(t, s)\big|_{s=0} = D_s^{(\alpha)}(R_s f)(t)\big|_{s=0}.$$

It remains to prove that R_s satisfy the condition of associativity. Let us first show that, for any formal power series $f(t)$, we have

$$\left(R_\alpha^s(R_s f)\right)(t) = (R_s(R_\alpha f))(t) \quad (\alpha \in \Gamma). \tag{1.13}$$

Denote by $F_\alpha(t, s)$ and $G_\alpha(t, s)$ the right-hand and left-hand sides of (1.13), respectively. It is clear that $F_\alpha(t, s)$ satisfies the system of equations

$$\left(L_\beta^s F_\alpha\right)(t, s) = \left(R_\beta^t F_\alpha\right)(t, s)$$

and

$$F_\alpha(t, 0) = (R_\alpha f)(t)$$

Let us show that $G_\alpha(t, s)$ is a solution of the same system of equations with the same initial condition. (By virtue of condition (iii), this would mean that relation (1.13) is satisfied.) Indeed,

$$\left(L_\beta^s G_\alpha\right)(t, s) = L_\beta^s\left(R_\alpha^s(R_s f)\right)(t) = R_\alpha^s\, L_\beta^s(R_s f)(t)$$

$$= R_\alpha^s\, R_\beta^t(R_s f)(t) = R_\beta^t\, R_\alpha^s(R_s f)(t) = \left(R_\beta^t G_\alpha\right)(t, s).$$

Further,

$$G_0(t, 0) = R_\alpha^s(R_s f)(t)\big|_{s=0} = D_s^{(\alpha)}(R_s f)(t)\big|_{s=0} = (R_\alpha f)(t).$$

Let us now prove that the operators R_s are associative. Denote

$$\Phi(t, s, r) = \left(R_r^s(R_s f)\right)(t) \quad \text{and} \quad \Psi(t, s, r) = \left(R_s^t(R_r f)\right)(t).$$

The function Φ satisfies the system of equations

$$L_\alpha^r \Phi = R_\alpha^s \Phi \quad (\alpha \in \Gamma) \tag{1.14}$$

with the initial condition $\Phi(t, s, 0) = (R_s f)(t)$. The function $\Psi(t, s, r)$ also satisfies this initial condition. By using (1.13), we obtain

$$R_\alpha^s \Psi = R_\alpha^s\left(R_s^t(R_r f)\right)(t) = R_s^t\left(R_\alpha^t(R_r f)\right)(t) = R_s^t\left(L_\alpha^r(R_r f)\right)(t) = L_\alpha^r \Psi,$$

i.e., the function $\Psi(t, s, r)$ also satisfies system (1.14) and, hence, by virtue of condition (iii) of the theorem, is equal to $\Phi(t, s, r)$.

■

Remark. The assertion of Theorem 1.3 remains true if, in its formulation, the space of formal power series is replaced by the space $C^\infty(U)$, where U is a domain in $\mathbb{R}^n$ which contains the point 0, and we require that the solution $u(t, s)$ of system (1.11) satisfying the initial condition (1.12) is an element of the space $C^\infty(U \times U)$.

The statement of Theorem 1.3 belongs to Levitan and is an analog of the first converse Lie theorem. In a series of works (see the bibliography in [Lev10]), he proved analogs of the second and third Lie theorems (direct and converse) for generalized translation operators whose first-order or second-order generators are degenerate and form a basis of a certain Lie algebra (for generators of the first order, the manifold Q is a local Lie group and the relevant generalized translation operators coincide with translations on this group). These results are contained in well-known monographs by Levitan ([Lev9] and [Lev10]) and, therefore, not presented here. We only note that Grabovskaya and S. Krein ([GrK1]–[GrK3]) gave an explicit description of the generators of generalized translation operators studied by Levitan as second-order integro-differential operators. Grabovskaya, Kononenko, and Osipov [GKO] constructed a representation of the Lie algebra in integro-differential operators of any order. These operators are then used to reconstruct generalized translation operators. As becomes clear in the sequel, in many well-known examples of generalized translation operators, generators satisfy much more general commutation relations than in the case of Lie algebras.

1.5. Algebraic Approach to the Infinitesimal Theory of Formal Generalized Translation Operators. In the present subsection, we describe another approach to the construction of (formal) generalized translation operators in terms of a given infinitesimal object $\mathcal{D}_e(Q)$. As earlier, we assume that $\mathcal{D}_e(Q)$ is formed by functionals whose supports are concentrated at the point e. As an infinitesimal object, we consider the space $\mathbb{C}[x_1, \dots, x_n]$ of polynomials in variables $x_1, \dots, x_n$ with associative multiplication (which does not necessarily coincide with the ordinary multiplication of polynomials) and the right identity $\varepsilon \equiv 1$. As a global object, we take the space $\mathbb{C}[[x_1, \dots, x_n]]$ of formal power series in the variables $x_1, \dots, x_n$ with comultiplication Δ defined by duality in terms of multiplication in $\mathbb{C}[x_1, \dots, x_n]$ (in Subsection 1.2, we demonstrated that comultiplication generates generalized translations).

Note that the *Laplace transformation*

$$\mathcal{D}_e(Q) \ni F \mapsto F(x_1, \dots, x_n) = \left\langle F, e^{x_1 t_1 + \dots + x_n t_n} \right\rangle$$

maps the space $\mathcal{D}_e(Q)$ into the space of polynomials, $\mathbb{C}[x_1, \dots, x_n]$ and, hence, the definition of infinitesimal objects introduced in Subsection 1.3 does not contradict the outlined procedure.

Thus, let $\mathbb{C}[[x]] = \mathbb{C}[[x_1, \dots, x_n]]$ be the space of formal power series and $\mathbb{C}[x] = \mathbb{C}[x_1, \dots, x_n]$ be the dual space of polynomials in the variables $x_1, \dots, x_n$. We have the following analog of Theorem 1.1:

Theorem 1.4. *The space $\mathbb{C}[x]$ is an algebra with respect to a certain associative multiplication (which does not necessarily coincide with the ordinary multiplication of polynomials) and the polynomial $P_0(x) \equiv 1$ is the right identity of this multiplication if and only if, in $\mathbb{C}[[x]]$, there exists a family of generalized translation operators*

$$R_s\colon \mathbb{C}[[x]] \to \mathbb{C}[[x, y]] = \mathbb{C}[[x]] \otimes \mathbb{C}[[y]]$$

such that the point 0 is the identity element of these operators. The generalized translation operators R_s and the corresponding operation of multiplication in $\mathbb{C}[x]$ are connected by the formula

$$\langle F * G, f\rangle = \langle F^x \otimes G^y, (R_y f)(x)\rangle \quad (F, G \in \mathbb{C}[x];\ f \in \mathbb{C}[[x]]). \tag{1.15}$$

Proof. *Necessity.* Associative multiplication in $\mathbb{C}[x]$ induces coassociative comultiplication

$$\Delta\colon \mathbb{C}[[x]] \to \mathbb{C}[[x]] \otimes \mathbb{C}[[y]] = \mathbb{C}[[x, y]].$$

Consider a formal series $(\Delta f)(x, y)$ $(f \in \mathbb{C}[[x]])$ and assume that the variables $y = (y_1, \dots, y_n)$ take numerical values $y = (y_1^0, \dots, y_n^0)$. As a result, we arrive at a formal series $(\Delta f)(x, y_0) \in \mathbb{C}[[x]])$. We set

$$R_{y_0} f(x) = (\Delta f)(x, y_0).$$

It is clear that the operator R_{y_0} is linear and acts in $\mathbb{C}[[x]]$. Let us show that the operators R_y $(y \in \mathbb{C}^n)$ form a family of generalized translation operators. It is not difficult to see that the axiom of associativity is equivalent to the coassociativity of comultiplication, namely,

$$(\mathrm{id} \otimes \Delta)\Delta f = (\Delta \otimes \mathrm{id})\Delta f \quad (f \in \mathbb{C}[[x]])$$

Let us show that $R_0 = \mathrm{id}$. We have

$\langle F, R_0 f\rangle = \langle F, (\Delta f)(x, 0)\rangle = \langle F, \langle 1^y, (\Delta f)(x, y)\rangle\rangle = \langle F \otimes 1, \Delta f\rangle = \langle F * 1, f\rangle = \langle F, f\rangle$

$$(f \in \mathbb{C}[[x]];\ F \in \mathbb{C}[x]).$$

The proof of *sufficiency* is evident and we leave it to the reader.

■

Let $\mathbb{C}[x] \ni F \mapsto 1 * F$ be a projection which maps $\mathbb{C}[x]$ onto a subalgebra $\tilde{\mathbb{C}}[x]$. As in Subsection 1.3, the subalgebra $\tilde{\mathbb{C}}[x]$ is regarded as an infinitesimal algebra of reduced generalized translation operators.

We fix a basis

$$e_\alpha = \frac{x^\alpha}{\alpha!}$$

in $\mathbb{C}[[x]]$, where $\alpha = (\alpha_1, \ldots, \alpha_n)$ is a multiindex,

$$x^\alpha = x_1^{\alpha_1} \ldots x_n^{\alpha_n} \quad \text{and} \quad \alpha! = \alpha_1! \ldots \alpha_n!.$$

In this basis, the operation of comultiplication can be written as

$$\Delta e_\alpha = c_\alpha^{\beta\gamma} e_\beta \otimes e_\gamma$$

The condition $R_0 = \mathrm{id}$ means that $c_\alpha^{\beta 0} = \delta_\alpha^\beta$.

According to Theorem 1.4, the corresponding operation of multiplication in $\mathbb{C}[x]$ in the dual basis e^α has the form $e^\beta * e^\gamma = c_\alpha^{\beta\gamma} e^\alpha$. It is clear that the ordinary multiplication of polynomials in $\mathbb{C}[x]$ corresponds to additive translations in $\mathbb{C}[[x]]$, namely,

$$\Delta x^\alpha = (x + y)^\alpha.$$

We denote

$$\langle e_\alpha \otimes e_\beta, e^\gamma\rangle = e_\alpha \delta_\gamma^\beta$$

and define right generators R_γ by the formula

$$R_\gamma e_\alpha = \langle \Delta e_\alpha, e^\gamma\rangle$$

The correspondence $\mathbb{C}[x] \ni e^\gamma \mapsto R_\gamma$ *is a representation of the algebra* $\mathbb{C}[x]$.

Indeed, the coassociativity of comultiplication in $\mathbb{C}[[x]]$ is equivalent to the equality

$$c_\gamma^{\mu\beta} c_\mu^{\sigma\alpha} = c_\gamma^{\sigma\mu} c_\mu^{\alpha\beta}.$$

On the other hand,

$$R_\alpha R_\beta e_\gamma = c_\gamma^{\mu\beta} c_\mu^{\sigma\alpha} e_\sigma$$

and

$$c_\sigma^{\alpha\beta} R_\sigma e_\gamma = c_\mu^{\alpha\beta} c_\gamma^{\sigma\mu} e_\sigma.$$

Thus, generators satisfy the same relations as elements of the basis e^α but some right generators may be degenerate.

■

In Example 1, the indicated procedure leads to universal enveloping algebras $U(\mathfrak{g})$ of Lie algebras. By the PBW theorem, $U(\mathfrak{g})$ is isomorphic to the algebra $\mathbb{C}[x_1, \ldots, x_n]$ with the following relations:

$$X_i X_j = X_j X_i + c_{ij}^k X_k,$$

where c_{ij}^k are structural constants of the Lie algebra $\mathfrak{g}$. This enables us to introduce the operation of group multiplication in $\mathbb{C}[[X_1, \ldots, X_n]]$.

In the general case, in order to construct generalized translation operators for a fixed set of relations

$$J = \{f_i(X_j) = 0\},$$

where $\gamma \in \Gamma$ is a set of indices, $f_i \in T^m$, and T^m is a free tensor algebra with m generators, it suffices to establish an isomorphism of linear spaces

$$T^m / I \to \mathbb{C}[x_1, \ldots, x_n],$$

under which X_γ is mapped into e^γ (i.e., to prove an analog of the PBW theorem). Here, I is an ideal in T^m generated by the system of relations J). If the required isomorphism τ is established, then the space $\mathbb{C}[x_1, \ldots, x_n]$ inherits the structure of an associative algebra with right identity 1 and, by virtue of Theorem 1.4, a family of generalized translation operators R_s whose right generators $R_\gamma\ (\gamma \in \Gamma)$ satisfy the relations J is defined in the space $\mathbb{C}[[X_1, \ldots, X_n]]$. In the general case, if the indicated isomorphism exists, the procedure of its construction is not exactly specified and remains somewhat arbitrary. To eliminate this situation, we introduce left generators L_α of generalized translation operators by setting

$$L_\alpha e_\gamma = \langle e^\alpha, \Delta e_\gamma \rangle,$$

where

$$\langle e^\alpha, e_\beta \otimes e_\gamma \rangle = e_\beta \delta^\alpha_\gamma.$$

It is easy to see that the correspondence $\mathbb{C}[x] \ni e^\alpha \mapsto L_\alpha$ is an antirepresentation of the algebra $\mathbb{C}[x]$ and the algebra generated by left generators is antiisomorphic to $\mathbb{C}[x]$.

If we now establish an isomorphism τ of the linear spaces T^m/I and $\mathbb{C}[x_1, \dots, x_n]$, where I is an ideal in the free tensor algebra T^m generated by the relations

$$J = \left\{ f_i(X_\gamma) = 0 \,\middle|\, \gamma \in \Gamma,\ f_i \in T^m \right\},$$

then, by setting

$$\tau(X)\tau(Y) = \tau(XY) \quad (X, Y \in T^m/I),$$

we equip $\mathbb{C}[x_1, \dots, x_n]$ with the structure of an algebra with identity 1. Left generators of the relevant generalized translation operators R_s in $\mathbb{C}[[x_1, \dots, x_n]]$ satisfy the relations J and the indicated generalized translation operators are determined to within an isomorphism.

Possible generalizations of the PBW theorem for algebras with quadratic relations are discussed in [Ver2]. For some classes of relations, PBW-type theorems are proved in the next section.

A useful method for proving PBW-type theorems was proposed by Bergman [Berg].

1.6. Some Facts from the Theory of Topological Vector Spaces. Let $\mathcal{E}$ be a locally convex space and let $\mathcal{E}'$ be its dual space.

The *weak topology* $\sigma(\mathcal{E}', \mathcal{E})$ in $\mathcal{E}'$ is defined by a family of seminorms

$$\mathcal{E}' \ni F \mapsto |\langle F, \varphi \rangle| \quad (\varphi \in \mathcal{E}).$$

Similarly, the *weak topology* in $\mathcal{E}$ is defined by seminorms

$$\mathcal{E} \ni \varphi \mapsto |\langle F, \varphi \rangle| \quad (F \in \mathcal{E}').$$

The *Mackey topology* $\tau(\mathcal{E}, \mathcal{E}')$ in $\mathcal{E}$ is the topology of uniform convergence on $\sigma(\mathcal{E}', \mathcal{E})$ – compact convex sets in $\mathcal{E}'$. Similarly, we define the Mackey topology $\tau(\mathcal{E}', \mathcal{E})$ in $\mathcal{E}'$.

The Mackey–Arens theorem states that any locally convex topology τ in $\mathcal{E}$ admits the space $\mathcal{E}'$ as a space of τ-continuous linear functionals if and only if it is not weaker than $\sigma(\mathcal{E}, \mathcal{E}')$ and not stronger than $\tau(\mathcal{E}, \mathcal{E}')$.

The *strong topology* $\beta(\mathcal{E}', \mathcal{E})$ in the space $\mathcal{E}'$ is defined as the topology of uniform convergence on bounded subsets of $\mathcal{E}$. The strong topology $\beta(\mathcal{E}', \mathcal{E})$ is, generally speaking, stronger than the Mackey topology.

A space $\mathcal{E}$ is called *reflexive* if the space $\mathcal{E}''$, $\beta(\mathcal{E}', \mathcal{E})$-dual to the space $\mathcal{E}'$ and equipped with the topology $\beta(\mathcal{E}'', \mathcal{E}')$, is topologically isomorphic to the original space $\mathcal{E}$.

A space $\mathcal{E}$ is called a *Mackey space* if its topology coincides with $\tau(\mathcal{E}, \mathcal{E}')$. A complete metrizable locally convex space is called a *Fréchet space.* Any Fréchet space is a Mackey space.

We are interested in the space $C^\infty(Q)$ of infinitely differentiable functions defined on a smooth manifold Q and equipped with the topology of uniform convergence with all their derivatives on compact subsets of Q. The space $\mathcal{D}(Q)$ dual to $C^\infty(Q)$ consists of generalized functions with compact supports. The space $C^\infty(Q)$ is a reflexive Fréchet space. Hence, the Mackey topology on $\mathcal{D}(Q)$ coincides with the strong topology $\beta(\mathcal{D}(Q), C^\infty(Q))$. In what follows, we always assume that $\mathcal{D}(Q)$ is equipped with the Mackey topology, unless otherwise stated.

Let $\mathcal{E}$ and $\mathcal{F}$ be locally convex spaces. In the algebraic tensor product $\mathcal{E}\otimes\mathcal{F}$, we consider the strongest locally convex topology in which the canonical bilinear mapping

$$\mathcal{E}\times\mathcal{F} \ni (\varphi, \psi) \mapsto \varphi\otimes\psi \in \mathcal{E}\otimes\mathcal{F}.$$

is continuous. This topology is called *projective.*

The completion of $\mathcal{E}\otimes\mathcal{F}$ in the projective topology is denoted by $\mathcal{E}\hat{\otimes}\mathcal{F}$. There exists a one-to-one correspondence between linear continuous mappings from $\mathcal{E}\hat{\otimes}\mathcal{F}$ into an arbitrary locally convex space $\mathcal{G}$ and continuous bilinear mappings from $\mathcal{E}\times\mathcal{F}$ into $\mathcal{G}$ (see [Gro]): every linear mapping $\mathcal{E}\hat{\otimes}\mathcal{F}\to\mathcal{G}$ is associated with its composition with the canonical mapping

$$\mathcal{E}\times\mathcal{F}\to\mathcal{E}\hat{\otimes}\mathcal{F}.$$

Moreover,

$$C^\infty(Q)\,\hat{\otimes}\,C^\infty(Q) = C^\infty(Q\otimes Q),$$

and the dual space $(C^\infty(Q) \hat{\otimes} C^\infty(Q))'$ equipped with the projective topology coincides with $\mathcal{D}(Q) \hat{\otimes} \mathcal{D}(Q)$ (see [Gro]).

If $\mathcal{E}$, $\mathcal{F}$, and $\mathcal{G}$ are Fréchet spaces and α is a separately continuous (in the weak topology) bilinear mapping from the space $\mathcal{E}' \times \mathcal{F}'$ into the space $\mathcal{G}'$ equipped with the weak topology, then α is a strongly continuous mapping from $\mathcal{E}' \times \mathcal{F}'$ into $\mathcal{G}'$ equipped with the topology $\beta(\mathcal{G}', \mathcal{G})$.

2. Analog of Lie Theory for Some Classes of Generalized Translation Operators

As shown in Section 1, the space of polynomials $\mathbb{C}[X_1, \dots, X_n]$ endowed with the structure of an arbitrary associative algebra with right identity 1 can be regarded as an infinitesimal object of a family of formal generalized translation operators in $\mathbb{C}[[X_1, \dots, X_n]]$. The same picture is observed in the case of duality $\langle C^\infty(Q), \mathcal{D}(Q)\rangle$. This means that the problem of construction of analogs of Lie theory for arbitrary families of generalized translation operators is too general. It is thus interesting to select some classes of generalized translation operators for which it is possible to construct meaningful analogs of Lie theory. In the present section, we consider three classes of generalized translation operators of this sort, namely, the Delsarte generalized translation operators, the Delsarte-type generalized translation operators, and generalized translation operators associated with the hypergroup $G//H$.

2.1. Infinitesimal Object for the Delsarte Generalized Translation Operators. Let G be a real or complex Lie group, let $\mathfrak{g}$ be its Lie algebra, and let $U(\mathfrak{g})$ be the universal enveloping algebra of the algebra $\mathfrak{g}$. We consider a compact subgroup $\Gamma \subset \operatorname{Aut} G$ of the group of automorphisms of the group G and denote by $d\gamma$ the normalized Haar measure of the group Γ. Recall that, for all $f \in C^\infty(Q)$, the Delsarte generalized translation operators are defined by the equality

$$(R_{g_2})(g_1) = \int_\Gamma f(g_1\gamma(g_2))\, d\gamma \quad (g_1, g_2 \in G).$$

Since any continuous automorphism $\gamma \in \Gamma$ is analytic, its differential $d\gamma_e$ in the identity element of the group G is an automorphism of the algebra $\mathfrak{g}$. It is easy to see that the *mapping*

$$\Gamma \ni \gamma \mapsto d\gamma_e \in \operatorname{Aut} \mathfrak{g}$$

is a *representation of the group* Γ.

This assertion immediately follows from the definition of the differential

$$d\gamma_e(X) f \circ \gamma = X(f \circ \gamma) \quad (X \in \mathfrak{g},\ \gamma \in \Gamma).$$

■

In what follows, for simplicity, we denote the action $d\gamma_e$ $(\gamma \in \Gamma)$ of the group Γ

on $\mathfrak{g}$ by the same letter, i.e.,

$$\gamma(X) = d\gamma_e(X) \quad (X \in \mathfrak{g},\ \gamma \in \Gamma).$$

Let $X_1, \ldots, X_n \in \mathfrak{g}$ be a basis in $\mathfrak{g}$. By the Poincare–Birkhoff–Witt theorem, the monomials

$$X^\alpha = X_1^{\alpha_1} \ldots X_n^{\alpha_n} \quad (\alpha = (\alpha_1, \ldots, \alpha_n);\ \alpha_i \in \mathbf{N}_0,\ i = 1, \ldots, n)$$

form a basis in $U(\mathfrak{g})$. We extend the action of the group Γ to $U(\mathfrak{g})$ by setting

$$\pi(\gamma)\, 1 = 1 \quad \text{and} \quad \pi(\gamma)(X^\alpha) = \left(\gamma(X_1)^{\alpha_1} \ldots \gamma(X_n)^{\alpha_n}\right)$$

and extending this operation to the entire $U(\mathfrak{g})$ by linearity.

The mapping

$$\pi(\gamma)\colon U(\mathfrak{g}) \to U(\mathfrak{g})$$

introduced above preserves the canonical filtration in $U(\mathfrak{g})$ *and is an automorphism of* $U(\mathfrak{g})$.

This statement is proved by induction on the degree of filtration of elements from $U(\mathfrak{g})$. For $n = 0, 1,$ the statement is evident. Suppose that the statement is true for all elements of $U(\mathfrak{g})$ whose degree of filtration does not exceed m. It suffices to prove that

$$\pi(\gamma)(X_i X^\alpha) = \pi(\gamma)(X_i)\pi(\gamma)(X^\alpha) \quad (|\alpha| \le m)$$

for all $X_i \in \mathfrak{g}$ $(i = 1, \ldots, n)$. For the sake of definiteness, we assume that $i = n$. Then, by the induction hypothesis, we obtain

$$\pi(\gamma)\left(X_n X_1^{\alpha_1} \ldots X_n^{\alpha_n}\right) = \pi(\gamma)\left(X_1 X_n X_1^{\alpha_1 - 1} \ldots X_n^{\alpha_n}\right)$$

$$+ \left[\gamma(X_n), \gamma(X_1)\right](\gamma(X_1))^{\alpha_1 - 1} \ldots (\gamma(X_n))^{\alpha_n} = \ldots$$

$$= \pi(\gamma)\left(X_1^{\alpha_1} \ldots X_n^{\alpha_n + 1}\right) + \left[\gamma(X_n), \gamma(X_1)\right](\gamma(X_1))^{\alpha_1 - 1} \ldots (\gamma(X_n))^{\alpha_n}$$

$$+ \ldots + (\gamma(X_1))^{\alpha_1} \ldots (\gamma(X_{n-1}))^{\alpha_{n-1} - 1}\left[\gamma(X_n), \gamma(X_{n-1})\right](\gamma(X_n))^{\alpha_n}$$

$$= (\gamma(X_1))^{\alpha_1}\ldots(\gamma(X_{n-1}))^{\alpha_{n-1}}\gamma(X_n)(\gamma(X_n))^{\alpha_n} + \ldots +$$

$$+ [\gamma(X_n), \gamma(X_1)](\gamma(X_1))^{\alpha_1 -1}\ldots(\gamma(X_n))^{\alpha_n}$$

$$= (\gamma(X_1))^{\alpha_1}\ldots(\gamma(X_{n-1}))^{\alpha_{n-1}-1}\gamma(X_n)\gamma(X_{n-1})(\gamma(X_n))^{\alpha_n}$$

$$+ (\gamma(X_1))^{\alpha_1}\ldots(\gamma(X_{n-1}))^{\alpha_{n-1}-2}\left[\gamma(X_n), \gamma(X_{n-1})\right]\gamma(X_{n-1})(\gamma(X_n))^{\alpha_n}$$

$$+ \ldots + [\gamma(X_n), \gamma(X_1)](\gamma(X_1))^{\alpha_1 -1}\ldots(\gamma(X_n))^{\alpha_n} = \ldots$$

$$= \pi(\gamma)(X_n)\,\pi(\gamma)(X^{\alpha}).$$

■

By using this assertion, one can easily show that the definition of the action of the group Γ on $U(\mathfrak{g})$ is independent of the choice of a basis in $\mathfrak{g}$. It is also clear that the mapping

$$\Gamma \ni \gamma \mapsto \pi(\gamma) \in \operatorname{Aut} U(\mathfrak{g})$$

is a representation of the group Γ.

Consider a linear operator $S\colon U(\mathfrak{g}) \to U(\mathfrak{g})$ defined by the formula

$$S(Y) = \int_{\Gamma} \pi(\gamma)\, Y\, d\gamma \quad (Y \in U(\mathfrak{g})). \tag{2.1}$$

Since the representation $\pi(\gamma)$ preserves the canonical filtration $(U_m(\mathfrak{g}))_{m\geq 0}$ (recall that the space $U_m(\mathfrak{g})$ is a linear span of all products $X_1 \ldots X_p$, where $X_1, \ldots, X_p \in \mathfrak{g}$ and $p \leq m$, and each space $U_m(\mathfrak{g})$ is finite-dimensional), the integral in (2.1) is the ordinary integral of the vector function

$$\Gamma \ni \gamma \mapsto \pi(\gamma)\, Y \in U_m(\mathfrak{g})$$

which takes values in the finite-dimensional vector space $U_m(\mathfrak{g})$, where m is the degree of filtration of elements $Y \in U_m(\mathfrak{g})$. A vector $Y \in U_m(\mathfrak{g})$ is called Γ-*invariant* if $\pi(\gamma)\, Y = Y$ for all $\gamma \in \Gamma$.

Lemma 2.1. *(i)* $S^2 = S$, *i.e.,* S *is a projector in* $U(\mathfrak{g})$.

(ii) the subspace $U(\mathfrak{g})_{\Gamma} = S(U(\mathfrak{g}))$ *is a subalgebra of the algebra* $U(\mathfrak{g})$ *and coincides with the set of all* Γ*-invariant vectors.*

Proof. We prove (i). For any $Y \in U(\mathfrak{g})$, in of view of the invariance of the Haar measure, we conclude that

$$S^2(Y) = \int_\Gamma \int_\Gamma \pi(\gamma_1\gamma_2) Y d\gamma_1 d\gamma_2 = \int_\Gamma \pi(\gamma_2) Y d\gamma_2 = S(Y).$$

To prove (ii), we first show that

$$S(Y)S(Z) = S(S(Y)Z) = S(YS(Z)) \quad (Y, Z \in U(\mathfrak{g})). \tag{2.2}$$

Indeed,

$$\begin{aligned} S(Y)S(Z) &= \int_\Gamma \int_\Gamma \pi(\gamma_1) Y d\gamma_1 \pi(\gamma_2) Z d\gamma_2 \\ &= \int_\Gamma \int_\Gamma \pi(\gamma_2\gamma_1) Y d\gamma_1 \pi(\gamma_2) Z d\gamma_2 \\ &= \int_\Gamma \pi(\gamma_2) \left(\int_\Gamma \pi(\gamma_1) Y d\gamma_1 Z \right) d\gamma_2 = S(S(Y)Z) \end{aligned}$$

$$(Y, Z \in U(\mathfrak{g})).$$

The second equality in (2.2) can be proved similarly. It follows from (2.2) that $U(\mathfrak{g})_\Gamma$ is a subalgebra of $U(\mathfrak{g})$.

Let us show that $U(\mathfrak{g})_\Gamma$ consists of all Γ-invariant vectors. It is clear that all Γ-invariant vectors are contained in $U(\mathfrak{g})_\Gamma$.

Conversely, let $Y \in U(\mathfrak{g})_\Gamma$. Then $SY = Y$ and

$$\pi(\gamma) Y = \pi(\gamma) SY = \int_\Gamma \pi(\gamma)\pi(\gamma_1) Y d\gamma_1 = \int_\Gamma \pi(\gamma_1) Y d\gamma_1 = Y.$$

■

Let $\mathcal{D}_e^R(G/\Gamma)$ be an infinitesimal algebra of reduced generalized translation operators.

Theorem 2.1. *The algebras* $U(\mathfrak{g})_\Gamma$ *and* $\mathcal{D}_e^R(G/\Gamma)$ *are isomorphic.*

Proof. In Section 1, it was shown that the algebra $\mathcal{D}_e^R(G)$ generated by all group generators

$$(H_\alpha f)(g) = D_h^{(\alpha)} f(gh)\big|_{h=e}$$

$$(g \in G,\ f \in C^\infty(G);\ \alpha = (\alpha_1, \dots, \alpha_n) \text{ is a multiindex})$$

is isomorphic to the algebra $\mathcal{D}_e(G)$ of distributions whose supports are concentrated in the identity of the group. By the Schwartz theorem, this algebra is isomorphic to $U(\mathfrak{g})$ and, consequently, $\mathcal{D}_e^R(G)$ is isomorphic to $U(\mathfrak{g})$.

Let $\kappa\colon U(\mathfrak{g}) \to \mathcal{D}_e^R(G)$ be the corresponding isomorphism. We define the action of Γ on $\mathcal{D}_e^R(G)$ by setting

$$\pi(\gamma) H_\alpha = \kappa\pi(\gamma)\kappa^{-1} H_\alpha.$$

Let us show that, for all $f \in C^\infty(G)$,

$$\pi(\gamma) H_\alpha f \circ \gamma = H_\alpha(f \circ \gamma) \quad (\gamma \in \Gamma;\ H_\alpha \in \mathcal{D}_e^R(G)). \tag{2.3}$$

Indeed, let $H_1, \dots, H_n$ be first order generators of the group G with respect to a fixed coordinate system in a neighborhood of the identity of the group G. Then

$$X_1 = \kappa^{-1} H_1, \dots, X_n = \kappa^{-1} H_n$$

is a basis in $\mathfrak{g}$. It is well-known that every generator H_α can be represented in the form of a polynomial P_α in generators of the first order, i.e., $H_\alpha = P_\alpha(H_1, \dots, H_n)$. Therefore,

$$\begin{aligned}
\pi(\gamma) H_\alpha f \circ \gamma &= [\kappa\pi(\gamma)\kappa^{-1}(H_\alpha)] f \circ \gamma \\
&= \kappa\pi(\gamma) P_\alpha(X_1, \dots, X_n) f \circ \gamma \\
&= P_\alpha(\pi(\gamma) H_1, \dots, \pi(\gamma) H_n) f \circ \gamma \\
&= P_\alpha(H_1, \dots, H_n)(f \circ \gamma) = H_\alpha(f \circ \gamma).
\end{aligned}$$

By using (2.3), for all $\alpha = (\alpha_1, \dots, \alpha_n)$, we can write

$$\begin{aligned}
\int_\Gamma (\pi(\gamma) H_\alpha) f(g)\, d\gamma &= \int_\Gamma (H_\alpha(f \circ \gamma))(\gamma^{-1}(g))\, d\gamma \\
&= \int_\Gamma D_h^{(\alpha)} f \circ \gamma\big(\gamma^{-1}(g) h\big)\big|_{h=e}\, d\gamma
\end{aligned}$$

$$= \int_{\Gamma} D_h^{(\alpha)} f(g\gamma(h))\big|_{h=e}\, d\gamma = (R_\alpha f)(g) \quad (g, h \in G,\ f \in C^\infty(G),$$

where R_α is a right generator of a Delsarte generalized translation operator, whence it follows that $\kappa S P_\alpha(X_1, \dots, X_n) = R_\alpha$ for all $\alpha = (\alpha_1, \dots, \alpha_n)$. Moreover, since the polynomials $P_\alpha(X_1, \dots, X_n)$ form a basis in $U(\mathfrak{g})$, the homomorphism κ maps the algebra $U(\mathfrak{g})_\Gamma$ onto the entire algebra $\mathcal{D}_e^R(G/\Gamma)$. Since κ is an isomorphism between $U(\mathfrak{g})$ and $\mathcal{D}_e^R(G)$, we have

$$\operatorname{Ker} \kappa \restriction U(\mathfrak{g})_\Gamma = 0$$

and κ is an isomorphism between $U(\mathfrak{g})_\Gamma$ and $\mathcal{D}_e^R(G)$.

■

Denote by $((X_1 \dots X_p))$ the symmetrization of the monomial $X_1 \dots X_p$, namely,

$$((X_1 \dots X_p)) = \frac{1}{p!} \sum_{\omega \in S(p)} X_{\omega(1)} \cdots X_{\omega(p)} \quad (X_1 \dots X_p \in \mathfrak{g})$$

Let $S(\mathfrak{g})$ be a symmetric algebra of the vector space $\mathfrak{g}$, let $(S_m(\mathfrak{g}))_{m \ge 0}$ be its canonical filtration ($S_m(\mathfrak{g})$ is a linear span of all products of the form $X_1 \dots X_p$, where $p \le m$ and $X_1 \dots X_m \in \mathfrak{g}$), and let

$$S(\mathfrak{g}) \ni X_1 \dots X_p \mapsto \omega(X_1 \dots X_p) = ((X_1 \dots X_n)) \in U(\mathfrak{g}).$$

be the canonical bijection of $S(\mathfrak{g})$ onto $U(\mathfrak{g})$. We define the action of Γ on $S(\mathfrak{g})$ in a natural way, i.e.,

$$\pi(\gamma)(X_1 \dots X_p) = (\pi(\gamma)X_1) \dots (\pi(\gamma)X_p) \quad (X_1 \dots X_p \in \mathfrak{g}).$$

Lemma 2.2. *The canonical bijection* $\omega : S(\mathfrak{g}) \to U(\mathfrak{g})$ *is an isomorphism of* Γ*-modules, i.e.,* $\omega \circ \pi(\gamma) = \pi(\gamma) \circ \omega$ *for all* $\gamma \in \Gamma$.

Proof. It is necessary to prove the equality

$$((\pi(\gamma)X_1 \dots \pi(\gamma)X_n)) = \pi(\gamma)((X_1 \dots X_n))$$

$$(X_1 \dots X_n \in \mathfrak{g};\ \gamma \in \Gamma,\ n \in \mathbb{N}_0)$$

We proceed by induction on the canonical filtration in $U(\mathfrak{g})$. For elements of $U_1(\mathfrak{g})$, the assertion of the lemma is obvious. To perform a step of induction, we apply the equality (see, e.g., [Dix3], p. 120),

$$X((X_1 \dots X_n)) = ((X X_1 \dots X_n)) - \frac{b_1}{1!} \sum_h (([X, X_h] X_1 \dots \hat{X}_h \dots X_n))$$

$$+ \frac{b_2}{2!} \sum_{h \neq l} (([[X, X_h], X_l] X_1 \dots \hat{X}_h \dots \hat{X}_l \dots X_n)) - \dots , \qquad (2.4)$$

where $b_i \in \mathbb{C}$, $X_1, \dots, X_n \in \mathfrak{g}$, $\hat{X}_h$ means that the relevant element $X_h \in \mathfrak{g}$ is omitted. By applying $\pi(\gamma)$ to relation (2.4) and using the induction hypothesis, we obtain

$$\pi(\gamma) X((\pi(\gamma) X_1 \dots \pi(\gamma) X_n)) = \pi(\gamma)((X X_1 \dots X_n))$$

$$- \frac{b_1}{1!} \sum_h (([\pi(\gamma) X, \pi(\gamma) X_h]$$

$$\times \pi(\gamma) X_1 \dots \widehat{\pi(\gamma) X_h} \dots \pi(\gamma) X_n))$$

$$+ \frac{b_2}{2!} \sum_{h \neq l} (([[\pi(\gamma) X, \pi(\gamma) X_h], \pi(\gamma) X_l]$$

$$\times \pi(\gamma) X_1 \dots \widehat{\pi(\gamma) X_h} \dots \widehat{\pi(\gamma) X_l} \dots \pi(\gamma) X_n)) - \dots .$$

On the other hand by virtue of (2.4), we have

$$\pi(\gamma) X((\pi(\gamma) X_1 \dots \pi(\gamma) X_n))$$

$$= ((\pi(\gamma) X \pi(\gamma) X_1 \dots \pi(\gamma) X_n))$$

$$- \frac{b_1}{1!} \sum_h (([\pi(\gamma) X, \pi(\gamma) X_h] \pi(\gamma) X_1 \dots \widehat{\pi(\gamma) X_h} \dots \pi(\gamma) X_n)) + \dots ,$$

and, hence, we arrive at the required equality

$$((\pi(\gamma) X \pi(\gamma) X_1 \dots \pi(\gamma) X_n)) = \pi(\gamma)((X X_1 \dots X_n))$$

■

Theorem 2.2. *Assume that* Γ *is a reductive algebraic group (i.e., a subgroup of* $GL(n, \mathbb{C})$ *every linear representation of which is completely reducible; in particular, finite groups and compact Lie groups are reductive). Then the algebra* $U(\mathfrak{g})_\Gamma$ *is finitely generated, i.e., has finitely many generators. If* Γ *is finite, then the algebra* $U(\mathfrak{g})_\Gamma$ *is generated by the elements*

$$J_\alpha = \frac{1}{|\Gamma|} \sum_{\gamma \in \Gamma} \pi(\gamma)((X^\alpha))$$

where $|\alpha| \le |\Gamma|$.

Proof. Let S_Γ be a ring of Γ-invariants of the algebra $S(\mathfrak{g})$, i.e.,

$$S_\Gamma = \{Z \in S(\mathfrak{g}) \mid \pi(\gamma)Z = Z\}.$$

By Lemma 2.2, $U(\mathfrak{g})_\Gamma = \omega(S_\Gamma)$. Since Γ is reductive, the ring of invariants S_Γ of the commutative algebra $S(\mathfrak{g})$ is finitely generated (see, e.g., [Kra]).

We fix a set of generating elements $I_1, \dots, I_m$ of S_Γ and show that the elements

$$J_1 = \omega(I_1), \dots, J_m = \omega(I_m)$$

generate $U(\mathfrak{g})_\Gamma$.

Since ω preserves filtration, i.e., $\omega(S_p(\mathfrak{g})) = U_p(\mathfrak{g})$, for any element

$$I_1^{\alpha_1} \dots I_m^{\alpha_m} \in S_p(\mathfrak{g}),$$

we have

$$\omega\left(I_1^{\alpha_1} \dots I_m^{\alpha_m}\right) = \left(\left(I_1^{\alpha_1} \dots I_m^{\alpha_m}\right)\right) = ((I_1))^{\alpha_1} \dots ((I_m))^{\alpha_m} + W = J_1^{\alpha_1} \dots J_m^{\alpha_m} + W,$$

where

$$J_1^{\alpha_1} \dots J_m^{\alpha_m} \in U_p(\mathfrak{g}),$$

and the degree of filtration of the element $W \in U(\mathfrak{g})$ is less than p.

Let $Y \in U(\mathfrak{g})_\Gamma$. Since S_Γ is finitely generated, we obtain

$$\omega^{-1}(Y) = P(I_1, \dots, I_m),$$

where P is a polynomial in the variables $I_1, \dots, I_m$. In view of the property of ω

proved above, we have

$$Y = P(J_1, \dots, J_m) + V$$

and the degree of filtration of the element $V \in U(\mathfrak{g})_\Gamma$ is lower than the degree of filtration of Y. The fact that $U(\mathfrak{g})_\Gamma$ is finitely generated can now be proved by induction on the degree of filtration of $U(\mathfrak{g})$ and by using the fact that S_Γ is finitely generated.

If Γ is a finite group, then the Noether theorem (see, e.g., [Kra]) implies that the invariants generating the ring S_Γ have the form

$$I_\alpha = \frac{1}{|\Gamma|} \sum_{\gamma \in \Gamma} \pi(\gamma) X^\alpha.$$

where $|\alpha| \le |\Gamma|$.

This yields the last assertion of the theorem.

■

Let $\mathcal{A}$ be an associative algebra over the field $\mathbb{C}$ with identity and finitely many generators $x_1, \dots, x_m$. For any n, we denote

$$A_0 = \mathbb{C} \quad \text{and} \quad A_1 = \text{l.s.}\{x_1, \dots, x_m\}.$$

Let A_n be the linear space spanned by all elements of the form $x_1 \dots x_p$ $(x_1, \dots, x_p \in \mathcal{A}_1;\ p \le n)$. Then $A_0 \subset A_1 \subset \dots$ is a filtration in $\mathcal{A}$. Denote $\mathcal{A}^{(k)} = \mathcal{A}_k / \mathcal{A}_{k-1}$.

The algebra $\mathcal{A}$ is called a *Poincare–Birkhoff–Witt algebra (PBW-algebra)* if

$$\dim \mathcal{A}^{(n)} = C^n_{n+m-1} \quad (n = 0, 1, \dots)$$

for any set of generators $x_1, \dots, x_m$. The algebra $\mathcal{A}$ is called a *PBW-type algebra* if, for any set of generators $x_1, \dots, x_m$, we have

$$\dim \mathcal{A}^{(n)} < C^n_{n+m-1} \quad (n = 0, 1, \dots)$$

(see [Ver2]).

Theorem 2.3. *Let $\mathcal{A}$ be a finitely generated subalgebra of a universal enveloping algebra $U(\mathfrak{g})$ of a Lie algebra $\mathfrak{g}$. Then $\mathcal{A}$ is a PBW-type algebra.*

Proof. Let $Z_1, \dots, Z_m$ be generators of the algebra $\mathcal{A}$. Let us show that symmetric polynomials in $Z_1, \dots, Z_m$ are linear generators of the algebra $\mathcal{A}$ (it is clear that the assertion of Theorem 2.3 immediately follows from this fact). Each monomial

$$\Phi = Z_{i_1} \dots Z_{i_p} \in \mathcal{A}$$

inherits the filtration from $U(\mathfrak{g})$. The PBW theorem for $U(\mathfrak{g})$ implies that the degree of filtration of the element

$$\Phi_1 = \frac{1}{p!} \sum_{\omega \in S(p)} Z_{i_{\omega(1)}} \dots Z_{i_{\omega(p)}} - \Phi$$

in $U(\mathfrak{g})$ is lower than the degree of filtration of the element Φ. The proof is completed by induction on the canonical filtration of $U(\mathfrak{g})$.

■

Corollary 2.1. *The algebra* $U(\mathfrak{g})_\Gamma$ *is an algebra of PBW type.*

In what follows, we study the simplest case of algebras $U(\mathfrak{g})_\Gamma$ with $|\Gamma| = 2$. Let $\Gamma = \{\text{id}, \gamma\}$, where γ is an involutive automorphism of the group G. The Delsarte generalized translation operators associated with this group are called $\mathbb{Z}_2$-Delsarte generalized translation operators.

We reduce the representation

$$\Gamma \ni \gamma \mapsto d\gamma_e \in \operatorname{Aut}\mathfrak{g}$$

of the group Γ in the Lie algebra $\mathfrak{g}$, i.e., choose a basis $X_1, \dots, X_m, X_{m+1}, \dots, X_n$ of the algebra $\mathfrak{g}$ such that $\gamma(X_k) = -X_k$ for $1 \le k \le m$ and $\gamma(X_j) = X_j$ for $m < j \le n$. The existence of a basis of this sort follows from the fact that any representation of a finite group is completely reducible and the commutative group Γ has only two one-dimensional irreducible representations.

Let

$$\mathfrak{g}_0 = \text{l.s.}\{X_1, \dots, X_m\}$$

and

$$\mathfrak{g}_1 = \text{l.s.}\{X_{m+1}, \dots, X_n\}.$$

Since $\gamma_e \in \operatorname{Aut}\mathfrak{g}$, $\mathfrak{g}_1$ is a subalgebra of the Lie algebra $\mathfrak{g}$ and, moreover, $[\mathfrak{g}_0, \mathfrak{g}_1] \subset \mathfrak{g}_0$ and $[\mathfrak{g}_0, \mathfrak{g}_0] \subset \mathfrak{g}_1$. In terms of the structural constants c_{ij}^k of the Lie algebra

$\mathfrak{g}$, this means that $c_{ij}^s = 0$ for $i, j > m$ and $s \le m$, $c_{ik}^t = 0$ for $t, i > m$ and $1 \le k \le m$, and $c_{kl}^t = 0$ for $1 \le k, l, t \le m$.

It is clear that

$$s \upharpoonright \mathfrak{g}_1 = \mathrm{id}$$

and

$$s \upharpoonright \mathfrak{g}_0 = 0.$$

By virtue of Theorem 2.2, the Lie algebra $U(\mathfrak{g})_\Gamma$ is generated by the elements $Z_i = X_i$ with $i > m$ and

$$Z_{kl} = \frac{1}{2}(X_k X_l + X_l X_k),$$

where $1 \le k, l \le m$. The generators Z_l $(m < i \le n)$ and Z_{kl} $(1 \le k, l \le m)$ are called *canonical*.

By direct calculations, we show that the canonical generators satisfy the following relations:

$$[Z_i, Z_j] = c_{ij}^t Z_t \quad (m < i, j \le n),$$

$$[Z_i, Z_{kl}] = c_{ik}^t Z_{tl} + c_{il}^t Z_{tk} \quad (m < i \le n,\ 1 \le k, l \le m) \tag{2.5}$$

$$\begin{aligned} Z_{kl} Z_{pq} - Z_{kp} Z_{lq} &+ \frac{1}{4}\left(c_{kl}^i c_{pq}^j - 2c_{lp}^i c_{kq}^j - c_{kp}^i c_{lq}^j\right) Z_j \\ &+ \frac{1}{2} c_{kl}^i Z_i Z_{pq} + \frac{1}{2} c_{pq}^i Z_{kl} Z_i - \frac{1}{2} c_{lp}^i Z_i Z_{kq} \\ &- \frac{1}{2} c_{lp}^i Z_{kq} Z_i - \frac{1}{2} c_{kp}^i Z_i Z_{lq} - \frac{1}{2} c_{lq}^i Z_{kp} Z_i \\ = \frac{1}{2} c_{lp}^i c_{iq}^s Z_{sk} &- \frac{1}{2} c_{lp}^i c_{ik}^s Z_{sq} + \frac{1}{2} c_{lp}^i c_{ki}^s c_{sq}^j Z_j \quad (1 \le k, l, p, q \le m) \end{aligned}$$

where summation is carried out over the repeating indices and $[Z_\lambda, Z_\mu]$ denotes the commutator $[Z_\lambda, Z_\mu] = Z_\lambda Z_\mu - Z_\mu Z_\lambda$. By applying the last relation in (2.5) several times and using the fact that $Z_{kl} = Z_{lk}$, we can obtain an expression for the commutator $[Z_{kl}, Z_{pq}]$.

Thus, for $|\Gamma| = 2$, $U(\mathfrak{g})_\Gamma$ is an algebra with quadratic relations (for $|\Gamma| > 2$, one can construct examples of relations whose order is higher than two). The total number of different relations in (2.5) is given by the formula

$$P_{nm} = 5C_4^m + 9C_3^m + 4C_2^m + \frac{1}{2}(n-m)(m^2+n-1)$$

$$= P_{mm} + \frac{1}{2}(n-m)(m^2+n-1).$$

Generally speaking, P_{nm} is greater than the total number of relations between the basis vectors in a Lie algebra of dimensionality

$$n + \frac{m(m-1)}{2}$$

(the last number is, in fact, the number of linearly independent canonical generators). For $m = 1$, relations (2.5) between the canonical generators are relations of the Lie type and have the form

$$[Z_i, Z_j] = c_{ij}^k Z_k \quad (2 \le i, j, k \le n) \quad \text{and} \quad [Z_i, Z_{11}] = 2c_{i1}^1 Z_{11}. \tag{2.6}$$

Thus, the elements $Z_{11}, Z_2, \ldots, Z_n$ form a basis of a new Lie algebra $\tilde{\mathfrak{g}}$. For $m \ge 2$, the collection of relations (2.5) always contains non-Lie relations.

One can easily see that relations (2.5) are invariant under nondegenerate linear transformations of the form

$$Z_i \to \overline{Z}_i = a_i^j Z_j \quad (m < i, j \le n) \quad \text{and} \quad Z_{kl} \to \overline{Z}_{kl} = b_k^t b_l^s Z_{st} \quad (1 \le k, l, t, s \le m),$$

where

$$\det\left(a_i^j\right)_{ij=m+1}^n \ne 0 \quad \text{and} \quad \det\left(b_k^t\right)_{kt=1}^m \ne 0.$$

Each transformation of canonical generators of this type is associated with a nondegenerate linear transformation of the Lie algebra $\mathfrak{g}$. This transformation can be written in the matrix form as

$$\begin{pmatrix} B & 0 \\ 0 & A \end{pmatrix},$$

where

$$B = \left(b_k^t\right)_{k,t=1}^m \quad \text{and} \quad A = \left(a_i^j\right)_{i,j=m+1}^n.$$

Namely, if

$$\overline{X}_k = b_k^t X_t \quad \text{and} \quad \overline{X}_i = a_i^j X_j \quad (1 \le k \le m,\ m < i \le n)$$

is a new basis in $\mathfrak{g}$, then

$$\overline{Z}_{kl} = \frac{1}{2}(\overline{X}_k \overline{X}_l + \overline{X}_l \overline{X}_k)$$

and

$$\overline{Z}_i = \overline{X}_i \quad (1 \le k,\ l \le m;\ m < i \le n).$$

Since the matrix of the automorphism γ is diagonal in the basis $\overline{X}_r$ $(1 \le r \le n)$, the generators $\overline{Z}_i$ $(m < i \le n)$ and $\overline{Z}_{kl}$ $(1 \le k, l \le m)$ are also canonical. The relationships between canonical generators written in the form (2.5) are called canonical.

The collection of generators Z_i $(i > m)$ and Z_{kl} $(k, l \le m)$ of the algebra $U(\mathfrak{g})_\Gamma$ satisfying relations (2.5) is called the *infinitesimal object* of $\mathbb{Z}_2$-Delsarte generalized translation operators.

Let $\mathfrak{g}$ and $\overline{\mathfrak{g}}$ be Lie algebras of the Lie groups G and $\overline{G}$, respectively, and let $\Gamma = (\mathrm{id}, \gamma)$ and $\overline{\Gamma} = (\mathrm{id}, \overline{\gamma})$ be the subgroups of $\operatorname{Aut} G$ and $\operatorname{Aut} \overline{G}$ determining the $\mathbb{Z}_2$-Delsarte generalized translation operators R_g $(g \in G)$ and $R_{\overline{g}}$ $(\overline{g} \in \overline{G})$. Two families of $\mathbb{Z}_2$-Delsarte generalized translation operators R_g $(g \in G)$ and $R_{\overline{g}}$ $(\overline{g} \in \overline{G})$ are called *isomorphic* if there exists a diffeomorphism $\varphi\colon G \to \overline{G}$ such that

$$\varphi(e) = \overline{e}, \quad \varphi(g^{-1}) = \varphi(g)^{-1},$$

and

$$\left(R_{g_1}(f \circ \varphi)\right)(g_2) = \left(R_{\varphi(g_1)} f\right)(\varphi(g_2))$$

for any $g, g_1, g_2 \in G$ and $f \in C^\infty(\overline{G})$ (cf. the definition in Section 4 of Chapter 1).

Denote by Φ the differential of the mapping φ at the identity element e of the group G. Since φ is a diffeomorphism, we have $\dim \mathfrak{g} = \dim \overline{\mathfrak{g}}$. For any function $f \in C^\infty(\overline{G})$, we can write

$$f(\varphi(g_1 g_2)) + f(\varphi(g_1 \gamma(g_2))) = f(\varphi(g_1)\varphi(g_2)) + f(\varphi(g_1)\overline{\gamma}(\varphi(g_2))) \tag{2.7}$$

$$(g_1, g_2 \in G).$$

By setting $g_1 = e$ in (2.7), we get $\varphi(\gamma(g_2)) = \bar{\gamma}(\varphi(g_2))$. This yields the relation $\Phi \circ \gamma = \bar{\gamma} \circ \Phi$, where γ and $\bar{\gamma}$ are the actions of Γ and $\bar{\Gamma}$ in $\mathfrak{g}$ and $\bar{\mathfrak{g}}$, respectively. Relation (2.7) also implies that, for any $g_1, g_2 \in G$, one of the following relations is true:

(a) $\varphi(g_1 g_2) = \varphi(g_1)\varphi(g_2)$,

(b) $\varphi(g_1 g_2) = \varphi(g_1)\varphi(\gamma(g_2))$.

Let G_e be the connected component of G which contains the identity.

Denote

$$H = \{h \in G_e \,|\, \gamma(h) = h\}.$$

Clearly, H is a closed subgroup of the group G. Hence, H is a topological Lie subgroup of the group G. Let $\mathfrak{g}_1$ be its Lie algebra. Relations (a) and (b) coincide on H. For any $h \notin H$, we denote

$$A_h = \{g \in G_e \,|\, \varphi(gh) = \varphi(g)\varphi(h)\}$$

and

$$B_h = \{g \in G_e \,|\, \varphi(gh) = \varphi(g)\varphi(\gamma(h))\}.$$

It is clear that A_h and B_h are closed subsets of the connected group G_e and, moreover, $A_h \cup B_h = G_e$ and $A_h \cap B_h = \varnothing$. Since $e \in A_h$ and $A_h \neq \varnothing$, we have $B_h = \varnothing$. Consequently, relation (a) holds for all $h \in G_e$, which means that φ is an isomorphism from G_e into $\bar{G}_e$. Hence, Φ is an isomorphism of the Lie algebras $\mathfrak{g}$ and $\bar{\mathfrak{g}}$.

Let $X_1, \ldots, X_m, X_{m+1}, \ldots, X_n$ be a basis in $\mathfrak{g}$ such that $\gamma X_i = X_i$ $(i > m)$ and $\gamma X_k = -X_k$ $(k \leq m)$. Since $\Phi \circ \gamma = \bar{\gamma} \circ \Phi$, we conclude that the basis

$$\bar{X}_1 = \Phi X_1, \ldots, \bar{X}_n = \Phi X_n$$

in $\bar{\mathfrak{g}}$ possesses the same properties relative to the action of the group $\bar{\Gamma}$.

Note that the Lie algebras $\mathfrak{g}$ and $\bar{\mathfrak{g}}$ are isomorphic and, therefore, their structural constants in the bases X_i and $\bar{X}_i$ $(i = 1, \ldots, n)$, coincide, i.e., $c_{ij}^k = \bar{c}_{ij}^k$ $(i, j, k = 1, \ldots, n)$. Hence, the relations between the canonical generators

$$\bar{Z}_i = \bar{X}_i \quad (i > m)$$

and

$$\overline{Z}_{kl} = \frac{1}{2}(\overline{X}_k \overline{X}_l + \overline{X}_l \overline{X}_k) \quad (1 \le k,\ l \le k)$$

have the form (2.5), i.e., the *infinitesimal objects of isomorphic $\mathbb{Z}_2$-Delsarte generalized translation operators coincide.*

The $\mathbb{Z}_2$-generalized translation operators R_g $(g \in G)$ and $R_{\overline{g}}$ $(\overline{g} \in \overline{G})$ are called *locally isomorphic* if there exists an isomorphism $\Phi : \mathfrak{g} \to \overline{\mathfrak{g}}$ between Lie algebras of the groups G and $\overline{G}$ such that $\Phi \circ \gamma = \overline{\gamma} \circ \Phi$, where γ and $\overline{\gamma}$ are the actions of the group $\mathbb{Z}_2$ in $\mathfrak{g}$ and $\overline{\mathfrak{g}}$, respectively.

The following theorem demonstrates that relations (2.5) are sufficient for the reconstruction of $\mathbb{Z}_2$-Delsarte generalized translation operators:

Theorem 2.4. *Let $\mathcal{A}$ be a finitely generated associative algebra whose generators $J_1, \dots, J_p$ are canonical generators for some $\mathbb{Z}_2$-Delsarte generalized translation operators and satisfy the canonical relations. If the quadratic part of at least one of these relations differs from the commutator, then the $\mathbb{Z}_2$-Delsarte generalized translation operators can be uniquely (to within a local isomorphism) reconstructed for given $\mathcal{A}$ in a certain neighborhood of the identity of the group G. Otherwise, unique reconstruction is, generally speaking, impossible.*

The proof of this theorem is quite cumbersome and we do not present it here.

Denote by V the linear span of the elements Z_i $(i > m)$ and Z_{kl} $(k, l \le m;$ we assume that $Z_{kl} = Z_{lk})$. Consider the tensor algebra T of the vector space V. Recall that

$$T = T^0 \oplus T^1 \oplus \dots,$$

where $T^0 = \mathbb{C}$ (or $\mathbb{R}$ if $\mathfrak{g}$ is real) and

$$T^n = V \otimes \dots \otimes V \quad (n \text{ factors}).$$

Multiplication in T is defined as ordinary tensor multiplication.

Let J be a two-sided ideal of the algebra T generated by tensors of the form

$$Z_i \otimes Z_j - Z_j \otimes Z_i - c_{ij}^t Z_t \quad (i, j > m),$$

$$Z_i \otimes Z_{kl} - Z_{kl} \otimes Z_i - c_{ik}^s Z_{sl} - c_{il}^s Z_{ks} \quad (i > m;\ k \le l \le m),$$

$$Z_{kl} \otimes Z_{pq} - Z_{kp} \otimes Z_{lq} + \frac{1}{4}\left(c_{kl}^i c_{pq}^j - 2c_{lp}^i c_{kq}^j - c_{kp}^i c_{lq}^j\right) Z_i \otimes Z_j$$

$$+ \frac{1}{2} c_{kl}^i Z_i \otimes Z_{pq} + \frac{1}{2} c_{pq}^i Z_{kl} \otimes Z_i - \frac{1}{2} c_{lp}^i Z_i \otimes Z_{kq}$$

$$- \frac{1}{2} c_{lp}^{i} Z_{kq} \otimes Z_i - \frac{1}{2} c_{kp}^{i} Z_i \otimes Z_{lq} - \frac{1}{2} c_{lq}^{i} Z_{kp} \otimes Z_i - \frac{1}{2} c_{lp}^{i} c_{iq}^{s} Z_{sk}$$

$$+ \frac{1}{2} c_{lp}^{i} c_{ik}^{s} Z_{sq} - \frac{1}{2} c_{lp}^{i} c_{ki}^{s} c_{sq}^{j} Z_j \quad (1 \le k, l, p, q \le m). \tag{2.8}$$

The associative algebra $U(V) = T/J$ is called an *algebra enveloping relations (2.5).*

The composition σ of the canonical mappings $V \to T \to U(V)$ is called the *canonical mapping* of V into $U(V)$.

Lemma 2.3. *Let σ be the canonical mapping of V into $U(V)$ and let $\tau: V \to A$ be a linear mapping from V into an associative algebra A with identity such that relations (2.5) hold for the elements*

$$T_i = \tau(Z_i) \quad (i > m) \quad \text{and} \quad T_{kl} = \tau(Z_{kl}) \quad (k, l \le m).$$

Then there exists a unique homomorphism τ': $U(V) \to A$ such that $\tau'(1) = 1$ and $\tau' \circ \sigma = \tau$.

Proof. The mapping τ' is unique because $U(V)$ is generated by the identity 1, $\sigma(Z_i)$, and $\sigma(Z_{kl})$ $(i > m;\ k, l \le m)$.

Denote by φ the homomorphism from the algebra T into A which extends τ and is such that $\varphi(1) = 1$. Since the elements

$$T_i = \tau(Z_i) \quad \text{and} \quad T_{kl} = \tau(Z_{kl}) \quad (i > m;\ k, l \le m)$$

satisfy relations (2.5), we have $\varphi(J) = 0$ and, hence, under factorization with respect to the ideal J, the homomorphism φ defines a homomorphism τ' from the algebra $U(V)$ into A such that $\tau'(1) = 1$ and $\tau \circ \sigma = \tau$.

∎

Let

$$\sigma(Z_i) = J_i \quad \text{and} \quad \sigma(Z_{kl}) = J_{kl} \quad (1 \le k \le l \le m;\ m < i \le n).$$

We introduce partial ordering in the set of indices φ of the form (k, l), i $(k \le l \le m;$ $i > m)$ by setting $(k, l) < (p, q)$ if $\max(k, l) \le \min(p, q)$, $(k, l) < i$ $(i > m)$ and consider the set $\{m+1, \ldots, n\}$ with natural ordering. With every finite sequence of indices $I = (\varphi_1, \ldots, \varphi_p)$ (each index φ_λ is either (k_λ, l_λ) or i_λ), we associate an element

$$Y_I = Y_{\varphi_1} \ldots Y_{\varphi_p}$$

of the algebra $U(V)$. A sequence $I = (\varphi_1, \ldots, \varphi_p)$ is called increasing whenever $\varphi_1 < \varphi_2 < \ldots < \varphi_p$.

Theorem 2.5. *Elements of the form* Y_I, *where* I *is a finite increasing sequence, form a basis in the vector space* $U(V)$.

Theorem 2.5 can be proved by using the Bergman diamond lemma (see [Berg]).

According to Theorem 2.5, one can construct a homomorphism $\omega : U(V) \to U(\mathfrak{g})_\Gamma$ by setting

$$\omega(Y_I) = Z_I = Z_{\varphi_1} \ldots Z_{\varphi_p}$$

for any sequence $I = (\varphi_1, \ldots, \varphi_p)$.

Since elements of the form Y_I, where I is an increasing sequence of indices, form a basis of the algebra $U(V)$, elements of the form Z_I, where I is an increasing sequence of indices, are linear generators of $U(\mathfrak{g})_\Gamma$. Each element of this sort can be written as

$$Z_I = Z_{k_1 k_1}^{\alpha_1} Z_{k_1 k_2}^{\delta_{1,2}} Z_{k_2 k_2}^{\alpha_2} Z_{k_2 k_3}^{\delta_{2,3}} \ldots Z_{k_p k_p}^{\alpha_p} Z_{i_1} Z_{i_2} \ldots Z_{i_q}, \tag{2.9}$$

where

$$1 \le k_1 < k_2 < \ldots < k_p \le m, m+1 \le i_1 \le \ldots \le i_q \le n,$$

$$\alpha_j = 0, 1, \ldots, \quad \delta_{j,j+1} \in \{0, 1\}.$$

By expressing the generators Z_{kl} and Z_i of the algebra $U(\mathfrak{g})_\Gamma$ in terms of the elements X_j of a basis of the Lie algebra $\mathfrak{g}$, we conclude that each element Z_I of the form (2.9) admits a representation

$$Z_I = X_{k_1}^{2\alpha_1 + \delta_{1,2}} X_{k_2}^{2\alpha_2 + \delta_{1,2} + \delta_{2,3}} \ldots X_{k_p}^{2\alpha_p + \delta_{p-1,p}} X_{i_1} X_{i_2} \ldots X_{i_q} + W, \tag{2.10}$$

where the degree of filtration of the element W is lower than that of Z_I.

By the PBW theorem, Z_I are linearly independent. Hence, the elements $Z_I \in U(\mathfrak{g})_\Gamma$ *given by* (2.9) *form a basis in* $U(\mathfrak{g})_\Gamma$ and the algebra $U(\mathfrak{g})_\Gamma$ is isomorphic to the algebra $U(V)$ which is an algebra enveloping relations (2.5). This means that all relations in $U(\mathfrak{g})_\Gamma$ follow from relations (2.5).

Example. Let $\mathfrak{g} = sl_2(\mathbb{R})$. In a basis

$$X = \begin{pmatrix} 0 & 1 \\ 1 & 0 \end{pmatrix}, \quad Y = \begin{pmatrix} 1 & 0 \\ 0 & -1 \end{pmatrix}, \quad K = \begin{pmatrix} 0 & 1 \\ -1 & 0 \end{pmatrix}, \tag{2.11}$$

we define the action of an involutive automorphism σ which determines the Cartan decomposition of $sl_2(\mathbb{R})$, i.e., $\sigma(X) = -X$, $\sigma(Y) = -Y$, and $\sigma(K) = K$, and consider the algebra $U(\mathfrak{g})_\Gamma$ with $\Gamma = \{\text{id}, \sigma\}$.

The generators of this algebra

$$Z_{11} = X^2, \quad Z_{12} = \frac{1}{2}(XY + YX), \quad Z_{22} = Y^2, \quad Z_3 = K$$

satisfy the relations which readily follow from (2.5) after evident transformations. Indeed,

$$[Z_3, Z_{11}] = [Z_{22}, Z_3] = 4Z_{12},$$

$$[Z_{12}, Z_3] = 2Z_{11} + 2Z_{22},$$

$$Z_{12}Z_{11} = Z_{11}Z_{12} + 4Z_3Z_{11} - 4Z_3,$$

$$Z_{22}Z_{12} = Z_{12}Z_{22} + 4Z_3Z_{22} + 8Z_{12} - 4Z_3,$$

$$Z_{11}Z_{22} = Z_{12}^2 + 3Z_3^2 - 3Z_3Z_{12} - Z_{12}Z_3 + 4Z_{22},$$

$$Z_{22}Z_{11} = Z_{12}^2 + 3Z_3^2 + 3Z_3Z_{12} + Z_{12}Z_3 + Z_{11}.$$

Remark. The infinitesimal object of Delsarte generalized translation operators can easily be described in the case where Γ consists of inner automorphisms of the group G. If Γ is isomorphic to a Lie subgroup H of G, then, clearly, $U(\mathfrak{g})_\Gamma$ coincides with the centralizer of the Lie algebra $\mathfrak{h}$ of the group H in $U(\mathfrak{g})$.

In the case where Γ is the group of all inner automorphisms of a compact Lie group G, the algebra $U(\mathfrak{g})_\Gamma$ (in this case, it coincides with the center $Z(\mathfrak{g})$ of the algebra $U(\mathfrak{g})$) was described by Gelfand [Gel2]. Moreover, Duflo proved that the algebra $Z(\mathfrak{g})$ is isomorphic to the subalgebra $S(\mathfrak{g})^G$ which consists of G-invariant elements of the symmetric algebra $S(\mathfrak{g})$ (note that, generally speaking, this isomorphism does not coincide with symmetrization because the latter does not preserve the operation of multiplication). If G is a complex semisimple Lie group, then, by virtue of the Chevalley theorem, there exists a one-to-one correspondence between $Z(\mathfrak{g})$ and the algebra of all polynomials over the Cartan subalgebra invariant with respect to the Weyl group. In this case, $Z(\mathfrak{g})$ has l independent generators, where $l = \operatorname{rank} \mathfrak{g}$.

2.2. Delsarte-Type Generalized Translation Operators and Generalized Lie Algebras. In the present subsection, we describe the infinitesimal object of Delsarte-type generalized translation operators introduced by Gurevich [Gur2].

Let G be a connected simply connected Lie group with Lie algebra $\mathfrak{g}$. In this case, the semigroup $\operatorname{End} G$ of endomorphisms of the group G is isomorphic to the semigroup $\operatorname{End} \mathfrak{g}$ of endomorphisms of the algebra $\mathfrak{g}$ and is an algebraic submanifold of the semigroup $\operatorname{End} \mathbb{R}^n$ $(\operatorname{End} \mathbb{C}^n)$.

Let $\Gamma \subset \operatorname{End} G$ be the semigroup (with identity) of endomorphisms of the group G such that Γ is a closed smooth (analytic) submanifold of the manifold $\operatorname{End} G$. Generalized translation operators R_h $(h \in G)$ acting in the space $C^\infty(G)$ are called *Delsarte-type* generalized translation operators if $R_e = I$ and

$$(R_h f)(g) = \langle f(A(g)B(h)), \mu_{A,B} \rangle \quad (g, h \in G;\ f \in C^\infty(G)), \tag{2.12}$$

where $\mu_{A,B} \in \mathcal{D}(\Gamma \times \Gamma)$. If $\Gamma \subset \operatorname{Aut} G$ is a compact group and $\mu_{A,B} = \delta_E \otimes m_\Gamma$, then, clearly, R_h are Delsarte generalized translation operators; here, E is the identity of the group Γ and m_Γ is its Haar measure.

The operators R_s defined by relation (2.12) are not necessarily generalized translation operators. In what follows, we present a sufficient condition for R_s to be generalized translations.

We say that the space $\Phi = C^\infty(\Gamma)$ is endowed with *algebraic structure* if Φ is an associative algebra with identity $1(A) \equiv 1$ $(A \in \Gamma)$ with respect to an operation of associative continuous multiplication

$$\Phi \times \Phi \ni (\varphi, \psi) \mapsto \varphi \circ \psi \in \Phi$$

invariant under left translations in the semigroup Γ, i.e.,

$$({}_A\varphi \circ {}_A\psi)(\cdot) = {}_A(\varphi \circ \psi)(\cdot) \quad (A \in \Gamma;\ A^\varphi(M) = \varphi(AM)).$$

Theorem 2.6. *Assume that the space $\Phi = C^\infty(\Gamma)$ is endowed with algebraic structure. Consider a functional $\mu_{A,B}$ defined in the space $C^\infty(\Gamma) \hat{\otimes} C^\infty(\Gamma)$ by the formula*

$$\langle \varphi(A)\psi(B), \mu_{A,B} \rangle = (\varphi \circ \psi)(E) \quad (\varphi, \psi \in C^\infty(\Gamma)). \tag{2.13}$$

Then the operators R_s $(s \in G)$ defined by relations (2.12), where $\mu_{A,B}$ is the functional introduced in (2.13), form a family of generalized translation operators.

Proof. Since $1(A)$ is the right identity of the operation $\circ$, we have

$$(R_e f) = \langle f(A(g)1(B)), \mu_{A,B}\rangle = (f(\cdot(g)\circ 1))(E) = f(E(g)) = f(g) \quad (g \in G).$$

In view of the fact that the operation of multiplication $\circ$ is left invariant in Φ, the law of multiplication can be written in the form

$$(\varphi \circ \psi)(C) = \langle \varphi(CA)\psi(CB), \mu_{A,B}\rangle \tag{2.14}$$

By virtue of (2.14) and the fact that the operation of multiplication in Φ is associative, we obtain

$$\begin{aligned}\langle\langle \varphi(MA)\psi(MCB)\theta(MBD), \mu_{C,D}\rangle, \mu_{A,B}\rangle \\ &= (\varphi\circ(\psi\circ\theta))(M) = ((\varphi\circ\psi)\circ\theta)(M) \\ &= \langle\langle \varphi(MAC)\psi(MAD)\theta(MB), \mu_{C,D}\rangle, \mu_{A,B}\rangle\end{aligned}$$

For $M = E$, we get

$$\langle\langle \varphi(A)\psi(BC)\theta(BD), \mu_{C,D}\rangle, \mu_{A,B}\rangle = \langle\langle \varphi(AC)\psi(AD)\theta(B), \mu_{C,D}\rangle, \mu_{A,B}\rangle \tag{2.15}$$

On the other hand, it follows from (2.12) that the axiom of associativity for the operators R_s takes the form

$$\begin{aligned}\left(R_q^h(R_h f)\right)(g) &= \langle\langle f(A(g)BC(h)BD(q)), \mu_{A,B}\rangle, \mu_{C,D}\rangle \\ &= \langle\langle f(AC(g)AD(h)B(q)), \mu_{A,B}\rangle, \mu_{C,D}\rangle = \left(R_h^g(R_q)\right)(g).\end{aligned}$$

This relation is indeed true by virtue of (2.15). Thus, R_s are generalized translation operators.

■

It is easy to see that, for any Γ, the ordinary operation of multiplication of functions on Γ is associated with operators of group translation. Delsarte generalized translation operators correspond to the operation of multiplication

$$(f\circ g)(C) = f(C)\int_\Gamma g(BC)dB,$$

where dB is the Haar measure on the compact group of automorphisms of Γ normalized by the condition

$$\int_{\Gamma} dB = 1.$$

Let us now construct an example of Delsarte-type generalized translation operators which are not Delsarte generalized translation operators. Let $G = SU(2)$ and let X, Y, Z be a basis in its Lie algebra $su(2)$ such that

$$[X, Y] = Z, \quad [Y, Z] = X, \quad \text{and} \quad [Z, X] = Y.$$

Consider two automorphisms $A, B \in \mathrm{Aut}(su(2))$ given in the basis X, Y, Z by the matrices $A = \mathrm{diag}(-1, 1, -1)$ and $B = \mathrm{diag}(1, -1, -1)$. The automorphisms A and B are involutive and generate a commutative group

$$\Gamma = \{E, A, B, AB = BA\}.$$

The characters $\chi_0 \equiv 1$, χ_1, χ_2, and χ_3 are determined by the relations

$$\chi_1(A) = \chi_2(B) = -1, \quad \chi_2(A) = \chi_1(B) = 1, \quad \text{and} \quad \chi_3 = \chi_1\chi_2$$

and form a basis in the space Φ of functions defined on Γ. In Φ, we introduce an algebraic structure by the relations

$$\chi_0 = \chi_1 \circ \chi_1 = \chi_2 \circ \chi_2, \quad \chi_1 \circ \chi_2 = -\chi_2 \circ \chi_1 = \chi_3,$$

$$\chi_0 \circ \chi_i = \chi_i \circ \chi_0 = \chi_i \quad (i = 0, 1, 2, 3).$$

The generalized translation operators associated with this algebra have the form

$$(R_s f)(t) = \frac{1}{2}\,[f(ts) + f(B(t)s) + f(tA(s)) - f(B(t)A(s))]$$

$$(t, s \in SU(2);\ f \in C^{\infty}(SU(2)))$$

By direct calculations, one can show that the first-order generators of these generalized translation operators X_1, X_2, and X_3 satisfy the relations

$$X_1X_2 + X_2X_1 = -X_3, \quad X_2X_3 + X_3X_2 = X_1, \quad X_3X_1 + X_1X_3 = X_2.$$

It is clear that, by using this procedure, one can construct many examples of nontrivial Delsarte-type generalized translation operators.

Let us establish the following properties of right generators R_α of arbitrary generalized translation operators R_s $(s \in G)$:

Lemma 2.4. *(i) If* $(R_\alpha f)(e) = a_\beta (D^{(\beta)} f)(e)$ *(summation is carried out over the repeating indices; all sums are finite), then either* $(R_\alpha f)(e) = (D^{(\alpha)} f)(e)$ *or generators are linearly dependent;*

(ii) if the generators R_α *from (i) are linearly independent, then the relation*

$$a^{\alpha\beta} R_\alpha R_\beta = c^\gamma R_\gamma \tag{2.16}$$

is equivalent to its restriction to the point e.

Proof. We prove (i). Since

$$\begin{aligned}(R_s R_\alpha f)(t) &= R_s D_r^{(\alpha)} R_r f(t)\big|_{r=e} \\ &= \left(D_r^{(\alpha)} R_s^t R_r f\right)(t)\big|_{r=e} = \left(D_r^{(\alpha)} R_r^s R_s f\right)(t)\big|_{r=e} = \left(R_\alpha^s R_s f\right)(t),\end{aligned}$$

we have

$$\begin{aligned}(R_\alpha f)(t) &= (R_s (R_\alpha f))(t)\big|_{s=e} \\ &= \left(R_\alpha^s (R_s f)\right)(t)\big|_{s=e} = a_\beta D_s^{(\beta)} (R_s f)(t)\big|_{s=e} = \sum_\beta a_\beta (R_\beta f)(t)\end{aligned}$$

for all $t \in Q$. This proves (i).

Let us prove (ii). It is necessary to show that the relation

$$a^{\alpha\beta} (R_\alpha (R_\beta f))(e) = c^\gamma (R_\gamma f)(e)$$

implies (2.16). Indeed, by using (i), we easily obtain

$$\begin{aligned}a^{\alpha\beta} (R_\alpha R_\beta f)(t) &= a^{\alpha\beta} (R_s R_\alpha R_\beta f)(t)\big|_{s=e} = a^{\alpha\beta} \left(R_\alpha^s R_\beta^s R_s f\right)(t)\big|_{s=e} \\ &= c^\gamma \left(R_\lambda^s R_s f\right)(t)\big|_{s=e} = \sum_\gamma c^\gamma D_s^{(\gamma)} (R_s f)(t)\big|_{s=e} = c^\gamma (R_\gamma f)(t).\end{aligned}$$

■

We now describe the infinitesimal object for Delsarte-type generalized translation operators. As in Subsection 2.1, let $\pi(B) \in \operatorname{End} U(\mathfrak{g})$ denote the action of an endomorphism $B \in \Gamma$ on the universal enveloping algebra $U(\mathfrak{g})$ of the Lie algebra $\mathfrak{g}$ of the group G. The mapping

$$\Gamma \ni B \mapsto \pi(B) \in \operatorname{End} U(\mathfrak{g})$$

is clearly a representation of the semigroup Γ.

Let X_α, $\alpha = (\alpha_1, \ldots, \alpha_n)$, be generators of right translations in the group G. By virtue of the Schwartz theorem, they form a basis in $U(\mathfrak{g})$.

Let $\pi_\alpha^\beta(B)$ be matrix elements of a representation π of the semigroup Γ in the basis X_α. By differentiating relation (2.12) $\alpha = (\alpha_1, \ldots, \alpha_n)$ times with respect to s and setting $s = e$, we obtain

$$(R_\alpha f)(t) = \langle \pi_\alpha^\beta(B)\, X_\beta f(A(t)),\, \mu_{A,B} \rangle,$$

whence, in particular, it follows that the generators R_α of the Delsarte-type generalized translation operators satisfy condition (i) of Lemma 2.4. Suppose that the generators R_α $(\alpha \in \Upsilon)$ are linearly independent for a collection of multiindices Υ. According to Lemma 2.4 (i),

$$R_\alpha\big|_e = X_\alpha\big|_e = D^{(\alpha)} \quad (\alpha \in \Upsilon).$$

Thus,

$$\begin{aligned} (R_\alpha R_\beta f)(e) &= D^{(\alpha)}(R_\beta f)(e) \\ &= \langle \pi_\beta^\delta(B)\, D_t^{(\alpha)} X_\delta f(A(t))_{t=e},\, \mu_{A,B} \rangle \\ &= \langle \pi_\beta^\delta(B) X_\alpha (X_\delta f \circ A)(e),\, \mu_{A,B} \rangle \\ &= \langle \pi_\beta^\delta(B)\, \pi_\alpha^\gamma(A)\, X_\gamma X_\delta f(e),\, \mu_{A,B} \rangle \\ &= \langle \pi_\beta^\delta(B)\, \pi_\alpha^\gamma(A),\, \mu_{A,B} \rangle (X_\gamma X_\delta f)(e) \\ &= \left(\pi_\alpha^\gamma \circ \pi_\beta^\delta\right)(E)\, X_\gamma X_\delta f(e) \qquad (\alpha, \beta \in \Upsilon). \end{aligned} \tag{2.17}$$

These relations enable us to formulate the following procedure of finding relations for generators of Delsarte-type generalized translation operators. If it is possible to solve relations (2.17), i.e., to express $(X_\gamma X_\delta f)(e)$ in terms of $(R_\alpha R_\beta f)(e)$, then, by using the known relations for right generators of the group G, one can find relations for generators of Delsarte-type generalized translation operators at the point $t = e$. Lemma 2.4 (ii) enables us to omit the requirement $t = e$. As a result, we arrive at relations for the generators R_α.

In what follows, we restrict ourselves to the case where $\Upsilon = \{\alpha \mid |\alpha| = 1\}$, i.e., we assume that the first-order generators of Delsarte-type generalized translation operators are nondegenerate and linearly independent. Instead of the multiindex $\alpha^{(k)} = (0, \ldots, 1, \ldots, 0) \in \Upsilon$, we write the number k of the nonzero component of $\alpha^{(k)}$.

Denote

$$R_{\alpha^{(k)}} = R_k, \quad X_{\alpha^{(k)}} = X_k, \quad \pi_{\alpha^{(k)}}^{\alpha^{(l)}} = \pi_k^l, \quad d_{kl}^{ij} = \left(\pi_k^i \circ \pi_l^j\right)(E)$$

$$(i, j, k, l = 1, \ldots, n; \; n = \dim \mathfrak{g}).$$

Relation (2.17) now turns into the equality

$$R_i R_j = d_{ij}^{kl} X_k X_l.$$

Theorem 2.7. *Assume that the first-order generators of Delsarte-type generalized translation operators* R_k $(k = 1, \ldots, n)$ *are nondegenerate and linearly independent and all matrices* $\left(d_{ij}^{kl}\right)$ *are invertible.*

Then the generators $R_1, \ldots, R_n$ *are generators of the algebra* $\mathcal{D}_e^R(G)_\Gamma$ *generated by all* R_α *and, moreover,*

$$R_i R_j - h_{ij}^{kl} R_k R_l = b_{ij}^m R_m \quad (i, j = 1, \ldots, n), \tag{2.18}$$

where

$$h_{ij}^{kl} = d_{ij}^{pq} d_{qp}^{-1kl}, \quad b_{ij}^m = d_{ij}^{pq} c_{pq}^m,$$

and c_{pq}^m *are structural constants of the Lie algebra* $\mathfrak{g}$ *in the basis* $X_1, \ldots, X_n$.

Proof. It is clear that the support of the convolution $F * G$ of any functionals F, $G \in \mathcal{D}^R(G)_\Gamma$ whose supports coincide with the identity element e of the group G also coincides with the identity element (we use the same notation as in Subsection 2.1). Thus, the generators R_α are linear generators of $\mathcal{D}_e^R(G)_\Gamma$.

Let $\{R_\alpha \mid \alpha \in \Xi\}$ be a basis in $\mathcal{D}_e^R(G)_\Gamma$. For any $\alpha \in \Xi$, one can indicate polynomials $P_\gamma(X_1, \ldots, X_n)$ such that

$$(R_\alpha f)(e) = \langle (\pi^\gamma_\alpha(B)\, X_\gamma f)(A(t)),\, \mu_{A,B}\rangle$$

$$= \langle (\pi^\gamma_\alpha(B)\, P_\gamma(X_1, \dots, X_n) f)(A(t)),\, \mu_{A,B}\rangle. \tag{2.19}$$

Since the matrices $\left(d^{kl}_{ij}\right)$ are nondegenerate, for any γ, one can find polynomials Q^k_γ and Q_γ (which may be equal to zero) such that

$$P_\gamma(X_1, \dots, X_n) = X_k Q^k_\gamma(R_1, \dots, R_n) + Q_\gamma(R_1, \dots, R_n).$$

By substituting the right-hand side of this equality for P_γ in (2.19) and setting $t = e$, by virtue of Lemma 2.4 (i), we obtain

$$(R_\alpha f)(e) = \left(c^\gamma_\alpha \left[\sum_{k=1}^{n} D^{\alpha^{(k)}} Q^k_\gamma(R_1, \dots, R_n) + Q_\gamma(R_1, \dots, R_n) \right] f \right)(e)$$

$$= (c^\gamma_\alpha\, [R_k Q^k_\gamma(R_1, \dots, R_n) + Q_\gamma(R_1, \dots, R_n)\,] f)(e). \tag{2.20}$$

By repeating the reasoning used in Lemma 2.4 (ii), we conclude that (2.20) holds for all $t \in G$, i.e., R_k generates $\mathcal{D}^R_e(G)_\Gamma$.

To prove (2.18), we pass in the equality

$$X_k X_l - X_l X_k = c^m_{kl} X_m$$

from the generators X_k of the Lie algebra $\mathfrak{g}$ to the generators R_k. As a result, we obtain

$$\left(d^{-1ij}_{kl} R_i R_j - d^{-1pq}_{lk} R_p R_q\right) f(e) = c^m_{kl} R_m f(e) \tag{2.21}$$

because

$$(R_m f)(e) = (X_m f)(e) = \left(D^{\alpha^{(m)}} f\right)(e)$$

by virtue of Lemma 2.4 (i). By multiplying relation (2.21) by $\left(d^{-1ij}_{kl}\right)$ and using Lemma 2.4 (ii), we arrive at relation (2.18).

■

A finite-dimensional linear space V is called a *generalized Lie algebra or an S-algebra* if it is an algebra with multiplication $C\colon V \otimes V \to V$ and there exists a tensor

$H: V \otimes V \to V \otimes V$ which is called *symmetry* such that the following conditions are satisfied:

(1) $H^2 = \text{id}$.

(2) $(H \otimes 1)(1 \otimes H)(H \otimes 1) = (1 \otimes H)(H \otimes 1)(1 \otimes H)$, i.e., H is a solution of the Yang–Baxter equation.

(3) $CH = -C$.

(4) $C(C \otimes 1) + C(C \otimes 1)(1 \otimes H)(H \otimes 1) + C(C \otimes 1)(H \otimes 1)(1 \otimes H) = 0$.

(5) $H(C \otimes 1) = (1 \otimes C)(H \otimes 1)(1 \otimes H)$.

Theorem 2.8. *The space* $V = \text{l.s.}\{R_1, \dots, R_n\}$ *is a generalized Lie algebra with multiplication* $C(R_i \otimes R_j) = b_{ij}^m R_m$ *and symmetry* $H = \left(h_{ij}^{kl}\right)$.

Proof. It is clear that $H^2 = \text{id}$, i.e., $h_{ij}^{pq} h_{pq}^{kl} = \delta_{ij}^{kl}$. Let us now show that H is a solution of the Yang–Baxter equation, i.e.,

$$h_{ij}^{kl} h_{lm}^{pq} h_{kp}^{st} = h_{jm}^{uv} h_{iu}^{sr} h_{rv}^{tq}.$$

To do this, we note that, by virtue of (2.14), one can write

$$\pi_\alpha^\gamma \circ \pi_\beta^\delta(C) = \langle \pi_\alpha^\gamma(CA)\, \pi_\beta^\delta(CB),\, \mu_{A,B} \rangle$$

$$= \langle \pi_\sigma^\gamma(A)\, \pi_\tau^\delta(B),\, \mu_{A,B} \rangle\, \pi_\alpha^\sigma(C)\, \pi_\beta^\tau(C)$$

$$= (\pi_\sigma^\gamma \circ \pi_\tau^\delta)(E)\, \pi_\alpha^\sigma(C)\, \pi_\beta^\tau(C) = d_{\sigma\tau}^{\gamma\delta}\, \pi_\alpha^\sigma(C)\, \pi_\beta^\tau(C) \quad (C \in \Gamma).$$

Consider the conjugate law of multiplication

$$\pi_\alpha^\gamma * \pi_\beta^\delta(C) = d_{\alpha\beta}^{\sigma\tau}\, \pi_\sigma^\gamma(C) \pi_\tau^\delta(C) \quad (C \in \Gamma) \tag{2.22}$$

(this type of multiplication can be expressed in terms of the original operation of multiplication by passing to conjugate quantities, namely, $f * g = (f^* \circ g^*)^*$). Multiplication $*$ is associative. By using the fact that ordinary multiplication is commutative and relation (2.22), we obtain

$$\pi_l^k * \pi_q^p = d_{lq}^{ij} d_{ji}^{-1ts} \pi_t^p * \pi_s^k = h_{lq}^{ts} \pi_t^p * \pi_s^k. \tag{2.23}$$

Condition (2) from the definition of generalized Lie algebras now follows from relation (2.23) and the associativity of multiplication $*$. All other conditions from the definition of generalized Lie algebras admit direct verification.

■

Note that the tensor H associated with a generalized Lie algebra corresponding to Delsarte-type generalized translation operators is equivalent to the tensor

$$H' = \left(h_{ij}^{\prime kl} = \delta_{ij}^{lk}\right),$$

associated with an ordinary Lie algebra. Actually, in this case, the tensor H can be represented in the form

$$h_{ij}^{kl} = d_{ij}^{pq} d_{qp}^{-1kl} = d_{ij}^{pq} \delta_{pq}^{st} d_{ts}^{-1kl}.$$

Note that finite-dimensional graded Lie algebras, including Lie superalgebras, are generalized Lie algebras (see [Moso]). For any generalized Lie algebra, one can construct a "generalized formal Lie group" ([Gur4]). It is interesting to describe generalized Lie algebras corresponding to Delsarte-type generalized translation operators.

2.3. Infinitesimal Algebra of the Hypercomplex System $L_1(G, H)$. Let G be a Lie group and let H be its compact subgroup. In Section 2 of Chapter 2, we studied the hypercomplex system $L_1(G, H)$. The generalized translation operators associated with this hypercomplex system have the form (see relation (2.1) in Chapter 2)

$$(R_{g_2} f)(g_1) = \int_H f(g_1 h g_2)\, dh,$$

where $g_1, g_2 \in G$ and $f \in C^\infty(G)$ is a biinvariant function. By differentiating this relation, we arrive at the following expression for the right generator R_α of the generalized translation operator R_g:

$$(R_\alpha f)(g) = \int_H (X_\alpha f)(gh)\, dh \quad (\alpha = (\alpha_1, \ldots, \alpha_n);\ g \in G), \tag{2.24}$$

where X_α is the generator of the right translations in the group G.

Since f is biinvariant, this yields

$$\begin{aligned}(R_\alpha f)(g) &= \int_H D_{g_1}^{(\alpha)} f(ghg_1)|_{g_1=e}\, dh \\ &= \int_H\int_H D_{g_1}^{(\alpha)} f(ghh_1g_1h_1^{-1})|_{g_1=e}\, dh dh_1 \\ &= \int_H\int_H ((\mathrm{Ad}_{h_1} X^{(\alpha)})f)(gh)\, dh dh_1 \quad (g\in G). \end{aligned} \tag{2.25}$$

Let $\mathfrak{g}$ and $\mathfrak{h}$ be the Lie algebras of the groups G and H, respectively, and let $U(\mathfrak{g})$ be the universal enveloping algebra of the Lie algebra $\mathfrak{g}$ realized by differential operators in $C^\infty(G)$. The generators X_α $(\alpha = (\alpha_1, \dots, \alpha_n))$ of the Lie group G form a basis in $U(\mathfrak{g})$. Let

$$U^H = \{X\in U(\mathfrak{g})\,|\, \mathrm{Ad}_h X = X,\ h\in H\},\quad I = \text{l.s.}\{\mathfrak{h}\cdot U(\mathfrak{g})\},\quad \text{and}\quad J = \text{l.s.}\{U(\mathfrak{g})\,\mathfrak{h}\}.$$

Clearly, U^H is a subalgebra of $U(\mathfrak{g})$ and I and J are, respectively, the left and right ideals of $U(\mathfrak{g})$. As in Subsection 2.1, one can easily show that the operator

$$PX = \int_H \mathrm{Ad}_h X\, dh \quad (X\in U(\mathfrak{g}))$$

is a projector and $PU(\mathfrak{g}) = U^H$. Denote $I^H = PI$ and $J^H = PJ$.

Lemma 2.5. *$I^H = I\cap U^H = J\cap U^H = J^H$ is a two-sided ideal in U^H.*

Proof. Let us show that $PI\subset I$. Consider an element $X\in I$ of the form $X = ZX_1$, where $Z\in\mathfrak{h}$ and $X_1\in U(\mathfrak{g})$. It is clear that $\mathrm{Ad}_h X = (\mathrm{Ad}_h X_1)$ $(h\in H)$. Since $\mathrm{Ad}_h Z\in\mathfrak{h}$, we have $\mathrm{Ad}_h X\in I$. Consequently, $\mathrm{Ad}_h I\subset I$, whence it follows that $PI\subset I$. Since $PI\subset U^H$, we have $PI = I\cap U^H$. Similarly, we prove that $PJ = J\cap U^H$. Let us show that $J^H = I^H$. Clearly, for this purpose, it suffices to establish the equality $P(ZX) = P(XZ)$ $(Z\in\mathfrak{h},\ X\in U(\mathfrak{g}))$. For any $Z\in\mathfrak{h}$ and $X\in U(\mathfrak{g})$, we have $ZX = XZ + \mathrm{ad}_Z X$ and, hence, it is necessary to show that $P(\mathrm{ad}_Z X) = 0$ or, equivalently, that $\mathrm{Im}(\mathrm{ad}_Z)\subset \mathrm{Ker}\,P$ for all $Z\in\mathfrak{h}$. Since P and ad_Z $(Z\in\mathfrak{h})$ preserve the filtration in $U(\mathfrak{g})$, it suffices to prove this inclusion for their restrictions $\mathrm{Im}(\mathrm{ad}_Z \restriction U_m(\mathfrak{g}))\subset \mathrm{Ker}(P\restriction U_m(\mathfrak{g}))$ to an arbitrary layer of the filtration $U_m(\mathfrak{g})$. Denote $P_m = P\restriction U_m(\mathfrak{g})$ and $\mathrm{ad}_Z^m = \mathrm{ad}_Z \restriction U_m(\mathfrak{g})$. Since $U_m(\mathfrak{g})$ is finite-dimensional, it can be equipped with a scalar product $(\cdot,\cdot)_0$ and, hence, with H – invariant scalar product

$$(X_1, X_2)_H = \int_H (\mathrm{Ad}_h X_1, \mathrm{Ad}_h X_2)_0\, dh \quad (X_1, X_2\in U_m(\mathfrak{g})).$$

One can easily see that, in this scalar product, $P = P^*$ and $(\mathrm{ad}_Z^m)^* = -\mathrm{ad}_Z^m$. By the definition of the projection P, we have $\mathrm{Im}\, P \subset \mathrm{Ker}\, \mathrm{ad}_Z$ for all $Z \in \mathfrak{h}$. Indeed, $\mathrm{Im}\, P = U^H$ but $\mathrm{Ad}_h \restriction U^H = \mathrm{id}$ and, therefore, $U^H \subset \mathrm{Ker}\, \mathrm{ad}_Z$ $(Z \in H)$. Thus, for any $X \in U_m(\mathfrak{g})$ and $\Phi \in \mathrm{Im}\, \mathrm{ad}_Z^m$, we have

$$(P\Phi, X)_H = (\Phi, PX)_H = 0$$

because $\mathrm{Im}\, \mathrm{ad}_Z^m \perp \mathrm{Ker}\, \mathrm{ad}_Z^m$.

■

Theorem 2.9. *The algebra $\mathcal{D}_e^R(G,H)$ generated by the generators R_α is isomorphic to the algebra U^H/J^H.*

Proof. By using (2.24), we construct a linear mapping $\rho : U(\mathfrak{g}) \to \mathcal{D}_e^R(G,H)$, i.e., we associate each generator X_α with a generator R_α of generalized translation operators R_s. By virtue of (2.25), the restriction of ρ to U^H is a surjection. Let us show that $\sigma = \rho \restriction U^H$ is a homomorphism of the algebra U^H onto $\mathcal{D}_e^R(G,H)$. Indeed, for any PX_α, $PX_\beta \in U^H$, by using the fact that the Haar measure is invariant and the function f is biinvariant, we obtain

$$\begin{aligned}
\rho(PX_\alpha\, PX_\beta) &= \int_H \int_H \int_H ((\mathrm{Ad}_{h_1} X_\alpha)(\mathrm{Ad}_{h_2} X_\beta) f)(gh)\, dh\, dh_1\, dh_2 \\
&= \int_H \int_H \int_H D_{g_1}^{(\alpha)}\, \mathrm{Ad}_{h_2} X_\beta\, f(ghh_1^{-1} g_1 h_1)|_{g_1 = e}\, dh_1\, dh_2\, dh \\
&= \int_H \int_H D_{g_1}^{(\alpha)} D_{g_2}^{(\beta)} f(ghg_1 h_1 g_2)|_{\substack{g_1 = e \\ g_2 = e}}\, dh_1\, dh = R_\alpha R_\beta f(g) \quad (g \in G).
\end{aligned}$$

To prove the theorem, it is necessary to show that the kernel of the homomorphism σ coincides with I^H. Let $f(g) \in C^\infty(G)$ be a right-invariant function. Then the function

$$f_1(g) = (L_{g_1} f)(g) = \int_H f(g_1 hg)\, dh$$

is biinvariant for all $g_1 \in G$. For $\Phi = c^\alpha X_\alpha \in \mathrm{Ker}\, \sigma$, this yields

$$0 = \int_H \Phi f_1(gh)\, dh = \int_H \sum_\alpha c^\alpha D_a^{(\alpha)} f_1(gha)|_{a=e}\, dh$$

$$= \int_H \sum_\alpha c^\alpha D_a^{(\alpha)} \int_H f(g_1 h_1 gha) dh_1 |_{a=e} dh$$

$$= \int_H \int_H \sum_\alpha c^\alpha D_a^{(\alpha)} f(g_1 h_1 ghah^{-1})|_{a=e} dh dh_1$$

$$= \int_H \int_H (\mathrm{Ad}_{h^{-1}} \Phi f)(g_1 h_1 g) dh dh_1 = \int_H (\Phi f)(g_1 hg) dh.$$

By setting $g = e$, for any $g_1 \in G$, we get

$$\int_H \Phi f(g_1 h) dh = 0 \quad (\Phi \in \operatorname{Ker} \sigma). \tag{2.26}$$

Let $\Phi = c^\alpha X_\alpha \in U^H$ and let $f \in C^\infty(G)$ be right-invariant. Then

$$(\Phi f)(g) = \sum_\alpha c^\alpha D_a^{(\alpha)} f(ga)|_{a=e} = \int_H \sum_\alpha c^\alpha D_a^{(\alpha)} f(ghh^{-1}ah)|_{a=e} dh$$

$$= \int_H (\mathrm{Ad}_{h^{-1}} \Phi f)(gh) dh = \int_H (\Phi f)(gh) dh, \tag{2.27}$$

i.e., the function $(\Phi f)(g)$ is right-invariant. If $\Phi \in \operatorname{Ker}\sigma$, then, by virtue of (2.26) and (2.27), $(\Phi f)(g) = 0$ for any right-invariant function. This means that $\Phi \in J$. Thus, if $\Phi \in \operatorname{Ker}\sigma \subset U^H$, then $\Phi \in J$, i.e., $\operatorname{Ker}\sigma \subset J^H$. The inverse inclusion is obvious. Consequently, $\operatorname{Ker}\sigma = J^H$.

■

In exactly the same way as in Subsection 2.1, we prove that $\mathcal{D}_e^R(G,H)$ is a finitely generated PBW-type algebra.

Example. Let $G = SO(4)$ be a group of rotations of the space $\mathbb{R}^4$ with an orthonormal frame (e_1, e_2, e_3, e_4) and let $H = O(2)$ be the complete group of rotations of the plane (e_3, e_4). We set $\alpha^{(k)} = \delta_k^i$, $\alpha^{(ki)} = \delta_k^i + \delta_l^i$ $(i = 1, \ldots, 6;$ δ_k^i is the Kronecker symbol), $X_k = X_{\alpha(k)}$, and $X_{ki} = X_{\alpha(kl)}$, where X_α are generators of the group $SO(4)$ in coordinates of the second kind. Similar notation is introduced for generators of the generalized translation operators corresponding to the pair $(SO(4), O(2))$. The generators R_1, R_{22}, R_{24}, R_{44} are linearly independent, generate the algebra $\mathcal{D}_e^R(G,H)$, and satisfy the commutation relations

$$[R_{22}, R_1] = 2R_{24} + R_1, \quad [R_{44}, R_1] = -2R_{24} - R_1, \quad [R_{24}, R_1] = R_{44} - R_{22},$$

$$[R_{44}, R_{22}] = 2R_1R_{24} + R_1^2 - R_{22} + R_{44},$$
$$[R_{24}, R_{22}] = R_1R_{22}, \quad [R_{44}, R_{24}] = R_{44}R_1; \tag{2.28}$$

$$R_{24}^2 = R_{22}R_{44}. \tag{2.29}$$

Denote $z_1 = R_1$, $z_2 = R_{22}$, $z_3 = R_{24}$, $z_4 = R_{44}$, and $V = \text{l.s.}\{z_1, z_2, z_3, z_4\}$. Relations (2.28) can now be rewritten as

$$[z_i, z_j]_H = z_i \otimes z_j - h_{ij}^{kl} z_k \otimes z_l = b_{ij}^p z_p, \tag{2.30}$$

where the tensors $H = \left(h_{ij}^{kl}\right)$ and $C = \left(b_{ij}^b\right)$ are determined from (2.28). Relation (2.29) means that

$$z_3 \otimes z_3 = z_2 \otimes z_4. \tag{2.31}$$

The linear space V endowed with the bracket $[\cdot, \cdot]_H$ is a generalized Lie algebra. This algebra possesses the ideal $V_1 = [V, V]_H$, $V_1 = \text{l.s.}\{z_1 + 2z_3, z_2 - z_4\}$. The algebra V_1 possesses the ideal $V_2 = [V_1, V_1]_H = [V_2, V_2]_H$, $V_2 = \text{l.s.}\{z_2 - z_4\}$.

Let $U(G, H) = T/J$ be an algebra enveloping relations (2.30); here, T is a free tensor algebra over V and J is the ideal generated by relations (2.30). Let J_n be elements of the nth tensor power in T and let β_n be the imbedding $\beta_n\colon J_n \to U_n(G, H)$. For $U(G, H)$, we have the following analog of the PBW theorem:

The elements 1 and $\beta_n\,(z_{i_1} \otimes \ldots \otimes z_{i_s})$ $(s \geq 1, i_1 \leq \ldots \leq i_s, i_k = 1, \ldots, 4)$ form a basis in $U(G, H)$. If $U_1(G, H) = T/J_1$, where the ideal J_1 is generated by relations (2.30) and (2.31), then the assertion formulated above remains true for $i_k \in \{1, 2, 3\}$ or $i_k \in \{1, 2, 4\}$. We also have

$$\dim U_1(G, H) = (n+1)^2 < C_3^{n+3},$$

i.e., $U_1(G, H)$ is a PBW-type algebra. The algebra $U_1(G, H)$ is isomorphic to the algebra $\mathcal{D}_e^R(G, H)$.

3. Duality of Generators of One-Dimensional Compact and Discrete Hypercomplex Systems

Let G be the group of motions of a circle. The functions e^{int} $(n \in \mathbb{Z},\ t \in [-\pi, \pi])$ are its characters; moreover, they are the eigenfunctions of the generator X of the group G. On the other hand, the characters e^{int} regarded as functions of n are generalized eigenfunctions of the shift operator $\hat{X}$ in $l_2(\mathbb{Z})$ (see, e.g., [Ber10] or [Ber11]). In a certain sense, the operator $\hat{X}$ can be regarded as the generator of the group dual to G and, hence, the operators X and $\hat{X}$ can be regarded as dual to each other. A similar situation is observed for some families of orthogonal polynomials satisfying a differential equation with respect to the continuous variable and a difference equation with respect to the discrete variable. To within a coefficient, these polynomials are characters of two Hermitian hypercomplex systems which are dual to each other (see Section 3 in Chapter 1). Thus, it is natural to consider the relevant differential and difference operators as dual to each other. The existence of such duality of generators was conjectured for the fist time by Berezin and Gelfand [BeG].

The generator of a hypercomplex system can be regarded as the Fourier image of the operator of multiplication by a function uniquely defined by the characters of this hypercomplex system. In Subsection 3.1, we show that, under certain restrictions imposed on generalized translation operators associated with a compact hypercomplex system, the generator thus defined coincides with the classical generator of generalized translation operators.

Generalized translation operators associated with hypercomplex systems can be uniquely reconstructed from the corresponding generator by using a formula of the type $(R_s f)(t) = (\chi(s, X)f)$ established by Vainerman for generalized translation operators constructed in the Sturm–Liouville problem by the method of Levitan and Povzner (see Section 4 in Chapter 2). In this case, $\varphi(s, \lambda)$ are eigenfunctions of the original Sturm–Liouville problem and X is the self-adjoint operator associated with this problem. In Subsection 3.2, we show that the formulas for the construction of generalized translation operators from self-adjoint operators remain applicable in more general situations. In Subsection 3.3, we establish and analyze relations between the generator of the hypercomplex system constructed in Example 2 (Section 1 in Chapter 1) and the Fourier image of the generator of the dual hypercomplex system.

3.1. Generators of One-Dimensional Compact and Discrete Hypercomplex Systems. Let $L_1 = L_1([a, b], m)$ be a normal commutative hypercomplex system (in a sense of Subsection 2.3 in Chapter 1), let a segment $[a, b]$ $(a, b < \infty)$ of the real axis be its basis, let $e \in [a, b]$ be its basis unity, and let m be a multiplicative measure. We assume that the structure measure $c(A, B, r)$ of this hypercomplex system is, generally speaking, nonpositive and the hypercomplex system satisfies the axioms of quantized hypercomplex systems. For this purpose, it is sufficient that

$$\int_a^b |c(A,B,r)|\,dm(r) \le C\,m(A)m(B) \quad (C \ge 1;\, A, B \in \mathcal{B}([a,b])),$$

and that the linear space $A(Q) = L_2 * L_2$ be an algebra with respect to the ordinary multiplication of functions $(fg)(t) = f(t)g(t)$ dense in $C(Q)$, where $L_2 = L_2([a,b], m)$. It follows from the theory of quantized hypercomplex systems that the dual hypercomplex system $l_1(\hat{Q},\hat{m})$ is discrete; moreover, it is a commutative $*$-algebra (but, generally speaking, not a Banach algebra). In this case, we also have analogs of the Plancherel theorem, inversion formula, and Pontryagin duality theory. Note that $\hat{m}(\chi) = \|\chi\|_2^{-2}$, where $\chi \in \hat{Q}$ is a character of the hypercomplex system L_1, $\hat{Q}$ is a basis of the dual hypercomplex system, and $\|\cdot\|_2 = \|\cdot\|_{L_2}$. For simplicity, we assume that the structure measure $\hat{c}(\chi, \psi, \theta)$ $(\chi, \psi, \theta \in \hat{Q})$ of the dual hypercomplex system is finite in θ for fixed χ and ψ and $|\chi(t)| \le 1$ $(\chi \in \hat{Q},\, t \in [a,b])$ (if the structure measure of at least one of hypercomplex systems L_1 or $l_1(\hat{m})$ is nonnegative, then the last condition is automatically satisfied).

Assume that the characters $\chi(t)$ are infinitely differentiable with respect to t and the function $f(t,n) = \chi^{(n)}(t)$ $(\chi \in Q,\, n \in N)$ is not identically equal to zero. Such hypercomplex systems are called *one-dimensional Lie hypercomplex systems.* In [Schw5], it was proved that every hypergroup $Q = [a,b]$ with smooth generalized translation operators is Hermitian. Thus, assume that $L_1\,[a,b]$ is a Hermitian hypercomplex system and that $e \in \partial[a,b]$. The following statement independently established by A. Scwartz under somewhat stronger restrictions [Schw5] justifies this assumption:

Assume that L_1 is an Hermitian hypercomplex system and the generalized translation operators R_s $(s \in [a,b])$ associated with this hypercomplex system map $C^\infty([a,b])$ into $C^\infty([a,b]\times[a,b])$. Then $e \in \partial[a,b]$.

Indeed, let $e \in (a,b)$. Denote by X_k the kth-order generator of the hypercomplex system L_1:

$$(X_k f)(t) = \frac{d^k}{ds^k}(R_s f)(t)\Big|_{s=e}, \qquad \mathcal{D}(X_k) = C^\infty([a,b]).$$

It is easy to see that X_k is an Hermitian operator in L_2. Indeed, since the hypercomplex system is Hermitian, we get

$$(X_k f, g)_2 = \frac{\partial^k}{\partial s^k} \int (R_s f)(t)\overline{g(t)}\, dm(t)|_{s=e}$$

$$= \frac{\partial^k}{\partial s^k} \int f(t)\overline{((R_s g)(t)}\, dm(t)|_{s=e} = (f, X_k g)_2.$$

Denote by $\tilde{X}_k$ the closure of X_k. Since $e \in (a, b)$, the operator $\tilde{X}_1$ is nondegenerate

$$\left(\tilde{X}_1 f\right)(e) = \frac{\partial}{\partial s}(R_s f)(e)|_{s=e} = \frac{\partial}{\partial s}(R_e f)(s)|_{s=e} = f'(e) \quad (f \in C^\infty([a, b])).$$

On the other hand, the characters of Hermitian hypercomplex systems are real, $\overline{\chi(t)} = \chi(t)$. It follows from the inequality $|\chi(t)| \le 1$ that $\chi'(e) = 0$ $\left(\chi \in \hat{Q}\right)$, whence

$$(X_1 \chi)(t) = \frac{\partial}{\partial s}(R_s \chi)(t)|_{s=e} = \frac{\partial}{\partial s}\chi(s)\chi(t)|_{s=e} = 0 \quad \left(\chi \in \hat{Q}\right).$$

Since the collection of characters form a basis in the space L_2, by approximating an arbitrary function $f \in L_2$ by linear combinations of characters and using the fact that $\tilde{X}_1$ is closed, we conclude that $\tilde{X}_1 f = 0$.

■

Denote $a^{\#} = b$ and $b^{\#} = a$. Thus, for instance, if $e = a$, then $e^{\#} = b$. We enumerate the characters $\chi \in \hat{Q}$ by using the set $\mathbb{N}_0 = \mathbb{N} \cup \{0\}$ so that $\chi_0(t) \equiv 0$ and denote $\chi(t, n) = \chi_n(t)$, i.e., the characters of the hypercomplex systems L_1 and $l_1(\hat{m})$ are regarded as functions of two variables $t \in [a, b]$ and $n \in \mathbb{N}_0$. Let

$$N = \min\left\{k \in \mathbb{N} \mid \chi^{(k)}(e) \neq 0 \text{ for at least one } \chi \in \hat{Q}\right\}.$$

Assume that $N < \infty$, i.e., there exists $k \in \mathbb{N}$ such that $\chi^{(k)}(e) \neq 0$. Consider the function

$$\hat{\lambda}(n) = \frac{\partial^N}{\partial t^N}\chi(t, n)|_{s=e}.$$

By expanding $\chi(t)$ in the Taylor series for at the point e and using the inequality $\chi(t) \le \chi(e)$, we conclude that $\hat{\lambda}(n)$ is nonnegative for $e = b$ and odd N and nonpositive, otherwise.

Suppose that (A) $\hat{\lambda}(n) \neq \hat{\lambda}(m)$ if $n \neq m$ and (B) there exists a number $n_0 \in \mathbb{N}$ such that $\chi(t, n_0)$ homeomorphically maps $[a, b]$ into its range.

Lemma 3.1. *For condition (B) to be satisfied, it is sufficient (C) that there exist a number $n_0 \in \mathbb{N}$ such that the problem*

$$(R_{n_0} f)(r) = Kf(r), \quad f(0) = 1 \quad (r \in \mathbb{N}_0), \tag{3.1}$$

where R_k $(k \in \mathbb{N}_0)$ are generalized translation operators associated with the hypercomplex system $l_1(\hat{m})$, is uniquely solvable for all $K \in [-1, 1]$.

Proof. By substituting $r = 0$ into (3.1), we get $K = f(n_0)$ and relation (3.1) turns into $(R_{n_0} f)(r) = f(n_0)f(r)$. Thus, the character $\hat{\chi}_K$ of the hypercomplex system $l_1(\hat{m})$ is a solution of problem (3.1). By the Pontryagin principle of duality for quantized hypercomplex systems, the multiplicative measure of the hypercomplex system dual to $l_1(\hat{m})$ is equal to m and the characters of $l_1(\hat{m})$ lying in supp m are parametrized by points of the interval $[a, b]$. Thus, if $\chi_K \in$ supp $\hat{m}$, then $\hat{\chi}_K(n) = \chi(t_K, n)$ and it is clear that $t_K (K \in [-1, 1])$ runs through the entire interval $[a, b]$. Consider the function $\psi(t) = \chi(t, n_0)$. It is obvious that $\psi(t)$ is a character of the hypercomplex system L_1. Let us show that $\psi(t)$ maps $[a, b]$ onto $[\psi(e^{\#}), 1]$ and is bijective. Since ψ is continuous, its surjectivity is obvious and injectivity follows from (C). Therefore, ψ is a homeomorphism.

■

It is easy to see that condition (C) is equivalent to the following condition: (C′) there exists a number n_0 such that any solution of the problem $(R_{n_0} f)(r) = Kf(r), f(0) = 0$ $(k \in [-1, 1])$ is trivial.

Consider the function

$$\lambda(t) = \frac{1}{|\hat{\lambda}(n_0)|}\chi(t, n_0) + \left(1 - \frac{1}{|\hat{\lambda}(n_0)|}\right),$$

where $\hat{\lambda}(n_0) \neq 0$ in view of condition (A), and denote by $M_{\hat{\lambda}}$ and M_λ the operators of multiplication by the functions $\hat{\lambda}(n)$ and $\lambda(t)$ in the spaces $l_2(\hat{m})$ and L_2, respectively. We define self-adjoint operators X and $\hat{X}$ as follows:

$$X = \mathcal{F}^{-1} M_{\hat{\lambda}} \mathcal{F}, \quad \hat{X} = \mathcal{F} M_\lambda \mathcal{F}^{-1}, \tag{3.2}$$

where $\mathcal{F}$ is the Fourier transform of L_2 on $l_2(\hat{m})$. Clearly, the spectrum $\sigma(X)$ of the operator X is equal to the closure of the set of values of the function $\hat{\lambda}(n)$ and $\sigma(\hat{X})$ is the set of values of the function $\lambda(t)$. In view of conditions (A) and (B), the spectra of X and $\hat{X}$ are simple.

The operator $\hat{X}$ *is bounded and given by the formula*

$$(\hat{X}f)(n) = \frac{1}{|\hat{\lambda}(n_0)|}(R_{n_0}f)(n) + \left[1 - \frac{1}{|\hat{\lambda}(n_0)|}\right]f(n) \quad (f \in l_2(\hat{m})). \tag{3.3}$$

Let $f \in l_1(\hat{m}) \cap l_2(\hat{m})$. Then

$$(\hat{X}f)(n) = \int_a^b \lambda(t)\overline{\chi(t,n)} \sum_{k=0}^{\infty} f(k)\chi(t,k)\,\hat{m}(k)\,dm(t),$$

and, in view of the fact that the series in the integrand is uniformly convergent, we get

$$\hat{X}f(n) = \frac{1}{|\hat{\lambda}(n_0)|}\sum_{k=0}^{\infty} f(k)\,m(k)\int_a^b \chi(t,n_0)\chi(t,n)\chi(t,k)\,dm(t)$$

$$+ \left(1 - \frac{1}{|\hat{\lambda}(n_0)|}\right)\sum_{k=0}^{\infty} f(k)\,\hat{m}(k)\int_a^b \chi(t,k)\chi(t,n)\,dm(t).$$

Since

$$\hat{c}(p,q,r) = \hat{m}(p)\hat{m}(q)\int_a^b \chi(t,p)\chi(t,q)\chi(t,r)\,dm,$$

$$(R_p f)(q) = \frac{1}{\hat{m}(p)\hat{m}(q)}\sum_r f(r)\,\hat{c}(p,q,r)\,\hat{m}(r),$$

and the characters in $L_2([a,b],m))$ are orthogonal, we arrive at relation (3.3).

∎

In what follows, we use the theory of expansion in generalized eigenvectors (see, e.g., [Ber10], [Ber11], and [BeKo]).

Lemma 3.2. *The chain*

$$\mathbb{R}^{\infty} \supset l_2(\hat{m}) \supset \mathbb{R}_0^{\infty} \tag{3.4}$$

is suitable for expansion in generalized eigenvectors of the operator $\hat{X}$, *i.e.,* $\mathbb{R}_0^{\infty}$ *is a nuclear space densely and topologically imbedded in* $l_2(\hat{m})$; *moreover, it is the projective limit of Hilbert spaces and* $\hat{X}$ *continuously maps* $\mathbb{R}_0^{\infty}$ *into* $\mathbb{R}_0^{\infty}$.

Proof. It is necessary to prove only the last assertion. In view of (3.3) and the fact that $\hat{c}(p, q, r)$ is finite, $\mathbb{R}_0^\infty$ is invariant under the action of $\hat{X}$. To show that $\hat{X}$ is continuous in $\mathbb{R}_0^\infty$, we fix a weight $\tau = (\tau_k)_{k=0}^\infty$ and denote $\|\cdot\|_{\tau\hat{m}} = \|\cdot\|_{l_2(\tau\hat{m})}$. Since $\mathbb{R}_0^\infty = \operatorname{pr}\lim l_2(\tau)$, it suffices to find a weight τ' such that $\left\|\hat{X}f\right\|_{\tau\hat{m}} \le C'\|f\|_{\tau'\hat{m}}$ $(f \in \mathbb{R}_0^\infty)$. For any $f \in \mathbb{R}_0^\infty$, we can write

$$\begin{aligned}\left\|\hat{X}f\right\|_{\tau\hat{m}}^2 &= \sum_k \left|(\hat{X}f)(k)\right|^2 \tau_k \hat{m}(k)\\ &\le \frac{2}{\left|\hat{\lambda}(n_0)\right|^2}\sum_k |(R_{n_0}f)(k)|^2\,\tau_k\hat{m}(k) + 2\left(1-\frac{1}{\left|\hat{\lambda}(n_0)\right|}\right)^2\sum_k |f(k)|^2\,\tau_k\hat{m}(k).\end{aligned}$$

Consider the first sum on the right-hand side of this inequality. We have

$$\sum_k |(R_{n_0}f)(k)|^2\,\tau_k\hat{m}(k) = \frac{1}{\hat{m}^2(n_0)}\sum_k \left|\sum_r f(r)\hat{c}(n_0,k,r)\hat{m}(r)\right|^2 \frac{\tau_k}{\hat{m}(k)}.$$

Let $N(k) = 1 + \max\{l \mid \hat{c}(n_0, k, r) \ne 0\}$. It is clear that

$$\sum_k |(R_{n_0}f)(k)|^2\,\tau_k\hat{m}(k) \le \frac{1}{\hat{m}^2(n_0)}\sum_k N(k)\left(\sum_r f^2(r)\hat{c}^2(n_0,k,r)\hat{m}^2(r)\right)\tau_k\frac{1}{\hat{m}(k)}.$$

Since f and $\hat{c}$ are compactly supported, both sums in this expression are finite. Therefore,

$$\sum_k |(R_{n_0}f)(k)|^2\,\tau_k\hat{m}(k) \le \frac{1}{\hat{m}^2(n_0)}\sum_r f^2(r)\left(\sum_k N(k)\hat{c}^2(n_0,k,r)\frac{\tau_k\hat{m}(r)}{\hat{m}(k)}\right)\hat{m}(r)$$

and

$$\begin{aligned}\left\|\hat{X}f\right\|_{\tau\hat{m}}^2 \le{}& \frac{2}{\left|\hat{\lambda}(n_0)\hat{m}(n_0)\right|^2}\sum_r f^2(r)\left(\sum_k N(k)\hat{c}^2(n_0,k,r)\frac{\tau_k\hat{m}(r)}{\hat{m}(k)}\right)\hat{m}(r)\\ &+ 2\left(1-\frac{1}{\left|\hat{\lambda}(n_0)\right|}\right)^2\sum_r f^2(r)\tau_r\hat{m}(r).\end{aligned}$$

Thus, we can take

$$\tau'(r) = \frac{1}{\left|\hat{\lambda}(n_0)\right|^2 \hat{m}^2(n_0)} \sum_k N(k)\hat{c}^2(n_0,k,r)\frac{\tau_k \hat{m}(r)}{\hat{m}(k)} + \left(1 - \frac{1}{\left|\hat{\lambda}(n_0)\right|}\right)^2 \tau(r)$$

as the required weight.

■

Remark 1. The operator $\hat{X}$ is self-adjoint in $l_2(\hat{m})$ and, hence, it is a Carleman operator (see, e.g., [Ber10]). Thus, almost all its generalized eigenvectors with respect to the spectral measure lie in the space $l_2\left(\tau^{-1}\hat{m}\right)$, where

$$\sum_{k=0}^{\infty} \frac{1}{\tau_k} < \infty.$$

If, in addition,

$$\sum_k \frac{1}{\tau_k} \hat{m}(k) < \infty,$$

then, as becomes clear somewhat later, necessarily $l_2\left(\tau^{-1}\hat{m}\right)$ contains all its generalized eigenvectors.

Remark 2. The condition that the structure measure $\hat{c}(p, q, r)$ of the dual hypercomplex system is compactly supported is necessary, in fact, only to take the standard space $\mathbb{R}_0^\infty$ as a space of test functions for the expansion of $\hat{X}$ in generalized eigenfunctions. This condition can be removed if we use $\hat{X}$ to construct a nuclear rigging of the space $l_2(\hat{m})$ (see, e.g., [BeKo]) and require that the series

$$\sum_{r=0}^{\infty} c(p, q, r)\hat{m}(r)$$

be absolutely convergent for fixed $p, q \in \mathbb{N}_0$.

Theorem 3.1.

(i) For any fixed n, the character $\chi(t, n)$ regarded as a function of t is an eigenfunction of the operator X corresponding to the eigenvalue $\hat{\lambda}(n)$.

(ii) For any fixed t, the character $\chi(t, n)$ regarded as a function of n is a generalized eigenvector of the operator $\hat{X}$ with respect to chain (3.4) and if

condition (C) is satisfied, then any generalized eigenvector of $\hat{X}$ *is a multiple of* $\chi(t, n)$ *for some* $t \in [a, b]$.

Proof. We prove (i). It is clear that $\chi(t, n) \in \mathcal{D}(X)$ and, hence,

$$(X\chi(\cdot, n))(t) = \sum_k \hat{\lambda}(k) \int_a^b \chi(s, k)\chi(s, n)\, dm(s)\chi(t, k)\, \hat{m}(k)$$

$$= \sum_k \hat{\lambda}(k)\, \delta_{k,n} \|\chi(\cdot, n)\|_2^2\, \chi(t, k)\, \hat{m}(k) = \hat{\lambda}(n)\, \chi(t, n).$$

We prove (ii). For any $\eta(p) \in \mathbb{R}_0^{\infty}$, we have

$$\big(\chi(t,\cdot), \hat{X}\eta\big)_{l_2(\hat{m})} = \frac{1}{\big|\hat{\lambda}(n_0)\big|} \sum_k \chi(t, k) R_{n_0}\eta(k)\, \hat{m}(k)$$

$$+ \left(1 - \frac{1}{\big|\hat{\lambda}(n_0)\big|}\right) \sum_k \chi(t, k)\eta(k)\, \hat{m}(k).$$

The first series on the right-hand side of this equality takes the form

$$\sum_{k=0}^{\infty} \chi(t, k) R_{n_0}\eta(k)\, \hat{m}(k) = \sum_r \frac{\eta(r)}{\hat{m}(n_0)} \sum_k \hat{c}(k, n_0, r)\hat{m}(r)\, \chi(t, k)$$

$$= \sum_r \frac{\eta(r)}{\hat{m}(n_0)} \sum_k \hat{c}(r, n_0, k)\chi(t, k)\hat{m}(k)$$

$$= \sum_r \eta(r)\chi(t, n_0)\chi(t, r)\, \hat{m}(r).$$

Here, we have used the fact that the dual hypercomplex system is normal and commutative, and the fact that, for compactly supported $\eta(r)$ and $\hat{c}(p, q, r)$, both series contain finitely many terms). As a result, we obtain

$$\big(\chi(t,\cdot), \hat{X}\eta\big)_{l_2(\hat{m})} = \left[\frac{1}{\big|\hat{\lambda}(n_0)\big|}\chi(t, n_0) + \left(1 - \frac{1}{\big|\hat{\lambda}(n_0)\big|}\right)\right] \sum_r \chi(t, r)\eta(r)\, \hat{m}(r)$$

$$= \lambda(r)\, (\chi(t,\cdot), \eta)_{l_2(\hat{m})}.$$

Let us now prove the last assertion in (ii). Let $f(n)$ be a nontrivial generalized eigenvector corresponding to the eigenvalue λ. For any $\eta \in \mathbb{R}_0^\infty$, we can write

$$0 = \left(f, \hat{X}\eta - \lambda\eta\right)_{l_2(\hat{m})}$$

$$= \sum_r f(k)\left[\frac{1}{\left|\hat{\lambda}(n_0)\right|}(R_{n_0}\eta)(k) + \left(1 - \frac{1}{\left|\hat{\lambda}(n_0)\right|} - \lambda\right)\eta(k)\right]\hat{m}(k).$$

Let $\eta = \delta_{k,r}m^{-1}(r)$. Then, for any $r \in \mathbb{N}_0$,

$$0 = f(r)\left(1 - \frac{1}{\left|\hat{\lambda}(n_0)\right|} - \lambda\right) + \frac{1}{\left|\hat{\lambda}(n_0)\right|}(R_{n_0}f)(r). \tag{3.5}$$

For $r = 0$, we get $f(n_0) + f(0)\left[(1-\lambda)\left|\hat{\lambda}(n_0)\right| - 1\right] = 0$, whence

$$\lambda = \frac{f(n_0)}{\left|\hat{\lambda}(n_0)\right|} + \left[1 - \frac{1}{\left|\hat{\lambda}(n_0)\right|}\right].$$

By substituting the value of λ in (3.5), we obtain system (3.1). Without loss of generality, we can assume that $f(0) = 1$ (if $f(0) = 0$, then, by virtue of (C′), $f \equiv 0$). The characters $\chi(t, n)$ of the hypercomplex system $l_1(\hat{m})$ are solutions of (3.1). To complete the proof, it remains to use condition (C).

■

Note that (ii) implies that all generalized eigenvectors of the operator $\hat{X}$ lie in the space $l_2\left(\tau^{-1}\hat{m}\right)$, where

$$\sum_k \frac{1}{\tau_k}\hat{m}(k) < \infty.$$

The domain $\mathcal{D}(X)$ of the operator X is an ideal with respect to multiplication $*$ in L_2 and, moreover,

$$(Xf * g)(t) = \left(X(f * g)\right)(t) \quad (f \in \mathcal{D}(X),\ g \in L_2), \tag{3.6}$$

$$\left(\hat{X}f \,\hat{*}\, g\right)(n) = \left(\hat{X}^+(f \,\hat{*}\, g)\right) \quad (f, g \in l_2(\hat{m})), \tag{3.7}$$

where $\hat{X}^+$ is the operator adjoint to the operator $X \restriction \mathbb{R}_0^\infty$ with respect to chain (3.4) (see [BeKo]).

Indeed, we have

$$(Xf*g)(t) = (\mathcal{F}^{-1} M_{\hat{\lambda}}\, \mathcal{F}f * \mathcal{F}^{-1}\mathcal{F}g) = (\mathcal{F}^{-1}(\hat{\lambda}\, \mathcal{F}f(\cdot)\, \mathcal{F}g(\cdot)))(t)$$

$$= (\mathcal{F}^{-1}(\hat{\lambda}(\cdot)\, \mathcal{F}(f*g))(\cdot)))(t) = \mathcal{F}^{-1} M_{\hat{\lambda}}\, \mathcal{F}(f*g)(t) = (X(f*g))(t),$$

Here, we have used the properties of Fourier transforms. Relation (3.7) is proved in exactly the same way.

■

Theorem 3.2. *The generalized translation operators* $(R_s f)(t)$ $(t, s \in [a, b])$ *and* $(R_p f)(q)$ $(p, q \in \mathbb{N}_0)$ *are uniquely reconstructed from the operators* X *and* $\hat{X}$ *by the formulas*

$$(R_s f)(t) = (\chi(s, \hat{\lambda}^{-1}(X))f)(t) \quad (t, s \in [a, b];\ f \in L_2), \tag{3.8}$$

$$(R_p f)(q) = (\chi(\lambda^{-1}(\hat{X}), p)f)(q) \quad (p, q \in \mathbb{N}_0; f \in l_2(\hat{m})) \tag{3.9}$$

Proof. Let us prove (3.8). We have

$$\mathcal{F}(R_s(\cdot))(p) = \int_a^b f(t)(R_s\chi)(t, p)\,dm(t) = \chi(s, p)(\mathcal{F}f)(p) \quad (f \in L_2([a, b], m)).$$

Note that

$$(Xf)(t) = \sum_k \hat{\lambda}(k)(\mathcal{F}f)(k)\chi(t, k)\,\hat{m}(k)$$

is, in fact, the spectral decomposition of the operator X. Therefore, for all $f \in L_2([a, b], m)$, we can write

$$\begin{aligned}(\chi(s, \hat{\lambda}^{-1}(X))f)(t) &= \sum_{k=0}^{\infty} \chi(s, \hat{\lambda}^{-1}(\hat{\lambda}(k))\hat{m}(k)\,\chi(t, k)(\mathcal{F}f)(k) \\ &= \sum_k \chi(s, k)\chi(t, k)(\mathcal{F}f)(k)\,\hat{m}(k) \\ &= \sum_k (\mathcal{F}(R_s f(\cdot)))(k)\chi(t, k)\,\hat{m}(k) \\ &= (\mathcal{F}^{-1}\mathcal{F}R_s)(t) = (R_s f)(t).\end{aligned}$$

The proof of (3.9) is similar. Indeed, it can easily be seen that

$$(\mathcal{F}^{-1}R_p f(\cdot))(t) = \chi(t,p)(\mathcal{F}^{-1}f)(t) \quad (f\in l_2(\hat{m})).$$

By using the spectral decomposition of the operator $\hat{X}$

$$\left(\hat{X}f\right)(k) = \int_{\{\kappa=\lambda(t)\}} \kappa(\mathcal{F}^{-1}f)(\lambda^{-1}(\kappa))\chi(\lambda^{-1}(\kappa),k)\,d\mu(\kappa),$$

where $d\mu(\kappa) = dm((\lambda^{-1}(\kappa))$, we obtain

$$\begin{aligned}(\chi(\lambda^{-1}(\hat{X}),p)f)(k) &= \int_a^b \chi(t,p)(\mathcal{F}^{-1}f)(t)\chi(t,k)\,dm(t) \\ &= \int_a^b \mathcal{F}^{-1}(R_p f(\cdot))\chi(t,k)\,dm(t) = (R_p f)(k).\end{aligned}$$

■

Note that the idea of reconstruction of generalized translation operators according to formulas of the types (3.8) and (3.9) belongs to Vainerman. Theorem 3.1 actually justifies the following definition: The operators X and $\hat{X}$ are called *generators* of the dual hypercomplex systems L_1 and $l_1(\hat{m})$, respectively, and regarded as *dual* to each other.

The following statement establishes the relationship between X and $\hat{X}$:

Theorem 3.3.

(i) $\operatorname{Ker} X = \{\nu \mid \nu = \text{const}\}$ *and* $\lambda(t)$ *is a unique solution of the problem*

$$(X\lambda)(t) = \operatorname{sign}\,\hat{\lambda}(n_0)\,\chi(t,n_0), \qquad \lambda(e) = 1; \tag{3.10}$$

(ii) $\{\nu \mid \nu = \text{const}\} \subset \operatorname{Ker}\left(\hat{X}^+ - 1\right)$ *(in the case where condition (C) is satisfied, the inclusion turns into the equality) and* $\hat{\lambda}(n)$ *is a solution of the problem*

$$\left(\hat{X}^+\hat{\lambda}\right)(n) = \hat{\lambda}(n) + \operatorname{sign}\hat{\lambda}(n_0), \qquad \hat{\lambda}(0) = 0, \tag{3.11}$$

where $\hat{X}^+$ *is the operator adjoint to* $\hat{X}\in\mathbb{R}_0^\infty$ *with respect to chain (3.4). Furthermore, this solution is unique if condition (C) is satisfied.*

Proof. We prove (i). Let $\nu = \text{const}$. Clearly, $X\nu = \nu(X\chi)(t,0) = \nu\hat{\lambda}(0) = 0$. On the other hand, if $f(t) \in \operatorname{Ker} X$, then $M_{\hat{\lambda}}\mathcal{F}f \equiv 0$ or

$$\hat{\lambda}(n)\int f(t)\chi(t,n)dm(t) = 0$$

for all $n \in \mathbb{N}_0$. Since $\hat{\lambda}(n) \neq 0$ for $n \neq 0$, we have

$$\int f(t)\chi(t,n)dm(t) = 0$$

for all $n \neq 0$, whence, since the family of characters is complete in $L_2([a,b],m)$, we conclude that $f(t) = \nu\chi(t,0) = \nu$. This implies that problem (3.10) is uniquely solvable. As follows from the explicit expression for $\lambda(t)$, this function satisfies relation (3.10).

Let us prove (ii). Its first statement is obvious. Suppose that condition (C) is satisfied and $\nu \in \operatorname{Ker}\left(\hat{X}^+ - 1\right)$. By using the same arguments as in Theorem 3.1 (ii), for all $f \in \mathbb{R}_0^\infty$, we obtain

$$0 = \left(\nu, \hat{X}f - f\right)_{l_2(\hat{m})} = \frac{1}{\left|\hat{\lambda}(n_0)\right|}\sum_r f(r)\,\hat{m}(r)\,[(R_{n_0}\nu)(r) - \nu(r)].$$

Since f is arbitrary, this enables us to conclude that $\nu(r)$ satisfies system (3.1) with $K = 1$. By virtue of (C), we have $\nu(r) = \nu(0) = \text{const}$. Let us prove (3.11). Since the structure measure is compactly supported, it follows from the definition of the number N that

$$\begin{aligned}
\left(\hat{X}^+\hat{\lambda}\right)(p) &= \frac{1}{\left|\hat{\lambda}(n_0)\right|}(R_{n_0}\hat{\lambda})(p) - \left(1 - \frac{1}{\left|\hat{\lambda}(n_0)\right|}\right)\hat{\lambda}(p) \\
&= \frac{1}{\left|\hat{\lambda}(n_0)\right|}\sum_r \hat{c}(n_0,p,r)\frac{\partial^N}{\partial t^N}\chi(t,r)|_{n=e}\,\frac{\hat{m}(r)}{\hat{m}(n_0)\hat{m}(p)} + \left(1 - \frac{1}{\left|\hat{\lambda}(n_0)\right|}\right)\hat{\lambda}(p) \\
&= \frac{1}{\left|\hat{\lambda}(n_0)\right|}\,\frac{\left.\dfrac{\partial^N}{\partial t^N}\left(\sum_r \hat{c}(n_0,p,r)\chi(t,r)\hat{m}(r)\right)\right|_{n=e}}{\hat{m}(n_0)\hat{m}(p)} + \left(1 - \frac{1}{\left|\hat{\lambda}(n_0)\right|}\right)\hat{\lambda}(p) \\
&= \frac{1}{\left|\hat{\lambda}(n_0)\right|}\frac{\partial^N}{\partial t^N}(\chi(t,n_0)\chi(t,p))|_{t=e} + \left(1 - \frac{1}{\left|\hat{\lambda}(n_0)\right|}\right)\hat{\lambda}(p) \\
&= \frac{1}{\left|\hat{\lambda}(n_0)\right|}\left[\hat{\lambda}(n_0) + \hat{\lambda}(p)\right] + \left(1 - \frac{1}{\left|\hat{\lambda}(n_0)\right|}\right)\hat{\lambda}(p) = \hat{\lambda}(p) + \operatorname{sign}\hat{\lambda}(n_0).
\end{aligned}$$

The unique solvability of problem (3.11) follows from the form of the kernel of the operator $X^+ - 1$.

■

It should be noted that, without loss of generality, one can assume that $N = 1$. We set $\varphi(t) = |t-e|^N/N!$. It is clear that $\varphi(t)$ is a homeomorphism of $[a, b]$ into $[0, (b-a)^N/N!]$ (at the same time, φ is not a diffeomorphism). In the segment $[0, (b-a)^N/N!]$, we construct an Hermitian hypercomplex system $L_1([0, (b-a)^N/N!], m_1)$ isomorphic to $L_1([a, b], m)$ by setting

$$(L_{\varphi(s)}f)(\varphi(t)) = (L_s(f\circ\varphi))(t),$$

$$e_1 = 0 \quad (t, s \in [a, b];\ f \in C[0, (b-a)^N/N!]),$$

$$\int_0^{(b-a)^N/N!} f(p)\,dm_1(p) = \int_a^b f\circ\varphi(t)\,dm(t).$$

The functions $\psi_n = \chi_n \circ \varphi^{-1}$ are, clearly, the characters of the hypercomplex system $L_1([0, (b-a)^N/N!], m_1)$, whence we conclude that $\psi_n'(e_1) = \chi^{(N)}(e) = \hat{\lambda}(n)$ $(n \in \mathbb{N}_0)$. Let $X_1 = \mathcal{F}_1^{-1} M_{\hat{\lambda}} \mathcal{F}_1$ be a generator of the hypercomplex system $L_1([0, (b-a)^N/N!], m_1)$, where $\mathcal{F}_1$ is the Fourier transformation with respect to the characters $\psi(t, n) = \psi_n(t)$. The operator $Ff(t) = (f\circ\varphi)(t)$ is a unitary operator from $L_2([0, (b-a)^N/N!], m_1)$ into $L_2([a, b])$. It is clear that $X_1 = F^{-1}XF$.

Let us show that, under certain restrictions imposed on the generalized translation operators R_s $(s \in [a, b])$, the operator X coincides with a generator of order N in the sense of (1.3), i.e., with

$$(X_N f)(t) = \frac{\partial^N}{\partial t^N} R_s f(t)|_{s=e}.$$

Without loss of generality, we can restrict ourselves to the case $N = 1$. Assume that the measure m is absolutely continuous with respect to the Lebesgue measure dt, that the relevant density is a sufficiently smooth function, and that, in a certain neighborhood U of the point e, one can always find smooth functions $\alpha(s)$ and $\beta(s)$, $\alpha \le \alpha(s)$, $\beta(s) \le b$, $\alpha(e) = e$, $\beta(e) = e^{\#}$, such that

$$(R_s f)(t) \in C^\infty([e, \alpha(s)] \times [e, s])$$

$$\cap\, C^\infty([\alpha(s), \beta(s)] \times [e, s]) \cap C^\infty([\beta(s), e^{\#}] \times [e, s]))$$

for any $s \in U$ and $f \in C^{\infty}([a, b])$ and we have

$$\sup_{s \in U} \frac{\partial}{\partial s} R_s f(t) = M(t) \in L_{\infty}([a, b], m).$$

In relations presented above, the first component of the products such as $I \times [e, s]$ corresponds to the variable t and the second component corresponds to the variable s. For $e = b$ and sufficiently small $|s - e|$, the expressions $[e, s]$, $[e, \alpha(s)]$, $[\alpha(s), \beta(s)]$, and $[\beta(s), e^{\#}]$, denote $[s, b]$, $[\alpha(s), b]$, $[\beta(s), \alpha(s)]$, and $[a, \beta(s)]$, respectively.

This quite cumbersome condition appears to be necessary because the "natural" requirement of smoothness of $R_s f(t)$ in both two variables contradicts the existing examples.

Theorem 3.4. *The generator X_1 is essentially self-adjoint and its closure coincides with the operator X defined by formula (3.2).*

Proof. Suppose that operator X_1 is Hermitian. Let us show that X_1 is essentially self-adjoint. Let $g(t) \in \operatorname{Ran}(X_1 + i)^{\perp}$. Then, for all $n \in \mathbb{N}_0$, we can write

$$0 = \int_a^b [(X_1\chi)(t, n) + i\chi(t, n)]\overline{g(t)}\mu(t)dt = \int_a^b (\chi'(e, n) + i)\chi(t, n)\overline{g(t)}\,\mu(t)dt.$$

At the same time, since the characters are real, we have $\chi'(e, n) \neq i$, whence it follows that $(g, \chi(\cdot, n))_2 = 0$ $(n \in \mathbb{N}_0)$. Since the system of characters is complete in L_2, we conclude that $\dim \operatorname{Ran}(X_1 + i)^{\perp} = 0$. Similarly, $\dim \operatorname{Ran}(X_1 - i)^{\perp} = 0$ and, consequently, X_1 is essentially self-adjoint.

Let $f \in \mathcal{D}(X_1)$. Then

$$\begin{aligned}(M_{\hat{\lambda}} \mathcal{F} f)(\chi_n) &= \int_a^b f(t) \frac{\partial}{\partial s}(R_s \chi_n)(t)|_{s=e} \mu(t)dt \\ &= \int_a^b f(t)(X_1\chi_n)(t)\mu(t)dt \\ &= \int_a^b (X_1 f)(t)\chi_n(t)\mu(t)dt = (\mathcal{F}(X_1 f))(\chi_n)\end{aligned}$$

because X_1 is Hermitian and $\chi_n(t) \in \mathcal{D}(X_1)$. This implies that $X_1 \subset \mathcal{F}^{-1} M_{\hat{\lambda}} \mathcal{F} = X$ and, hence, in view of the essential self-adjointness of X_1, its closure coincides with X.

To prove that X_1 is Hermitian, we need the following formula:

$$\frac{\partial}{\partial s}\int_{\alpha(s)}^{\beta(s)} f(t,s)\,dt = \int_{\alpha(s)}^{\beta(s)} \frac{\partial f(t,s)}{\partial s} + \beta'(s)f(\beta(s),s) - \alpha'(s)f(\alpha(s),s), \qquad (3.12)$$

where $f(t,s)$, $\alpha(s)$, and $\beta(s)$ are sufficiently smooth functions, and dt is the Lebesgue measure.

Lemma 3.3. *The generator*

$$(X_1 f) = \frac{\partial}{\partial s}(R_s f)(t)|_{s=e}$$

is an Hermitian operator.

Proof. By using the Lebesgue theorem and relation (3.12), we obtain

$$(Xf,g)_2 = \int_a^b \frac{\partial}{\partial s}(R_s f)(t)|_{s=e}\overline{g(t)}\,\mu(t)\,dt$$

$$= \lim_{s\to e}\int_{\alpha(s)}^{\beta(s)} \frac{\partial}{\partial s}(R_s f)(t)|_{s=e}\overline{g(t)}\,\mu(t)\,dt$$

$$= \lim_{s\to e}\Bigg[\frac{\partial}{\partial s}\int_a^b (R_s f)(t)\overline{g(t)}\mu(t)\,dt - \frac{\partial}{\partial s}\int_a^{\alpha(s)} (R_s f)(t)\overline{g(t)}\mu(t)\,dt$$

$$- \frac{\partial}{\partial s}\int_{\beta(s)}^b (R_s f)(t)\overline{g(t)}\mu(t)\,dt - \beta'(s)(R_s f)(\beta(s))\,\overline{g(\beta(s))}\,\mu(\beta(s))$$

$$+ \alpha'(s)(R_s f)(\alpha(s))\overline{g(\alpha(s))}\mu(\alpha(s))\Bigg]$$

$$= (f*\overline{g})'(e) - \lim_{s\to e}\left[\int_a^{\alpha(s)} \frac{\partial}{\partial s}(R_s f)(t)\overline{g(t)}\mu(t)\,dt + \int_{\beta(s)}^b \frac{\partial}{\partial s}(R_s f)(t)\overline{g(t)}\mu(t)\,dt\right].$$

For $s \in U$, in view of the fact that $\alpha(s)$ is continuous, we get

$$\left| \int_a^{\alpha(s)} \frac{\partial}{\partial s}(R_s f)(t)\overline{g(t)}\mu(t)dt \right|$$

$$\leq \int_a^{\alpha(s)} \sup_{s\in U}\left|\frac{\partial}{\partial s}(R_s f)(t)\right| |g(t)|\mu(t)dt = \int_a^{\alpha(s)} M(t)|g(t)|\mu(t)dt$$

$$\leq \|M\|_{L_\infty([a,b],m)} \int_a^{\alpha(s)} |g(t)|\mu(t)dt \to 0 \text{ as } s \to e \quad (f, g \in C^\infty([a,b])).$$

A similar estimate holds for

$$\int_{\beta(s)}^{b} \frac{\partial}{\partial s}(R_s f)(t)\overline{g(t)}\mu(t)dt,$$

whence

$$(X_1 f, g)_2 = (f * \bar{g})'(e) \quad (f, g \in C^\infty([a,b])) \tag{3.13}$$

but $(f * \bar{g})'(e) = (g * \bar{f})'(e)$. By using (3.13) and the fact that the generalized translation operators are real, we prove that the operator X_1 is Hermitian. Indeed,

$$(X_1 f, g)_2 = (\bar{f} * g)'(e) = (X_1 \bar{g}, f)_2 = \overline{(X_1 g, f)_2} = (f, X_1 g)_2 \quad (f, g \in C^\infty([a,b])).$$

■

Example 1. Let $[a, b] = [0, \pi]$. Consider an Hermitian hypercomplex system generated by generalized translation operators

$$(R_s f)(t) = \frac{1}{2}[f(|t-s|) + f(\pi - |\pi - t - s|)] \tag{3.14}$$

with the Lebesgue measure as a multiplicative measure. The functions $\chi(t, n) = \cos nt$ are characters of this hypercomplex system, $\hat{\lambda}(n) = -n^2$, and $\lambda(t) = \cos t$. The generator X is a differential operator generated by the second derivative d^2/dt^2 and the boundary conditions $f'(0) = f'(\pi) = 0$ and the operator $\hat{X}$ has the form

$$(\hat{X} f)(n) = \frac{1}{2}[f(n+1) + f(|n-1|)].$$

It is clear that the generalized translation operators from (3.14) satisfy the smoothness condition from Theorem 3.4, where $\alpha(s) = s$, $\beta(s) = \pi - s$, and X coincides with the closure of the second-order generator restricted to $f'(0) = f'(\pi) = 0$. It is easy to see that the mapping $[0, \pi] \ni t \mapsto \cos t = x$ establishes an isomorphism between the hypercomplex system $L_1([0, \pi], dt)$ constructed from the generalized translation operators (3.14) and the hypercomplex system

$$L_1\left([-1, 1], \frac{dx}{\sqrt{1-x^2}}\right)$$

constructed from the Chebyshev polynomials of the first kind. Under this isomorphism, the second-order generator X_2 of the hypercomplex system $L_1([0, \pi], dt)$ transforms into the first-order generator

$$X_1' = -(1-x^2)\frac{d^2}{dx^2} + x\frac{x}{dx}$$

of the hypercomplex system

$$L_1\left([-1, 1], \frac{dx}{\sqrt{1-x^2}}\right).$$

Example 2. Consider the Hermitian hypercomplex system $L_1([-1, 1], (1-t)^{\alpha}(1+t)^{\beta}dt)$ constructed from the Jacobi polynomials $P_n(t) = P_n^{(\alpha,\beta)}(t)$, $(t \in [-1, 1]$, $\alpha \geq \beta$, $\alpha + \beta \geq -1)$ orthonormal with respect to the measure $dm(t) = (1-t)^{\alpha}(1+t)^{\beta}dt$ (see Subsection 3.5 in Chapter 2). Generally speaking, the structure measure of this hypercomplex system may take negative values (see [Gas3]).

The functions $\chi(t, n) = P_n(t)/P_n(1)$ $(t \in [-1, 1], n \in \mathbb{N}_0)$ are characters of this hypercomplex system. The dual hypercomplex system $l_1(\hat{m})$ $(\hat{m}(n) = P_n^2(1))$ is described in Section 3 of Chapter 2. Both hypercomplex systems $L_1([-1, 1], m)$ and $l_1(\hat{m})$ are Banach algebras (see Section 3 in Chapter 2); moreover, the structural constants of the hypercomplex system $l_1(\hat{m})$ are nonnegative for $\alpha \geq \beta$ and $\alpha + \beta + 1 \geq 0$. Hence, all conditions formulated at the beginning of this subsection are satisfied. It is easy to see that

$$\hat{\lambda}(n) = \frac{d}{dt}\chi(t, n)\Big|_{t=1} = \frac{n(n+\alpha+\beta+1)}{2(\alpha+1)}$$

(see [Sze], p. 75) and $\lambda(t) = t$. Clearly, the generator X is generated by the differential equation for Jacobi polynomials

$$X = -\frac{1}{2(\alpha+1)}\left\{(1-t^2)\frac{d^2}{dt^2} - [\alpha-\beta+(\alpha+\beta+2)t]\frac{d}{dt}\right\}.$$

(The equality $(X\chi)(t,n) = \chi'(1,n)\chi(t,n)$ is evident and it remains to use the fact that the family of of characters is complete in $L_1([-1,1],m)$.) Let us show that

$$(\hat{X}f)(n) = \frac{1}{P_n(1)}L(f(\cdot)P_\bullet(1))(n), \tag{3.15}$$

where

$$(Lu)(n) = \alpha_n u(n+1) + \beta_n u(n) + \alpha_{n-1}u(n-1) \quad (\alpha_n > 0,\ \beta_n \in \mathbb{R},\ \alpha_{-1} = 0,\ u \in l_2)$$

is the difference operator in l_2 for which Jacobi polynomials are polynomials of the first kind. Indeed, for $f \in l_2$, we can write

$$(\hat{X}f)(n) = \frac{1}{P_n(1)}\int tP_n(t)\sum_{k=0}^{\infty} f(k)P_k(t)P_k(1)\,dm(t)$$

$$= \frac{1}{P_n(1)}[\alpha_n f(n+1)P_{n+1}(1) + \beta_n f(n)P_n(1) + \alpha_{n-1}f(n-1)P_{n-1}(1)].$$

■

Thus, the generator $\hat{X}$ is unitary equivalent to L.

3.2. General Case of the Construction of Generalized Translation Operators from a Generator. In the present subsection, we show that relations (3.8) and (3.9) remain true in a more general case where it is, generally speaking, impossible to construct a hypercomplex system from the collection of generalized translation operators R_s (or R_p).

Consider an unbounded self-adjoint operator X with simple spectrum and complete system of eigenvectors acting in $L_2([a,b],m)$, where m is a regular positive Borel measure. Assume that X is semibounded, $\operatorname{Ker} X = \{\lambda \mid \lambda = \text{const}\}$, and the eigenfunctions of X are continuous and real. We enumerate the points of the discrete spectrum so that $\lambda_0 = 0$, and set $e = a$ if X is semibounded from below and $e = b$ if X is semibounded from above. Let $\chi(t,n)$ be the eigenfunction of the operator X corresponding to an eigenvalue λ_n and normalized by the condition $\chi(e,n) = 1$. Suppose that $|\chi(t,n)| \le C$. Denote by λ^{-1} the function inverse to the function $\mathbb{N}_0 \ni n \mapsto \lambda_n \in \mathbb{R}$.

Theorem 3.5. *The operators* $(R_s f)(t) = (\chi(s,\lambda^{-1}(X))f)(t)$ $(f \in L_2)$ *form a commutative family of generalized translation operators with the following properties:*

(i) $(R_s\bar{f})(t) = \overline{(R_s f)(t)}$ *m-almost everywhere in* t;

(ii) $(R_s f)(t) = (R_t f)(s)$ $m \times m$*-almost everywhere;*

(iii) $\| R_s \| \leq C$;

(iv) the operator valued function $[a, b] \ni s \mapsto R_s$ *is weakly continuous;*

(v) $\int_a^b (R_s f)(t) dm(t) = \int_a^b f(t) dm(t)$ $(f \in C[a, b])$ *and the measure* m *is strongly invariant;*

(vi) $(R_s \chi)(t, n) = \chi(s, n)\chi(t, n)$.

Proof. It is clear that $\chi(t, 0) = 1$. To prove that R_s are commutative generalized translation operators, it is necessary to check the following properties:

(a) $R_e = I$;

(b) $(R_s f)(t) \in L_2([a, b], m)$ m-almost everywhere in t;

(c) $(R_r^s(R_s f)((t) = (R_s^t(R_r f))(t)$ $m \times m$-almost everywhere;

(d) $(R_r^t R_s f)(t) = (R_s^t(R_r f))(t)$ $m \times m$-almost everywhere.

Since $\chi(t, k)$ is a complete set of eigenvectors, we conclude that

$$(Xf)(t) = \sum_{k \in \mathbb{N}_0} \lambda_k (f, \chi(\cdot, k))_2 \chi(t, k) \|\chi(\cdot, k)\|_2^{-2}$$

$$= \sum_{\lambda_k} \lambda_k (f, \chi(\cdot, \lambda^{-1}(\lambda_k))_2 \chi(t, \lambda^{-1}(\lambda_k)) \left\|\chi(\cdot, \lambda^{-1}(\lambda_k))\right\|_2^2$$

is the spectral decomposition of the operator X. Hence,

$$(R_s f)(t) = \chi(s, \lambda^{-1}(X)) f)(t) = \sum_k \chi(s, k)\chi(t, k)(f, \chi(\cdot, k))_2 \|\chi(\cdot, k)\|_2^{-2}. \quad (3.16)$$

This immediately implies property (iii) (because $\| R_s \| = \| \chi(s, \lambda^{-1}(X)) \| < C$) and property (ii). Condition (b) follows from (ii), and (d) follows from the fact that R_r and

R_s are functions of single operator. It is also clear that $R_e = I$. Let us check the associativity of generalized translation operators, condition (c). Denote

$$C_k = (f, \chi(\cdot, k))_2 \, \|\chi(\cdot,k)\|_2^{-2}.$$

Since the eigenfunctions of the operator X are orthogonal, we get

$$(\chi(r, \lambda^{-1}(X))\,\chi(s, \lambda^{-1}(X))f)(t)$$

$$= \sum_p \chi(r,p) \sum_k C_k(\chi(\cdot, k), \chi(\cdot, p))_2 \chi(s,k)\chi(t,k)$$

$$= \sum_k (f, \chi(\cdot, k))_2 \chi(r,k)\chi(s,k)\chi(t,k),$$

whence we readily arrive at (c). Property (i) follows from the fact that the eigenfunctions of X are real.

We check (iv). Since $\chi(s, \lambda^{-1}(X))_2 \le C\|f\|_2 \|g\|_2$, by using the Lebesgue theorem, we find that

$$(\chi(s, \lambda^{-1}(X))f, g)_2 = \sum_k C_k \chi(s,k) \int_a^b \chi(t,k)\, \overline{g(t)}\, dm(t)$$

$$= \sum_k (f, \chi(\cdot, k))_2 \, \overline{(g, \chi(\cdot,k))_2} \, \|\chi(\cdot,k)\|_2^{-2} \chi(s,k) \qquad (f, g \in L_2)$$

This yields (iv) because the last series is uniformly bounded and, hence, uniformly convergent. It follows from (3.16) that

$$(R_s\chi)(t,n) = \sum_k \chi(t,k)\chi(s,k)(\chi(\cdot, n), \chi(\cdot, k))_2 \, \|\chi(\cdot,k)\|_2^{-2} = \chi(t,n)\chi(s,n).$$

Finally, by using the relation

$$\|R_s f\|_{L_1} \le m^{1/2}([a,b])\, \|R_s f\|_2 \le C_1 \|f\|_2,$$

and the Lebesgue theorem, we conclude that the measure m is invariant, i.e.,

$$\int_a^b \sum_k (f, \chi(\cdot, k))_2 \chi(t,k)\chi(s,k) \, \|\chi(\cdot, k)\|_2^{-2}\, dm(t)$$

$$= \sum_k (f, \chi(\cdot, k))_2 \chi(s, k) \|\chi(\cdot, k)\|_2^{-2} \int_a^b \chi(t, k) dm(t)$$

$$= (f, \chi(\cdot, 0))_2 = \int_a^b f(t) dm(t).$$

Since $R_s^* = R_s$, the measure m is strongly invariant.

■

A similar theorem is true in the discrete case.

Consider a bounded self-adjoint operator $\hat{X}$ $\left(\|\hat{X}\| = 1\right)$ acting in the space $l_2(\hat{m})$, where $\hat{m}(k) > 0$, with simple spectrum homeomorphic to the segment $[a, b] \subset \mathbb{R}$. Assume that this homeomorphism is determined by a function $[a, b] \ni t \mapsto \hat{\lambda}(t) \in \sigma(\hat{X})$ such that $\sup \lambda(t) = \lambda(e) = 1$. (Clearly, $e \in \partial[a, b]$.) The operator $\hat{X}$ as a self-adjoint operator in $l_2(\hat{m})$, is a Carleman operator and, hence, for decomposition in its generalized eigenvectors, we can take the chain

$$l_2\left(\frac{1}{p}\hat{m}\right) \supset l_2(\hat{m}) \supset l_2(p\hat{m}), \tag{3.17}$$

where $p = (p_n)_{n=0}^{\infty}$ is an arbitrary weight satisfying the condition

$$\sum_{k=0}^{\infty} \frac{1}{p_k} < \infty.$$

Without loss of generality, we can assume that

$$\sum_{k=0}^{\infty} \frac{\hat{m}(k)}{p_k} < \infty.$$

Let us now extend rigging (3.17) by using a nuclear space $\mathcal{D} \subset l_2(p\hat{m})$ with the following properties:

1) $\hat{X}$ continuously maps $\mathcal{D}$ into $\mathcal{D}$;

2) for any $\lambda(t) \in \sigma(\hat{X})$, the dimensionality of the space $N(\lambda(t))$ generated by generalized eigenvectors correspond to an eigenvalue $\lambda(t)$ is equal to one.

Denote by $\chi(t, n) \in N(\lambda(t))$ a generalized eigenvector normalized by the condition $\chi(t, 0) = 1$. Let $\chi(t, n)$ be continuous real functions of t for fixed n. Assume that $\chi(e, n) \equiv 1$ and there exists a number $C \geq 1$ such that $|\chi(t, n)| \leq C$. As in Theorem 3.5, we denote by λ^{-1} the function inverse to the function $[a, b] \ni t \mapsto \lambda(t) \in \sigma(\hat{X})$.

Theorem 3.6. *The operators* $R_p\chi(q) = \chi\big(\lambda^{-1}(\hat{X}), p\big) f(q)$ *form a commutative family of generalized translation operators with the following properties:*

(i) $\big(R_q \bar{f}\big)(p) = \overline{\big(R_q f\big)(p)}$;

(ii) $(R_p f)(q) = (R_q f)(p)$;

(iii) $\| R_p \| \leq C$;

(iv) the eigenvectors $\chi(t, n)$ *regarded as functions of* t *are orthogonal with respect to the measure* $d\mu(t) = \|\chi_t\|_-^{-2} d\rho(\lambda(t))$, *where* $\|\cdot\|_-$ *is the norm in* $l_2(p^{-1}\hat{m})$ *and* ρ *is the trace measure of the operator* $\hat{X}$.

If, in addition, the functions $\chi(t, n)$ *are complete in* $L_2([a, b], \mu)$, *then*

(v) the measure $\hat{m}$ *is invariant, i.e.,*

$$\sum_q (R_p f)(q)\, \hat{m}(q) = \sum_q f(q)\, \hat{m}(q) \quad (f \in l_1(\hat{m}) \cap l_2(\hat{m}));$$

The last condition is, in particular, satisfied if the linear span of the functions $\chi(t, n)$ *is uniformly dense in* $C([a, b])$.

Proof. As in the proof of Theorem 3.5, it is necessary to check not only conditions (i)–(v) but also (a)–(d). It is clear that $R_0 = I$ and the operators R_p are commuting. By using formula (3.5) from Chapter 3 in [BeKo], we obtain

$$(\chi(\lambda^{-1}(\hat{X})), p) f)(q) = \int_a^b \chi(t, p)\chi(t, q)(\chi(t, \cdot), f)_0 \|\chi_t\|_-^{-2} d\rho(\lambda(t)),$$

where $(\cdot, \cdot)_0 = (\cdot, \cdot)_{l_2(\hat{m})}$,

$$\varphi(\lambda(t)) = \frac{\chi(t, k)}{\|\chi(t, \cdot)\|_-} \quad \text{and} \quad \mathbf{J}f(k) = \frac{1}{\sqrt{p_k}} f_k.$$

This immediately implies (i)–(iii) and (b). The relation

$$(\chi(\lambda^{-1}(\hat{X})), p)_q(\chi(\lambda^{-1}(\hat{X})), q)f)(r)$$

$$= (\chi(\lambda^{-1}(\hat{X})), p)(\chi(\lambda^{-1}(\hat{X})), r)f)(q)$$

$$= \int_a^b \chi(t,p)\chi(t,r)\chi(t,q)(\chi(t,\cdot),f)_0 \|\chi(t,\cdot)\|_-^{-2} d\rho(\lambda(t))$$

implies associativity of the generalized translation operators R_p.

Let us prove (iv). By using formula (3.7) from [BeKo, Chapter 3], for all $f, g \in l_2(\hat{m})$, we obtain

$$(f,g)_0 = (E([a,b])f,g)_0 = \int_a^b (f,\chi(t,\cdot))_0(g,\chi(t,\cdot))_0 \|\chi(t,\cdot)\|_-^{-2} d\rho(\lambda(t))$$

We set $f(k) = \delta_p^k$ and $g(k) = \delta_q^k$, where δ_j^i is the Kronecker symbol. Then

$$\int_a^b \chi(t,p)\chi(t,q) d\rho(\lambda(t)) = \delta_p^q \frac{1}{\hat{m}(p)}$$

which immediately implies (iv) and the equality $\|\chi(\cdot,p)\|_{L_2(\mu)}^{-2} = \hat{m}(p)$.

To prove the last assertion of theorem, it suffices to consider the case $f(k) = \delta_p^k$. It is clear that

$$\sum_k R_q \delta_p^k \hat{m}(k) = \sum_k \int_a^b \chi(t,p)\chi(t,q)\chi(t,k) d\mu(t)\, \hat{m}(p)\hat{m}(k).$$

On the other hand, since the functions $\chi(t,k)$ are complete in $L_2(\mu)$, we have

$$\chi(t,p)\chi(t,q) = \sum_k \int_a^b \chi(t,p)\chi(t,q)\chi(t,k) d\mu(t) \chi(t,k) \|\chi(\cdot,k)\|_{L_2}^{-2}. \tag{3.18}$$

almost everywhere with respect to the measure m. Since the functions $\chi(t,n)$ are continuous in t and uniformly bounded, both sides of relation (3.18) are continuous in t. By setting $t = e$ in (3.18) and using the fact that $\hat{m}(k) = \|\chi(\cdot,k)\|_{L_2}^{-2}$, we obtain

$$\sum_k \int_a^b \chi(t,p)\chi(t,q)\chi(t,k)d\mu(t)\,\hat{m}(k) = 1,$$

whence it follows that

$$\sum_k R_q^k(\delta_p^k)\hat{m}(k) = \hat{m}(p) = \sum_k \delta_p^k\hat{m}(k).$$

■

The conditions of Theorems 3.5 and 3.6 are satisfied by the generators X and $\hat{X}$ of one-parameter Lie hypercomplex systems, by the operators appearing in the regular Sturm–Liouville problem, and by singular Sturm–Liouville operators defined on an interval studied in [AcT1].

We now analyze the case of regular Sturm–Liouville problem in more details. Consider a differential expression

$$L = -\frac{d^2}{dt^2} + q(t), \quad q \in C([0,\pi]).$$

Let $\mu(t)$ be a solution of the Cauchy problem

$$-\mu'' + qm = 0, \quad \mu(0) = 1, \quad \mu'(0) = h\mu(0) \quad (h \neq \infty).$$

Denote

$$\frac{\mu'(\pi)}{\mu(\pi)} = H.$$

Under the assumption that $H \neq \infty$, we consider the Sturm–Liouville problem

$$Ly = -y'' + qy = -\lambda y, \tag{3.19}$$

$$y'(0) = hy(0), \quad y'(\pi) = Hy(\pi). \tag{3.20}$$

Let $\varphi(t, \lambda_n)$ be eigenfunctions of problem (3.19)–(3.20), i.e., solutions of problem (3.19)–(3.20) satisfying the condition $\varphi(0, \lambda_n) = 1$. It is known that the eigenvalues λ_n are distinct and have the following asymptotic behavior in n:

$$\sqrt{\lambda_n} = n + \frac{C}{n} + O\left(\frac{1}{n^2}\right)$$

and

$$\varphi(t, \lambda_n) = \cos \lambda_n t + O\left(\frac{1}{\sqrt{\lambda_n}}\right)$$

uniformly in $t \in [0, \pi]$ (see, e.g., [LeS]). Clearly, $\lambda_0 = 0$ and $\varphi(t, \lambda_0) = \mu(t)$. According to the Sturm theorem on zeros of solutions, we conclude that $\mu(t) > 0$ $(t \in [0, \pi])$. Consider a unitary operator

$$L_2([0, \pi], dt) \ni u(t) \mapsto \frac{u(t)}{\mu(t)} = Vu(t) \in L_2([0, \pi], \mu^2(t)dt)$$

and denote $X = VLV^{-1}$, where L is a self-adjoint operator in $L_2([0, \pi], dt)$ associated with problem (3.19)–(3.20). It is easy to see that X satisfies all conditions of Theorem 3.5. Its eigenfunctions have the form

$$\chi(t, n) = \frac{\varphi(t, \lambda_n)}{\mu(t)}.$$

Thus, it is possible to construct generalized translation operators satisfying conditions (i)–(vi) for the regular Sturm–Liouville problem. Note that these generalized translation operators can also be constructed by the method of Povzner (see Section 4 in Chapter 2). It is thus interesting to determine the classes of potentials $q(t)$ for which it is possible to construct commutative quantized hypercomplex systems from the generalized translation operators associated with the Sturm–Liouville problem.

3.3. Analog of the Canonical Commutation Relations for the Delsarte Generalized Translation Operators. First, we recall necessary facts about canonical commutation relations. A pair of self-adjoint operators A, B acting in a Hilbert space $\mathcal{H}$ is called a representation of *canonical commutation relations* (CCR) if

$$AB - BA = -iI. \tag{3.21}$$

The situation where both A and B are bounded is impossible. Moreover, relation (3.21) should be understood on a dense subset of analytic vectors of the operators A and B (recall that a vector $\varphi \in \mathcal{H}$ is called a *joint analytic vector* of a finite family of operators $\{A_i\}_{i=1}^{n}$ if there exists $s > 0$ such that

$$\sum_{k=0}^{\infty} \frac{1}{k!} \sum_{1 \le i_1, \ldots, i_k \le n} \left\| A_{i_1} \ldots A_{i_k} \varphi \right\| s^n < \infty),$$

or, which is the same, in the Weyl form $e^{itA}e^{isB} = e^{its}e^{isB}e^{itA}$. If there are no nontrivial subspaces of $\mathcal{H}$ invariant under the action of all operators e^{itA} and e^{isB} $(t, s \in \mathbb{R}^1)$, then the considered representation of CCR is called irreducible. According to the von Neumann uniqueness theorem, there exists a unique (to within unitary equivalence) irreducible representation of CCR, namely,

$$\mathcal{H} = L_2(\mathbb{R}^1, dt), \quad A = \frac{1}{i}\frac{d}{dt}, \quad B = M_t,$$

where M_t is the operator of multiplication by t. This representation is called the Schrödinger representation. Note that the operator $A = \frac{1}{i}\frac{d}{dt}$ is a generator of the group $\mathbb{R}$ and $B = M_t$ is the Fourier image of a generator of the dual group which is also equal to $\mathbb{R}$. In this case, $R_t = e^{itA}$ is an operator of translation in $\mathbb{R}$ and $T_s = e^{isB}$ is the Fourier image of an operator of translation in the dual group. Our aim is to replace the group $\mathbb{R}$ by a one-dimensional hypercomplex system, the operator A by a generator of this hypercomplex system, and the operator B by the Fourier image of a generator of the dual hypercomplex system (Fourier transformation is carried out over the characters of this hypercomplex system) and to establish relationships between these operators. These relations can be regarded as an analog of CCR.

First, we note that the constructions realized in Subsection 3.1 can easily be generalized to the case of one-dimensional hypercomplex systems with locally compact bases. Actually, as shown in [Schw5], to within an isomorphism of hypercomplex systems, one can assume that the semiaxis $[0, \infty)$ is a basis of a one-dimensional Hermitian Lie hypercomplex system and $e = 0$ is its basis unit. Assume that a basis of the dual hypercomplex system also coincides with the semiaxis $\hat{Q} = [0, \infty)$ and, hence, its characters are given by a smooth function $\chi(t, \tau)$ $(t \in Q, \tau \in \hat{Q})$. The functions $\lambda(t)$ and $\hat{\lambda}(\tau)$ $(t \in Q, \tau \in \hat{Q})$ determining the generators $\hat{X}$ and X, respectively, can be found by differentiating the function $\chi(t, \tau)$, namely,

$$\lambda(t) = \frac{\partial^k}{\partial \tau^k}\chi(t, \tau)|_{\tau=0} \quad \text{and} \quad \hat{\lambda}(\tau) = \frac{\partial^l}{\partial t^l}\chi(t, \tau)|_{t=0},$$

where

$$k = \min\left\{ n \colon \left.\frac{\partial^n \chi}{\partial^n \tau}\right|_{\tau=0} \neq 0 \;\text{ for some }\; t \in Q \right\}$$

and

$$l = \min\left\{ n \colon \left.\frac{\partial^n \chi}{\partial t^n}\right|_{t=0} \neq 0 \;\text{ for some }\; \tau \in \hat{Q} \right\}.$$

All results of Subsections 3.1 and 3.2 remain true in this case. We omit the proofs of relevant assertions and consider the following simple example:

Let $L_1([0,\infty), dt)$ be the hypercomplex system constructed in Example 2 in Section 2 of Chapter 1. The corresponding Delsarte generalized translation operators have the form

$$R_s f(t) = \frac{1}{2}[f(t+s) + f(|t-s|)], \quad f \in C([0,\infty)).$$

Recall that the characters of this hypercomplex system are $\chi(t,\tau) = \cos\tau t$ $(t, \tau \geq 0)$. It is clear that the dual hypercomplex system $L_1(\hat{Q}, \hat{m})$ has the basis $\hat{Q} = [0,\infty)$ and the multiplicative measure $d\hat{m}(\tau) = d\tau$. Moreover, the generalized translation operators associated with this hypercomplex system are also Delsarte generalized translation operators

$$(R_\tau f)(\mu) = \frac{1}{2}[f(\tau+\sigma) + f(|\tau-\sigma|)] \quad (f \in C([0,\infty))).$$

This means that the hypercomplex system $L_1([0,\infty), dt)$ is self-dual. The generator $X = \mathcal{F} M_{-\tau^2} \mathcal{F}$ of this hypercomplex system is generated by the differential expression $\frac{d^2}{dx^2}$ and the initial condition $f'(0) = 0$ $(f \in C^\infty([0,\infty))$, where $\mathcal{F}$ is the Fourier cosine transformation, $M_{-\tau^2}$ is the operator of multiplication by $-\tau^2$ in $L_2([0,\infty), d\tau)$, and the generalized translation operators have the form $R_s = \cos s\sqrt{X}$ $(s \in ([0,\infty))$. The generator of the dual hypercomplex system $\hat{X}$ is unitarily equivalent to the operator of multiplication by $-t^2$ in the space $L_2([0,\infty), dt)$. Denote by T_τ the Fourier images of the generalized translation operators R_τ $(\tau \in [0,\infty) = \hat{Q})$, i.e., $T_\tau = \mathcal{F}^{-1} R_\tau \mathcal{F}$. It is easy to see that $(T_\tau f)(t) = \cos\tau t f(t)$ $(f \in L_2)$.

Let $A = -X$ and $B = M_{t^2} = -\mathcal{F}^{-1}\hat{X}\mathcal{F}$. It is clear that there are no quadratic relations between A and B. At the same time, *there are (exactly) two linearly independent relations of the third order for A and B*:

$$[A,[B,A]] = 8A, \tag{3.22}$$

$$[B,[A,B]] = 8B. \tag{3.23}$$

This statement can be checked by direct calculation. For the present, relations (3.22)–(3.23) are regarded as formal relations on a set of smooth functions.

Let us show that the *operators $A = -X$ and $B = M_{t^2}$ have a dense set of joint analytic vectors*. Indeed, it is clear that the functions $v_{2k}(t) = H_{2k}(t)e^{-t^2/2}$ $(k \in \mathbb{N}_0)$, where $H_{2k}(t)$ are orthonormal Hermite polynomials, form a basis in the space $L_2([0,\infty), dt)$. Let us show that $\mathcal{D} = \text{l.s.}\{v_{2k}(t) \mid k \in \mathbb{N}_0\}$ is a space of joint analytic vectors of the operators A and B. It is evident that $\mathcal{D} \subset \mathcal{D}(A) \cap \mathcal{D}(B)$ and is invariant under the action of A and B. By using the recurrence relations for Hermite polynomials, we obtain

$$t^2 v_n(t) = \frac{1}{2}\sqrt{(n+1)(n+2)}\, v_{n+2} + \left(n+\frac{1}{2}\right) v_n + \frac{1}{2}\sqrt{n(n-1)}\, v_{n-2},$$

$$\begin{aligned}\frac{d^2 v_n}{dt^2} &= -(2n+1)\, v_n + t^2 v_n \\ &= \frac{1}{2}\sqrt{(n+1)(n+2)}\, v_{n+2} - \left(n+\frac{1}{2}\right) v_n + \frac{1}{2}\sqrt{n(n-1)}\, v_{n-2} \quad (n = 2k).\end{aligned}$$

By induction on $k = 0, 1, \ldots,$ we prove that, for all $n \geq 2$,

$$\| X_1 \ldots X_k v_n \| \leq 2^k (s+2k)^k \quad \left(s = n + \frac{1}{2}\right), \tag{3.24}$$

where, for any $i = 1, \ldots, k,$ the operator X_i is equal either to A or to B. For $k = 0$ and $k = 1$, the assertion is evident. By the induction hypothesis, we obtain

$$\begin{aligned}\| X_1 \ldots X_{k-1} X_k v_n \| &\leq \frac{1}{2}\sqrt{(n+1)(n+2)}\, \| X_1 \ldots X_{k-1} v_{n+2} \| \\ &\quad + \left(n+\frac{1}{2}\right) \| X_1 \ldots X_{k-1} v_n \| + \frac{1}{2}\sqrt{n(n-1)}\, \| X_1 \ldots X_{k-1} v_{n-2} \| \\ &\leq \left[\frac{1}{2}(s+1)(s+2k)^{k-1} + s(s+2k-2)^{k-1} \right. \\ &\quad \left. + \frac{1}{2}(s-1)(s+2k-4)^{k-1}\right] 2^{k-1} \\ &\leq 2^{k-1}(s+2k)^{k-1}\left(\frac{1}{2}(s+1) + s + \frac{1}{2}(s-1)\right) < 2^k (s+2k)^k.\end{aligned}$$

Here, we have used the fact that

$$\sqrt{(n+1)(n+2)} < \frac{1}{2}(2n+3) \quad \text{and} \quad s + 2k - 4 > 0 \quad \text{for all} \quad n \geq 2.$$

By using (3.24), one can easily show that, for all $n \geq 2$ and $t < (4e(n+3))^{-1}$, the series

$$\sum_{k=1}^{\infty} \frac{t^k}{k!} \sum_{i_1, \ldots, i_k} \left\| X_{i_1} \ldots X_{i_k} v_n \right\| \leq \sum_{k=1}^{\infty} \frac{t^k}{k!} \left(2 + \sum_{i=1}^{k} (k-i)i \right) 2^k (s+2k)^k$$

$$\leq \sum_{k=1}^{\infty} \frac{e^{-k}}{\sqrt{2\pi k}} \left(\frac{s}{k} + 2 \right)^k \left[2 + \frac{1}{6}(k-1) \right]$$

is convergent. The case $n = 0$ is investigated analogously (recall that we consider only even n).

■

We can now give the following definition: A pair of positive self-adjoint operators A and B acting in a separable Hilbert space $\mathcal{H}$ such that relations (3.22)–(3.23) hold on a dense set of joint analytic vectors of A and B is called a representation of relations (3.22)–(3.23).

A representation of relations (3.22)–(3.23) is called irreducible if the space $\mathcal{H}$ has no nontrivial closed subspace $\mathcal{H}_0$ such that $E_A(\Delta)\mathcal{H}_0 \subset \mathcal{H}_0$ and $E_B(\Delta)\mathcal{H}_0 \subset \mathcal{H}_0$ ($\Delta \in ([0, \infty))$, where E_A and E_B are the resolutionsof the identity for the operators A and B, respectively. It is easy to see that this is equivalent to the condition that the space $\mathcal{H}$ has no nontrivial closed subspace $\mathcal{H}_0 \subset \mathcal{H}$ invariant under the action of $U(t) = \cos t\sqrt{A}$ and $V(s) = \cos s\sqrt{B}$ or, which is the same, that any bounded operator Λ commuting with all $U(t)$, $V(s)$ ($t, s \in ([0, \infty))$ is a multiple of the identity operator, i.e., $\Lambda = \lambda I$.

Let us show that the *representation of (3.22)–(3.23) by the operators* $Au = -u''$ ($u'(0) = 0$) *and* $B = M_{t^2}$ *is irreducible.*

Indeed, assume that a bounded operator Λ commutes with T_τ and R_s ($t, s \in ([0, \infty))$. Then, by using the relation $[\Lambda, T_\tau] = 0$, we conclude that Λ commutes with the operator of multiplication by the Fourier (cosine) image of an arbitrary function $f \in L_1([0, \infty), dt)$. Since $\mathcal{F}(L_1([0, \infty), dt))$ is dense in $C_\infty([0, \infty))$ and the operator Λ is continuous, Λ commutes with the operator of multiplication by an arbitrary function from $L_\infty([0, \infty), dt)$. By Riesz–Neumann theorem, Λ coincides with the operator of multiplication by a function $f \in L_\infty([0, \infty), dt)$. The condition $[\Lambda, R_s] = 0$ now takes the form

$$u(t+s)[f(t) - f(t+s)] + u(|t-s|)[f(t) - f(|t-s|)] = 0 \quad (t, s \in ([0, \infty)),$$

whence it follows that $f(t+s) = f(t)$ for almost all $(t, s) \in [0, \infty) \times [0, \infty)$, i.e., the function $f(t)$ is constant almost everywhere. Hence, the representation under consideration is irreducible.

■

Theorem 3.7. *If $A \neq 0$, then there are no representations of relation (3.22) by bounded self-adjoint operators. The same assertion is true for relation (3.23).*

Proof. We consider only relation (3.22) because relation (3.23) is analyzed in exactly the same way. The proof is based on the use of the identity

$$4n^2 A^{2n-1} = A^n B A^n - \frac{1}{2}(A^{2n}B + BA^{2n}) \quad (n = 1, 2, \dots) \tag{3.25}$$

established below. If A and B are bounded operators, then identity (3.25) implies that, for any $n = 1, 2, \dots,$

$$4n^2 \|A\|^{2n-1} \le \|A^n B A^n\| + \frac{1}{2}(\|A^{2n}B\| + \|BA^{2n}\|) \le 2\|A\|^{2n}\|B\|,$$

whence it follows that $\|A\|\|B\| \ge 2n^2$ $(n = 1, 2, \dots)$ in contradiction with the assumption that both operators A and B are bounded.

Let us now prove (3.25). Denote

$$T_k^n = \frac{1}{2}(A^{2n-k}BA^k + A^k BA^{2n-k}) \quad (k = 0, \dots, 2n).$$

Then relation (3.22) can be rewritten as $T_1^1 = 4A + T_0^1$. It is obvious that

$$T_k^n = T_{2n-k}^n \quad \text{and} \quad A^s T_k^n A^s = T_{k+s}^{n+s} \quad (k = 0, \dots, n;\ s = 0, 1, \dots).$$

If $p + q = 2m$, then we have

$$A^p T_k^n A^p + A^q T_k^n A^q = T_{k+p}^{m+n} + T_{k+q}^{m+n}.$$

By using these properties of T_k^n, we arrive at the following recurrence formula:

$$T_k^n = 4A^{2n-1} + \frac{1}{2}\left(T_{k+1}^n + T_{k-1}^n\right), \tag{3.26}$$

Indeed,

$$T_k^n = \frac{1}{2}\left[A^{2n-k-1}T_1^1 A^{k-1} + A^{k-1}T_1^1 A^{2n-k-1}\right]$$

$$= \frac{1}{2}\left[A^{2n-k-1}(4A + T_1^1)A^{k-1} + A^{k-1}(4A + T_1^1)T^{2n-k-1}\right]$$

$$= 4A^{2n-1} + \frac{1}{2}\left(T^n_{2n-k-1} + T^n_{k-1}\right) = 4A^{2n-1} + \frac{1}{2}\left(T^n_{k+1} + T^n_{k-1}\right).$$

The proof of the theorem is completed by the following lemma:

Lemma 3.4. *The equality*

$$T^n_n = 4kA^{2n-1} + T^n_{n-k} \tag{3.27}$$

holds for all $k = 0, \dots, n$.

Proof. We proceed by induction on k. According to (3.26),

$$T^n_n = 4A^{2n-1} + \frac{1}{2}\left(T^n_{n+1} + T^n_{n-1}\right) = 4A^{2n-1} + T^n_{n-1},$$

i.e., equality (3.27) holds for $k = 1$. By using the induction hypothesis, we write

$$T^n_n = 4k^2A^{2n-1} + T^n_{n-k} = 4k^2A^{2n-1} + 4A^{2n-1} + \frac{1}{2}\left(T^n_{n-k+1} + T^n_{n-k-1}\right)$$

$$= 4(k^2+1)A^{2n-1} + \frac{1}{2}\left(T^n_{n-(k+1)} + T^n_{n-(k+1)}\right).$$

By applying the induction hypothesis once again, we get

$$T^n_{n-(k-1)} = T^n_n - 4(k-1)^2A^{2n-1},$$

whence it follows that

$$T^n_n = 4(2k^2 + 2 - (k-1)^2)A^{2n-1} + T^n_{n-(k+1)} = 4(k+1)^2A^{2n-1} + T^n_{n-(k+1)}.$$

■

By setting $k = n$ in relation (3.27), we obtain (3.25).

■

To establish all irreducible representations of relations (3.22)–(3.23), one can use the method of many-dimensional dynamical systems (see [VaS1], [VaS2]). For the sake of brevity, we omit the proofs and present only the final result (it was obtained by Vaisleb).

Let $(v_n)_{n=0}^{\infty}$ be an orthonormal basis in a Hilbert space $\mathcal{H}$. Then any irreducible representation of relations (3.22)–(3.23) by positive operators is unitarily equivalent to one of the following representations:

(a) $$Av_n = \sqrt{n(n+\lambda-1)}\, v_{n-1} + (2n+\lambda)\, v_n + \sqrt{(n+1)(n+\lambda)}\, v_{n+1},$$

$$Bv_n = -\sqrt{n(n+\lambda-1)}\, v_{n-1} + (2n+\lambda) v_n - \sqrt{(n+1)(n+\lambda)}\, v_{n+1} \quad (\lambda \geq 0);$$

(b) $$Av_n = -\sqrt{n(n+\lambda-1)}\, v_{n-1} + (2n+\lambda) v_n - \sqrt{(n+1)(n+\lambda)}\, v_{n+1},$$

$$Bv_n = \sqrt{n(n+\lambda-1)}\, v_{n-1} + (2n+\lambda) v_n + \sqrt{(n+1)(n+\lambda)}\, v_{n+1} \quad (\lambda \geq 0);$$

(c) $\dim \mathcal{H} = 1,\ A = B = 0$.

Thus, the von Neumann theorem is no longer true for relations (3.22)–(3.23). The representation

$$A = -\frac{d^2}{dt^2} \quad \text{and} \quad B = M_{t^2}$$

analyzed above is unitarily equivalent to representation (b) with $\lambda = 1/2$. One can easily prove this fact by choosing a basis in $L_2([0, \infty], dt)$ in the form $v_k = H_{2k}(t)e^{-t^2/2}$, where H_{2k} are Hermite polynomials. Note that, in case (a) with $\lambda > 0$, the operator A is defined by the Jacobian matrix and the corresponding polynomials of the first kind are the Laguerre polynomials $L_n(x, \alpha)$ $(\lambda = \alpha + 1, \alpha > -1)$. Thus, A is unitarily equivalent to the operator M_t of multiplication by the variable t in the space $L_2([0, \infty),$ $t^\alpha e^{-t} dt)$. Under this equivalence, the operator B turns into the operator

$$-4ty'' + 4(t-\alpha-1)y' - ty + 2(\alpha+1)y \quad (y \in C^\infty([0, \infty))).$$

It is easy to see that the unitary operator

$$Uf(x) = \sqrt{2} f(x^2) e^{-x^2/2} \quad (f \in L_2([0, \infty), t^\alpha e^{-t} dt)$$

transforms B into the Bessel operator

$$\tilde{B} = -\frac{d^2}{dx^2} - \frac{2\alpha+1}{x}\frac{d}{dx}$$

and A into the operator $\tilde{A}$ of multiplication by x^2 in the space $L_2([0, \infty), x^{2\alpha+1} dx)$. For $\alpha \geq -1/2$, the Bessel operator $\tilde{B}$ can be associated with generalized translation operators which preserve positivity and, by using these operators, one can construct the hypercomplex system $L_1([0, \infty), t^{2\alpha+1} dx)$. The functions

$$\chi(x, \lambda) = 2^{\alpha}\Gamma(\alpha+1)J_{\alpha}(\lambda x)/(\lambda x)^{\alpha} \quad (x \geq 0,\ \lambda \geq 0),$$

where J_α is a Bessel function, are characters of this hypercomplex system. The operator $\tilde{A}$ is the Fourier image of a generator of the dual hypercomplex system. Hence, although the uniqueness theorem is not true for relations (3.22) and (3.23), the class of hypercomplex systems whose generators satisfy relations (3.22) and (3.23) is not broad.

If the requirement that the operators A and B in representations of (3.22) and (3.23) must be positive is omitted, one can also count all possible representations. The number of representations of this sort is greater than the number of representations considered above.

Remark. By setting

$$X = \frac{i}{2}A, \quad Y = \frac{1}{2i}B, \quad Z = [X, Y],$$

we conclude that relations (3.22) and (3.23) can be reduced to relations between elements of the basis X, Y, Z of the Lie algebra $sl_2(\mathbb{R})$. By choosing, in the collection of unitarizable irreducible representations of $sl_2(\mathbb{R})$ (it is well known that such representations admit a complete classification (to within unitary equivalence)), representations for which the operators A and B are positive, we can obtain the collection of irreducible representations of (3.22), (3.23) without using the indicated method of dynamical systems.

Supplement. Hypercomplex Systems and Hypergroups: Connections and Distinctions

The supplement consists of two parts. The first part (Sections 1 and 2) is devoted to the presentation of the results obtained by Berezansky and S. Krein in 1950–1953. These results were published in 12 papers ([BKr1]–[BKr3] and [Ber1]–[Ber9]; see also a brief survey [BKr4]). The second part (Sections 3–5) contains the discussion of some recent results and the analysis of the relationship between hypercomplex systems and the theory of hypergroups.

1. Hypercomplex Systems with Locally Compact Basis. Definition and Properties

1.1. Finite-Dimensional Case. A finite-dimensional hypercomplex system over the field of complex numbers is defined a finite-dimensional algebra A with a fixed basis $Q = \{e_1, \dots, e_d\}$ $(d \in \mathbb{N})$. Hence, $f = \sum_{p=1}^{d} f_p e_p$ for all $f \in A$. To introduce the operation of multiplication $*$ in A, it suffices to define it for elements of the basis, i.e.,

$$e_p * e_q = \sum_{r=1}^{d} c_{pq}^{r} e_r \qquad (p, q = 1, \dots, d), \tag{S.1}$$

where $\{c_{pq}^r\}_{p,q,r=1}^{d}$ is a cubic matrix of structural constants $c_{pq}^r \in \mathbb{C}$. One can easily write relations which make multiplication in A associative.

We identify the vector e_p with p (hence, $Q = \{1, \dots, d\}$) and interpret the coordinate f_p of a vector $f \in A$ as the value of the function at the point $p = e_p$. With this interpretation, the hypercomplex system A coincides with the space of complex-valued functions f defined on the basis Q: $Q \ni p \mapsto f(p) \in \mathbb{C}$ with ordinary linear operations and multiplication defined by the formula which follows from (S.1), namely, for any f and g

$$(f * g)(r) = \sum_{p,q=1}^{d} f(p) g(q) c_{pq}^{r} = \sum_{p,q \in Q} f(p) g(q) c(p, q, r) \tag{S.2}$$

$$(r \in Q;\ c(p, q, r) = c_{pq}^{r} = (p * q)(r)).$$

The general idea is to replace the basis $\{1, \dots, d\}$ by a general locally compact space Q. In this case, the elements of the hypercomplex system are regarded as complex-valued functions on Q, the matrix of structural constants is replaced by a function $Q \times Q \times Q \ni (p, q, r) \mapsto c(p, q, r) \in \mathbb{C}$, and multiplication (convolution) is defined by analogy with (S.2) as

$$(f * g)(r) = \int_Q \int_Q f(p) g(q) c(p, q, r)\, dm(p)\, dm(q) \quad (r \in Q), \tag{S.3}$$

where m is a suitable measure.

However, in some important cases (e.g., for the ordinary convolution of functions on $\mathbb{R}$), multiplication in (S.3) makes sense only if $c(p, q, r)$ is a generalized function. Therefore, it is reasonable to define not a "structure function" $c(p, q, r) = (p * q)(r)$, but a "structure measure" $c(A, B, r)$ which is equal to the convolution of two indicators $\kappa_A(\cdot)$ and $\kappa_B(\cdot)$ of sets $A, B \subset Q$. Thus, instead of "a continuous analog" of the structural constants $c_{pq}^{r} = c(p, q, r)$ of finite-dimensional hypercomplex systems, we consider their matrix analog

$$c(A, B, r) = \sum_{p \in A, q \in B} c(p, q, r) \quad (A, B \subset Q;\ r \in Q). \tag{S.4}$$

1.2. Definition of Hypercomplex Systems with Locally Compact Basis. Let Q be a complete separable locally compact metric space of points $p, q, r, \dots$, let $\mathcal{B}(Q)$ be the σ-algebra of Borel subsets, and let $\mathcal{B}_0(Q)$ be the subring of $\mathcal{B}(Q)$ which consists of sets with compact closure. We consider the Borel measures, i.e., positive regular measures on $\mathcal{B}(Q)$ finite on compact sets. We denote by $C(Q)$ the space of continuous functions on Q and let $C_b(Q)$, $C_\infty(Q)$, and $C_0(Q)$ be spaces consisting of bounded, approaching zero at infinity, and compactly supported functions from $C(Q)$, respectively.

A hypercomplex system with basis Q is defined by its *structure measure* $c(A, B, r)$ $(A, B \in \mathcal{B}(Q);\ r \in Q)$. For fixed B and r (A and r), the structure measure $c(A, B, r)$ is a Borel measure in A (in B) with the following properties:

(H1) the function $c(A, B, r)$ lies in $C_0(Q)$ for all $A, B \in \mathcal{B}_0(Q)$,;

(H2) the following associativity relation holds for all $A, B \in \mathcal{B}_0(Q)$ and $s \in Q$:

$$\int_Q c(A, B, r)\, d_r c(E_r, C, s) = \int_Q c(B, C, r)\, d_r c(E_r, C, s). \tag{S.5}$$

The structure measure is called *commutative* if

$$\text{(H3)} \quad c(A, B, r) = c(B, A, r) \quad (A, B \in \mathcal{B}_0(Q)). \tag{S.6}$$

A measure m is called a *multiplicative measure* whenever

$$\int_Q c(A, B, r)\, dm(r) = m(A)m(B) \ (A, B \in \mathcal{B}_0(Q)). \tag{S.7}$$

(H4) We suppose the existence of a multiplicative measure.

Theorem 1.1. *For any $f, g \in L_1(Q, m)$, the convolution*

$$(f * g)(r) = \int_Q f(p)\, d_p\left(\int_Q g(q)\, d_q c(E_p, E_q, r)\right), \tag{S.8}$$

is well defined. This formula generalizes the definition of convolution (S.3). The space $L_1(Q, m)$ with this convolution is a Banach algebra (under condition (H3), the algebra is commutative). This Banach algebra is called a hypercomplex system with basis Q.

It is obvious that $c(A, B, r) = (\kappa_A * \kappa_B)(r)$ $(A, B \in \mathcal{B}_0(Q))$. One may construct hypercomplex system both with and without unity. If the unity is not included in $L_1(Q, m)$, then it is convenient to join it formally to $L_1(Q, m)$. The unital algebra obtained as a result is denoted by $\mathcal{L}_1(Q, m)$.

We especially emphasize two points in the axiomatics of hypercomplex systems: the nonnegativity of the structure measure (hence, finite-dimensional hypercomplex systems have nonnegative structural constants) and continuity and finiteness of the convolution $(\kappa_A * \kappa_B)(\cdot)$ $(A, B \in \mathcal{B}_0(Q))$. These properties of hypercomplex systems are essential.

Let us now present an example of a hypercomplex system. A large part of the theory of hypercomplex systems is constructed by analogy with this example. We are speaking about the group algebra of a unimodular locally compact group G with the basis $Q = G$. The structure measure is defined by the equality

$$c(A, B, r) = \rho(B^{-1} rA) \quad (A, B \in \mathcal{B}(Q);\ r \in Q), \tag{S.9}$$

where ρ is the Haar measure on G. The multiplicative measure m is ρ. It is obvious that the ordinary convolution of functions summable with respect to the Haar measure on the group can be rewritten in the form (S.8). Therefore, the group algebra $L_1(G, \rho)$ is an example of a hypercomplex system with the locally compact basis $Q = s$.

In Section 1, we consider only commutative hypercomplex systems, i.e., we always assume that (H3) is satisfied. For such hypercomplex systems, by using the nonnegativity of the structure measure and by imposing some additional restrictions, we can prove the existence of a multiplicative measure (see Subsections 1.5 and 3.1).

1.3. Characters and Maximal Ideals. A nonzero function $Q \ni p \mapsto \chi(p) \in \mathbb{C}$ measurable and bounded almost everywhere is called a *character* of the hypercomplex system if, for any $A, B \in \mathcal{B}_0(Q)$,

$$\int_Q c(A,B,r)\chi(r)\,dm(r) = \chi(A)\chi(B)\chi(C) = \int_C \chi(r)\,dm(r), \quad C \in \mathcal{B}_0(Q). \tag{S.10}$$

The set of characters is in one-to-one correspondence with the set of maximal ideals of the Banach algebra $L_1(Q,m)$ (or the algebra $\mathcal{L}_1(Q,m)$, where we consider only proper ideals of $L_1(Q,m)$). This correspondence between a character χ and a maximal ideal M is described by the equality

$$f(M) = \int_Q f(p)\chi(p)\,dm(p) \quad (f \in L_1(Q,m)). \tag{S.11}$$

We denote by X the set of characters endowed with the topology of the space of maximal ideals. The set X is a locally compact metric space if $L_1(Q,m)$ is an algebra without unity and it is a compact space if $L_1(Q,m)$ is an algebra with unity. Relation (S.7) now implies that the function $f(p) \equiv 1$ $(p \in Q)$ is always a character of a hypercomplex system. It follows from (S.11) that $\|\chi\|_{L_\infty(Q,m)} = \|M\| \le 1$.

2.4. Normal Hypercomplex Systems with Basis Unity. Consider a class of hypercomplex systems for which, by analogy with group algebra, one can construct meaningful harmonic analysis. For these systems, there exists an analog of the inverse elements for points of the basis and there exists a point of the basis which plays the role of "unity located in the basis", (i.e., the identity element in the case of group algebras).

(H5) A hypercomplex system is called *normal,* if there exists an involutive homeomorphism $Q \ni p \mapsto p^* \in Q$ such that $m(A) = m(A^*)$,

$$c(A,B,C) = c(C,B^*,A), \quad \text{and} \quad c(A,B,C) = c(A^*,C,B) \quad (A, B \in \mathcal{B}_0(Q)),$$

where $c(A,B,C) = \int_C c(A,B,r)\,dm(r)$.

It is worth noting that, for normal hypercomplex systems, the mapping $L_1(Q,m) \ni f(p) \mapsto f^*(p) = \overline{f(p^*)} \in L_1(Q,m)$ is an involution in the Banach algebra $L_1(Q,m)$. Consider some properties of normal hypercomplex systems.

Theorem 1.2. *The multiplicative measure in a normal hypercomplex system is unique. Characters of this system are continuous. The radical (i.e., the intersection of all maximal ideals) coincides with the set of annihilators (i.e., elements* $a \in L_1(Q, m)$ *such that* $a * f = 0$ *for any* $f \in L_1(Q, m)$*).*

A character χ of a normal hypercomplex system is called *Hermitian* if it corresponds to a symmetric maximal ideal $M^* = M$. In other words, χ is Hermitian if $\chi(p^*) = \overline{\chi(p)}$ $(p \in Q)$. Denote the collection of all Hermitian characters by $X_h \subset X$.

In normal hypercomplex systems, one can introduce generalized translation operators T_p $(p \in Q)$ as follows:

$$(T_p f)(q) = \lim_{n\to\infty} \frac{1}{m(B_n(p))m(B_n(q))} \int_Q c(B_n(p), B_n(q), r) f(r) dm(r), \qquad \text{(S.12)}$$

where $B_n(s)$ is a ball in Q of radius $1/n$ centered at the point s. We can prove that the limit in (S.12) exists for $f \in L_\infty(Q, m)$. Convolution (S.9) can be rewritten as relation (3) in Introduction. Roughly speaking, if the function $c(p, q, r)$ is as in (S.3), operator (S.12) takes the form

$$(T_p f)(q) = \int_Q c(p, q, r) f(r)\, dm(r) \qquad (p \in Q).$$

In the case of group algebras, we have $(T_p f)(q) = f(qp)$ $(p, q \in Q = G)$.

In normal hypercomplex systems, $c(A, B, r) \le m(A)$ $(A, B \in \mathcal{B}_0(Q);\ r \in Q)$. Hence, there exists the Radon–Nikodym derivative $dc(\cdot, B, r)/dm(\cdot)(p) = c'(p, B, r)$ defined for m-almost all $p \in Q$. It can be interpreted as the convolution of the δ-function δ_p with the indicator $\kappa_B(\cdot)$, i.e., $c'(p, B, r) = (\delta_p * \kappa_B)(r) \in [0, 1]$. This convolution is a measurable bounded compactly supported function.

(H6) A point $e \in Q$ is called the *basis unity* of a normal hypercomplex system if

$$c(A, B, e) = m(A^* \cap B) \quad (A, B \in \mathcal{B}_0(Q)). \qquad \text{(S.13)}$$

In the case of group algebras, the identity element of the group is the basis unity of the hypercomplex system $L_1(G, \rho)$. If Q is discrete, then $u(p) = 1/m(e)\delta_e^p$ is the unity of the algebra $L_1(Q, m)$.

Theorem 1.3. *A normal hypercomplex system with basis unity is semisimple.*

A point $e \in Q$ *is a basis unity if and only if* $\chi(e) = 1$ *for all characters and the system of characters is complete in* $C_0(Q)$ *in a sense of* $L_1(Q, m)$ *(i.e., if* $\int g(r)\chi(r)\, dm(r) = 0$ *for* $g \in C_0(Q)$ *and all characters* χ*, then* $g = 0$*).*

We develop harmonic analysis for normal hypercomplex systems with basis unity, including some generalizations of harmonic analysis on commutative compact and discrete groups (the Plancherel theorem, positive definite functions, duality, and almost periodic functions). In the remaining part of Section 1, these results are mainly presented for hypercomplex system with compact or discrete bases.

1.5. Harmonic Analysis for Hypercomplex Systems with Compact Basis. Let Q be a compact basis. We have the following theorem on the existence of a multiplicative measure:

Theorem 1.4. *Assume that the structure measure $c(A, B, r)$ of the general commutative hypercomplex system with compact basis Q satisfies the condition $c(Q, Q, r) > 0$ for any open set $O \subset Q$ and any $r \in Q$. Then there exists a multiplicative measure (i.e., in this case, (H4) follows from (H1), (H2), and (H3)).*

For normal hypercomplex systems, the following theorem on completeness of characters is true:

Theorem 1.5. *Let $L_1(Q, m)$ be a normal hypercomplex system with basis unity and compact basis Q. The space of characters X is discrete and countable, i.e., $X = \{\chi(p)\} = (\chi_j(p))_{j=1}^{\infty}$ (if Q is a finite set, then X is finite). All characters are Hermitian and form a complete orthogonal system in $L_2(Q, m)$. Denote $\hat{Q} = X_h = X$. This set is called the dual basis. The Parseval equality takes the following form:*

$$\int_Q f(p)\overline{g(p)}\, dm(p) = \sum_{\chi \in \hat{Q}} \hat{f}(\chi)\overline{\hat{g}(\chi)}\hat{m}(\chi) \quad \textit{for any} \quad f, g \in L_2(Q, m), \tag{S.14}$$

where

$$\hat{f}(\chi) = \int_Q f(p)\overline{\chi(p)}\, dm(p) \quad \textit{and} \quad \hat{m}(\chi) = \left(\int_Q |\chi(p)|^2\, dm(p)\right)^{-1},$$

and $\hat{m}$ is a Plancherel measure.

Corollary 1.1. *We have*

$$c(A, B, r) = \sum_{\chi \in \hat{Q}} \chi(A)\chi(B)\overline{\chi(r)}\, \hat{m}(\chi) \quad (A, B \in \mathcal{B}(Q);\ r \in Q). \tag{S.15}$$

Note that it follows from (S.14) and (S.10) that $(c(A, B, \cdot)\hat{m}(\chi)) = \overline{\chi(A)\chi(B)}$. By reconstructing $c(A, B, r)$ from $(c(A, B, \cdot)\, \hat{m}(\chi))$ and using the positivity of $c(A, B, r)$, we obtain (S.15).

A continuous bounded function $Q \ni p \mapsto \varphi(p) \in \mathbb{C}$ is called *positive definite* if the inequality

$$\sum_{j,k=1}^{n} (T_{p_j}\varphi)(p_k^*)\xi_j \bar{\xi}_k \geq 0. \tag{S.16}$$

holds for all $p_1, \ldots, p_n \in Q$ and $(\xi_1, \ldots, \xi_n) \in \mathbb{C}^n$ $(n \in \mathbb{N})$.

Theorem 1.6. *A function* $\varphi(p)$ *is positive definite if and only if*

$$\varphi(p) = \sum_{\chi \in \hat{Q}} \chi(p)\sigma(\chi), \quad \sigma(\chi) \geq 0, \quad \textit{and} \quad \sum_{\chi \in \hat{Q}} \sigma(\chi) < \infty, \tag{S.17}$$

where the numbers $\sigma(\chi) = \hat{\varphi}(\chi)\ \hat{m}(\chi)$ *are uniquely determined by* φ.

This theorem admits the following equivalent formulation: If the Fourier coefficients $\hat{\varphi}(\chi)$ of the function $\varphi \in L_\infty(Q, m)$ are nonnegative, then the expansion of φ in a series in χ is absolutely and uniformly convergent.

1.6. Harmonic Analysis for Hypercomplex Systems with Discrete Basis. Consider a normal hypercomplex system $L_1(Q, m)$ with a countable discrete basis Q and basis unity (in this case, it is the unity of the algebra $L_1(Q, m)$). Hence,

$$L_1(Q, m) = l_1(m) = \left\{ Q \ni p \mapsto f(p) \in \mathbb{C} \,\middle|\, \sum_{p \in Q} |f(p)|\, m(p) < \infty \right\}.$$

The structure measure of this hypercomplex system is given by (S.4) in the form of an infinite structure matrix $c(p, q, r) = (\delta_p * \delta_q)(r) \geq 0$ $(p, q, r \in Q)$. It is now convenient to weaken condition (H1): *we do not require, that* $c(p, q, r)$ *is compactly supported for fixed* $p, q \in Q$. The existence of a multiplicative measure (i.e., (H4)) is supposed. The definition of characters and the normality condition now take the form

$$\sum_{r \in Q} c(p, q, r)\chi(r)m(r) = \chi(p)m(p)\chi(q)m(q),$$
$$c(p, q, r)m(r) = c(p, q^*, r)m(r). \tag{S.18}$$

The theorem on completeness of characters now admits the following formulation:

Theorem 1.7. *The spaces* X *and* X_h *of characters and Hermitian characters of a normal hypercomplex system with discrete basis and basis unity are compact. There* *exists a Borel measure* $\hat{m}$ *on* X_h *(Plancherel measure) such that the following Parseval equality holds for all* $f, g \in L_2(Q, m) = l_2(m)$:

$$\sum_{p \in Q} f(p)\overline{g(p)}m(p) = \int_{\hat{Q}} \hat{f}(\chi)\overline{\hat{g}(\chi)}d\hat{m}(\chi),$$

$$\hat{f}(\chi) = \sum_{p \in Q} f(p)\overline{\chi(p)}m(p), \tag{S.19}$$

where $\hat{Q} = \operatorname{supp} \hat{m}$ *is the dual basis.*

By setting $g = \delta_p$ in (S.19), we obtain the following inversion formula for f:

$$f(p) = \int_{\hat{Q}} \hat{f}(\chi)\chi(p)d\hat{m}(\chi). \tag{S.20}$$

Corollary 1.2. *We have*

$$c(p, q, r) = m(p)m(q)\int_{\hat{Q}} \chi(p)\chi(q)\overline{\chi(r)}d\hat{m}(\chi) \quad (p, q, r \in Q). \tag{S.21}$$

Positive definite sequences $Q \ni p \mapsto \varphi(p) \in \mathbb{C}$ are defined by (S.16). The Bochner–Herglotz theorem for sequences of this sort can be formulated as follows:

Theorem 1.8. *A function* $\varphi(p)$ *is positive definite if and only if*

$$\varphi(p) = \int_{X_h} \chi(p)d\sigma(\chi) \quad (p \in Q), \tag{S.22}$$

where the positive Borel measure σ *on* X_h *is uniquely determined by* φ.

1.7. Construction of Hypercomplex Systems with Compact and Discrete Bases by Using Orthogonal Systems of Functions. The formulas presented in Subsections 1.5 and 1.6 enable one to construct hypercomplex systems with compact and discrete bases.

Theorem 1.9. *Let* Q *be a compact space with finite Borel measure* σ *positive on open subsets of* Q, *let* $\hat{Q}$ *be a countable set and let* $\Sigma = (\varphi_j(p))_{j=\hat{Q}}$ *be an orthogonal (with respect to* σ*) system of continuous functions on* Q *which is complete in* $L_2(Q, m)$. *Suppose that*

(i) if $\varphi_j \in \Sigma$, *then* $\overline{\varphi_j} \in \Sigma$;

(ii) one can indicate points $e \in Q$ *and* $\hat{e} \in \hat{Q}$ *such that* $\varphi_j(e) = 1$ *for all* $j \in \hat{Q}$ *and* $\varphi_{\hat{e}}(p) = 1$ *for all* $p \in Q$.

Then the following formulas similar to (S.15) and (S.21)

$$c(A, B, r) = \sum_{j \in \hat{Q}} \varphi_j(A)\varphi_j(B)\overline{\varphi_j(r)}\, m(j), \tag{S.23}$$

$$m(j) = \left(\int_Q |\varphi_j(p)|^2 d\sigma(p) \right)^{-1} \quad (A, B \in \mathcal{B}(Q);\ r \in Q),$$

$$c(p, q, r) = m(p)m(q) \int_Q \varphi_p(t)\varphi_q(t)\overline{\varphi_r(t)}\, d\sigma(t) \quad (p, q, r \in \hat{Q}) \tag{S.24}$$

define the structure measures of normal hypercomplex systems with basis unity in the cases of a compact basis and a discrete basis, respectively. It should be additionally assumed that the sums in (S.23) in the first case and the integrals in (S.24) in the second case are nonnegative.

A system of functions φ_j with properties outlined above is called an *A-system.* If we it is necessary to construct a single hypercomplex system with the structure measure (S.23) or (S.24), the requirements imposed on the measure σ and a system of functions can be weakened.

1.8. Duality for Hypercomplex Systems with Compact and Discrete Basis. The results presented in Subsections 1.5–1.7 enable one to construct duality theory for normal hypercomplex systems with compact and discrete bases. It can be regarded as a generalization of the classical duality theory for commutative compact and discrete groups.

Let $L_1(Q, m)$ be a hypercomplex system with compact basis, let $\hat{Q}$ be the dual countable basis (the collection of all characters χ, φ, $\psi, \ldots$), and let $\hat{m}$ be a Plancherel measure of the form (S.14). The space $L_1(\hat{Q}, \hat{m}) = l_1(\hat{m})$ becomes a hypercomplex system with discrete basis if we define the dual structure measure $\hat{c}$ by the formula

$$\hat{c}(\chi, \varphi, \psi) = \hat{m}(\chi)\,\hat{m}(\varphi) \int_Q \chi(p)\varphi(p)\overline{\psi(p)}dm(p) \quad (\chi, \varphi, \psi \in \hat{Q}) \tag{S.25}$$

and assume that the integrals in (S.25) are nonnegative. This dual hypercomplex system is normal if we set $\chi^* = \overline{\chi}$ and possesses the basis unity $\hat{e} \equiv 1$.

Every point $p \in Q$ defines a function on $\hat{Q}$: $\hat{Q} \ni \chi \mapsto \chi(p) \in \mathbb{C}$. It is easy to see that this function is an Hermitian character of the hypercomplex system $l_1(\hat{m})$. Therefore, $Q \subset \hat{\hat{Q}}$, where $\hat{\hat{Q}}$ is the collection of Hermitian characters of the dual hypercomplex system $l_1(\hat{m})$. We say that hypercomplex systems are in *duality* if $Q = \hat{\hat{Q}}$.

Theorem 1.10. *Let $L_1(Q, m)$ be a normal hypercomplex system with compact basis and basis unity. The discrete set $\hat{Q}$ of characters is a basis of the dual normal hypercomplex system $l_1(\hat{Q}, \hat{m})$ if and only if the pointwise product $\varphi_1(t)\varphi_2(t)$ of any two positive definite functions $\varphi_1(t)$, $\varphi_2(t)$ $(t \in Q)$ is positive definite. The set $\hat{\hat{Q}}$ of Hermitian characters of the dual hypercomplex system coincides with the original compact set Q if and only if the pointwise product $\hat{\varphi}_1(\chi)\hat{\varphi}_2(\chi)$ of any two positive definite sequences $\hat{\varphi}_1(\chi)$ and $\hat{\varphi}_2(\chi)$ $(\chi \in \hat{Q})$ is positive definite. In this case, the Plancherel measure $\hat{\hat{m}}$ coincides with the original measure m.*

Note that, since $\operatorname{supp} m = Q$, we have $\operatorname{supp} \hat{\hat{m}} = \hat{\hat{Q}}$. Hence, $\hat{\hat{Q}} = Q$ is the dual basis for $l_1(\hat{m})$ in a sense of Theorem 1.7.

A similar theorem holds if we start with a hypercomplex system $l_1(m)$ with a discrete basis Q, the structure measure $c(p, q, r)$ $(p, q, r \in Q)$, and characters $\chi, \varphi, \psi, \ldots$. In this case, for the dual compact basis $\operatorname{supp} \hat{m} = \hat{Q}$, we can construct the hypercomplex system $L_1(\hat{Q}, \hat{m})$ with the structure measure such that

$$\hat{c}(A, B, \chi) = \sum_{p \in Q} p(A) p(B) \overline{\chi(p)}\, m(p), \quad p(A) = \int_A \psi(p)\, d\hat{m}(\psi) \tag{S.26}$$

for all $A, B \in \mathcal{B}(\hat{Q})$ and $\chi \in \hat{Q}$.

The set of characters of the hypercomplex system $L_1(\hat{Q}, \hat{m})$ coincides with Q and the Plancherel measure of the dual hypercomplex system coincides with m.

Note that the original proof of Theorem 1.10 was based on the use of another definition of normality in the discrete case (see Subsection 1.10). For the standard definition of normality, the last condition of positive definiteness of pointwise products of sequences can be omitted.

1.9. Hypercomplex Systems with Complex-Valued Measures. Hypercomplex systems of this sort can be regarded as a generalization of the notion of finite-dimensional hypercomplex systems in a sense that the structural constants c_{pq}^r may be complex. All definitions are similar to the definitions in Subsections 1.2–1.4; $c(A, B, r)$ is, in this case, a complex Borel measure (or charge) of A (B) for fixed B (A) and r. The multiplicative measure is a nonnegative Borel measure m such that the following inequality holds (instead of (S.7)):

$$\int_Q v(A, B, r)\, dm(r) \le m(A) m(B) \qquad (A, B \in \mathcal{B}(Q);\ r \in Q), \tag{S.27}$$

where $v(A, B, r)$ is the total variation on $A \times B$ of the charge c_r defined on $\mathcal{B}(Q \times Q)$ by the formula $c_r(A \times B) = c(A, B, r)$. It follows from inequality (S.27) that the convolution (S.8) exists for all $f, g \in L_1(Q, m)$ such that

$$\| f \times g \|_{L_1(Q,m)} \leq \| f \|_{L_1(Q,m)} \| g \|_{L_1(Q,m)}.$$

The basis Q can be, in particular, compact or discrete.

After appropriate reformulation, the major part of the results presented in Subsections 1.2–1.7 remain true for hypercomplex systems with complex measures (It is clear that the existence of a multiplicative measure is supposed). However, it is quite difficult to construct the relevant duality theory because inequality (S.27) for the dual hypercomplex system is violated (for more details, see Section 5).

1.10. References. The results presented in Subsections 1.2–1.4 for the case of a compact basis were obtained in [BKr1]–[BKr5] and [Ber9] (the locally compact case was considered in [Ber6]). The results of Subsections 1.6–1.9 with another definition of normality in the discrete case $(c(p, q, r) = c(r, q^*, p)$ instead of (S.18)) were established in [Ber2]–[Ber9]. The theory of almost periodic functions in hypercomplex systems is discussed in Subsection 2.6.

2. *Examples of Hypercomplex Systems*

2.1. Group Algebra of a Locally Compact Group. This example was mentioned in Subsection 1.2. The structure measure is defined by (S.9). The facts presented in Subsections 1.2–1.8 imply the relevant facts from harmonic analysis on such groups.

2.2. Center of the Group Algebra of a Commutative Compact Group. Let $L_1(G, \rho)$ be the group algebra of a compact group G (ρ is the Haar measure on G). The center of $L_1(G, \rho)$ is formed by functions which are constant on the conjugacy classes $p = \{h^{-1}gh \mid h \in G\}$ $(g \in G)$. Let Q be the collection of such classes endowed with the quotient topology relative to the mapping $G \ni g \mapsto \tilde{g} = p \in Q$. The center of the group algebra of the commutative compact group G gives an example of a normal hypercomplex system $L_1(Q, m)$ with basis unity and compact basis. The structure measure is defined by the equality $c(A, B, r) = \rho(\tilde{A}g \cap \tilde{B})$, where $\tilde{A} = \{h \in G \mid \tilde{h} \in A\}$ and $g \in r$, the multiplicative measure m is the image of ρ, the basis unity coincides with the group unity e, and $p^* = p^{-1}$. The characters of the hypercomplex system $L_1(Q, m)$ have the form $\chi(p) = X(g)n^{-1}$, where $g \in p$ and X is a character of an irreducible representation of G, $n = \dim X$. The results of Subsections 1.2–1.8 yield the representation theory and the harmonic analysis of compact groups, including the Tanna-ka–M. Krein duality principle.

2.3. Harmonic Analysis of Spherical Functions. Let G be a compact group, let H be a closed subgroup of G, let $Q = G/H$ be the space of double cosets $p = \{h_1 g h_2 \mid h_1, h_2 \in G\}$ $(g \in G)$, and let $G \ni g \mapsto \tilde{g} = p \in Q$ be a natural mapping. The functions in the group algebra $L_1(G, \rho)$ constant on classes the p form a subalgebra of $L_1(G, \rho)$. We suppose that this subalgebra is commutative (i.e., (G, H) is a Gelfand pair). Then we can introduce a normal hypercomplex system with compact basis Q and basis unity. Thus, the structure and multiplicative measures are defined by the same formulas as in Subsection 2.2 but with the new natural mapping, $p^* = p^{-1}$, and the basis unity $e = \{H\}$. Characters of such hypercomplex systems coincide with spherical functions. The results presented in Subsections 1.2–1.8 are, in fact, basic results of harmonic analysis on Gelfand pairs.

2.4. Hypercomplex systems Associated with the Sturm–Liouville Operator. We consider the Sturm–Liouville expression on a semiaxis

$$\mathcal{L}y = -y'' + a(p)y, \quad p \in [0, \infty), \tag{S.28}$$

where $a(p)$ is a real continuous function of bounded variation on $[0, \infty)$. Let $\tilde{a}(p)$ be a nonnegative nonincreasing function such that

$$|a(p') - a(p'')| \le \tilde{a}(p') - \tilde{a}(p'') \quad (0 \le p' < p'' < \infty)$$

(e.g., $\tilde{a}(p) = \mathrm{Var}_p^\infty a$). Consider the equation $-y'' + \tilde{a}(p) = 0$. Denote by $\mu(p)$ the solution of this equation such that $\mu(0) = 1$ and $\mu'(0) = 0$. By using the well-known method of Levitan and Povzner, we can define the convolution of functions on $[0, \infty)$.

We extend the function $a(p)$ to the entire real line as an even function and consider the following hyperbolic partial differential equation:

$$\mathcal{L}_p[u(p, q)] = \mathcal{L}_q[u(p, q)].$$

For a two times differentiable function $f(p)$, we set

$$(T_p f)(q) = u(p, q)/\mu(p)\mu(q) \quad (p, q \in [0, \infty)),$$

where $u(p, q)$ is a solution of the last equation satisfying the initial conditions $u(p, 0) = f(p)\mu(p)$ and $u_q'(p, 0) = 0$ (the function $f(p)$ is supposed to be extended to the entire real line as an even function). By using the Riemann method for hyperbolic equations, we can extend the operator T_p to a broader class of functions $f(p)$.

We define convolution by the formula

$$(f * g)(r) = \int_0^\infty (T_r f)(q) g(q) \mu^2(q) dq \quad (r \in [0, \infty)). \tag{S.29}$$

This operation generates the hypercomplex system $L_1(Q, m)$ with the basis $s = [0, \infty)$ and the multiplicative measure $dm(p) = \mu^2(p)dp$. The real structure measure is given by $s(A, B, r) = (\kappa_A * \kappa_B)(r)$. The hypercomplex system $L_1(Q, m)$ is normal $(p^* = p)$ and possesses the basis unity $e = 0$.

Assume that $\tilde{a}(p) \to 0$ as $p \to \infty$. The characters of the hypercomplex system coincide with $\varphi(t, \lambda)/\mu(t)$, where $\varphi(t, \lambda)$ are eigenfunctions of the Sturm–Liouville operator defined by (S.28) with the initial condition $y'(0) = 0$ and λ belongs to the spectrum of this operator. The hypercomplex system $L_1(Q, m)$ possesses the nonnegative structure measure (which satisfies (S.7)) if the function $a(p)$ is nonincreasing. In this case, $\tilde{a} = a$.

2.5. Hypercomplex Systems Associated with Orthogonal Polynomials. Let E be a compact subset of the real line, let σ be a finite Borel measure on E positive on open sets, and let $P_0(t)$, $P_1(t), \ldots$ be the relevant system of orthonormal polynomials. Let $b = \sup E$. It is obvious that $P_n(b) \neq 0$. According to Subsection 1.7, by using this system of polynomials, one can construct a hypercomplex system with the discrete basis $Q = \{0, 1, \ldots\}$ by setting

$$c(p, q, r) = \frac{P_b(b)P_q(b)}{P_r(b)} \int_E P_p(t)P_q(t)P_r(t)d\sigma(t)$$

$$= m(q) \int_E \varphi_p(t)\varphi_q(t)\varphi_r(t)d\sigma(t), \tag{S.30}$$

$$\varphi_p(t) = \frac{P_p(t)}{P_p(b)}, \quad m(p) = |P_p(b)|^2 = \left(\int_Q |\varphi_p(t)|^2 d\sigma(t)\right)^{-1}$$

for all $p, q, r \in Q$. We suppose that the integrals in (S.30) are nonnegative. This hypercomplex system is normal $p^* = p$ and possesses the basis unity $e = 0$.

The procedure of construction of hypercomplex systems associated with orthogonal polynomials can be generalized to the case of real $c(p, q, r)$ as in Subsection 1.9. We can also present the procedure for choosing constants $m(p)$ $(p \in Q)$ guaranteeing the validity of inequality (S.27) in the case of discrete bases. Moreover, for such hypercomplex systems, one can prove a theorem similar to the theorem on continuity of transmutation operators. Indeed, if we can construct a hypercomplex system for a measure $d\sigma(t)$, then, under certain restrictions imposed on the function $h(t) > 0$ $(t \in E)$, we can construct a hypercomplex system for the measure $h(t)d\sigma(t)$. Some special systems of orthogonal polynomials were investigated.

If (a) there exists $e \in E$ such that

$$P_r(e) = \max_{t \in E} |P_r(t)| \quad (r = 0, 1, \ldots);$$

(b) there is a multiplication formula

$$R_n(t)R_n(s) = \int_E R_n(z)\,d\mu_{t,s}(z) \qquad (t, s \in E),$$

where $d\mu_{t,s}(z)$ is a nonnegative bounded measure and $R_n(t) = P_n(t)/P_n(e)$, then, by using (S.23), we can construct a normal hypercomplex system with the compact basis E. Here, $t^* = t$ and the point e is the basis unity.

2.6. Almost Periodic Functions and Sequences. Let $L_1(Q, m)$ be a normal hypercomplex system with locally compact basis and basis unity and let T_p be the corresponding generalized translation operators (S.12). Following Delsarte and Levitan, a continuous function $Q \ni p \mapsto f(p) \in \mathbb{C}$ is called *almost periodic* if the family of functions $\{T_p f\}_{p\in Q}$ is precompact in the topology of uniform convergence (i.e., the closures of all these functions are compact).

We can construct the general theory of almost periodic functions of this sort. Namely, we can define an average $M(f)$ of an arbitrary function $f(p)$ by

$$M(f) = \operatorname*{Lim}_{N\to\infty} \frac{1}{m(Q_N)} \int_{Q_n} f(p)\,dm(p),$$

where Lim denotes the Banach limit and Q_N $(N = 1, 2, \dots)$ is a sequence of compact subsets of Q such that $Q_N \subset Q_{N+1}$ and $\cup\, Q_N = Q$. Almost periodic functions satisfy the following Parseval equality:

$$(f,f) = (f_a, f_a) + \sum_{\chi} |(f,\chi)|^2 (\chi,\chi)^{-1}, \qquad (f,g) = M(f\bar{g}), \qquad \text{(S.31)}$$

where summation is carried out over the countable set of characters of the hypercomplex system and f_a is the projection of f onto a space of "annihilators". We can also establish the conditions under which f_a is equal to zero and relation (S.31) turns into the classical Parseval equality. One can also prove a theorem on uniform approximation of almost periodic functions by linear combinations of characters.

This theory is applicable to hypercomplex systems associated with the Sturm–Liouville operator, to hypercomplex systems associated with orthogonal polynomials with measure $h(t)(1-t)^{\alpha}(1+t)^{\beta}$ $(t \in [-1, 1],\ \alpha, \beta \ge -1/2,\ h(t) > 0)$, and to almost periodic even functions defined on commutative locally compact groups. For these examples, some results of Levitan and Marchenko were extended and improved.

2.7. References. The results of Subsection 2.2 can be found in [Ber1] and [Ber9], For the results of Subsections 2.3 and 2.4, see [Ber2], [Ber9] and [Ber7], respectively.

It is clear that all results in Subsection 2.3 remain valid if we replace a compact group G by a locally compact group. Subsection 2.5 contains a presentation of the results obtained in [Ber2], [Ber4], and [Ber9] modified according to the style of [BeK5]. The results of Subsection 2.6 were established in [Ber3], [Ber5], [Ber6], [Ber8], and [Ber9].

3. Harmonic Analysis in the Locally Compact Case

In the present section, we describe harmonic analysis for a general noncommutative hypercomplex system $L_1(Q, m)$ with locally compact basis Q and basis unity $e \in Q$.

3.1. Fourier Transformation and the Plancherel Theorem. First, we consider the problem of existence of a multiplicative measure in the locally compact case.

Theorem 3.1. *Suppose that the structure measure is commutative (i.e., (H3) is satisfied) and has the following properties:*

(i) $c(Q, A, r) \in C_b(Q)$ *for any* $A \in \mathcal{B}_0(Q)$;

(ii) $\lim_{r \to \infty} c(Q, A, r) = c(Q, A, \infty)$ *exists for all* $A \in \mathcal{B}_0(Q)$ *and the function* $\mathcal{B}_0(Q) \ni A \mapsto c(Q, A, \infty)$ *is a regular Borel volume;*

(iii) the inequalities

$$c(Q, Q, r) > 0 \quad (r \in Q) \quad \text{and} \quad c(Q, Q, \infty) > 0$$

hold for any open $O \in \mathcal{B}_0(Q)$.

Then, for this structure measure, the multiplicative measure exists (but is not necessarily unique).

In the noncommutative case we suppose that the multiplicative measure exists. In normal hypercomplex systems, the multiplicative measure is unique.

All results of harmonic analysis established in the cases of compact and discrete bases can be generalized to the case of noncommutative hypercomplex systems with locally compact bases. Any hypercomplex system of this sort is associated a family of (left) generalized translation operators given by the formula $(L_s f, g^*)_{L_2} = (f * g)(s)$ $(f, g \in L_2)$. These operators can be extended to uniformly bounded operators acting in L_∞.

A dense linear subspace s of a Hilbert space H with two pairs of operations: multiplication $a \cup b$ and involution $a^{\cup}$ (first pair) and multiplication $a \cap b$ and involution $a^{\cap}$ (second pair) is called a *Hilbert bialgebra* if

1) D is a Hilbert algebra for each pair of operations;

2) the *operator of connection* W in $H \otimes H$ defined by the equality

$$(a \otimes b'^{\cap}, W(a'^{\cup} \otimes b))_{H \otimes H} = (a \cup a', b \cup b')_H \qquad (a, a', b, b'),$$

is continuous.

Theorem 3.2. *Let* $L_1\ (Q, m)$ *be a normal hypercomplex system with basis unity. Then the linear subspace* $D = L_1 \cap L_\infty$ *has the structure of a Hilbert bialgebra with the operations* $f \cap g = f * g$ *and* $f^{\cup}(r) = \overline{f(r^*)}$ *(first pair) and* $f \cap g(r) = f(r)g(r)$ *and* $f^{\cap}(r) = \overline{f(r)}$ *(second pair).*

For the Hilbert bialgebra $D = L_1 \cap L_\infty$ from Theorem 3.2, let $\mathcal{L}_{\cup}$ be the W^*-algebra associated with the Hilbert algebra $(D, *)$. By $\mathcal{L}_{\cup}^{+}$ we denote the set of positive elements of $\mathcal{L}_{\cup}$ and by φ the canonical exact semifinite normal trace on $\mathcal{L}_{\cup}^{+}$, defined as $\varphi(A) = (a, a)_{L_2}$ if $A^{1/2} = \mathcal{L}_{\cup}(a)$ $a \in D$, where $\mathcal{L}_{\cup}(a)$ is the operator of left convolution with a, i.e., $\mathcal{L}_{\cup}(a) b = a * b$ $(a, b \in D)$. We extend φ to $\mathcal{L}_{\cup}$ by linearity. From the central decomposition, we obtain

$$L_2 = \int_Z H_\lambda \, d\rho(\lambda) \qquad \text{and} \qquad \mathcal{L}_{\cup} = \int_Z \mathcal{L}_{\cup}(\lambda) \, d\rho(\lambda),$$

where Z is the quasispectrum of $\mathcal{L}_{\cup}$.

The *Fourier transform* of the function $a \in D = L_1 \cap L_\infty$ is defined as an operator-valued function

$$\hat{a}(\lambda) = \int_Q a(s) L_s(\lambda) \, dm(\lambda) \qquad (\lambda \in Z), \tag{S.32}$$

where $L_s(\lambda)$ are the components of the central decomposition of the left generalized translation operator $L_s \in \mathcal{L}_{\cup}$ associated with the hypercomplex system.

Theorem 3.3 (analog of the Plancherel theorem and inversion formula)**.** *The Fourier transforms (S.32) satisfy the following Plancherel and inversion formulas:*

$$(a, b) = \int_Z \varphi_\lambda(L_\lambda(b)^* L_\lambda(a)) \, d\rho(\lambda), \tag{S.33}$$

$$a(s) = \int_Z \varphi_\lambda(L_s(\lambda) \, \hat{a}(\lambda)) \, d\rho(\lambda), \tag{S.34}$$

where φ_λ *and* $L_\lambda(a)$ *are the components of the central decomposition of the trace* φ *and the operator* $\mathcal{L}_U(a)$ *respectively. The Plancherel measure* ρ *in (S.33) and (S.34) is unique to within equivalence. It follows from the Gelfand–Naimark–Segal construction that the Fourier transform (S.32) and the inversion formula (S.34) establish a unitary isomorphism between* $L_2(Q, m)$ *and the Hilbert space* H_φ *determined by the trace* φ *on* $\mathcal{L}_U$.

Denote by $\mathcal{N}$ the intersection of $L_1(Q, m)$ with the linear span of positive definite functions on Q. If the hypercomplex system $L_1(Q, m)$ is commutative, then the Fourier transform can be written as

$$\hat{a}(\chi) = \int_Q a(r)\overline{\chi(r)}\, dm(r) \qquad (\chi \in X_h).$$

Theorem 3.3 yields the following assertion:

Corollary 3.1. *There exists a unique regular Borel measure* $\hat{m}$ *on the space of Hermitian characters* X_h *such that, for any function* $x \in \mathcal{N}$, *its Fourier transform* $\hat{x}$ *lies in* $L_1(X_h, \hat{m})$ *and* $x(r) = \int_{X_h} \hat{x}(\chi)\chi(r)\, d\chi$. *The Fourier transformation can be continuously extended to a unitary operator* $\mathcal{F}\colon L_2(Q, m) \to L_2\big(\hat{X}_h, \hat{m}\big)$.

In the commutative case, we have also an analog of the Bochner theorem.

Theorem 3.4. *A function* $\varphi \in C_b(Q)$ *is positive definite if and only if there exists a finite positive regular measure* μ *on the space* X_h *such that*

$$\varphi(r) = \int_{X_h} x(r)\, d\mu(x) \qquad (r \in Q).$$

3.2. Duality of Commutative Hypercomplex Systems. Let $L_1(Q, m)$ be a commutative hypercomplex system Denote by $\hat{Q}$ the support of the Plancherel measure $\hat{m}$. For any $\Phi, \Psi \in \mathcal{B}(\hat{Q})$ and $\chi \in \hat{Q}$, we set

$$\hat{c}(\Phi, \Psi, \chi) = \int_Q \overline{\chi(r)} \int_\Phi \varphi(r)\, d\hat{m}(\varphi) \int_\Psi \psi(r)\, d\hat{m}(\psi)\, dr. \tag{S.35}$$

Our idea is to take the function $\hat{c}(\Phi, \Psi, \chi)$ as the structure measure of the dual hypercomplex system but, generally speaking, it does not satisfy all axioms (H1)–(H3) (axioms of positivity and finiteness are not true for $\hat{c}(\Phi, \Psi, \chi)$). Nevertheless, under certain additional restrictions this function turns into a structure measure.

Theorem 3.5. *$\hat{c}(\Phi, \Psi, \chi)$ is a structure measure if and only if it belongs to $C_0(\hat{Q})$ for any fixed $\Phi, \Psi \in \mathcal{B}(\hat{Q})$ and the pointwise product of any two characters $\varphi, \psi \in \hat{Q}$ is a positive definite function.*

Under the conditions of Theorem 3.5, we can construct the *dual hypercomplex system* $L_1(\hat{Q}, \hat{m})$ associated with the structure measure $\hat{c}(\Phi, \Psi, \chi)$. The Plancherel measure $\hat{m}$ is multiplicative for $\hat{c}(\Phi, \Psi, \chi)$. The dual hypercomplex system is a normal hypercomplex system with involution $\chi^*(r) = \overline{\chi(r)}$ and basis unity $\hat{e} \equiv 1(r)$. For any point $r \in Q$, we can construct an Hermitian character $\hat{r}(\chi) = \chi(r)$ of the dual hypercomplex system. Denote by $\hat{\hat{m}}$ the Plancherel measure corresponding to the dual hypercomplex system and let $\hat{\hat{Q}} = \operatorname{supp} \hat{\hat{m}}$. We say that these hypercomplex systems are in *duality* whenever $Q = \hat{\hat{Q}}$.

Let L_s be the left generalized translation operators associated with the hypercomplex system $L_1(Q, m)$. The hypercomplex system $L_1(Q, m)$ satisfies the condition of *separate continuity* if, for any $f \in C_0(Q)$, the function $(L_s f)(t)$ is separately continuous.

Theorem 3.6 (duality theorem). *Let $L_1(Q, m)$ be a commutative normal hypercomplex system with basis unity and with the following properties:*

(i) $\hat{e} \equiv 1 \in \hat{Q}$;

(ii) the pointwise product $\chi(r)\psi(r)$ of characters $\chi, \psi \in \hat{Q}$ is a positive definite function and the support of the measure corresponding to this function by Theorem 3.4 lies in $\hat{Q}$;

(iii) the function $\hat{c}(\Phi, \Psi, \chi)$ lies in $C_0(\hat{Q})$;

(iv) the hypercomplex system satisfies the condition of separate continuity.

Then the dual hypercomplex system $L_1(\hat{Q}, \hat{m})$ also satisfies (i)–(iv) and duality takes place.

Let us now describe an analog of the Fourier algebra for hypercomplex systems. Assume that a commutative hypercomplex system $L_1(Q, m)$ satisfies the conditions of Theorem 3.6. Denote $A(Q) = L_2(Q, m) * L_2(Q, m)$. It is easy to see that $A(Q) \subset C_\infty(Q)$. Moreover, $A(Q)$ is a Banach space with the norm

$$\|f_1 * f_2\|_{A(Q)} = \|(f_1 * f_2)^\wedge\|_{L_1(\hat{Q}, \hat{m})}.$$

Theorem 3.7. *If a commutative normal hypercomplex system with basis unity satisfies the conditions of Theorem 4.6, then $A(Q)$ is a Banach algebra with the ordinary* *pointwise multiplication of functions. Moreover, the algebra* $A(Q)$ *is isomorphic to* $L_1(\hat{Q}, \hat{m})$. *It is called the Fourier algebra of the hypercomplex system* $L_1(Q, m)$.

3.3. Representations of Hypercomplex Systems and Approximation Theorem. Let $L_1(Q, m)$ be a normal (not necessarily commutative) hypercomplex system with basis unity. A family of bounded operators $U = (U_p)_{p \in Q}$ in a separable Hilbert space H is called a *∗-representation of hypercomplex system* if

(a) $U_e = 1$;

(b) $U_p^* = U_{p^*}$ $(p \in Q)$;

(c) for any $\xi \in H$, the vector-valued function $Q \ni p \mapsto U_p \xi \in H$ is continuous in the weak topology;

(d) $\int\limits_Q c(A, B, r)\, U_r\, dm(r) = \int\limits_A U_p\, dm(p) \int\limits_B U_q\, dm(q)$ for any $A, B \in \mathcal{B}_0(Q)$.

A ∗-representation U_p is called *bounded* if the function $Q \ni p \mapsto \| U_p \|$ is bounded. Bounded ∗-representations of commutative hypercomplex systems coincide with Hermitian characters.

It is easy to see that the formula

$$U_x = \int\limits_Q x(p) U_p\, dm(p) \quad (x \in L_1(Q, m))$$

establishes a one-to-one correspondence between bounded ∗-representations and nondegenerate representations of the Banach algebra $L_1(Q, m)$. If a normal hypercomplex system with basis unity satisfies the condition of separate continuity, then

$$L_s(U_r \xi, \psi)_H = (U_s U_r \xi, \psi)_H \quad (r, s \in Q;\ \xi, \psi \in H),$$

where L_s are left generalized translations associated with the hypercomplex system, for any bounded ∗-representation U_r. It is worth noting that this formula can be used as a definition of representations of hypercomplex systems.

The system of all bounded ∗-representations of a normal hypercomplex system with basis unity is complete. There is a bijective correspondence between the set of positive definite functions on Q and the set of unitary equivalence classes of bounded cyclic ∗-representations of hypercomplex systems. A positive definite function $\varphi(r)$ and the corresponding ∗-representation U_r are connected by the formula $\varphi(r) = (U_r \xi_0, \xi_0)_H$, where ξ_0 is a cyclic vector.

A positive definite function $\varphi(r) \not\equiv 0$ is called *elementary* if all positive definite functions φ_1 such that $\varphi - \varphi_1$ is positive definite have the form $\varphi_1 = \lambda\varphi$ $(0 \le \lambda \le 1)$. There is a bijective correspondence between the set of elementary positive definite functions on Q and the set of unitary equivalence classes of irreducible bounded $*$-representations of the hypercomplex system.

Theorem 3.8 (approximation theorem). *If a normal hypercomplex system $L_1(Q, m)$ with basis unity satisfies the condition of separate continuity, then the linear span of elementary positive definite functions is dense in $C(Q)$ in the topology of uniform convergence on compact subsets.*

Corollary 3.2. *The linear span of Hermitian characters of a commutative normal hypercomplex system with basis unity satisfying the condition of separate continuity is dense in $C(Q)$ in the topology of uniform convergence on compact subsets.*

3.4. References. Theorem 3.1 for the compact case appeared in [BKr3] and, for the locally compact case, in [Kal1]. Theorem 3.3 can be found in [VaK2]. This theorem follows from the results of [VaL]. The other results in Subsections 3.1 and 3.2 are given in the present book (see also [BeK1]). The approximation theorem (Theorem 3.8) for the compact case was proved in [Vre1] and, for the commutative case, in [Ros2]. As far as we know, the general case of this theorem is published in the present book for the first time.

4. Hypergroups and Hypercomplex Systems

In this section, we analyze the relationship between hypercomplex systems and hypergroups.

4.1. Hypergroups. At the beginning of the 70-s, Dunkl [Dun2], Spector [Spe1] and Jewett [Jew] independently introduced a concept of hypergroups. For convenience, we now briefly recall the relevant definition.

Let $M_b(Q)$ denote the set of bounded Radon measures and let $M^1(Q)$ be the set of probability Radon measures in a locally compact space Q with the weak topology. The locally compact space Q is called a *hypergroup* if

(G1) there exists a separately continuous mapping (convolution) $Q \times Q \ni (p, q) \mapsto \delta_p * \delta_q \in M_1(Q)$; $M_b(Q)$ is an associative Banach algebra with convolution given by the formula

$$\langle \mu * \nu, f \rangle = \langle \langle \delta_p * \delta_q, f \rangle, \mu \otimes \nu \rangle \quad (\mu, \nu \in M_b(Q),\ f \in C_0(Q));$$

(G2) supp $(\delta_p * \delta_q)$ is compact;

(G3) there is an involutive homeomorphism $Q \ni p \mapsto p^* \in Q$ such that $(\delta_p * \delta_q)^* = \delta_{q^*} * \delta_{p^*}$, where μ^* is defined by the equality

$$\langle \mu^*, f \rangle = \int_Q f(p^*)\, d\mu(p));$$

(G4) there exists a point $e \in Q$ such that $\delta_e * \delta_p = \delta_p * \delta_e = \delta_p$ $(p \in Q)$;

(G5) $e \in \operatorname{supp}(\delta_p * \delta_q)$ if and only if $p = q^*$;

(G6) the mapping $(p, q) \mapsto \operatorname{supp}(\delta_p * \delta_q)$ is separately continuous in the Michael topology (in the collection of compact subsets $\mathcal{K}(Q)$ of Q, this topology is generated by the subbasis $\{\mathcal{K} \in \mathcal{K}(Q) \mid K \cap U \neq \varnothing, K \subset V\}$, where U and V are open subsets of Q.

A locally compact space Q is called a *weak hypergroup* if conditions (G1)–(G4) are satisfied. Suppose that the left Haar measure m exists on the hypergroup Q, i.e., $\delta_p * m = m$ for all $p \in Q$ (in the compact, discrete, and commutative cases, the existence of Haar measure was proved in [Jew] and [Spe4]). If $\delta_p * m = m * \delta_p = m$, (i.e., the left Haar measure coincides with the right Haar measure), the hypergroup is called *unimodular.* There exists a family of generalized translation operators $(L_p f)(q) = \langle \delta_p * \delta_q, f \rangle$ $(f \in C_\infty(Q))$ associated with the hypergroup Q. The Banach space $L_1(Q, m)$ is a Banach algebra with convolution given by the formula

$$(f * g)(p) = \int_Q (L_p f)(q)\, g(q^*)\, dm(q).$$

The next theorems is an evident consequence of (G1)–(G6).

Theorem 4.1. *If Q is a unimodular hypergroup, then $L_1(Q, m)$ is a normal hypercomplex system with basis unity and condition of separate continuity.*

It follows from Proposition 4.1 that the class of normal hypercomplex systems is broader than the class of unimodular hypergroups. Hence, all results established for hypercomplex systems of this sort are true for unimodular hypergroups.

The assertion converse to Theorem 4.1 is not true because there is an example of a normal hypercomplex system with basis unity associated with generalized Chebyshev polynomials which is not a hypergroup (the authors are deeply grateful to A. Schwartz for this example). Nevertheless, it follows from Theorem 4.2 that hypergroups and hypercomplex systems are very closely related and, roughly speaking, they are the same algebraic objects.

Theorem 4.2. *If the normal hypercomplex system* $L_1(Q, m)$ *with basis unity satisfies the condition of separate continuity, then* Q *is a weak hypergroup. If* Q *is discrete, then the definitions of hypergroup and normal hypercomplex system with basis unity coincide.*

From the standpoint of applications, hypercomplex systems are useful in duality theory (see Theorems 3.5 and 3.6), while hypergroups are useful for the investigation of more detailed structures (e.g., for the analysis of subhypergroups, etc.). Generally speaking, this distinction is connected with the fact that conditions (G5) and (G6) are not satisfied for hypercomplex systems.

4.2. References. The fundamentals of the theory of hypergroups are exposed in [Jew], [Ros1], and [Hey2]. There is also a forthcoming book by Bloom and Heyer. The relationship between hypercomplex systems and hypergroups was analyzed in [BeK1].

5. *Generalizations*

In the present section, we briefly discuss the existing generalizations of the concept of hypercomplex systems. One generalization of this sort (hypercomplex systems with complex-valued structure measures) was described in Subsection 1.9. The major part of the results presented in Section 3 are true for hypercomplex systems with complex-valued measures (with the exception of duality theory, because the Plancherel measure is not multiplicative for such hypercomplex systems). Another class of compact and discrete hypercomplex systems with real structure measures, good duality, and meaningful harmonic analysis was proposed by Vainerman [Vai2].

For this class of hypercomplex systems, the unity, in general, does not belong to the basis. In what follows, we describe the so-called "quantized hypercomplex systems".

It is well known that the duality theory of locally compact noncommutative groups can be formulated in terms of Kac algebras [Kac] or, equivalently, in terms of a special class of Hilbert bialgebras. Thus, the functions defined on a group with ordinary multiplication (commutative operation) and convolution (noncommutative operation) constitute a bialgebra of the indicated type. The transition from special bialgebras of this sort to general bialgebras with two noncommutative operations may be regarded as "quantization" of the group. In this case, we have the following principle of conformity: Under certain conditions, the commutativity of one operation of the bialgebra enables one to realize this bialgebra as a bialgebra of functions on the group. Quantized hypercomplex systems appear if one wishes to realize a similar construction for hypercomplex systems. The axioms of quantized hypercomplex systems are simpler than the axioms of Kac bialgebras. If one of the operations is commutative, quantized hypercomplex systems can be realized as objects close to hypercomplex systems with locally compact bases.

5.1. Principle of Conformity. Let $\mathcal{D}$ be a Hilbert bialgebra such that $\mathcal{L}_{\cup}$ and $\mathcal{L}_{\cap}$ are left W^*-algebras associated with the Hilbert algebras $(\mathcal{D}, \cup)$ and $(\mathcal{D}, \cap)$, let $\mathcal{L}_{\cup^*}$

and $\mathcal{L}_{\cap *}$ be the preduals, and let $E_\cap$ $(E_\cup)$ be the closure of the linear span of the operators $L_\cap(a \cup b)$ $(L_\cup(a \cap b))$ $(a, b \in \mathcal{D})$ in the uniform norm.

A Hilbert bialgebra $\mathcal{D}$ is called a *quantized hypercomplex system* if

(A1) $(a^\cup)^\cap = (a^\cap)^\cup$, $(a \cap b)^\cap = a^\cap \cup b^\cap$, $(a \cap b)^\cup = a^\cup \cup b^\cup$ $(a, b \in \mathcal{D})$;

(A2) $\mathcal{L}_{\cap *}$ is subalgebra of $E_\cup$ and $\mathcal{L}_{\cup *}$ is subalgebra of $E_\cap$.

Theorem 5.1 (principle of conformity). *Suppose that the operation* $\cap$ *in a quantized hypercomplex system* $\mathcal{D}$ *is commutative. Then there exist a locally compact space* Q, *nonnegative measure* m, *and involutive homeomorphism* $Q \ni t \mapsto t^* \in Q$ *such that* 1) $m^* = m$; 2) $E_\cap$ *is isomorphic to* $C_\infty(Q)$; 3) *this isomorphism can be extended to isometries of* H *onto* $L_2(Q, m)$ *and* $\mathcal{L}_{\cap *}$ *onto* $L_1(Q, m)$.

It follows from Theorem 5.1 that $L_1(Q, m)$ is endowed with the structure of $*$-algebra from $\mathcal{L}_{\cap *}$. Thus, $L_1(Q, m)$ is an object close to a hypercomplex system with locally compact basis. Theorem 3.3 remains true for this algebra [VaK2].

By interchanging the operations $\cup$ and $\cap$ in a quantized hypercomplex system, we obtain a new quantized hypercomplex system $\mathcal{D}'$ (the dual quantized hypercomplex system). It is obvious that $\mathcal{D}'' = \mathcal{D}$. If one of the operations in the quantized hypercomplex system is commutative, the proposed method allows us to construct an object dual to, in general, a noncommutative hypercomplex system with locally compact basis. If both operations are commutative, by virtue of Theorem 5.1, we can realize this dual object as an object $L_1(\hat{Q}, \hat{m})$ close to a hypercomplex system with locally compact basis. Here, $\hat{Q}$ consists of $*$-characters of the $*$-algebra $L_1(Q, m)$. As a result, we arrive at a generalization of the Pontryagin duality principle for hypercomplex systems.

5.2. Positivity. If we wish to establish such properties as the existence of the basis unity and multiplicative property for the measure m, it is necessary to introduce the following additional axiom:

(A3) the mappings $\mathcal{L}_{\cup *} \ni \omega \mapsto \langle I, \omega \rangle$ and $\mathcal{L}_{\cap *} \ni \omega \mapsto \langle I, \omega \rangle$, where I is the identity operator, can be extended to characters of the C^*-algebras $E_\cap$ and $E_\cup$.

Proposition 5.1. *Assume that a quantized hypercomplex system satisfy (A3) and the operation* $\cap$ *is commutative. Then*

(i) there exists a point $e \in Q$ *(basis unity) such that*

$$(f * g)(e) = \int_Q f(t) g(t) dm(t) \qquad (f, g \in L_2(Q, m));$$

(ii) the measure m is multiplicative

$$\int_Q (f * g)(t)\,dm(t) = \int_Q f(t)\,dm(t) \int_Q g(s)\,dm(s) \qquad (f, g \in L_1(Q, m)).$$

We now introduce an axiom under which (in case where one operation in a quantized hypercomplex system is commutative) the convolution of positive functions in $L_1(Q, m)$ is positive.

An operator $A \in \mathcal{L}_{\cup}$ is called $\cap$-*positive definite* if $\langle A, \omega \cap \omega^{\cap} \rangle \geq 0$ for all $\omega \in \mathcal{L}_{\cup^*}$. Operators that are $\cup$-*positive definite* can be defined in exactly the same way.

We introduce the axioms

(A4) if A, B are $\cap$-positive definite, then AB is $\cap$-positive definite;

(A4)′ if A, B are $\cup$-positive definite, then AB is $\cup$-positive definite.

Theorem 5.2. *Assume that a quantized hypercomplex system satisfy (A3), (A4) and that the operation $\cap$ is commutative. Then $L_1(Q, m)$ is a Banach $*$-algebra and the convolution of positive functions is positive.*

Hence, the quantized hypercomplex systems satisfying the conditions of Theorem 5.2 have all properties of normal hypercomplex systems with locally compact basis and basis unity, except that the convolution of functions from $C_0(Q)$ is not necessarily compactly supported. We need axiom (A4)′ in order for a class of such hypercomplex systems to be closed under duality.

The class of quantized hypercomplex systems contains the class of Kac bialgebras:

If the connecting operator W of a quantized hypercomplex system $\mathcal{D}$ is unitary and

$$(W \otimes I)(I \otimes W)(W^* \otimes I) = (I \otimes W)(\sigma \otimes I)(I \otimes W)(\sigma \otimes I),$$

then $\mathcal{D}$ is a Kac bialgebra (here, σ is the operator of permutation).

5.3. References. All results of this section can be found in [VaK2] and [Kac]. In [Vai7], the construction of quantized hypercomplex systems was generalized to the nonunimodular case. An object close to quantized hypercomplex systems, namely, a pair of $*$-algebras in duality was studied in [Ver1] and [Ker2]. For related results, see [Kir]. In the finite-dimensional case, quantized hypercomplex systems were also studied (in another setting) in [Vai1].

6. Remarks on Terminology

It seems reasonable to standardize the terminology used in the theories of generalized translations, hypercomplex systems, hypergroups, convolution algebras, etc.

We propose to do the following:

In our opinion, the terminology should be based on the notion of hypergroups. A (finite) hypergroup is a basis Q of a finite-dimensional algebra (of a hypercomplex system) with nonnegative structural constants. Its points $p, q, r, \ldots$ can be multiplied as in a group but, as a result of this multiplication, $p * q$ we obtain not a point of Q but a collection of points with various nonnegative values which can be regarded as a vector with nonnegative coordinates. This vector gives the measure of each point in the product

$$p * q = \sum_{r \in Q} c_{pq}^{r}\, r. \tag{S.36}$$

The meaning of the word "hypergroup" is connected with the fact that we construct something over a group and this is reflected by the prefix "hyper". This term also reflects the opinion of J. Delsarte who suggested to use it in his constructions. It should be noted that the old term "hypercomplex systems" used for the objects of the type (S.36) is somewhat arbitrary because they are quite distant from the system of complex numbers and are cannot be regarded as its direct generalizations.

Special cases of the introduced notion are evident: commutative hypergroups, hypergroups with unity (lying, clearly, in Q; when the unity of an algebra is not in the basis, we may say that this is a hypergroup with generalized unity), and so forth.

If Q is infinite (countable and discrete), one may say that hypergroups are infinite (the use of series (S.36) does not cause any difficulties because the coefficients c_{pq}^{r} are nonnegative).

If Q is infinite and equipped with a nontrivial topology, we can consider compact, locally compact, and other types of hypergroups depending on the type of topological space Q. In this case, the algebras of measures, functions, or distributions on Q can be used as the original algebra. By analogy with the case of groups, this algebra should be called a hypergroup algebra. In our opinion, there should be no preferences in introducing topologies in hypergroup algebras. This is important because, depending on expected results and the desire to include a great number of examples, the researchers may wish to use different systems of axioms. Thus, to construct duality theory, it is more convenient to use hypercomplex systems. On the other hand, hypergroups are preferable for the investigation of subhypergroups, induced representations, and so on. It seems that the only requirement necessary for a hypergroup algebra is that the product of positive elements of the hypergroup algebra must be positive.

The objects introduced by using different special systems of axioms should be distinguished by prefixes and parentheses. Thus, a hypercomplex system may be called a locally compact L_1-hypergroup. This emphasizes the fact that the Banach algebra $L_1(Q, m)$ is a hypergroup algebra. Hypergroups may be called M-hypergroups due to the fact that the convolution of measures is defined.

If the structural constants are not positive, one may say that hypergroups are real, complex, etc. Hence, hypercomplex systems with complex structural constants should be called complex L_1-hypergroups.

For finite hypergroups, generalized translation operators acting upon the functions $Q \ni q \mapsto f(q) \in \mathbb{C}$ can be introduced by the formula

$$(T_p f)(q) = \sum_{r \in Q} c_{pq}^r f(r) \qquad (p, q \in Q). \tag{S.37}$$

For topological hypergroups, this formula should be appropriately modified (cf. (S.12)). Certainly, these objects are very important but it does not seem reasonable to use them as a basic concept. Here, the algebraic side of the construction is more important than the operators (S.37) associated with this construction (thus, it is not convenient to define the group $\mathbb{R}$ by using the ordinary translation operators, although this is, of course, possible).

BIBLIOGRAPHICAL NOTES

We suppose that the reader is familiar with general ideas of the books mentioned in this paragraph. The theory of measure on locally compact spaces is presented in [Bou2] and the theory of linear topological spaces in [Sch]. Elements of the theory of groups and Lie algebras necessary for our presentation can be found in [Hel1], [Hel2], and [Dix3]. For the theory of normalized algebras and harmonic analysis on locally compact groups, see [Dix2], [Nai1], or [HeR]. Necessary facts from the theory of W^*-algebras can be found in [Dix1] or [BrR]. The theory of rigged Hilbert spaces and the spectral projection theorem are presented in [Ber10] and [BeKo].

Chapter 1

Section 1. Commutative Banach algebras $L_1(Q, m)$ with respect to generalized convolutions of functions constructed for a given class of generalized translation operators were studied in early works by Levitan [Lev2]–[Lev5]. For these algebras, he established the Bochner and Plancherel theorems, the inversion formula, and a version of the Pontryagin duality. In a somewhat different form, discrete infinite-dimensional hypercomplex systems appeared in [Haa]. Normal hypercomplex systems with continual basis were introduced by Berezansky and S.Krein [BKr1]–[BKr4]. In these works, an important condition of positivity of the structural measure guaranteeing the possibility of construction of nontrivial harmonic analysis appeared for the first time. Commutative hypercomplex systems with compact and discrete bases were studied by Berezansky [Ber1]–[Ber9] and Berezansky and S.Krein [BKr1]–[BKr4]. The case of general locally compact bases was only outlined. The results of this series of works were summarized in the survey [BKr4]. The theory of commutative hypercomplex systems with locally compact basis was constructed by Berezansky and Kalyuzhnyi [BeK1]–[BeK5]. Its elements were presented in the monograph [BeKo]. The theorem on existence of a multiplicative measure in the case of a compact basis was proved by Berezansky and S.Krein [BKr3] and, in the locally compact case, by Kalyuzhnyi {Kal1]. The conditions of the indicated theorem in the case of a compact basis were weakened by Bakhtin [Bak]. The concept of Hilbert bialgebras was introduced by Kac [Kac]. All results presented in Section 1 belong to the authors except for Theorem 1.7 which was in fact established by Jewett [Jew].

Section 2. At present, the Delsarte–Levitan theory of generalized translation operators is very well developed and, hence, it is practically impossible to give a survey of all relevant results within the framework of these notes. We refer the reader to the monographs by Levitan ([Lev9] and [Lev10]) and reviews by Vainerman [Vai5] and Litvinov [Lit3] (In [Lit3], Litvinov, instead of the term "generalized translation operators," uses the term "hypergroup" introduced by J.Delsarte.) Fundamentals of the theory of hypergroups can be found in the work by Jewett [Jew] and surveys by Ross [Ros1] and Heyer [Hey2]. The existence of invariant measures on compact semihypergroups was established by Onipchuk [Oni]. It is worth noting that, parallel with topological hypergroups, serious attention is given to the investigation of abstract hypergroups. Each hypergroup of this sort is a set Q with associative binary operation on the set of its subsets. The theory of abstract hypergroups was developed by Corsini [Cor1], [Cor2], McMullen [McM1], McMullen and Price [McP1], [McP2], and other researchers (see also [Ros1]).

All statements presented in Section 2, with the exception of the general properties of generalized translation operators and Theorem 2.2 established for hypergroups by Jewett [Jew], belong to the authors. (Theorem 2.1 proved by Kalyuzhnyi is published for the first time.) The formalization of the process of reduction was proposed by Litvinov [Lit3].

Let us now give a brief description of other objects related to hypercomplex systems. First of all, one must mention the association schemes studied by P.Delsarte [DeP] (see also [BaI]). Each of these schemes is associated with a finite-dimensional hypercomplex system (a so-called Bose–Mesner algebra). Further, there are numerous works in the theory of probability on semigroups of measures with respect to generalized convolution ([Vol1], [Vol2], [Dav], [Ken], [Kenn], [Kin], [Lam], [Ost], [OTr], [Tru1]–[Tru5], [RoU], [Urb], [Cvd], and many others). We want to especially mention the paper [Kin], where, apparently, for the first time, the principal ideas of probability theory on hypergroups were formulated. The relationship between hypercomplex systems and Urbanic algebras [Urb] was analyzed by Olshanetskii [Ols1]. Generalized convolutions were also studied in [Dim], [GhM], [Pym1], and [Pym2].

Section 3. In a somewhat different axiomatics, Theorems 3.1–3.3 were established in terms of generalized translation operators by Levitan [Lev3], [Lev4]. For hypercomplex systems, they were proved by Berezansky and S.Krein [BKr4] and Berezansky and Kalyuzhnyi [BeK1]. The duality of commutative hypercomplex systems in the compact and discrete cases was established by Berezansky [Ber9]. In the locally compact case, it was established by Berezansky and Kalyuzhnyi [BeK1]. For hypergroups, duality of the same type was established by Jewett [Jew] and Spector [Spe1], [Spe2] (see also [Bür]) and, for real hypercomplex systems, by Vainerman [Vai2], [Vai3] (see also bibliographical notes to Section 5 in Chapter 1 below). Unlike hypercomplex systems, in the case of hypergroups, one fails to establish conditions under which the dual object is a hypergroup. Examples of hypergroups whose dual objects are not hypergroups were presented, e.g., in [Jew] (see also Sections 2 and 3 in Chapter 2).

The results obtained by Voit [Voi1] essentially complement Theorem 3.5. Namely, for any commutative hypergroup $(Q, *)$, there exists a unique positive essential character α_0. If $\alpha_0 \neq 1$, then the law of multiplication in Q can be "corrected" by setting

$$\langle \delta_t \circ \delta_s, f \rangle = \frac{\langle \delta_t * \delta_s, \alpha_0 f \rangle}{\alpha_0(t)\alpha_0(s)}$$

in such a way that $(Q, \circ)$ is also a hypergroup and its unit character is essential. Under certain conditions imposed on $(Q, \circ)$, the support of the Plancherel measure coincides with the set of all its characters. Thus, by "correcting" the hypergroup $SL_2(\mathbb{C}) // SU(2)$ as indicated above, we obtain the hypergroup $I(3) // SO(3)$, where $I(3)$ is the group of isometries of $\mathbb{R}^3$. In order that the unit character be essential, it is necessary and sufficient that there exist a sequence of functions $\varphi_n * \varphi_n^*(t)$ $(\varphi_n \in C_0(Q))$ approximating the unit element uniformly on compact sets ([GaG3]). As far as the subsequent development of the technique of positive characters and its applications are concerned, see the works by Voit [Voi3]–[Voi8].

The concept of A–systems was introduced by Berezansky in his candidate's degree thesis [Ber9] and, independently, by A.Schwartz [Schw3]. Representations of hypergroups and generalized translation operators were studied by Jewett [Jew], Vrem [Vre1], Bloom and Heyer [BlH2], Levitan [Lev9], [Lev10], Litvinov [Lit1]–[Lit3], Vainerman [Vai2]– [Vai3], Berezansky and Kalyuzhnyi [BeK3], [BeK4]. Theorem 3.11 belongs to Vrem [Vre1] and Theorem 3.12 belongs to Jewett [Jew].

The theorem on approximation (Theorem 3.13) for compact hypergroups was proved by Vrem [Vre1]. For commutative hypergroups, under certain additional restrictions, this theorem was proved by Ross [Ros2]. As far as we know, for noncommutative hypercomplex systems with locally compact basis, the theorem on approximation is published in the present book for the first time. Vainerman generalized this theorem to the case of real hypercomplex systems with compact basis [Vai2], [Vai3]. The space of test and generalized functions were introduced and studied in the language of hypercomplex systems by Berezansky and Kalyuzhnyi [BeK2], [BeK4] (for the analysis of representations of commutative hypercomplex systems). Vrem [Vre1] extended the Peter–Weyl theory to the case of compact hypergroups and Vainerman [Vai2], [Vai3] extended this theory to the case of real hypercomplex systems with compact basis. It is worth noting that elements of the Peter–Weyl theory for compact generalized translation operators were established Levitan [Lev9], [Lev10] (see also [Mat]). Hauenschild, Kaniuth and Kumar investigated representations of central hypergroups, i.e., hypergroups Q such that Q/Z is a compact set, where Z is the intersection of the maximal subgroup with the center of the hypergroup Q [HKK] (see also Section 4).

Various problems in the theory of functions almost periodic with respect to generalized translation operators were studied in [Del3], [Lev5], [Ber3], [Ber5], [Ber6], [Ber8], [Las1], and [Wol2]. Spectral synthesis on hypergroups and generalized translation operators was analyzed in [Vog1], [Vog2], [KuS1], [Lit1], [Lit3], [Chi], [ChK1], [ChK2], and [ChR]. Let us also mention some works devoted to the investigation of various problems of harmonic analysis on hypergroups: [BlH2], [Vre2, Vre3], [GeK], [Dun1]–[Dun3], [DuR1], [DuR2], [Jew], [Las3], [Las4], [LanW1], [LanW2], [Pla1], [Pla2], [Ras], [Ros1], [Ros2], and [Khi]. Numerous works are devoted to the analysis of various problems of probability theory on hypergroups. These are [Blo1], [Blo2], [BlH1]–

[BlH5], [Voi2], [Voi3], [Gal1], [Gal2], [GaG1]–[GaG3], [GaR], [Geb1], [Geb2], [Ze3], [Ze4], [Las5], [Las7], [Let1], [Ols2], and [Hey1]–[Hey6]. For surveys of the results obtained in this field, see [Hey1], [Hey2], and [Let1].

Section 4. The concept of subhypergroups was introduced by Jewett [Jew]. All results in Subsections 4.1 and 4.2, except Theorem 4.2 proved by Vrem [Vre1], belong to Jewett. Subhypergroups were also studied by Hauenschild [Hau] and Vrem [Vre4] and [Vre5]. If $\mu \in M_b^+(Q)$ is an idempotent measure on a hypergroup Q, then its support is a compact subhypergroup and Q is its normalized Haar measure [Jew]. Idempotent measures on commutative hypergroups were studied by Bloom [Blo1] (see also [BlH2]). The definition of morphisms of hypercomplex systems was given by Vainerman [Vai2], [Vai3]. Jewett [Jew] used a notion of orbital morphisms for the construction of new examples of hypergroups. He also mentioned that orbital morphisms are not necessarily homomorphisms. The major part of results presented in Subsection 4.3 belongs to Kalyuzhnyi. They are published for the first time. The concept of joins of hypergroups was introduced by Jewett [Jew] and studied by Vrem [Vre4] (see also [DuR2]).

Section 5. The results presented in Subsections 5.1–5.3 belong to Vainerman and Kalyuzhnyi [VaK1], [VaK2]. Theorem 5.3 was proved by Vainerman and Litvinov [VaL]. In [Kac], it was indicated that Kac algebras admit a generalization in terms of Hilbert bialgebras (in connection with generalized translation operators). At the same time, the relationships between their operations were not analyzed. A similar object, namely, a pair of dual $*$-algebras, was investigated by Vershik [Ver1] Kerov [Ker1], [Ker2], and Kirchberg [Kir]. The finite-dimensional case was studied by Vainerman [Vai1], [Vai4]. Harmonic analysis on Kac algebras is developed in the monograph [EnS]. In recent years, many researchers work in the field of quantization of groups in terms of deformations of algebras. The relationship between quantized groups and Kac algebras is discussed in [Dri] and [Vai10]. The results presented in Subsection5.4 belong to Vainerman [Vai2], [Vai3], [Vai5]. Another generalization of locally compact hypergroups was suggested by Gebuhrer [Geb2], [Geb3].

Chapter 2

Section 1. The center of the group algebra of a compact group and the theory of characters belong to the most developed branches of harmonic analysis and the theory of representations (see [HeR], [Nai1], [Nai2], etc.). The structure of hypercomplex systems in $ZL_1(G)$ was discovered by Berezansky [Ber1] (see also [Gel1] and [BeG]). Iltis [Ilt] and Hewitt and Ross [HeR] regarded $\hat{G}$ as an algebraic hypergroup (see also [McM1] and [Rot]). The structure of a hypergroup on $\hat{G}$ was studied by Kawada [Kaw], Berezin and Gelfand [BeG], and Helgason [Hel3]. Rider [Rid1], [Rid3] used the hypergroup structure on $\hat{G}$ for the characterization of idempotent measures on the classes of conjugate elements (see also [Rag1]). Gallardo [Gal1], Gallardo and Ries [GaR], and Eymard and Roynette [EyR] studied some problems of probability theory on $\widehat{SU(2)}$. The work of Blyumin [Bly] is devoted to the application of the hypergroup structure on $\hat{G}$ to the problem of digital analysis of signals. Theorem 1.1 belongs to Berezansky [Ber1]. The detailed presentation of the theory of duality of Tannaka and M.Krein can be found in [HeR]. The proof of the second part of the Tannaka–Krein duality theorem based on the duality theorem for hypercomplex systems is published in our book for the first time. In Subsection 1.6, we present well-known results from the theory of semisimple groups and Lie algebras (see, e.g., [Nai2], [Hel1], and [Hel2]). In Subsections 1.7 and 1.8, we present a part of the work by Berezin and Gelfand [BeG].

Section 2. Positive definite functions on a sphere were studied by Schoenberg [Scho]. The commutativity of the algebra of biinvariant functions associated with Riemannian symmetric spaces was established by Gelfand [Gel1] (see also [BeG]). M.Krein [Kre1] independently investigated another class of Gelfand pairs. Hypercomplex systems connected with these Gelfand pairs were studied by Berezansky [Ber9]. The unimodularity of the group G was discovered by Berg [Ber]. Harmonic analysis on Gelfand pairs was constructed by Godement [God] (see also [Mau]). Theorems 2.1–2.2 can be found in the well-known books by Dieudonne [Die1] and S.Lang [Lan]. Theorem 2.3 belongs to Gelfand [Gel1] and Berezansky [Ber9]. Theorem 2.4 was established by Gelfand and Naimark [GeN] (see also [BeG]). Spherical functions on symmetric Riemannian spaces of compact and noncompact types were described by Harish-Chandra [Har]. Note that, for Riemannian symmetric spaces of these types, the explicit form of the Plancherel measure is known (for the rigorous formulation of this result and relevant bibliography, see [Hel2]).

The problems of harmonic analysis on Gelfand pairs were studied in [Gan], [Die1], [Die2], [Koo1], [Kri1], [Kri2], [Kuc1], [Kuc2], [Kus], [Car], [McD1], [May], [Miz1], [Miz2], [Nus], [Ras], [Rag2], [Rag3], [RaR], [Rid2], [Far], [FlK1], [Hel1], [Hel2], and [Khi]. Dijk [Dij] and Thomas [Tho1] considered generalized Gelfand pairs. Problems of probability theory on Gelfand pairs were studied in [Bin1], [Bin2], [Let1], [Let2], [Hey1], [Hey5], and [Hey6] (see also the bibliography contained therein).

Section 3. Generalized translation operators associated with orthogonal polynomials were introduced by Levitan [Lev6] and then studied by many authors (see, e.g., [Zhy], [Raf1]–[Raf6], and [Fed]). Discrete hypercomplex systems constructed in terms of orthogonal polynomials appeared for the first time in the works by Berezansky [Ber4], [Ber5], and [Ber8]. He proved Theorem 3.1, 3.2, and 3.8–3.11. Insignificant corrections were made by Berezansky and Kalyuzhnyi [BeK5]. Hypercomplex systems of this sort were also studied by Lasser [Las2]–[Las4], [Las6] and A.Schwartz [Schw3], [Schw4]. Their generalizations were investigated by Vainerman [Vai9]. Theorem 3.3, in fact, belongs to A.Schwartz [Schw4] (see also [BlS]). Relations (3.12)–(3.15) and Theorem 3.5 were established by Lasser [Las2]. Theorem 3.6 was proved by Askey [Ask1] (see also

[Kus], [Gas1]–[Gas3], [Hyl], and the survey [Ask2]). The results of Subsection 3.5 are well-known. In addition to the works mentioned in the body of Section 3, examples of hypercomplex systems and generalized translation operators associated with orthogonal polynomials were constructed and studied in [Bav], [Voi2], [Voi4], [Vog1], [Wol2], [GöM], [Gla1], [Gla2], [GlR], [DuR1], [Dib], [Kenn], [Koo1], [McC], [Mar3], [Soa], [AsW1], [AsW2], and [AsG2]. Note that the availability of multiplication formulas in many of these examples is explained by the fact that the corresponding orthogonal polynomials are matrix elements of representations of compact quantum groups (see, e.g., [Koo3], [NoM1], [NoM2], and [VaW]) or algebras with special quadratic relations ([GrZ1] and [GrZ2]). Generalized translation operators associated with matrix-valued orthogonal polynomials were studied in by Osilenker [Osi2].

Section 4. An example of generalized translation operators associated with the Bessel differential operator was constructed by J.Delsarte [Del1]. Levitan [Lev1] constructed generalized translation operators for a differential operator

$$\pm\frac{d^2}{dx^2}+q(x)$$

given on a semiaxis in the case where the potential $q(x)$ is analytic. Povzner generalized this result to the case of continuous potentials. The conditions imposed on the potential were then weakened by Leblanc [Leb] and Hutson and Pym [HuP2], [HuP3]. The classification of generalized translation operators corresponding to the Sturm–Liouville operator was given by Gurevich [Gur1] (see also [Leb], [CoS3], [CoS4], [Zeu2], and [Zeu3]). The conditions guaranteeing that generalized translation operators preserve positivity and admit the construction of hypercomplex systems were established by Berezansky [Ber7] (see also [Boc1], [Boc2], and [Wei]). Vainerman [Vai7] studied generalized translation operators constructed for a Hamiltonian system of two differential equations on a semiaxis (in particular, for the Dirac system). Examples of generalized translation operators and convolution algebras associated with partial differential equations can be found in [Tri3], [Hul1], [Hul2], [Ste1], and [Ste2]. The results presented in Subsections 4.2–4.4 belong to Berezansky [Ber7]. Spectral analysis and synthesis on hypercomplex systems associated with a regular Sturm–Liouville operator are presented in the work by Sinyavskii [Sin].

Chapter 3

Section 1. The fundamentals of Lie theory for generalized translation operators were laid by J.Delsarte [Del1]–[Del4] and Levitan [Lev7], [Lev9], and [Lev10]. In our presentation, we mainly follow the survey by Litvinov [Lit3]. He gave the general definition of infinitesimal objects. All results in Subsections 1.1 and 1.2 and Theorem 1.2

belong to Litvinov. The interpretation of $\mathfrak{D}_e(Q)$ as a space tangent to the manifold $\mathfrak{D}(Q)$ at the point δ_e published here for the first time was suggested by Berezansky. All results in Subsection 1.4 belong to Levitan [Lev9] and [Lev10]. The results of Subsection 1.5 belong to Gurevich, except Theorem 1.4 proved by Litvinov [Lit3]. A fairly general Lie theory of generalized translation operators was constructed by Karasev [Kar1]–[Kar3] by using asymptotic methods (see also [Mas], [KaM1], and [KaM2]). Note that, in the framework of this theory, the axiom of associativity for generalized translation operators and relationships between their right generators hold asymptotically. Among applications of the Lie theory of generalized translation operators, one should especially mention algebraic topology (see [Buk] and [BKh]; concerning the theory of multi-valued groups, see also [Gur3]).

Section 2. Theorems 2.1 and 2.2, in fact, belong to J.Delsarte [Del4]. All other results in Subsection 2.1 belong to Kalyuzhnyi [Kal4]. In Subsection 2.2, we present a part of the paper by Gurevich [Gur2]. The results in Subsection 2.3 belong to Podkolzin [Pod1]–[Pod4]. In these works, Podkolzin also constructed the adjoint action of the hypergroup $G//H$ and its infinitesimal algebra and applied his results to the exact integration of nonlinear differential equations. Note that, in the case Riemannian symmetric spaces, the infinitesimal algebra of the hypercomplex system $L_1(G, H)$ was described by Gelfand [Gel1], [Gel2] (for a more detailed presentation, see the books [Hel1] and [Hel2]).

Section 3. The results of Section 3 belong to Kalyuzhnyi [Kal2], [Kal3]. A close point of view on the generator of the family of commutative generalized translation operators can be found in [HuP1], [Fis1], and [Fis2]. Vainerman applied formulas analogous to (3.2) to the construction of the family of unbounded generalized translation operators for a given Jacobian matrix [Vai9]. Formulas similar to (3.2) were established by Bloom and Heyer [BlH4] for the generator of a contracting semigroup of operators associated with a one-parameter semigroup of probability measures on a hypergroup. For the description of one-dimensional hypergroup, see [Schw5], [Zeu2], and [CoS3].

REFERENCES

Achour, A., and Trimeche, K.

[AcT1] *Operateurs de translation generalisee associes a un operateur differentiel singulier sur un interval borne,* C. R. Acad. Sci., Ser. A **288** (1979), 399–402.

[AcT2] *La g-fonction de Littlewood–Paley associee a un operateur differentiel singulier sur* $(0, \infty)$, Ann. Inst. Fourier **33** (1983), No.4, 203–226.

Annabi, H., and Trimeche, K.

[AnT] *Convolutions generalisees sur de la disque unite,* C. R. Acad. Sci., Ser. A **278** (1974), No.1, 21–24.

Arnaud, I. P.

[Arn] *Fonctions spheriques et fonctions definies positive sur l'arbre homogene,* C. R. Acad. Sci., Ser. A **290** (1980), 99–101.

Askey, R.

[Ask1] *Linearization of the product of orthogonal polynomials,* in: R.C.Gunning(ed.), Problems in Analysis, 1970, pp.131–138.

[Ask2] *Orthogonal polynomials and positivity,* Studies in Applied Math. (1970), No.6, 64–85.

Askey, R., and Gasper, G.

[AsG1] *Linearization of the product of Jacobi polynomials.* III, Canad. J. Math. **23** (1971), No.2, 332–338.

[AsG2] *Convolution structure for Laguerre polynomials,* J. D'Analyse Math. **31** (1977), 48–68.

Askey, R., and Wainger, S.

[AsW1] *A convolution structure for Jacobi series,* Amer. J. Math. **91** (1969), 463–483.

[AsW2] *A dual convolution structure for Jacobi polynomials,* in: Proceedings of the Conference on Orthogonal Expansions and Their Continuous Analogues, 1967, pp.25–36.

Attori, P., and Picardello, M.

[AtP] Spherical Functions on Symmetric Graphs, Lecture Notes in Math. **992** (1982.)

Bakhtin, I. A.

[Bak] *On positive linear operators and hypercomplex systems,* Ukr. Mat., Zh. **1** (1965), No.4, 3–11 *(in Russian).*

Bannai, E., and Itô, T.

[BaI] Algebraic Combinatorics. I. Association Schemes, Benjamin/Cumnings, Menlo Park, California, 1984.

Bapat, R. B., and Sunder, V. S.

[BaS] *On hypergroups of matrices,* Lin. Multilin. Algebra **29** (1991), 125–140.

Barrucand, P., and Dickinson, P.

[BaD] *On the associated Legendre polynomials,* in: Orthogonal Expansions and Their Continuous Analogues, Southern Illinois University Press, Edwardsville, 1968, pp.43–50.

Bavinck, H.

[Bav] *A special class of Jacobi series and some applications,* J. Math. Anal. Appl. **37** (1972), 767–797.

Berezansky, Yu. M.

[Ber1] *On the center of the group ring of a compact group,* Dokl. Akad. Nauk SSSR **72** (1950), No.5, 825–828 *(in Russian).*

[Ber2] *Hypercomplex systems with discrete basis,* Dokl. Akad. Nauk. SSSR **81** (1951), No.3, 329–332 *(in Russian).*

[Ber3] *To the theory of Levitan's almost periodic functions and sequences,* Dokl. Akad. Nauk. SSSR **81** (1951), No.4, 493–496 *(in Russian).*

[Ber4] *On some normed rings associated with orthogonal polynomials,* Ukr. Mat. Zh. **3** (1951), No.4, 412–432 *(in Russian).*

[Ber5] *On generalized almost periodic functions and sequences,* Ukr. Mat. Zh. **4** (1952), No.3, 340–341 *(in Russian).*

[Ber6] *To the theory of functions almost periodic with respect to shift in hypercomplex systems,* Dokl. Akad. Nauk SSSR **85** (1952), No.1, 9–12 *(in Russian).*

[Ber7] *On hypercomplex systems associated with the Sturm–Liouville equation on a semiaxis,* Dokl. Akad. Nauk SSSR **91** (1953), No.6, 1245–1248 *(in Russian).*

[Ber8] *On generalized almost periodic functions and sequences associated with differential and difference equations,* Mat. Sb. **32** (1953), No.1, 157–194 *(in Russian).*

[Ber9] Hypercomplex Systems with Compact and Discrete Bases, Candidate Degree Thesis (Physics and Mathematics), Institute of Mathematics, Ukrainian Academy of Sciences, Kiev, 1950 *(in Russian).*

[Ber10] Expansions in Eigenfunctions of Selfadjoint Operators, Naukova Dumka, Kiev, 1965; ***English translation:*** Amer. Math. Soc. Transl., vol. **17**, Providence, R.I., 1968.

[Ber11] Selfadjoint Operators in Spaces of Functions of Infinitely Many Variables, Naukova Dumka, Kiev, 1978; ***English translation:*** Amer. Math. Soc. Transl., vol. **63**, Providence, R.I., 1986.

Berezansky, Yu. M., and Kalyuzhnyi, A. A.

[BeK1] Hypercomplex Systems with Locally Compact Basis, Preprint No.40, Institute of Mathematics, Ukrainian Academy of Sciences, Kiev 1982; ***English translation:*** Sel. Math. Sov. **4** (1985), No.2, 151–200.

[BeK2] *Nuclear space of functions on the basis of a hypercomplex system,* Ukr. Mat. Zh. **35** (1983), No.1, 9–17 *(in Russian).*

[BeK3] *Representations of hypercomplex systems with locally compact bases,* Ukr. Mat. Zh. **36** (1984), No.4, 417–421 *(in Russian).*

[BeK4] *Spectral decompositions of representations of hypercomplex systems,* in: Spectral Theory of Operators and Infinite-Dimensional Analysis, Institute of Mathematics, Ukrainian Academy of Sciences, Kiev, 1984, pp.4–19 *(in Russian).*

[BeK5] *Hypercomplex systems associated with orthogonal polynomials,* Ukr. Mat. Zh. **38** (1986), No.3, 275–284 *(in Russian).*

Berezansky, Yu. M., and Kondrat'ev, Yu. G.

[BeKo] *Spectral Methods in Infinite–Dimensional Analysis,* Naukova Dumka, Kiev, 1988; ***English translation:*** Kluwer AP, Dordrecht, 1995.

Berezansky, Yu. M., and Krein, S. G.

[BKr1] *Continual Algebras,* Dokl. Akad. Nauk SSSR **72** (1950), No.1, 5–8 *(in Russian).*

[BKr2] *Some classes of continual algebras,* Dokl. Akad. Nauk SSSR **72** (1950), No.2, 237–240 *(in Russian).*

[BKr3] *Hypercomplex system with compact basis,* Ukr. Mat. Zh. **3** (1951), No.2, 184–204 *(in Russian).*

[BKr4] *Hypercomplex system with continual basis,* Usp. Mat. Nauk **12** (1957), No.1, 147–152 *(in Russian).*

Berezin, F. A., and Gelfand, I. M.

[BeG] *Some remarks to the theory of spherical functions on symmetric Riemann manifolds,* Tr. Mosk. Mat. Obshch. **5** (1956), 311–351 *(in Russian).*

Berg, C.

[Ber] *Dirichlet forms on symmetric spaces,* Ann. Inst. Fourier (Grenoble) **23** (1973), 135–156.

Bergman, G. M.

[Berg] *The diamond lemma for ring theory,* Adv. Math. **29** (1978), 178–218.

Bingham, N. H.

[Bin1] *Random walks on spheres,* Z. Wahrs. verw Gebiete **22** (1972), 169–192.

[Bin2] *Positive definite functions on spheres,* Proc. Camb. Philos. Soc. **73** (1973), 145–156.

[Bin3] *Factorisation theory and domains of attraction for generalized convolution algebras,* Proc. Lond. Math. Soc. **23** (1971), 16–30.

[Bin4] *Integrability theorems for convolutions,* J. Lond. Math. Soc. **18** (1978), No.3, 502–510.

Bloom, W. R.

[Blo1] *Infinitely divisible measures on hypergroups,* Lecture Notes in Math. **928** (1982), 1–15.

[Blo2] *Indempotent measures on commutative hypergroups,* Lecture Notes in Math. **1210** (1986), 13–23.

Bloom, W. R., and Heyer, H.

[BlH1] *Convergence of convolution products of probability measures on hypergroups,* Rend. Math. Serie VII **2** (1982), No.3, 547–563.

[BlH2] *The Fourier transform for probability measures for hypergroups,* Rend. Math. Serie VII **2** (1982), No.7, 317–334.

[BlH3] *Convolution semigroups and resolvent families of measures on hypergroups,* Math. Z. **188** (1985), No.4, 449–474.

[BlH4] *Non-symmetric translation invariant Dirichlet forms on hypergroups,* Bull. Austral. Soc. **36** (1987), 61–72.

[BlH5] *Continuity of convolution semigroups on hypergroups,* J. Theoret. Probab. **1** (1988), No.3, 271–286.

[BlH6] *Characterisation of potential kernels of transient convolution semigroups on a commutative hypergroup,* Lecture Notes in Math. **1379** (1989), 21–35.

Bloom, W. R., and Selvanathan, S.

[BlS] *Hypergroup structures on the set of natural numbers,* Bull. Austral. Math. Soc. **33** (1986), No.1, 89–102.

Blyumin, S. L.

[Bly] *Convolution systems related to compact groups,* Avtomat. Telemekh. (1978), No.12, 48–53 *(in Russian).*

Bochner, S.

[Boc1] *Positive zonal functions on spheres,* Proc. Nat. Acad. Sci. USA **40** (1954), No.10, 1141–1147.

[Boc2] *Sturm–Liouville and heat equations whose eigenfunctions are ultraspherical polynomials or associated Bessel functions,* in: Proceedings of the Conference on Differential Equations, University of Maryland, 1955, pp.23–48.

Bourbaki, N.

[Bou1] Topological vector spaces, Inostrannaya Literatura, Moscow, 1959 *(Russian translation).*

[Bou2] Integration. Measures on Locally Compact Spaces. Extension of a Measure. Integration of Measures. Measures on Separable Spaces, Nauka, Moscow, 1977 *(Russian translation).*

Braaksma, B. L. J.

[Bra] *A singular Cauchy problem and generalized translations,* in: H.A. Antosewicz (ed.), Differential Equations, Los Angeles, 1975, pp.40–52.

Braaksma, B. L. J., and Snoo, H. S. V. de

[BrS] *Generalised translation operator associated with a singular differential operator,* Lecture Notes in Math. **415** (1974), 63–77.

Bragiel, K.

[Bra] *On the spherical and zonal spherical functions on a compact quantum group,* Lett. Math. Phys. **22** (1991), 195–201.

Bratteli, O., and Robinson, D.

[BrR] Operator Algebras and Quantum Statistical Mechanics, Mir, Moscow, 1982 *(Russian translation).*

Bressoud, D. M.

[Bre] *Linearization and related formulas for q-ultraspherical polynomials,* SIAM J. Math. Anal. **12**, No.2, 161–168.

Bukhshtaber, V. M.

[Buk] *Characteristic classes in cobordisms and topological applications of the theory of single-valued and two-valued formal groups,* VINITI Series in Contemporary Problems in Mathematics **11** (1978), 5–178 *(in Russian).*

Bukhshtaber, V. M., and Kholodov, A. N.

[BKh] *Topological constructions related to multi-valued formal groups,* Izv. Akad Nauk SSSR, Ser. Mat. **46** (1982), No.1, 3–27 *(in Russian).*

Bürger, R.

[Bur] *Contributions to duality theory on groups and hypergroups,* in: Proceedings of the Seminar on Topology and Modern Harmonic Analysis (Torino–Milano, May–June 1982), vol. **2**, Roma, 1983, pp.1055–1070.

Bustoz, J., and Ismail, M. E. H.

[BuI] *The associated ultraspherical polynomials and their q-analogues,* Canad. J. Math. **34** (1982), 718–736.

Cary, R.

[Car] *Invariant polynomials and zonal spherical functions,* Boletin Soc. Brasil. Math. **11** (1980), No.1, 57–70.

Chapovsky, Yu. A., and Vainerman, L. I.

[ChV] *Hypergroup structures associated with a pair of quantum groups* $(SU_q(n), U_q(n-1))$, in: Methods of Functional Analysis in Problems of Mathematical Physics, Institute of Mathematics, Ukrainian Academy of Sciences, Kiev, 1992, pp.47–68.

Chebli, H.

[Che1] *Sur la positivite des operateurs de "translation generalisee" associes a un operateur de Sturm–Liouville sur* $[0, \infty[$, C. R. Acad. Sci., Ser.A **275** (1972), 601–604.

[Che2] *Sur un theoreme de Paley–Wiener associe a la decomposition spectrale d'un operateur de Sturm–Liouville sur* $[0, \infty]$, J. Func. Anal. **17** (1974), 447–461.

[Che3] *Operateurs de translation generalisee et semigroups de convolution,* Lecture Notes in Math. **404** (1974), 35–59.

[Che4] *Theoreme de Paley–Iener associe a un operateur singuliere sur* $(0, \infty)$, J. Math. Pures et Appl. **58** (1979), No.1, 1–19.

Chilana, A. K.

[Chi] *Harmonic analysis and hypergroups,* in: Proceedings of the Symposium on Recent Developments in Mathematics and Its Applications, Allahabad, 1978, pp.93–120.

Chilana, A. K., and Kumar, A.

[ChK1] *Ultrastrong Ditkin sets in hypergroups,* Proc. Amer. Math. Soc. **77** (1979), 353–358.

[ChK2] *Spectral synthesis in Segal algebras on hypergroups,* Pacif. J. Math. **80** (1979), No.1, 59–76.

Chilana, A. K., and Ross, K. A.

[ChR] *Spectral synthesis in hypergroups,* Pacif. J. Math. **76** (1978), No.2, 313–328.

Cholewinsky, F. M., and Haimo, D. T.

[ChH] *Classical analysis and generalized heat equation,* SIAM Rev. **10** (1968), 67–80.

Chu Van Dong

[Cvd] *Representation theorem for multiply self-decomposable probability measures on generalized convolution algebra,* Probab. Math. Statist. **6** (1985), No.1, 57–63.

Connett, W. C., and Schwartz, A. L.

[CoS1] *Convolution structures for eigenfunction expansions arising from regular Sturm–Liouville problems,* in: P.L.Butzer, R. L.Stehs, and B.Sz.-Nagy (eds.), Anniversary Volume on Functional Analysis and Approximation, Birkhäuser, Basel, 1984, pp.437–447.

[CoS2] *Product formulas for hypergroups and Jacobi polynomials,* Bull. Amer. Math. Soc. **22** (1990), No.1, 91–97.

[CoS3] *Analysis of a class of probability preserving measure algebras on compact intervals,* Trans. Amer. Math. Soc. **320** (1990), No.1, 371–393.

[CoS4] *A Hardy–Littlewood maximal inequality for Jacobi type hypergroups,* Proc. Amer. Math. Soc. **107** (1989), 137–143.

[CoS5] *Positive product formulas and hypergroups associated with singular Sturm–Liouville problems on a compact interval,* Colloq. Math. **60/61** (1990), No.2, 525–535.

Corsini, P.

[Cor1] *Ipergruppi semiregolari e regolary,* Rend. Semin. Mat. Univers. Politech. Torino **40** (1982), No.3, 35–46.

[Cor2] *Recenti reultati in teoria degli ipergruppi,* Bolletino Untone Mat. Ital. **A2** (1983), No.2, 133–138.

Davidson, R.

[Dav] *Arithmetic and other properties of certain Delphic semigroups,* Z. Wahrs und Verw. Gebiete **10** (1968), 120–145, 146–172.

Delsarte, J.

[Del1] *Sur une extension de la formule de Taylor,* J. Math. Pures et Appl. **17** (1938), No.3, 213–231.

[Del2] *Sur certaines transformation fonctionelles relative aux equations linearies aux derivees partiels du seconde ordre,* C. R. Acad. Sci., Ser. A **17** (1938), 178–182.

[Del3] *Une extension nouvelle de la theorie de fonction presque periodiques de Bohr,* Acta Math. **69** (1939), 259–317.

[Del4] *Hypergroupes et operateuers de permutation et de transmutation,* in: Colloques Internat., Nancy, 1956, pp.29–44.

Delsarte, J., and Lions, J.-L.

[DeL] *Transmutations d'operateurs differentiel's dans le domain complexe,* Comm. Math. Helv. **32** (1957), No.2, 113–128.

Delsarte, P.

[DeP] Algebraic Approach to Association Schemes in the Theory of Coding, Mir, Moscow, 1976 *(Russian translation).*

Dib, H.

[Dib] *Polynomes de Laguerre d'un argument matriciel,* C. R. Acad. Sci., Ser. A **304** (1987), No.1, 111–114.

Dieudonne, J.

[Die1] Grundzuge der Modernen Analysis, Bd 5/6, Vieweg, 1979.

[Die2] *Gelfand pairs and spherical functions,* Internat. J. Math. and Math. Sci. **2** (1979), 153–162.

Dijk, G. van

[Dij] *On generalized Gelfand pairs,* Proc. Jap. Acad. **A60** (1984), No.1, 30–34.

Dimovski I.

[Dim] Convolutional Calculus, Bulgarian Mathematical Monographs, Vol.2, Sofia, 1982.

Dixmier, J.

[Dix1] Les Algebres D'Operateurs dans L'Espace Hilbertien, Gauthier Villar, Paris, 1969.

[Dix2] C^*-Algebras and Their Representation, Gauthier-Villars, Paris, 1969.

[Dix3] Universal Enveloping Algebras, Mir, Moscow, 1978 *(Russian translation).*

Drinfeld, V. G.

[Dri] *Quantum groups,* Zapiski Nauch. Sem. LOMI **155** (1986), 18–49 *(in Russian).*

Dunkl, C. F.

[Dun1] *Operators and harmonic analysis on the sphere,* Trans. Amer. Math. Soc. **125** (1965), 250–263.

[Dun2] *Structure hypergroups for measure algebras,* Pacif. J. Math. **47** (1973), 413–425.

[Dun3] *The measure algebra of a locally compact hypergroup,* Trans. Amer. Math. Soc. **179** (1973), 331–348.

Dunkl, C. F., and Ramires, P. E.

[DuR1] *Krawtchouk polynomials and the symmetrisation of hypergroups,* SIAM J. Math. Anal. **5** (1974), No.3, 351–366.

[DuR2] *A family of countably compact P-hypergroups,* Trans. Amer. Math. Soc. **202** (1975), 334–356.

Ehrenpreis, L.

[Ehr] *Theory of distributions for locally compact spaces,* Mem. Amer. Math. Soc. **26** (1956).

Enock, M., and Schwartz, J.-M.

[EnS] Kac Algebras and Duality for Locally Compact Groups, Springer Verlag, 1992.

Eymard, P.

[Eym] *L'alegebre de Fourier d'un group localement compact,* Bull. Soc. Math. France **92** (1974), No.2, 181–235.

Eymard, P., and Roynette, B.

[EyR] *Marches aleatories sur le dual de SU (2),* Lecture Notes Math. **497** (1975), 108–152.

Faraut, J.

[Far] *Un theoreme de Paley–Wiener pour la transformation de Fourier sur un espace Riemannien symmetrique de rang un,* J. Func. Anal. **49** (1982), No.2, 230–268.

Faraut J., and Piardello, M. A.

[FaP] *The Plancherel measure for symmetric graphs,* Ann. Mat. Pura Appl. **198** (1982), 151–155.

Fedorov, V. M.

[Fed] *Approximation by algebraic polynomials with Chebyshev–Hermite weight,* Izv. Vyssh. Uchebn. Zaved., Ser. Mat. (1984), No.6, 55–63 *(in Russian).*

Fell, J., and Thoma E.

[FeT] *Einige Bemerkungen über vollsymmtrische Banachsche Algebren,* Archiv. Math. **12** (1961), No.1, 69–70.

Finet, C.

[Fin] *Ensembles de Riesz por les groupes et les hypergroups compacts,* C. R. Acad. Sci., Ser. A **310** (1990), No.7, 509–512.

Fishel, B.

[Fis1] *Generalized translations associated with an unbounded selfadjoint operator,* Math. Proc. Cambr. Phil. Soc. **99** (1986), No.3, 519–528.

[Fis2] *The Fourier–Plancherel transform as a spectral representation of a rigged Hilbert space,* Math. Proc. Cambr. Phil. Soc. **101** (1987), No.3, 567–572.

Flensted-Jensen, M.

[Fle] *Paley–Wiener type theorems for a differential operator connected with symmetric spaces,* Arkiv Math. **10** (1972), No.1, 143–162.

Flensted-Jensen, M., and Koorwinder T.

[FlK1] *The convolution structure for Jacobi function expansions,* Arkiv Math. **11** (1973), 245–262.

[FlK2] *Positive definite spherical functions on a non-compact, rank one symmetric space,* Lecture Notes in Math. **739** (1979), 249–282.

[FlK3] *Jacobi functions: the addition formula and the positivity of the dual convolution structure,* Arkiv Math. **17** (1979), 139–151.

Fourier, J. J., and Ross, K. A.

[FuR] *Random Fourier series on compact Abelian hypergroups,* J. Austral. Math. Soc. **37** (1984), 45–81.

Gallardo, L.

[Gal1] *Theoreme limites pour les marches aleatoires sur le dual de* $SU(2)$, Ann. Scient de L'Univ. de Clermont **67** (1969), 63–68.

[Gal2] *Exemples d'hypergroupes transients,* Lecture Notes in Math. **1210** (1986), 68–76.

Gallardo, L., and Gebuhrer, O.

[GaG1] *Lois de probabililite infinitement divisibles sur les hypergroups communitatifs, discrets, denombrables,* Lect. Notes in Math. (1984), 116–130.

[GaG2] *Marches aleatories et hypergroups,* Expos. Math. **5** (1987), 41–73.

[GaG3] Marches Aleatories et Analyse Harmonique sur les Hypergroupes Localement Compacts, Prepublication Univ. de Strasbourg I. R. M. A., 1985.

Gallardo, L., and Ries V.

[GaR] *La loi des grands nombres pour les marches aleatories sur le dual de $SU(2)$,* Studia Math. **66** (1979), 93–105.

Gangolli, R.

[Gan] *On the Plancherel formula and the Paley–Wiener theorem for spherical functions on semi-simple Lie groups,* Ann. Math. **93** (1971).

Gasper, G.

[Gas1] *Linearisation of the product of Jacobi polynomials* II, Canad. J. Math. **22** (1970), No.3, 532–543.

[Gas2] *Positivity and the convolution structure for Jacobi series,* Ann. Math. **93** (1971), 112–118.

[Gas3] *Banach algebras for Jacobi series and positivity of a kernel,* Ann. Math. **95** (1972), No.2, 261–280.

Gebuhrer, O.

[Geb1] *Quelques properites du noyau potential f'une marche aleatore sur les hypergroupes de type Kunze–Stein,* Lecture Notes in Math. **1210** (1986), 77–83.

[Geb2] *Algebres de Banach commutatives de measures de Radon bornees et theoremes ergodiques de type L^1,* C. R. Acad. Sci., Ser. A **306** (1988), 67–70.

[Geb3] Analyse Harmonique sur les Espaces de Gelfand–Levitan et Applicationa a la Theorie des Semigroupes de Convolution, Publ. Inst. Rech. Math. Avan. No.397 (1989).

[Geb4] *Multiplicateurs radiaux sur le groupe de Heisenberg,* C. R. Acad. Sci., Ser. A **313** (1991), 363–364.

Gebuhrer, O., and Kumar, A.

[GeK] *The Wiener property for a class of discrete hypergroups,* Math. Z. **202** (1989), No.2, 271–274.

Gelfand, I. M.

[Gel1] *Spherical functions on symmetric Riemann spaces,* Dokl. Akad. Nauk SSSR **70** (1950), No.1, 3–6 *(in Russian).*

[Gel2] *Center of an infinitesimal group ring,* Mat. Sb. **26** (1950), No.1, 103–112 *(in Russian).*

Gelfand, I. M., and Naimark, M. A.

[GeN] *Unitary representations of a unimodular group containing the identity representation of a unitary group,* Tr. Mosk. Mat. Obshch. **1** (1952), 423–475 *(in Russian).*

Gelfand, I. M., Raikov, D. A., and Shilov, G. E.

[GRS] Commutative Normed Rings, Fizmatgiz, Moscow, 1960 *(in Russian).*

Geronimus, J.

[Ger] *On a set of polynomials,* Ann. Math. **31** (1930), 681–686.

Ghahramani, F., and Medgalchi, A. R.

[GhM] *Compact multipliers on weighted hypergroup algebras,* Math. Proc. Cambr. Phil. Soc. I **98** (1985), No.3, 493–500; II **100** (1986), No.1, 145–149.

Glaeske, H.-J.

[Gla1] *Die Laguerre–Pinnej transformation,* Aequationes Math. **22** (1981), 73–85.

[Gla2] *On a convolution structure for a generalized Hermite transformation,* Serdica **9** (1983), No.2, 223–229.

Glaeske, H.-J., and Runst, T.

[GlR] *The discrete Jacobi transformation of generalized functions,* Math. Nachr. **132** (1987), 239–251.

Glazmann, I. M.

[Gla] Direct Methods of Qualitative Spectral Analysis of Singular Differential Operators, Fizmatgiz, Moscow, 1963 *(in Russian)*.

Godement, R.

[God] Introduction aux Travaux de A. Selberg, Seminaire Bourbaki, Expose 144, Paris, 1957.

Görlich, E., and Markett, C.

[GöM] *A convolution structure for Laguerre series*, Indag. Math. **44** (1982), 161–171.

Grabovskaya, R. Ya, Kononenko V. I., and Osipov, V. B.

[GKO] *On one family of generalized translation operators*, Izv. Akad. Nauk SSSR, Ser.Mat. **41** (1977), No. 2, 280–283 *(in Russian)*.

Grabovskaya, R. Ya., and Krein, S. G.

[GrK1] *On the formulae of permutation of operators representing a Lie algebra*, Funkts. Anal. Prilozh. **7** (1973), No.3.

[GrK2] *On one representation of the algebra of differential operators and related differential equations*, Dokl. Akad. Nauk SSSR **212** (1973), No.2, 280–283 *(in Russian)*.

[GrK3] *Second-order differential equations with operators generating a Lie algebra representation*, Math. Nachr. **75** (1976), 9–29.

Granovskii, Ya. I., and Zhedanov, A. S.

[GrZ1] *Orthogonal polynomials on Lie algebras*, Izv. Vyssh. Uchebn. Zaved., Ser. Fiz. (1986), No.5, 60–66 *(in Russian)*.

[GrZ2] Exactly Solvable Problems and Quadratic Algebras, Preprint DonFTI–89–7, Donetsk, 1989 *(in Russian)*.

Grosser, S., Mosak, R., and Moskowitz, M.

[GMM] *Duality and harmonic analysis on central topological groups I and II*, Indag. Math. **35** (1973), No.2, 65–91.

Grosser, S., and Moskowitz, M.

[GrM] *Harmonic analysis on central topological groups,* Trans. Amer. Math. Soc. **156** (1971), 419–454.

Grothendieck, A.

[Gro] *Produits tensoriels et espaces nucleaires,* Mem. Amer. Math. Soc. **16** (1955).

Gurevich, D. I.

[Gur1] *Generalized translation operators with right infinitesimal Sturm–Liouville operator,* Mat. Zametki **25** (1979), No.3, 393–408 *(in Russian).*

[Gur2] *Generalized translation operators on Lie groups,* Izv. Akad. Nauk Arm. SSR **18** (1983), No.4, 305–317 *(in Russian).*

[Gur3] *On multi-valued Lie groups,* Usp. Mat. Nauk **39** (1984), No.6 *(in Russian).*

[Gur4] *Yang–Bakster equations and generalizations of the formal Lie theory,* Dokl. Akad. Nauk SSSR **288**, No.4, 767–801 *(in Russian).*

[Gur5] *Elements of the formal Lie theory and the Poincare Birkhoff–Witt theorem for generalized translation operators,* Funkts. Anal. Prilozh. **20** (1986), No.4, 72–73 *(in Russian).*

Haar, A.

[Haa] *Über die Multiplicationstabelle der orthogonalen Functionen,* Math. Z. **31** (1930), 767–798.

Harish-Chandra

[Har] *Spherical functions on a semi-simple Lie group. I, II,* Amer. J. Math. **80** (1958), 241–310, 533–613.

Hartmann, K.

[Hart] $[FIA]_B^{\pm}$*-Gruppen und Hypergruppen,* Monatsch. Math. **89** (1980), 9–17.

Hartmann, K., Henrichs, R. W., and Lasser, R.

[HHL] *Duals of orbit spaces in groups with relatively compact inner authomorphism group are hypergroups,* Monatsch. Math. **88** (1979), 229–238.

Hauenschild, W.

[Hau] *Subhypergroups and normal subgroups,* Math. Ann. **256** (1981), No.1, 1–18.

Hauenschild, W., Kaniuth, E., and Kumar, A.

[HKK] *Harmonic analysis on central hypergroups and induced representations,* Pacific J. Math. **110** (1984), No.1, 83–112.

Helgason, S.

[Hel1] Differential Geometry and Symmetric Spaces, Academic Press, New York, 1962.

[Hel2] Groups and Geometric Analysis, Academic Press, New York, 1984.

[Hel3] *Lacunary Fourier series on noncommutative groups,* Proc. Amer. Math. Soc. **9** (1958), No.5, 782–790.

Hewitt, E., and Ross, K. A.

[HeR] Abstract Harmonic Analysis, Springer Verlag, Berlin, 1963, 1970.

Heyer, H.

[Hey1] *Convolution semigrpoup of probability measures on Gelfand pairs,* Exos. Math. **1** (1983), No.1, 3–45.

[Hey2] *Probability theory on hypergroups: a survey,* Lecture Notes in Math. **1064** (1984), 481–550.

[Hey3] *Convolutions semigroups of measures on Sturm–Liouville structures,* Symposia Math. **29** (1987), 131–162.

[Hey4] *Generalized translation structures and invariant Feller semigroups,* Math. Statist. Probab. Theory (1987), vol. A, 134–194.

[Hey5] *Convolution semigroups of local type on a commutative hypergroup,* Hokkaido Math. J. **18** (1989), No.2, 321–337.

[Hey6] *Infinitely divisible probability measures on discrete Gelfand pair,* Nagoja Math. J. **116** (1989), 43–62.

[Hey7] *Das einbettungs problem der Wahrscheinlichkeitstheorie,* Osterreichische Zeitschrift fur Statistik un Informatik (ZSI) **19** (1989), No.2, 191–213.

[Hey8] *Convolution semigroups and potential kernels on a commutative hypergroup,* in: K.H. Hofmann, J.P. Lawson, and J.S. Pym (eds.), The Analytical and Topological Theory of Semigroups, 1990, pp.279–312.

Hirschmann, I. I, Jr.

[Hir] *Sur les polynomes ultraspheriques,* C. R. Acad. Sci., Ser. A **242** (1956), 2212–2214.

Hulanicki, A.

[Hul1] *On positive functionals on a group algebra multiplicative on a subalgebra,* Studia Math. **37** (1970/1971), 163–171.

[Hul2] *Subalgebra of $L_1(G)$ associated with Laplacian on a Lie group,* Colloq. Math. **31** (1974), No.2, 259–287.

Hutson, V., and Pym, J. S.

[HuP1] *Generalized translations associated with an operator,* Math. Ann. **187** (1970), 241–258.

[HuP2] *Generalized translations associated with a differential operator,* Proc. Lond. Math. Soc. **24** (1972), No.3, 548–576.

[HuP3] *Measure algebra associated with a second-order differential operator,* J.Func. Anal. **12** (1973), 68–96.

Hylleras, E.

[Hyl] *Linearization of product of Jacobi polynomials,* Math. Scand. **10** (1962), 189–200.

Iltis, R.

[Ilt] *Some algebraic structure in the dual compact groups,* Canadian J. Math. **20** (1968), 1499–1510.

Ionescu Tulcea, C., and Simon, A. B.

[IoS] *Spectral representations and unbounded convolution operators,* Proc. Nat. Acad. Sci. **45** (1959), No.12, 1765–1767.

Jewett, R. I.

[Jew] *Spaces with an abstract convolution of measures,* Adv. Math. **18** (1975), No.1, 1–101.

Kac, G. I.

[Kac] *Ring groups and duality principle,* Tr. Mosk. Mat. Obshch., I **12** (1963), 259–301; II **13** (1965), 84–113 *(in Russian).*

Kalyuzhnyi, A. A.

[Kal1] *A theorem on the existence of multiplicative measure,* Ukr. Mat. Zh. **35** (1983), No.3, 369–371 *(in Russian).*

[Kal2] *Duality of generators of compact and discrete hypercomplex systems with one-dimensional bases,* Dokl. Akad. Nauk Ukr. SSR, Ser. A (1989), No.4, 20–24 *(in Russian).*

[Kal3] *Duality of generators of one-dimensional compact and discrete hypercomplex systems,* in: Methods of Functional Analysis in Problems of Mathematical Physics, Institute of Mathematics, Ukrainian Academy of Sciences, Kiev, 1990, pp.79–102 *(in Russian).*

[Kal4] *Infinitesimal object for Delsarte generalized translation operators,* in: Application of Method of Functional Analysis in Mathematical Physics, Institute of Mathematics, Ukrainian Academy of Sciences, Kiev, 1991, pp.65–70 *(in Russian).*

Kanjin, I.

[Kan] *Banach algebra related to disk polynomials,* Tohoku Math. J. **37** (1985), No.3, 395–404.

Kantorovich, L. V., Vulikh, B. Z., and Pinsker, A. G.

[KVP] Functional Analysis in Semiordered Spaces, Gostekhizdat, Moscow, Leningrad, 1950 *(in Russian).*

Karasev, M. V.

[Kar1] *Operators of regular representation for a class of non-Lie commutation relations,* Funkts. Anal. Prilozh. **13** (1979), No.3, 89–90 *(in Russian).*

[Kar2] *Analogs of objects of the theory of Lie groups for nonlinear Poisson brackets,* Izv. Akad. Nauk SSSR, Ser. Mat. **50** (1986), No.3, 508–538 *(in Russian).*

[Kar3] *Poisson Algebras of Symmetries and Asymptotics of Spectral Series,* Funkts. Anal. Prilozh. **20** (1986), No.1, 21–32 *(in Russian).*

Karasev, M., V. and Maslov, V. P.

[KaM1] *Algebras with general commutation relations and their applications. II. Operator unitarily-nonlinear equations,* VINITI Series in Contemporary Problems in Mathematics (1979), 149–267 *(in Russian).*

[KaM2] Nonlinear Poisson Bracket. Geometry and Quantization, Nauka, Moscow, 1991 *(in Russian).*

Kawada, Y.

[Kaw] *Üper den Dualitätsaatz der Charaktere nichtkommutativer Gruppen,* Proc. Phys.-Math. Soc. Japan **24** (1942), 97–109.

Kendall, D. G.

[Ken] *Delphic semigroups, infinitely divisible regenerative phenomena and the arithmetic of p-functions,* Z. Wahrsch. verw. Gebiete **9** (1968), 163–195.

Kennedy, M.

[Kenn] *A stochastic process associated with ultrasperical polynomials,* Proc. Roy. Irish Acad. **61** (1961), 89–100.

Kerov, S. V.

[Ker1] *Dual algebras of functions on a finite group,* Zap. Nauch. Sem. Lomi **39** (1974), 180–185 *(in Russian).*

[Ker2] *Duality of finite–dimensional ∗-algebras,* Vestn. Leningr. Univ. (1974), No.7, 23–29 *(in Russian).*

Khelemskii, A. Ya.

[Khe] Banach and Polynormed Algebras: General Theory, Representations, Homologies, Nauka, Moscow, 1989 *(in Russian).*

Khinkis, L. A.

[Khi] *On spaces of functions on semisimple Lie groups invariant under translation,* Dokl. Akad. Nauk Azerb. SSR **41** (1985), No.10, 10–11 *(in Russian).*

Kingman, J. E.

[Kin] *Random walks with spherical symmetry,* Acta Math. **109** (1963), No.1, 11–53.

Kirchberg, E.

[Kir] Darstellungen coinvolutiver Hopf-W^*-algebren und ihre Anwendung in der nicht-Abelschen Dualitatstheorie localkompakter Gruppen, Acad. der Wiss. der DDR, Berlin, 1977.

Koornwinder, T. H.

[Koo1] *A new proof of a Paley–Wiener type theorem for Jacobi transform,* Arkiv Math. **13** (1975), 145–159.

[Koo2] *Positivity proofs for linerization and connection coefficients of orthogonal polynomials satisfying an addition formula,* J. Lond. Math. Soc. **18** (1978), No.1, 101–114.

[Koo3] · *Orthogonal polynomials in connections with quantum groups,* in: P.Nevai (ed.), Orthogonal polynomials, Theory and Practice, 1990, pp.257–292.

[Koo4] *Positive convolution structures associated with quantum groups,* in: Probability Measures on Groups, X Plenum (1991), 249–268.

Kostant, B.

[Kos] *On the existence and irreducibility of certain series of representation,* in: I.M. Gelfand (ed.) Lie Groups and Their Representations, Halsted, New York, 1975, pp.231–329.

Kraft, H.

[Kra] Geometrische Methoden in der Invariantentheorie, Friedr. Vieweg, Braunschweig–Wiesbaden, 1985.

Krein, M. G.

[Kre1] *Hermitian positive kernels on homogeneous spaces,* I, II, Ukr. Mat. Zh. **1** (1949), No.4, 64–98; **2** (1950), No. 1, 10–59 *(in Russian).*

[Kre2] *Duality principle for a bicompact group and quadratic block-algebra,* Dokl. Akad. Nauk. SSSR **69** (1949), No.6, 725–728 *(in Russian).*

Krein, M. G., and Rutman, M. A.

[KrR] *Linear operators preserving a cone in the Banach space,* Usp. Mat. Nauk **3** (1948), 3–95 *(in Russian).*

Krier, R.

[Kri1] *Sur les ensembles de syntese se l'algebre de Banach commutative* $L^1(K\backslash G/K)$, C. R. Acad. Sci., Ser. A **278** (1974), 417–418.

[Kri2] *Sur les ideaux de l'algebre de Banach des donctions integrables biinvariantes,* Osaka J. Math **12** (1975), 745–765.

Kuchment, P. A.

[Kuc1] *Representations of solutions of invariant differential operators in some symmetric spaces,* Dokl. Akad. Nauk SSSR **259** (1981), No.3, 532–535 *(in Russian).*

[Kuc2] *On mean periodic functions on symmetric spaces,* Funkts. Anal. Prilozh. **16** (1982), No.3, 68–69 *(in Russian).*

Kumar, A., and Singh, A. I.

[KuS1] *Spectral synthesis in products and quotients of hypergroups,* Pacif. J. Math. **94** (1981), No.1, 177–192.

[KuS2] *Counterexamples in spectral synthesis on hypergroups,* Rend. Math. Appl. **8** (1988), No.3, 329–338.

[KuS3] *A dichotomy theorem for random walks on hypergroups,* Lecture Notes in Math. **1379** (1989), 179–184.

Kushner, H. B.

[Kus] *The linearization of the product of two zonal polynomials,* SIAM J. Math. Anal. **19** (1988), No.3, 687–717.

Laine, Th. P.

[Lai] *The product formula and convolution structure for the generalized Chebyshev polynomials,* SIAM J. Math. Anal. **11** (1980), No.1, 133–146.

Lamperti, J.

[Lam] *The arithmetic of certain semigroups of positive operators,* Proc. Cambr. Phil. Soc. **64** (1968), 161–166.

Lang, S.

[Lan] $SL_2(R)$, Addison-Wesley, London, 1975.

Lang, W.

[LanW1] *The structure of hypergroup measure algebras,* J. Austral. Math. Soc. **46** (1989), No.2, 319–342.

[LanW2] *A hypergroup measure algebra that admits no analogue of the structure theorem of J. L. Taylor,* Commutative Harmon. Analysis (Canton, NY, 1987): Contemp. Math. **91** (1989), 151–156.

Lasser, R.

[Las1] *Almost periodic functions on hypergroups,* Math. Ann. **250** (1980), 183–196.

[Las2] *Orthogonal polynomials and hypergroups,* Rend. Math. **3** (1983), No.7, 185–209.

[Las3] *Bochner theorems for hypergroups and their applications to orthogonal polynomial expansions,* J. Appr. Theory **87** (1983), 311–325.

[Las4] *Fourier–Stieltjes transforms on hypergroups,* Analysis 2 (1982), 281–303.

[Las5] *On the Levy–Hinchin formula for commutative hypergroups,* Lecture Notes in Math. **1064** (1984), 298–308.

[Las6] *On the problem of modified moments,* Proc. Amer. Math. Soc. **90** (1984), No.3, 360–362.

[Las7] *Convolution semigroups on hypergroups,* Pacif. J. Math. **127** (1987), No.2, 353–373.

Leblanc, N.

[Leb] *Classification des algebres de Banach associees aux operateurs differentiels de Sturm–Liouville,* J. Func. Anal. **2** (1968), 52–72.

Letac, G.

[Let1] *Problems classiques de probabilite sur un couple de Gelfand,* Lecture Notes in Math. **861** (1981), 93–120.

[Let2] *Les fonction spheriques d'es couple de Gelfand symetrique et les chalines de Markow,* Adv. Appl. Prob. **14** (1982), 272–294.

Levin, B. Ya.

[LeY] *Transformations of Fourier–Laplace type with the use of solutions of a differential equation of the second order,* Dokl. Akad. Nauk SSSR **106** (1956), No.2, 187–190 *(in Russian).*

Levitan, B. M.

[Lev1] *Generalization of shift operation in connection with almost periodic functions,* Mat. Sb. **7** (1940), No.3, 449–478 *(in Russian).*

[Lev2] *Normed rings generated by generalized translation operation,* Dokl. Akad. Nauk SSSR **47** (1945), No.1, 3–6 *(in Russian).*

[Lev3] *Theorems on representation of positive definite functions for generalized translation operation,* Dokl. Akad. Nauk SSSR **47** (1945), No.3, 163–165 *(in Russian).*

[Lev4] *Plancherel theorem for generalized translation operators,* Dokl. Akad. Nauk SSSR **47** (1945), No.5, 323–326 *(in Russian).*

[Lev5] *Duality law for generalized translation operation,* Dokl. Akad. Nauk SSSR **47** (1945), No.6, 401–403 *(in Russian).*

[Lev6] *Generalization of translation operation and infinite-dimensional hypercomplex systems,* Mat. Sb. I, **16** (1945), No.3, 259–280; II **17** (1945), No.1, 9–44; III **17** (1945), No.2, 163–192 *(in Russian).*

[Lev7] *Application of generalized translation operators to linear differential equations of the second order,* Usp. Mat. Nauk **4** (1949), No.1, 3–112 *(in Russian).*

[Lev8] *Lie theorems for generalized translation operators,* Usp. Mat. Nauk **16** (1961), No.4, 3–30 *(in Russian).*

[Lev9] Generalized Translation Operators and Some Their Applications, Nauka, Moscow, 1962; ***English translation:*** Israel Program for Scientific Translations, 1964.

[Lev10] Theory of Generalized Translation Operators, Nauka, Moscow, 1973 *(in Russian).*

Levitan, B. M., and Povzner, A. Ya.

[LeP] *Differential Equations on Semiaxis and Plancherel Theorem,* Dokl. Akad. Nauk SSSR **52** (1946), No.3, 483–486 *(in Russian).*

Levitan, B. M., and Sargsyan, A. Ya.

[LeS] Sturm–Liouville and Dirac Operators, Nauka, Moscow, 1988 *(in Russian).*

Litvinov, G. L.

[Lit1] *On generalized translation operators and their representations,* Tr. Sem. Vekt. Anal. MGU **18** (1978), 345–371 *(in Russian).*

[Lit2] *On dual topological algebras and topological Hopf algebras,* Tr. Sem. Vekt. Anal. MGU **18** (1978), 372–375 *(in Russian).*

[Lit3] *Hypergroups and hypergroup algebras,* VINITI Series in Contemporary Problems in Mathematics **26** (1985), 57–106 *(in Russian).*

Liukkonen, J., and Mosak, R.

[LiM] *Harmonic analysis and centers of group algebras,* Trans. Amer. Math. Soc. **195** (1974), 147–163.

Maltese, G.

[Mal1] *Spectral representations for solutions of certain abstract functional equations,* Comp. Math **15** (1962), No.1, 1–22.

[Mal2] *Spectral representations for some unbounded normal operators,* Trans. Amer. Math. Soc. **110** (1964), No.1, 79–87.

Markett, C.

[Mar1] *Product formulas for Bessel, Whittaker and Jacobi functions via the solutions of an associated Cauchy problem,* Internat. Series Numer. Math. **65** (1984), 449–462.

[Mar2] *Norm estimates for generalized transform operators associated with a singular differential operator,* Proc. Kon. Ned. Acad. Wet. Ser. A **87** (1984), No.3, 299–313.

[Mar3] *A new proof of Watson's product formula for Laguerre polynomials via a Cauchy problem associated with a singular differential operator,* SIAM J. Math. Anal. **17** (1986), No.4, 1010–1033.

[Mar4] *Product formulas and convolution structure for Fourier–Bessel series,* Constr. Appr. **5** (1989), 384–404.

Maslov, V. P.

[Mas] *Application of the method of ordered operators for finding exact solutions,* Teor. Mat. Fiz. **33** (1977), No.2, 185–207 *(in Russian).*

Matevosyan, A. K.

[Mat] *On decomposition in matrix elements of linear representations of generalized translation operators,* Izv. Akad. Nauk. Arm. SSR **22** (1987), No.3, 309–314 *(in Russian).*

Mautner, F. I.

[Mau] *Fourier analysis and symmetric spaces,* Proc. Nat. Acad. Sci. USA **37** (1951), No.8, 529–533.

Mayer-Lindenberg, F.

[May] *Zur Dualitats theory Symmetrisher Paare,* J. Reine Angew. Math. **321** (1981), 35–52.

McCully, J.

[McC] *The Laguerre transform,* Siam Rev. **2** (1960), 185–191.

McDonald, J.

[McD1] *Commuting differential operators and zonal spherical functions,* Lecture Notes in Math. **1271** (1987), 189–200.

[McD2] Symmetric Functions and Hall Polynomials, Clarendon Press, Oxford, 1979.

McMullen, J. R.

[McM1] *On the dual object of a compact connected group,* Math. Z. **185** (1984), 539–552.

[McM2] *An algebraic theory of hypergroups,* Bull. Austral. Math. Soc. **20** (1979), 35–55.

McMullen, J. R., and Price, J. F.

[McP1] *Reversible hypergroups,* Rend. Sem. Mat. Fis. Milano **47** (1977), 67–85.

[McP2] *Duality for finite Abelian hypergroups over spliting fields,* Bull. Austral. Math. Soc. **20** (1979), 57–70.

Meaney, C.

[Mea] *Spherical functions and the Fourier algebra,* Rend Sem. Mat. Fis. Milano **54** (1984), 127–132.

Michael, E. A.

[Mic] *Topologies on spaces of subsets,* Trans. Amer. Math. Soc. **71** (1955), 152–182.

Mizony, M.

[Miz1] *Contribution d'analyse harmonique spherique,* Publ. Dep. Math. Lyon **12** (1975), 1, 61–108.

[Miz2] *Algebres et noyaux de convolution sur le dual spherique d'n groupe de Lie semi-simple non compact et de rang 1,* Publ. Dep. Math. Lyon **13** (1976), 1, 1–14.

Mosak, R. D.

[Mos] *L^1 and C^* -algebras of $[FIA]_B^{\pm}$-groups and their representations,* Trans. Amer. Math. Soc. **163** (1972), 277–310.

Mosolova, M. V.

[Moso] *On functions of noncommutative operators generating a graduated Lie algebra,* Mat. Zametki **29** (1981), No.1, 35–44 *(in Russian).*

Naimark, M. A.

[Nai1] Normed Rings, Nauka, Moscow, 1968 *(in Russian).*

[Nai2] Theory of Representations of Groups, Nauka, Moscow, 1976 *(in Russian).*

Noumi, M., Jamada, H., and Mimachi, K.

[NJM] Finite-Dimensional Representations of the Quantum Group $GL_q(N, C)$ and the Zonal Spherical Functions on $U_q(n-1)/U_q(n)$, Preprint, 1990.

Noumi, M., and Mimachi, K.

[NoM1] *Askey–Wilson polynomials and the quantum group* $SU_q(2)$, Proc. Jap. Acad. Ser. A **66** (1990), No.6, 146–149.

[NoM2] *Quantum 2-spheres and big Jacobi polynomials,* Comm. Math. Phys. **128** (1990), 521–531.

Nussbaum, A. E.

[Nus] *Extension of positive definite functions and representations of functions in terms of spherical functions in symmetric spaces of noncompact type of rank 1,* Math. Ann. **215** (1975), 97–116.

Ocneanu, A.

[Ocn] *Quantized groups, string algebras and Galois theory for operator algebras,* in: Operator Algebras and Applications, vol. **2** (Warwik, 1987), LMS Lecture Notes Ser. **136** (1988), 119–172.

Olshanetskii, I. D.

[Ols1] *On construction of a hypercomplex system by the Urbanik algebra,* Ukr. Mat. Zh. **38** (1986), No.6, 801–804 *(in Russian).*

[Ols2] *On the Khinchin formula for compact hypergroups,* Ukr. Mat. Zh. **41** (1989), No.8, 1127–1130 *(in Russian).*

Onipchuk, S. V.

[Oni] *Idemponent measures on compact semihypergroups,* Ukr. Mat. Zh. **41** (1989), No.9, 1244–1247 *(in Russian).*

Osilenker, B. P.

[Osi1] *Generalized translation operators and convolution structure for orthogonal polynomials,* Dokl. Akad. Nauk SSSR **298** (1988), No.5, 1072–1076 *(in Russian).*

[Osi2] *Operators of generalized translation in orthogonal matrix polynomials,* Dokl. Akad. Nauk SSSR **318** (1991), No.2, 280–282 *(in Russian).*

[Osi3] *The representation of the trilinear kernel in general orthogonal polynomials and some applications,* J. Appr. Theory **67** (1991), No.1, 93–114.

Ostrovskii, I. V.

[Ost] *Description of the class I_0 in one special semigroup of probability measures,* Mathematical Physics and Functional Analysis, Issue IV, FTINT, Ukrainian Academy of Sciences, Kharkov, 1973, pp.3–12 *(in Russian).*

Ostrovskii, I. V., and Trukhina, I. P.

[OTr] *On the arithmetic of Schoenberg–Kennedy semigroups,* in: Problems of Mathematical Physics and Functional Analysis, Naukova Dumka, Kiev, 1976, pp.11–19 *(in Russian).*

Platonov, S. S.

[Pla1] *Subspaces invariant under generalized translations,* Mat. Zametki **39** (1986), No.4, 576–585 *(in Russian).*

[Pla2] *Subspaces invariant under generalized translations,* Mat. Zametki **47** (1990), No.6, 91–101 *(in Russian).*

Podkolzin, G. B.

[Pod1] *Infinitesimal algebras for a hypercomplex system with the basis* $SO(4)//SO(2)$, Ukr. Mat. Zh. **42** (1990), No.3, 427–429 *(in Russian).*

[Pod2] *Scheme of solution of nonlinear equations on the hypercomplex system* $SO(4)//SO(2)$, Ukr. Mat. Zh. **42** (1990), No.7, 971–974 *(in Russian).*

[Pod3] *Infinitesimal object for a hypercomplex system with the basis* $SO(4)//SO(2)$, Dokl. Akad. Nauk Ukr. SSR (1990), No.3, 27–30 *(in Russian).*

[Pod4] *Infinitesimal object for a hypercomplex system generated by dual cosets and nonlinear differential equations,* Funkts. Anal. Prilozh. **25** (1991), No.1, 91–93 *(in Russian).*

Povzner, A. Ya.

[Pov1] *On equations of Sturm–Liouville type and positive functions,* Dokl. Akad. Nauk SSSR **43** (1944), No.9, 387–391 *(in Russian).*

[Pov2] *On differential equations of Sturm–Liouville type on a semiaxis,* Mat. Sb. **23** (1948), 3–52 *(in Russian).*

Pym, J. S.

[Pym1] *Weakly separately continuous measure algebras,* Math. Ann. **175** (1968), 207–219.

[Pym2] *Dual structures for measure algebras,* Proc. Lond. Math. Soc. **19** (1969), No.3, 625–660.

Rafalson, S. Z.

[Raf1] *On approximation of functions by Fourier–Jacobi sums,* Izv. Vyssh. Uchebn. Zaved., Ser. Mat. (1968), No.4, 54–62 *(in Russian).*

[Raf2] *On approximation of functions in the mean by Hermite–Fourier sums,* Izv. Vyssh. Uchebn. Zaved., Ser. Mat. (1968), No.7, 78–94 *(in Russian).*

[Raf3] *On approximation of functions from classes* Lip_α *by Legendre sums,* Izv. Vyssh. Uchebn. Zaved., Ser. Mat. (1975), No.5, 109–112 *(in Russian).*

[Raf4] *Generalized translation operators related to the theory of orthogonal polynomials,* in: Proceedings of the International Conference on Constructive Theory of Functions (Varna, 1981), Sofia, 1983, pp.139–143 *(in Russian).*

[Raf5] *On norms of operators which are Fourier–Legendre quotients,* Izv. Vyssh. Uchebn. Zaved., Ser. Mat. (1983), No.8, 54–59 *(in Russian).*

[Raf6] *On approximation of functions in the mean by Fourier–Jacobi sums,* Izv. Vyssh. Uchebn. Zaved., Ser. Mat. (1984), No.5, 75–78 *(in Russian).*

Ragozin, D. L.

[Rag1] *Central measures on compact simple Lie groups,* J.Func. Anal. **10** (1972), 212–219.

[Rag2] *Zonal measure algebras on isotropy irreducible homogeneous spaces,* J.Func. Anal. **17** (1974), No.4, 355–356.

[Rag3] *Rotation invariant measure algebras on Euclidean space,* Indiana Univ. Math. J. **23** (1974), 1139–1154.

Ragozin, O. L., and Rothschild, L. P.

[RaR] *Central measures on semisimple Lie groups have essentialy compact support,* Proc. Amer. Math. Soc. **32** (1972), 585–589.

Rahman, M.

[Rah1] *The linerization of the product of continuous q-Jacobi polynomials,* Canad. J. Math. **33** (1981), 961–987.

[Rah2] *A product formula for the continuous q-Jacobi polynomials,* J. Math. Anal. Appl. **118** (1986), 309–322.

Raikov, D. A.

[Rai] *Harmonic analysis on commutative groups with Haar measure and the theory of characters,* Tr. Mat. Inst. Akad. Nauk SSSR **14** (1945), 1–86 *(in Russian).*

Rashevskii, P. K.

[Ras] *Description of closed invariant subspaces in some functional spaces,* Tr. Mosk. Mat. Obshch. **38** (1979), 139–185 *(in Russian).*

Reed, M. and Simon, B.

[ReS] Methods of Modern Mathematical Physics, vol.**1**, Academic Press, New York–San Francisco–London, 1972.

Rham, J. de

[Rha] Differentiable Manifolds, Inostrannaya Literatura, Moscow, 1956 *(Russian translation).*

Rider, D.

[Rid1] *Central idempotent measures on unitary group,* Canad. J. Math. **22** (1970), 719–725.

[Rid2] *Central idempotent measures on SIN groups,* Duke Math. J. **38** (1971), 187–191.

[Rid3] *Central idempotent measures on compact groups,* Trans. Amer. Math. Soc. **186** (1973), 459–479.

Roberts, P. H., and Ursell, H. D.

[RoU] *Random walks on a sphere and on a Riemannian manifold,* Phil. Trans. Roy. Soc. Lond., Ser. A **252** (1960), 317–356.

Ross, K. A.

[Ros1] *Hypergroups and centers of measure algebras,* Symposia Math. **22** (1977), 189–203.

[Ros2] *Centers of hypergroups,* Trans. Amer. Math. Soc. **243** (1978), 251–269.

Roth, R. L.

[Rot] *Character and conjugacy class hypergroups of a finite group,* Ann. Mat. Pura et Appl. **55** (1975), 295–311.

Schoefer, H. H.

[Sch] Topological Vector Spaces, MacMillan, New York, 1966.

Schoenberg, I. J.

[Scho] *Positive definite functions of spheres,* Duke Math. J. **9** (1942), 96–108.

Schwartz, A.

[Schw1] *The structure of the algebra of Hankel transforms and the algebra of Hankel–Stieltjes transforms,* Canad. J. Math. **23** (1971), 236–246.

[Schw2] *On the ideal structure of the algebra of radial functions,* Proc. Amer. Math. Soc. **26** (1970), 621–624.

[Schw3] *Generalized convolutions and positive definite functions associated with general orthogonal series,* Pacif. J. Math. **55** (1974), 565–582.

[Schw4] *l_1-Convolution algebras: representation and factorization,* Z. Wahrsh. verw. Gebiete **41** (1977), 161–176.

[Schw5] *Classification of one-dimensional hypergroups,* Proc. Amer. Math. Soc. **103** (1988), No.4, 1073–1081.

Schwartz, L.

[Schw] Analyse Mathématique, vol **1**, Hermann, 1967.

Shapiro, R. L.

[Sha1] *Irreducible representations of the group $SU(n)$ of class I with respect to $SU(n-1)$*, Usp. Mat. Nauk **22** (1967), No.1, 180–182 *(in Russian).*

[Sha2] *Special functions related to the representations of the group $SU(n)$ of class I with respect to $SU(n-1)$*, Izv. Vyssh. Uchebn. Zaved., Ser. Mat. (1968), No.4, 97–107 *(in Russian).*

Shilov, G. E.

[Shi] *On extension of maximal ideals,* Dokl. Akad. Nauk. SSR **39** (1940), No.2, 83–85 *(in Russian).*

Sinyavskii, V. V.

[Sin] *Regularity condition for the Banach algebra related to the Sturm-Liouville equation,* Dokl. Akad. Nauk Ukr. SSR, Ser. A (1977), No.11, 988–990 *(in Russian).*

Soardi, P. M.

[Soa] *Limit theorems for random walks on discrete semigroups related to nonhomogenuous trees and Chebyshef polynomials,* Math. Z. **200** (1989), 313–327.

Spector, R.

[Spe1] *Apercu de la theorie des hypergroupes,* Lecture Notes in Math. **497** (1975), 643–673.

[Spe2] *Dualite des hypergroupes commutatifs,* C. R. Acad. Sci., Ser. A **281** (1975), 27–30.

[Spe3] *Theorie axiomatique des hypergroupes,* C. R. Acad. Sci., Ser. A **280** (1975), 1743–1744.

[Spe4] *Measures invariantes sur les hypergroupes,* Trans. Amer. Math. Soc. **239** (1978), 147–165.

Stempak, K.

[Ste1] *On convolution products of radial measures on Heisenberg group,* Colloq. Math. **50** (1985), 125–128.

[Ste2] *An algebra associated with the generalized sublaplacian,* Stud. Math. **88** (1988), No.3, 245–256.

Sunder, V. S.

[Sun] II_1 *factors, their bimodules and hypergroups,* Trans. Amer. Math. Soc. **330** (1992), No.1, 227–256.

Szegö, G.

[Sze] Orthogonal Polynomials, Fizmatgiz, Moscow, 1962 *(Russian translation)*.

Tannaka, T.

[Tan] *Über de Dualitätssatz der nichtkommutativen topologischen Gruppen,* Tohoku Math. J. **45** (1939), 1–12.

Thomas, E. G. F.

[Tho1] The Theorem of Bochner–Schwartz–Godement for Generalized Gelfand Pairs, ZW-Rept. Math. Inst. Rijksuniv. Groningen, S. a. No.8312.

[Tho2] An Infinitesimal Characterisation of Gelfand Pairs, W-Rept. Math. Inst. Rijksuniv. Groningen, S. a. N.8313.

Trimeche, K.

[Tri1] *Convergence des series de Taylor generalisees,* C. R. Acad. Sci., Ser. A **281** (1975), 1015–1017.

[Tri2] *Probabilities infiniment divisibles et theorie de la limite centrale pour une convolution generalisee sur la demi-droite,* C. R. Acad. Sci., Ser. A **286** (1978), No.1, 63–66.

[Tri3] *Operateuers de permutation et analyse harmonique associes a des operateurs aux derivees partielles,* J. Math. Pures et Appl, **70** (1991), 1–73.

[Tri4] Transmutation Operators and Mean-Periodic Functions Associated with Differential Operators, Hardwood AP, Chur–London–Paris–NewYork–Melburne, 1988.

Trukhina, I. P.

[Tru1] *On one Lamperty problem,* Teor. Ver. Primen. **23** (1978), No.4, 869–870 *(in Russian)*.

[Tru2] *On one problem related to the arithmetic of probability measures on a sphere,* Zap. Nauch. Sem. LOMI **87** (1979), 143–158 *(in Russian)*.

[Tru3] *On semigroups of functions representable by series in Jacobi polynomials,* Dokl. Akad. Nauk Ukr. SSR, Ser. A (1977), No.1, 21–24 *(in Russian)*.

[Tru4] *Arithmetic of spherical symmetric measures in Lobachevsky spaces,* Teor. Funkts. Funkts. Anal. Prilozh. **34** (1980), 136–146 *(in Russian)*.

[Tru5] *Arithmetic of special semigroups of probability measures,* in: Proceedings of IV International Conference on Probability Theory and Mathematical Statistics, vol. III, 1985, pp.206–208 *(in Russian)*.

Urbanic, K.

[Urb] *Generalized convolutions,* Studia Math. I **23** (1964), No.3, 217–245; II **45** (1973), No.1, 57–70; III **80** (1984), No.2, 167–189; IV **83** (1986), No.1, 57–95; V **91** (1988), No.2, 153–178.

Vainerman, L. I.

[Vai1] *Finite Kac hypercomplex systems,* Funkts. Anal. Prilozh. **17** (1983), No.2, 68–69; ***English translation:*** Func. Anal. Appl. (1983), 135–136.

[Vai2] *Hypercomplex systems with compact and discrete bases,* Dokl. Akad. Nauk SSSR **278** (1984), No.1, 16–20; ***English translation:*** Soviet Math. Dokl. **30** (1984), No.2, 305–308.

[Vai3] *Harmonic analysis on hypercomplex systems with compact and discrete bases,* in: Spectral Theory of Operators and Infinite-Dimensional Analysis, Institute of Mathematics, Ukrainian Academy of Sciences, Kiev, 1984, pp.19–32 *(in Russian).*

[Vai4] *Duality principle for finite hypercomplex systems,* Izv. Vyssh. Uchebn. Zaved., Ser. Mat. (1985), No.2, 12–21 *(in Russian).*

[Vai5] *Duality of algebras with involution and generalized translation operators,* VINITI Series in Mathematical Analysis **24** (1986), 165–205; ***English translation:*** J. Sov. Math. **42**, No.6, 2113–2137 (1988) .

[Vai6] *Generalized translation operators generated by differential equations,* Dokl. Akad. Nauk SSSR **302** (1988), No.1, 14–17 *(in Russian).*

[Vai7] *Nonunimodular quantized hypercomplex systems,* Dokl. Akad. Nauk Ukr. SSR, Ser. A (1989), No.4, 7–10 *(in Russian).*

[Vai8] *On the abstract Plancherel formula and inversion formula,* Ukr. Mat. Zh. **41** (1989), 1041–1047 *(in Russian).*

[Vai9] *Hypercomplex systems and finite–difference operators,* in: Methods of Functional Analysis in Problems of Mathematical Physics, Institute of Mathematics, Ukrainian Academy of Sciences, Kiev, 1990, pp.42–50 *(in Russian).*

[Vai10] *Relations between compact quantum groups and Kac algebras,* Adv. Sov. Math. **9** (1992), 151–160.

[Vai11] *Two structures associated with actions of quantum groups,* in: Proceedings of the International Conference on Operator Algebras, Orleans, 1992, pp.157–167.

Vainerman, L. I., and Kac, G. I.

[VaK] *Nonunimodular ring groups and Hopf–von Neumann algebras,* Mat. Sb. **94** (1974), No.2, 194–225 *(in Russian).*

Vainerman, L. I., and Kalyuzhnyi, A. A.

[VaK1] *Quantized hypercomplex systems,* in: Boundary-Value Problems for Differential Equations, Institute of Mathematics, Ukrainian Academy of Sciences, Kiev, 1988, pp.13–29 *(in Russian).*

[VaK2] *Quantized hypercomplex systems,* Funkts. Anal. Prilozh. **23** (1989), No.4, 77–78 *(in Russian).*

Vainerman, L. I., and Litvinov, G. L.

[VaL] *Plancherel formula and inversion formula for generalized translation operators,* Dokl. Akad. Nauk SSSR **257** (1981), No.4, 792–795; ***English translation:*** Soviet Math. Dokl. **23** (1981), No.2, 333–337.

Vainerman, L.I., and Weiss, Y.

[VaW] *On a hypergroup associated with dual cosets of the quantum group* $SU_q(2)$, in: Application of Methods of Functional Analysis in Mathematical Physics, Institute of Mathematics, Ukrainian Academy of Sciences, Kiev, 1991, pp.13–29 *(in Russian).*

Vaisleb, E. E., and Samoilenko, Yu. S.

[VaS1] *On representation of the relations* $AB = BF(A)$ *by unbounded self-adjoint and unitary operators,* in: Boundary-Value Problems for Differential Equations, Institute of Mathematics, Ukrainian Academy of Sciences, Kiev, 1988, pp.30–52 *(in Russian).*

[VaS2] *Representation of operator relations by unbounded operators and many-dimensional dynamical systems,* Ukr. Mat. Zh. **42** (1990), No.8, 1011–1019 *(in Russian).*

Vershik, A. M.

[Ver1] *Geometrical theory of states, von Neumann boundary, and the duality of* C^**-algebras,* Zap. Nauch. Sem. LOMI **29** (1972), 147–154; ***English translation:*** J. Sov. Math. **3** (1975), No.6.

[Ver2] *Algebras with quadratic relations,* in: Spectral Theory of Operators and Infinite-Dimensional Analysis, Institute of Mathematics, Ukrainian Academy of Sciences, Kiev, 1984, pp.32–57 *(in Russian).*

Vilenkin, N. Ya.,

[Vil] Special Functions and Theory of Representations, Nauka, Moscow, 1965 *(in Russian).*

Vogel, M.

[Vog1] *Spectral synthesis on algebras of orthogonal polynomial series,* Math. Z. **194** (1987), No.1, 99–116.

[Vog2] *Harmonic analysis and spectral synthesis in central hypergroups,* Math. Ann. **281** (1988), No.3, 369–385.

Voit, M.

[Voi1] *Positive characters on commutative hypergroups and some applications,* Math. Z. **198** (1988), No.3, 405–421.

[Voi2] *Positive and negative definite functions on the dual space of a commutative hypergroup,* Analysis **9** (1989), No.4, 371–387.

[Voi3] *Central limit theorems for random walks on N_0 that are associated with orthogonal polynomials,* J. Multivar. Analysis **34** (1990), 290.

[Voi4] *Negative definite functions on commutative hypergroups,* Lecture Notes in Math. **1379** (1989), 378–388.

[Voi5] *A law of the iterated logarithm for a class of polynomial hypergroups,* Mh. Math. **109** (1990), 311–326.

[Voi6] *Factorization of probability measures on symmetric hypergroups,* J.Austral. Math. Soc. (Ser. A) **50** (1991), 407–417.

[Voi7] *On the Fourier transformation of positive definite measures on commutative hypergroups and dual convolution structure,* Manuscripta Math. **72** (1991), 141–153.

[Voi8] *On the dual space of a commutative hypergroup,* Arch. Math. **56** (1991), 380–385.

Volkovich, V. E.

[Vol1] *On analytic description of Urbanic algebras,* Izv. Akad. Nauk Uzb. SSR, Ser. Fiz.-Mat. **5** (1979), 12–17 *(in Russian).*

[Vol2] *On normed rings generated by Urbanic algebras,* Teor. Ver. Primen. **25** (1980), No.2, 432–433 *(in Russian).*

Vrem, R. C.

[Vre1] *Harmonic analysis on compact hypergroups,* Pacific J. Math. **85** (1979), 239–251.

[Vre2] *Lacunarity on compact hypergroups,* Math. Z. **164** (1978), 93–104.

[Vre3] *Continuous measures and lacunarity on hypergroups,* Trans. Amer. Math. Soc. **269** (1982), No.2, 549–556.

[Vre4] *Hypergroup joins and their dual objects,* Pasif. J. Math. **111** (1984), No.2, 483–495.

[Vre5] *Connectivity and supernormality results for hypergroups,* Math. Z. **195** (1987), No.3, 419–428.

[Vre6] *L^p-improving measures on hypergroups,* Lecture Notes in Math. **1379** (1989), 389–397.

[Vre7] *Torsion and torsion-free hypergroups,* Commutative Harmonic Analysis (Canton, NY, 1987), Contemp. Math., **91** (1989) 281–289.

[Vre8] *Independent sets and lacunarity for hypergroups,* J. Austral. Math. Soc. **50** (1991), No.2, 171–188.

Vretare, L.

[Vret] *Elementary spherical functions on symmetric spaces,* Math. Scand. **39** (1976), 343–358.

Weinberger, A.

[Wei] *A maximum property of Cauchy problem,* Ann. Math. **64** (1956), No.3, 505–512.

Wolfenstetter, S.

[Wol1] *Weakly almost periodic functions on hypergroups,* Monat. Math. **96** (1983), No.1, 67–79.

[Wol2] *Spectral synthesis on algebras of Jacobi polynomial series,* Arch. Math. (Basel) **43** (1984), No.4, 364–369.

Yosida, K

[Yos] Functional Analysis, Mir, Moscow, 1967 *(Russian translation).*

Zeuner, H. M.

[Zeu1] *On hyperbolic hypergroups,* Lect. Notes **1210** (1986), 216–224.

[Zeu2] *One-dimensional hypergroups,* Adv. Math. **76** (1989), No.1, 1–18.

[Zeu3] *The central limit theorem for Chebli–Trimeche hypergroups,* J. Theor. Probab. **2** (1989), No.1, 51–63.

[Zeu4] *Laws of large numbers for hypergroups on* R_+, Math. Ann. **283** (1989), 657–678.

[Zeu5] *Properties of the cosh hypergroups,* Lecture Notes in Math. **1379** (1989), 425–434.

Zhidkov, V. G.

[Zhi] *Constructive characterization of one class of nonperiodic functions,* Dokl. Akad. Nauk SSSR **169** (1966), No.5, 1002–1005 *(in Russian).*

Selected Papers Close to the Subject of the Monograph Which Appeared after the Publication of the Russian Edition of the Book

Berezansky, Yu. M.

1. *Nuclear spaces of test functions connected with hypercomplex systems and representations of such systems,* Contemp. Math. **183** (1995), 15–20.
2. *A connection between the theory of hypergroups and white noise analysis,* Rep. Math. Phys. **36** (1995), 215–234.
3. *A generalisation of white noise analysis by means of the theory of hypergroups,* Rep. Math. Phys. (1996) (to appear).
4. *Infinite-dimensional non-Gaussian analysis and generalized translation operators,* Funkts. Anal. Prilozh. **30** (1996), No.4, 61–65.
5. *Infinite-dimensional analysis associated with generalized translation operators,* Ukr. Mat. Zh. (to appear).

Berezansky, Yu. M., and Kalyuzhnyi, A. A.

1. *Hypercomplex systems and hypergroups: connections and distinctions,* Contemp. Math. **183** (1995), 21–44.

Berezansky, Yu. M., and Kondrat'ev, Yu. G.

1. *Non-Gaussian analysis and hypergroups,* Func. Anal. Appl. **29** (1995), No.3, 51–55.
2. *Biorthogonal systems in hypergroups: an extension of non-Gaussian analysis,* Meth. Func. Anal. Topol. **2** (1996), No.2.

Bloom, W. R., and Heyer, H.

1. Harmonic Analysis of Probability Measures on Hypergroups, De Gruyter, Berlin–New York, 1995.

Chapovsky, Yu. A.

1. *Gelfand pair associated with a Hopf algebra and a coideal,* Ukr. Mat. Zh. **46** (1994), No.8, 1055–1067.

Chapovsky, Yu. A., and Vainerman, L. I.

1. *Gelfand pair of compact quantum groups,* Func. Anal. Prilozh. **29** (1995), No.2, 67–71.

Kalyuzhnyi, A. A.

1. *Algebras with non-quadratic relations associated with Bessel hypergroups,* Meth. Func. Anal. Topol. (to appear).

Podkolzin, G. B.

1. *An infinitesimal algebra of the hypergroup generated by double cosets and nonlinear differential equations,* Contemp. Math. **183** (1995), 287–297.

Podkolzin, G. B., and Vainerman, L. I.

1. Quantum Stiefel Manyfold and Double Cosets of Quantum Unitary Group, Preprint Universite d'Orleans, 1996.

Vainerman, L. I.

1. *Gelfand pairs of quantum groups, hypergroups and q-special special functions,* Contemp. Math. **183** (1995), 373–394.
2. *Hypergroup structures associated with Gelfand pairs of compact quantum groups,* Asterisque **1995** (1995), 231–242.
3. *On the Gelfand pair associated with the quantum groups of motions of the plane and q-Bessel functions,* Rep. Math. Phys. (to appear).

SUBJECT INDEX

Other *Mathematics and Its Applications* titles of interest:

G. Gaeta: *Nonlinear Symmetries and Nonlinear Equations*. 1994, 258 pp. ISBN 0-7923-3048-X

V.A. Vassiliev: *Ramified Integrals, Singularities and Lacunas*. 1995, 289 pp. ISBN 0-7923-3193-1

N.Ja. Vilenkin and A.U. Klimyk: *Representation of Lie Groups and Special Functions*. Recent Advances. 1995, 497 pp. ISBN 0-7923-3210-5

Yu. A. Mitropolsky and A.K. Lopatin: *Nonlinear Mechanics, Groups and Symmetry*. 1995, 388 pp. ISBN 0-7923-3339-X

R.P. Agarwal and P.Y.H. Pang: *Opial Inequalities with Applications in Differential and Difference Equations*. 1995, 393 pp. ISBN 0-7923-3365-9

A.G. Kusraev and S.S. Kutateladze: *Subdifferentials: Theory and Applications*. 1995, 408 pp. ISBN 0-7923-3389-6

M. Cheng, D.-G. Deng, S. Gong and C.-C. Yang (eds.): *Harmonic Analysis in China*. 1995, 318 pp. ISBN 0-7923-3566-X

M.S. Livšic, N. Kravitsky, A.S. Markus and V. Vinnikov: *Theory of Commuting Nonselfadjoint Operators*. 1995, 314 pp. ISBN 0-7923-3588-0

A.I. Stepanets: *Classification and Approximation of Periodic Functions*. 1995, 360 pp. ISBN 0-7923-3603-8

C.-G. Ambrozie and F.-H. Vasilescu: *Banach Space Complexes*. 1995, 205 pp. ISBN 0-7923-3630-5

E. Pap: *Null-Additive Set Functions*. 1995, 312 pp. ISBN 0-7923-3658-5

C.J. Colbourn and E.S. Mahmoodian (eds.): *Combinatorics Advances*. 1995, 338 pp. ISBN 0-7923-3574-0

V.G. Danilov, V.P. Maslov and K.A. Volosov: *Mathematical Modelling of Heat and Mass Transfer Processes*. 1995, 330 pp. ISBN 0-7923-3789-1

A. Laurinčikas: *Limit Theorems for the Riemann Zeta-Function*. 1996, 312 pp. ISBN 0-7923-3824-3

A. Kuzhel: *Characteristic Functions and Models of Nonself-Adjoint Operators*. 1996, 283 pp. ISBN 0-7923-3879-0

G.A. Leonov, I.M. Burkin and A.I. Shepeljavyi: *Frequency Methods in Oscillation Theory*. 1996, 415 pp. ISBN 0-7923-3896-0

B. Li, S. Wang, S. Yan and C.-C. Yang (eds.): *Functional Analysis in China*. 1996, 390 pp. ISBN 0-7923-3880-4

P.S. Landa: *Nonlinear Oscillations and Waves in Dynamical Systems*. 1996, 554 pp. ISBN 0-7923-3931-2

Other *Mathematics and Its Applications* titles of interest:

A.J. Jerri: *Linear Difference Equations with Discrete Transform Methods.* 1996, 462 pp. ISBN 0-7923-3940-1

I. Novikov and E. Semenov: *Haar Series and Linear Operators.* 1997, 234 pp. ISBN 0-7923-4006-X

L. Zhizhiashvili: *Trigonometric Fourier Series and Their Conjugates.* 1996, 312 pp. ISBN 0-7923-4088-4

R.G. Buschman: *Integral Transformation, Operational Calculus, and Generalized Functions.* 1996, 246 pp. ISBN 0-7923-4183-X

V. Lakshmikantham, S. Sivasundaram and B. Kaymakcalan: *Dynamic Systems on Measure Chains.* 1996, 296 pp. ISBN 0-7923-4116-3

D. Guo, V. Lakshmikantham and X. Liu: *Nonlinear Integral Equations in Abstract Spaces.* 1996, 350 pp. ISBN 0-7923-4144-9

Y. Roitberg: *Elliptic Boundary Value Problems in the Spaces of Distributions.* 1996, 427 pp. ISBN 0-7923-4303-4

Y. Komatu: *Distortion Theorems in Relation to Linear Integral Operators.* 1996, 313 pp. ISBN 0-7923-4304-2

A.G. Chentsov: *Asymptotic Attainability.* 1997, 336 pp. ISBN 0-7923-4302-6

S.T. Zavalishchin and A.N. Sesekin: *Dynamic Impulse Systems.* Theory and Applications. 1997, 268 pp. ISBN 0-7923-4394-8

U. Elias: *Oscillation Theory of Two-Term Differential Equations.* 1997, 226 pp. ISBN 0-7923-4447-2

D. O'Regan: *Existence Theory for Nonlinear Ordinary Differential Equations.* 1997, 204 pp. ISBN 0-7923-4511-8

Yu. Mitropolskii, G. Khoma and M. Gromyak: *Asymptotic Methods for Investigating Quasiwave Equations of Hyperbolic Type.* 1997, 418 pp. ISBN 0-7923-4529-0

R.P. Agarwal and P.J.Y. Wong: *Advanced Topics in Difference Equations.* 1997, 518 pp. ISBN 0-7923-4521-5

N.N. Tarkhanov: *The Analysis of Solutions of Elliptic Equations.* 1997, 406 pp. ISBN 0-7923-4531-2

B. Riečan and T. Neubrunn: *Integral, Measure, and Ordering.* 1997, 376 pp. ISBN 0-7923-4566-5

N.L. Gol'dman: *Inverse Stefan Problems.* 1997, 258 pp. ISBN 0-7923-4588-6

S. Singh, B. Watson and P. Srivastava: *Fixed Point Theory and Best Approximation: The KKM-map Principle.* 1997, 230 pp. ISBN 0-7923-4758-7

A. Pankov: G-*Convergence and Homogenization of Nonlinear Partial Differential Operators.* 1997, 263 pp. ISBN 0-7923-4720-X

Other *Mathematics and Its Applications* titles of interest:

S. Hu and N.S. Papageorgiou: *Handbook of Multivalued Analysis*. Volume I: Theory. 1997, 980 pp. ISBN 0-7923-4682-3 (Set of 2 volumes: 0-7923-4683-1)

L.A. Sakhnovich: *Interpolation Theory and Its Applications*. 1997, 216 pp. ISBN 0-7923-4830-0

G.V. Milovanović: *Recent Progress in Inequalities*. 1998, 531 pp. ISBN 0-7923-4845-1

V.V. Filippov: *Basic Topological Structures of Ordinary Differential Equations*. 1998, 530 pp. ISBN 0-7293-4951-2

S. Gong: *Convex and Starlike Mappings in Several Complex Variables*. 1998, 208 pp. ISBN 0-7923-4964-4

A.B. Kharazishvili: *Applications of Point Set Theory in Real Analysis*. 1998, 244 pp. ISBN 0-7923-4979-2

R.P. Agarwal: *Focal Boundary Value Problems for Differential and Difference Equations*. 1998, 300 pp. ISBN 0-7923-4978-4

D. Przeworska-Rolewicz: *Logarithms and Antilogarithms*. An Algebraic Analysis Approach. 1998, 358 pp. ISBN 0-7923-4974-1

Yu. M. Berezansky and A.A. Kalyuzhnyi: *Harmonic Analysis in Hypercomplex Systems*. 1998, 493 pp. ISBN 0-7923-5029-4

MIX
Papier aus verantwortungsvollen Quellen
Paper from responsible sources
FSC® C105338

If you have any concerns about our products,
you can contact us on
ProductSafety@springernature.com

In case Publisher is established outside the EU,
the EU authorized representative is:
Springer Nature Customer Service Center GmbH
Europaplatz 3, 69115 Heidelberg, Germany

Printed by Libri Plureos GmbH
in Hamburg, Germany